History of Mechanism and Machine Science

Volume 39

Series Editor

Marco Ceccarelli, Department of Industrial Engineering, University of Rome Tor Vergata, Rome, Italy

This bookseries establishes a well-defined forum for Monographs and Proceedings on the History of Mechanism and Machine Science (MMS). The series publishes works that give an overview of the historical developments, from the earliest times up to and including the recent past, of MMS in all its technical aspects.

This technical approach is an essential characteristic of the series. By discussing technical details and formulations and even reformulating those in terms of modern formalisms the possibility is created not only to track the historical technical developments but also to use past experiences in technical teaching and research today. In order to do so, the emphasis must be on technical aspects rather than a purely historical focus, although the latter has its place too.

Furthermore, the series will consider the republication of out-of-print older works with English translation and comments.

The book series is intended to collect technical views on historical developments of the broad field of MMS in a unique frame that can be seen in its totality as an Encyclopaedia of the History of MMS but with the additional purpose of archiving and teaching the History of MMS. Therefore. the book series is intended not only for researchers of the History of Engineering but also for professionals and students who are interested in obtaining a clear perspective of the past for their future technical works. The books will be written in general by engineers but not only for engineers. The series is promoted under the auspices of International Federation for the Promotion of Mechanism and Machine Science (IFToMM).

Prospective authors and editors can contact Mr. Pierpaolo Riva (publishing editor, Springer) at: pierpaolo.riva@springer.com

Indexed by SCOPUS and Google Scholar.

More information about this series at http://www.springer.com/series/7481

Danilo Capecchi

Epistemology and Natural Philosophy in the 18th Century

The Roots of Modern Physics

Springer

Danilo Capecchi
Ingegneria Strutturale e Geotecnica
Sapienza Università di Roma
Roma, Italy

ISSN 1875-3442 ISSN 1875-3426 (electronic)
History of Mechanism and Machine Science
ISBN 978-3-030-52851-5 ISBN 978-3-030-52852-2 (eBook)
https://doi.org/10.1007/978-3-030-52852-2

This Springer imprint is published by the registered company Springer Nature Switzerland AG
The registered company address is: Gewerbestrasse 11, 6330 Cham, Switzerland

To my wife

Preface

At the beginning of the modern era, what is now called science was spread out among various disciplines: theology, handcrafts, magic, alchemy, astrology, medicine, natural philosophy, and mixed mathematics (that is optics, astronomy, mechanics, music, etc). In the 18th century, these disciplines had already broken up and became recomposed into an organization of science of the natural world that was similar to the modern one. The most important transformation was that affecting natural philosophy, which was considered, at least in the academic world, the most noble and by far the most important form of knowledge. It ceased to be a philosophy in the canonical sense, merging with other forms of knowledge and giving raise to disciplines such as physics, natural history, chemistry, medicine, and engineering, names which, while not initially fully shared, were established in the 19th century and are still used today. The new philosophers of nature were no longer canonical philosophers, although they continued to think about philosophy. They were rather mathematicians in the broad sense, namely, scholars who were interested in more than just pure mathematics. In some cases, this is particularly true for the new branches of physics such as electricity and magnetism, and they were also simply educated gentlemen, gifted with intelligence and curiosity.

The present book aims to document this process of transformation, concentrating on the 18th century, a century that in the past had been considered uninteresting for the history of science. It would represent the transition from the age of genius and the birth of modern science, in the 17th century, to the age of prodigious development, in the 19th century. This view does not stand up to thorough analysis. The 18th century, the century of Enlightenment and reason, as will be clear from the present book, was rather a century of great ferment and novelty.

To make the narrative practicable for a single individual, no great emphasis has been placed on the precise genesis of the various concepts and methods developed in scientific enterprises, except when this was necessary to make them clear. I have been content to take snapshots of situations by taking a look at discrete intervals of time. In several situations, reference is made to the authors who are famous today, such as Newton, the Bernoullis, Euler, d'Alembert, Lagrange, Lambert, Volta, etc. Not so much because they were the most creative and original minds, but mainly because their writings represent a synthesis of contemporary and previous studies. The above names should, therefore, be considered more labels of a period than references to real historical characters. The history of science was not made up of isolated heroes, but by an entire community and its legacy, by teachers, anonymous collaborators of celebrities, playing this role often only as a result of their social status. It is true that in the 18th century scientific research was carried out by a handful of scholars; though small, it was still an army, however, whose generals are only representative of the victories carried out by the soldiers. Referring only to the great characters of science has in any case an advantage that there is no need entering the merits of their acquisitions because well documented by historians, thus leaving room for other aspects.

This book intends to answer these three fundamental questions:

1. Was the transformation of ancient natural philosophy into (modern) science due to an internal evolution or an external appropriation?
2. What was the role of mechanical and experimental philosophies in this transformation?
3. What was the role of the newly born infinitesimal mathematics (*Calculus*)?

The answers come as follows:

1. It was a conquest from the outside by mathematicians (broader sense). They were able to extend the approach of mixed mathematics to the study of most phenomena.
2. Mechanical philosophy was crucial. It aroused interest in natural philosophy in many not canonical philosophers—including mathematicians—because its argumentations were much simpler than those offered by canonical philosophy, imbued with metaphysics. A similar argument holds good for experimental philosophy.
3. The role of Calculus was twofold. On the one hand, its great fecundity made it possible to solve very complex problems, thus giving mathematics more appeal in the approach of philosophy of nature. As regards to applications, on the other hand, the need to refer to regular functions defined on continua meant that nature was seen through glasses with thick lenses, which influenced its interpretation.

Before going further, a nomenclature should be established because many of the terms in use in the past and still in use today have changed their meaning, and thus a stipulation is necessary:

- *Canonical natural philosophy*. Study of nature under the concepts of matter, cosmos, and causation. Examples of canonical natural philosophies are the Aristotelean, the Platonic, the mechanicistic; but not Newton's and the approach to nature after him, even though the term natural philosophy is retained also in such cases.
- *Mixed mathematics*. Mathematics related to physical problems—that somehow joins quality to quantity—as they were established in the early modern era, such as astronomy (physical or positional), surveying, fortification, ballistic,

mechanics, hydrodynamics, pneumatics, and so on. A category quite distinct from that of subalternate sciences of Aristotelian mould that sometimes are considered as the canon of mixed mathematics.

- *Canonical philosophers* (or simply philosophers). Scholars that besides natural philosophy considered also and in a systematic way metaphysics and either logic, or ethic, or theology.
- *Mathematical practitioners* (or mathematicians broad meaning or more simply mathematicians). People with a more or less important training in theoretical mathematics who were also involved in practical activities (note that the idea of pure mathematicians is quite modern; until at least the end of the 18th century more or less all mathematicians were involved in practical activity or at least wrote about practical use of mathematics). The term *Mathematical practitioners* was introduces as a historical category in 1954 by the English geographer and historian of science Eva Germaine Rimington Taylor (1879–1966), but it is used here quite freely.

Of course one can envision a spectrum between canonical philosophers and mathematical practitioners. The columns of the following table show the possible main combinations, ranging from canonical philosophers with a very limited interest in mathematics (first column) to skilled artisans who knew little of philosophy and mathematics (last columns). H means high involvement and L means low or medium involvement.

Philosophy	H	H	L	L
Mathematics	L	H	H	L

16th Century. The birth of Greek rationality represented a fundamental step for the change in the form of the western knowledge of the natural world. But in the 16th century the change was possibly more radical when mathematicians began to widen the fields of classical mixed mathematics—which had flourished in Hellenistic Greece and remained vital in the Middle Ages—certainly pressed by demands of technology from a rapidly expanding society, by appropriating of parts of the canonical natural philosophy.

17th Century. The process of appropriation became more evident as the century progressed. Traditional mixed mathematics, optics, astronomy, music, mechanics, flanked by many other disciplines that in previous centuries had been studied only in natural philosophy, such as acoustics, meteorology, and hydrodynamics, were given a new and fundamental acquisition: dynamics (term introduced by Leibniz). Success obtained in this field by the mathematical (and experimental) approach due to the contribution of many scholars, including Galileo, Torricelli, Cavalieri, Huygens, Mariotte, Roberval, Descartes, Borelli, Leibniz, and eventually Newton, had a fundamental role in the process of erosion of the old philosophy of nature by (mixed) mathematicians. Because of this success, with the greater interest in experimentation to clarify and discover new "facts", and the interest in applications, natural philosophy was seen differently than in the past. Less attention was paid to traditional issues concerning nature, essence, and properties appropriate to all bodies—the so-called *physica generalis*—and more attention was paid to the examination and discussion of the particular bodies—the so-called *physica particularis*. Traditional explanations, both Aristotelian and Cartesian, appeared in some way sterile. Indeed the theories put forward did not have any element of objectivity; the same phenomenon could be explained in one way by one scholar, in another way by another, without an agreement being reached. New approaches to the study of nature became so appealing because, despite providing answers in a more restricted range, they had some kind of objectivity and were able to lead to a consensus of opinion. Furthermore, together with the explanations, the new approaches also provided for the prediction of phenomena, which had great utilitarian value, leading to technological applications.

Skepticism toward canonical natural philosophy led to the birth of *experimental philosophy*. It developed in different ways on the Continent and in England, so that it is stipulated by some historians that the label experimental philosophy be applied to English experimentations only, sponsored by the Royal society. The way of relating to experimentation is attributed by these historians to the conception of the role of the demiurge and magic in the Creation and, therefore, in what is called the ordinary course of nature. In England, more freedom would have been granted to the demiurge than on the Continent, and this would have given greater freedom to the English, who were not obliged to subsume experiments under general necessary laws. As a result, English scholars would have developed a science that favored phenomenological aspects while continental academics would have paid more attention to causation.

18th Century. The appropriation of natural philosophy by mathematicians became substantially complete. This process was aided by the spreading of Calculus, which made it possible to address many of yet unsolved problems. The received viewpoint considers the change in the approach to philosophy of nature to be largely a consequence of the technical results Newton achieved in the *Philosophiae naturalis principia mathematica* and *Opticks* and his empirical view. Since the 19th century, Newton has generally been seen as the founder of modern science, in particular of classical mechanics, in the form it has today. This point of view was also prevalent among modern science historians, at least until the mid-20th century. The assessment of Newton by contemporaries was more balanced. He was recognized as a great mathematician; his results in astronomy were considered extraordinary, but few saw him as a bringer of revolutionary results, not even in the fundaments of mechanics. In fact, he was considered just one of the many. Before him, Huygens, Wallis, and Hooke had obtained results of no less importance than his. Modern historiography introduces great variations in the received point of view, giving to Newton only what is Newton's.

In mechanics, the old mixed mathematics of Hellenistic origin, together with statics and Galilean dynamics, changed into a theory that had the same ambitions of the canonical natural philosophy, namely, to give a global response to the nature of motion based, however, only on a mathematical approach. The Bernoullis, Varignon, Euler, and a few others, explicitly introduced algebra and differential calculus, operating a non-trivial transformation that invested the very role of mathematics in the science of nature. Even for the study of the motion of the free mass point, the treatment of 18th-century mathematicians became unrecognizable compared to that of Newton. From geometric it became analytic. Relations of proportionality were replaced by equalities, overcoming the difficulties of dealing with non-homogeneous quantities, with a reinterpretation of the concept of physical magnitudes, already started by Descartes. Technological applications, as well as the curiosity of the mathematicians, required the solution of more complicated problems than those faced by Newton (essentially the motion of the free material point) such as those related to systems of material points, rigid bodies, and deformable continua.

For the other sections of physics, such as optics, electricity, magnetism, the raising thermology, and meteorology, sometimes named Baconian sciences, but also some disciplines today classified as mechanical, such as hydraulics, pneumatics, acoustics, the qualitative reasoning remained important. The objective was to identify the phenomena under observation with the greatest clarity. To achieve this goal it became necessary for mathematicians to take an interest in natural philosophy, more than it was necessary in mechanics.

Even in these Baconian sciences, however, there was a division, already found in the 17th century, between people devoted to pure experimentation and those who preferred a more theoretical approach. The supporters of a mainly experimental and phenomenological type of physics made use of refined instruments and provided, in some cases, very accurate measurements, as regards thermal and meteorological phenomena, for instance, with attempts being made at formulating mathematical laws; measurements for electricity and magnetism were more difficult. There was considerable progress in measuring devices. Some were built directly by the scholars who used them. There was thus, as often in the past, interaction between the mathematician and the technologist. The most theoretical experimenters often sought causal explanations. They required the appropriation of natural philosophy, which was made possible primarily by its transformation into a mechanicistic form, certainly much more understandable for people who were well educated but who were not canonical philosophers. They proposed explanatory models guided by experience. In some cases, it was a mechanicism *à la* Descartes or *à la* Boyle, where everything was referred to in terms of interactions of particles by contact. After Newton, interaction between particles could also take place at a distance and this type of mechanicism became prevalent.

The 18th century, especially in the second half, saw an impressive development of technology, named industrial revolution in the 19th century. The intertwining between this development and that of science, population growth, agricultural revolution, spread of scientific culture, and birth of the entrepreneurial and capitalist bourgeoisie is complex and even extremely interesting, but it is not treated in this book that is limited to examining the relations between the development of science and technology, with particular reference to epistemological aspects.

Seen from this perspective, the key figure of the technological development is that of the engineer. A figure that at the beginning of the 18th century had blurred connotations that gradually became more defined, and also thanks to the foundation of schools and associations that took care of his formation. The engineer who emerged in the end was that of an educated technician who mastered all areas of science with a particular attention to mathematics (especially on the Continent).

Summary of Chapters

Chapter 1. Newton's epistemology is analyzed as the prototype of the view at the turn of the 18th century, both by studying his official statements in the *Principia*, *Opticks*, in other published works and letters, and his actual way of making science. Newton in mechanics and optics was linked to tradition. He was the last of the great traditional mixed mathematicians; with slight exaggeration, one could say, the last of the ancients. He was a very complex character, involved in many issues, including alchemy and theology, spending most of his time in activities that today could hardly be classified as scientific. The chapter discusses how his being a mathematician influenced all his apparently non-mathematical interests, canonical natural philosophy included. The received point of view, that Newtonian science was already mature at its inception, is also disputed. This, however, has already been attested in recent historiography. Indeed, it is well established that the description of 18th-century classical mechanics as Newtonian mechanics should be considered an anachronism. It was the developments of 18th-century algebra and Calculus that produced Newtonian mechanics as we know it today, which was therefore primarily a creation of 18th-century mathematicians. In the examination of Newton's epistemology, I made no use of the modern categories of the philosophy of science to highlight its most intimate contradictions. I instead tried to frame Newton historically, giving him, for example, the label of empiricist, which he attributed to himself and which many of his contemporaries attributed to him, even if this label does not stand up to the criticism of a modern epistemology analysis.

Chapter 2. *Physica generalis* and/or *physica particularis* of Aristotelian mould became physics in a nearly modern sense with the use of a quantitative, experimental method to discover laws governing the inorganic world. The chapter explains how mathematicians were successful in replacing canonical philosophers nearly completely in the study of natural philosophy, both in research and academic contexts and how they invented an academic discipline that was called simply physics, concerned only with the study of inanimate matter, excluding alchemy. The new conception of physics for at least the whole of the 18th century still continued to be called natural philosophy, and even maintained some of the characteristics of old physics. A typical example is the explanation and acceptance of the principle of conservation of living forces and minimum action, typically 18th-century themes. Both were explained by the metaphysical principles of economy and simplicity of nature or by God, a type of explication not allowed in today's physics. Following the spread of mechanical philosophy in European universities and colleges, the theoretical explanations of philosophy of nature were accompanied by experiments, mainly concerning mechanics, hydraulics, pneumatics (and electricity). Later, especially in France, teaching began to be supported by mathematics. This led to tension between the mathematical-physics disciplines (such as mechanics, for instance), where the use of algebra and Calculus was possible and massive, and the other disciplines such as electricity, magnetism, and thermology, where this possibility did not yet exist. While in teaching, physics was often unitary, in research a split originated among people mainly interested in mathematics (the *geometers*), and others who were keener on experimentation (the *physicists*). However, many of them believed in the myth of the exact measurement, which was considered the prerequisite for mathematical treatment and, therefore, according to current ideas of the true scientific method. The complex relationship between experimental and mechanical philosophies (and the heuristic role of theories) is also addressed. In principle, experimental philosophy did not require the knowledge of mechanical philosophy. The latter, however, was helpful because it suggested explanatory models and made it possible to make predictions, which if sometimes proved to be false were, however, a starting point. For this reason, many experimental philosophers supported mechanical philosophy.

Chapter 3. The evolution of *mechanics* toward what is now called *classical mechanics* is explored in depth. Though 18th century mechanics saw great successes in theoretical astronomy, namely, the study of the solar system using the modern techniques of Calculus and the assumption of the law of the inverse square of distance for the gravitational forces, it also made tremendous progress in terms of rationalization and completion. The exposition of fundaments and applications in various treatises and papers are studied, especially those of continental scientists and, in particular, the syntheses by Euler and d'Alembert, who described the way mechanics could be transformed into a rational discipline, like mathematics, based on Calculus. Great attention was paid to the justification of the foundations of mechanics, which required a substantial involvement of metaphysics and epistemology to discuss fundamental notions: the nature of space and time, the constitution and the properties of bodies and the nature of motion. However, this effort was pursued not with the classical and organic approach of canonical philosophy but with the pragmatism of mathematicians. Sometimes this philosophical approach was made explicit and expounded in specific treatises or preliminary parts of scientific works. At other times, it remained implicit and could only be brought to light through careful reading. The complete mathematization of mechanics only occurred at the end of the 18th century with the Lagrangian synthesis, which kept considerations about the philosophy of nature to a minimum.

Chapter 4. D'Alembert called *physics in general* disciplines such optics, acoustics, positional astronomy, cosmology, magnetism, and electricity. Optics and positional astronomy were object of in-depth researches. A good deal of the optical works concerned with the theories of propagation of light, especially with those of an undulatory nature. They required complex mathematical treatments and the use of partial differential equations, and became fertile ground for mathematical physics. The experience with which the theories were compared was mostly based on experiments conducted in the 17th century by Newton and Huygens. Relevant new experimental work, on a quantitative basis, was carried out only relatively to what is today known as photometry with Bouguer and Lambert. Positional astronomy saw a great development made possible by the use of reflecting or refracting telescopes with the introduction of achromatic lenses, mistakenly considered impossible by Newton and theorized instead by Euler, as efficient instruments of astronomical observation. Magnetism caught the attention of many mathematicians and experimental philosophers, but it revealed a too complex topic for the times.

Electricity was the emerging science of the 18th century. In the chapter, it plays a paradigmatic role for the development of experimental sciences. Experimenting in electricity had several advantages over experimenting in other fields of physics. Once research began in earnest, experimenters quickly discovered new electrical phenomena, making their work rewarding. The number and quality of experiments grew dramatically, especially after the 1750s, when the discovery of the Leyden jar made it possible to accumulate large charges. After a brief mention of the situation in the 17th century, the chapter goes on to take a look at the English experimenters, including Francis Hauksbee, Stephen Gray, John Canton, and those from the Continent, including Jean Antoine Nollet, Pieter van Musschenbroek, Ulrich Theodor Aepinus, Giambattista Beccaria, and others.

Chapter 5. Modern technology historians identify the birth of a new figure in the 18th century, the scientific engineer. His goal was the rationalization of design and implementation of processes. For this purpose, he used hypotheses and experimentations, as in the (mathematical) physical sciences. With its dizzying growth, science revealed the possibility of applications to areas never thought of before. However, scientists were dealing with general problems. Their solutions did not provide for an immediate application of science. Thus, there was the basic need for an intermediate figure between the scientist and the final user. More precisely, there was a need for a sufficiently large body of qualified engineers.

The preparation of these technicians was different from country to country. In England, one moved with private associations, and in France, instead, the state was interested in. After general considerations on the relationship between mathematics, natural philosophy, new physics, and technology, the chapter goes on to characterize the engineers of the 18th century. For reasons of space, I will dwell only on the situation in England where many skilled engineers operated after the 1750s; I refer here to John Smeaton and James Watt, not so much because they are the most famous, but because in them one can observe at the higher degree the interaction between the scientist and the engineer.

Editorial Considerations

Most figures are redrawn to allow a better reading. They are, however, as much possible close to the original. Symbols of formulas are those of the authors, except cases easily identifiable. Translations of texts from various languages are as much as possible close to the original. For the Latin, Italian of the XVI and XVII centuries a critical transcription has been preferred. In the critical Latin transcription, some shortenings are resolved, "v" is modified in "u" and vice versa where necessary, ij in ii, following the modern rule; moreover, the use of accents is avoided. In the Italian critical transcription, some shortenings are resolved, "v" is modified in "u" and vice versa, and a unitary way of writing words is adopted. Books and papers are always reproduced in the original spelling. For the name of the different characters, the spelling of their native language is generally preferred, excepting for the ancient Greeks, for which the English spelling is assumed, and some medieval people, for which the Latin spelling is assumed, following the common use.

Through the text I searched to avoid modern terms and expressions as much as possible while referring to ancient theories. In some cases, however, I transgressed this resolution for the sake of simplicity.

Roma, Italy
May 2020

Danilo Capecchi

Contents

Chapter 1
Epistemology and Science at the Turn of the 18th Century

Abstract After a brief excursion of precursors, Descartes, Huygens and a few others, Newton's epistemology is analyzed as the prototype of the view at the turn of the 18th century, both by studying his official statements in the *Principia*, *Opticks*, in other published works or letters and his actual way of making science. Newton was the last of the great traditional mixed mathematicians; with slight exaggeration, one could say, the last of the ancients. He was a very complex character, involved in many issues, including alchemy, theology and management; spending most of his time in activities that today could hardly be classified as scientific. The chapter discusses how his being a mathematician influenced all his apparently not mathematical interests, canonical natural philosophy included.

1.1 The Heritage of the 17th Century

Most people involved in the early modern science that came after Galileo and before Newton in the 17th century accepted the mechanical philosophy. Mechanical philosophy was very useful to the development of science; it made easier the applications to natural philosophy of geometric models. For example, a series of spherical or cubic bodies perfectly hard or elastic. By means of the laws of mechanics a simulation, a prediction, could be obtained of the phenomenon under study. If the simulation was in agreement with reality it could be said that the proposed model is correct and a causal explanation has been possibly found.

The most meaningful characters of the early modern science after Galileo and before Newton are usually identified in Descartes and Huygens. This is however a simplification not necessarily the best one; the transformations that concerned the scientific knowledge were the result of collective work, largely due to mathematicians involved in mixed mathematics. A short list of the most known of them is referred to in the following Table 1.1.

To them one should add many others teachers of the universities and members of the new born scientific academies. The purpose of the present book is not however to document the spread of the process of transformation of the scientific knowledge but rather to exemplify the way it occurred. Thus, for the sake of simplicity I will

D. Capecchi, *Epistemology and Natural Philosophy in the 18th Century*,
History of Mechanism and Machine Science 39,
https://doi.org/10.1007/978-3-030-52852-2_1

Table 1.1 Mathematicians operating in the second half of 17th century

Giovanni Battista Baliani (1582–1666)	Italy
Giovanni Alfonso Borelli (1608–1679)	Italy
Bonaventura Francesco Cavalieri (1598–1647)	Italy
Giovanni Domenico Cassini (1625–1712)	Italy
Evangelista Torricelli (1608–1647)	Italy
Geminiano Montanari (1633–1687)	Italy
Vincenzo Viviani (1622–1703)	Italy
Gilles Personne de Roberval (1602–1675)	France
Jacques Rohault (1618–1672)	France
Blaise Pascal (1623–1662)	France
Ignace Gaston Pardies (1636–1673)	France
Honoré Fabri (1608–1688)	France
Edme Mariotte (1620–1684)	France
Robert Hooke (1635–1703)	England
John Wallis (1616–1703)	England
Isaac Barrow (1630–1677)	England
Christopher Wren (1632–1723)	England
Robert Boyle (1627–1691)	England

refer, besides Newton, mainly to Descartes and Huygens; the former is usually taken as an example of a philosopher who dealt with mixed mathematics, while the latter of a mixed mathematician who dealt with philosophy; this is however a crude simplification. A few words will however be devoted to the French Jesuit Ignace Gaston Pardies.

1.1.1 *Descartes's Purely Deductive Mixed Mathematics*

René Descartes (1596–1650) played an important role in the development of modern science. He contributed in a fundamental way to the development of mathematics, especially for what concerns algebra and analytical geometry. He contributed significantly to the development of mechanicism also, that was one of the main support to the mixed mathematics of the second half of the 17th century. He also gave important technical contributions in statics, hydrostatics, dynamics (modern meaning), optics, music. But in substance he kept extraneous to the fundamental idea of the mixed mathematics: the development of a theory on the basis of a deductive approach starting from more or less complex principles of empirical nature that were not questioned.

Descartes remained a canonical natural philosopher, although his philosophy was strongly modeled on mathematics. His search for explanations of the world leaded him on a path that at present appears to be blind, or at least was not followed by the scholars that we qualify today as scientists. He formulated a method that strived to explain natural phenomena based on possibly simple and evident notions and/or observations, drawn from rational reflection on concepts or from everyday experience, about the most fundamental aspects of the world. These provide the requisite (metaphysical) foundation for his physics. In other words, he proceeded from *clear* and *distinct* knowledge of general metaphysical objects, such as the material substance, to derive particular laws. Descartes's method of conducting science was thus quite different from the modern approach, where scholars do not first engage in a metaphysical search for first principles. Yet, the lack of this phase is exactly the criticism Descartes leveled at Galileo's physics, that is at modern physics: "I find in general that he philosophizes much better than the vulgar, in that he leaves the most he can the errors of the Schools [...] but without having considered the first causes of nature, [Galileo] has merely looked for the explanations of a few particular effects, and he has thereby built without foundations [28].[1]

Descartes was involved in mixed mathematics in different period of his life; the interest was prominent in his early phase however, before he developed his mature physics and metaphysics. Determinant for Descartes's move was his acquaintance with Isaac Beeckman (1588–1637) who involved him in his studies on hydraulics and acoustics. The first Descartes's published work on mixed mathematics was the *Musicae compendium* [26]. Written in 1618, it was published posthumous in The Netherlands in 1650; there was a second edition in 1653 to which other editions followed. To be noted that, differently from Mersenne, who largely wrote on music, Descartes never had acquaintance in any musical circle and possibly he did know of music only at La Flche and through Beeckman.

Other Descartes's works on mixed mathematics belong to his mature period. They were about optics, with the *Dioptrique* of 1637; statics, with the letter to Constantijn Huygens of 1637 entitled *Explication des engins par l'ayde desquels on peut, avec une petite force, lever un fardeau fort pesant* [12][2]; dynamics, with studies on the oscillation of composed pendulums, spread on letters to Mersenne in 1646 [13]. Below, for the sake of space, I will deal only with the study on statics and optics. In these works Descartes assumed principles based on reason and every day experience, while contrived experiments played no role.

1.1.1.1 Statics. The Principle of Virtual Work

The conception of statics of Descartes is spread in his correspondence with Mersenne, but its complete synthesis appeared in the letter of 1637 to Constantijn Huygens already cited [12].[3] The whole statics according to Descartes is based on the following 'obvious'principle:

---

[1] Vol. 2, letter to Mersenne of 11th October 1638, p. 380.

[2] pp. 164–173.

[3] pp. 164–169.

> The same force that can lift a weight, for example of 100 pounds to a height of two feet, it can also raise a weight of 200 pounds to one foot, or a weight of 400 pounds to the height of 1/2 foot, etc. [28].[4] (A.1)

Namely, the force (fatigue, work) needed to raise the weight p to the height h is the same as that required to raise $p/2$ to $2h$. It is a restricted formulation of the principle of virtual work.

Descartes's principle of statics is proved without any explicit reference to sensible experience, in a simply and 'convincing' way. The proof is associated to a thought experiment. Assume a weight of 200 pounds; it can ideally be decomposed into two weights of 100 pounds. The thought experiment makes it clear that either raising a weight of 100 pounds to the height of 2 feet or rising two weights of 100 pounds (that is a weight of 200 pounds) to the hight of 1 foot requires the same 'fatigue' or force by an operator, a man for instance. In fact lifting a weight of 100 pounds to the height of 2 feet can be imagined into two steps. First bring the weight to the height of 1 foot, then move it again (as if it were another body) from the height of 1 foot to that of 2 feet. But this last operation requires the same effort to lift 100 pound from 0 to 1 foot. Thus lifting 100 pounds to the eight of 2 feet is the same as to lift two weights of 100 pounds (and thus 200 pounds) to the height of 1 foot.

However, the above justification, coinciding with that of Galileo, not mentioned by Descartes, referred to in [12],[5] though ingenious, does not withstand critical analysis as noted by Mach [73].[6] Indeed, the admission that to raise 100 pounds in two stages is equivalent to raise 200 in one stage only, although intuitive, is not logically necessary and it is not contradictory to imagine that it is not true. The thought experiment makes the proof convincing because it incorporates, in an implicit way, empirical arguments. These are however consideration of a modern epistemologist; Descartes most probably would not have shared this opinion and would have considered the proof to be completely a priori.

1.1.1.2 Optics. The Law of Refraction

The fundamental considerations about optics by Descartes, apart from some letters, are referred to in the *Le monde ou le traité de la lumière* published posthumously in 1664 [27][7] and in the *Dioptrique* of 1637, a relatively short treatise of about 150 pages, divided into 10 chapters (Discourses), planned and written long before its actual publication [24].[8]

The most celebrated part of the *Dioptrique* is the second chapter where the law of refraction is 'deduced'. It is preceded by a preliminary chapter on the nature of light

[4] Vol. 2, pp. 435–436.

[5] p. 129.

[6] p. 84.

[7] Chapter XIII.

[8] The *Dioptrique* was published as an appended treatise to the *Discours de la methode* followed by the *Metéores* and the *Géometrie*, in that order.

and followed by some chapters of philosophy of nature concerning the anatomy of the eye, the role of the brain and the aspects determining vision. The treatise ends with three chapters, compressively of about 80 pages, of technical content connected to the use and construction of lenses. Here I will stress the phenomenon of refraction as described in the second chapter only.

In the first pages of the treatise Descartes had declared that his scope was not to inquire the 'true' nature of light but only to use some of its properties, such for instance refraction and colors: "*In this I will be imitating the astronomers* [emphasis added], who, although their assumptions are almost all false or uncertain, nonetheless, because they agree with many observations that they have made, never cease to allow the derivation of many very true and well-assured consequences" [28].[9] In any case the true nature of light could not be the object to empirical verification; it should be decided a priori on the basis of certain metaphysical reasonings.

According to Descartes light is nothing but, in luminous bodies, than the propagation of a tendency to motion through particles of the second of his three elements because of an impulse received by particles of the first of his elements, from the sun for instance. Remember that Descartes distinguished three elements of matter: luminous matter made up of very thin particles which could assume any shape so as to exactly fill all the angles they find in the bodies they meet; aetherial matter made up of small rounded particles (notice that some commentators call aetherial the first element) and terrestrial matter, which, because of its greater dimension, cannot be moved like the others [25].[10]

Because the particles of aether are perfectly rigid, light propagates instantaneously. It moves toward our eyes through the air and other transparent media in the same fashion as the resistance of bodies encountered by a blind person pass to his hand by the intermediary of the stick. This example will support—for Descartes and not a modern physicist—that light can extend its rays in an instant from the sun to us, that is its speed is infinite [28].[11] Moreover, light can be imagined to propagate along rays which could always be assumed to be completely straight, when they pass through an uniform transparent body. But when the rays encounter some other bodies, they are subject to being deflected by them, or weakened in the same way as the motion of a ball or a stone thrown in the air is weakened by the bodies it encounters, as empirical evidence shows. This give raise to the phenomenon of refraction [28].[12]

To explain refraction Descartes considered himself authorized to simplify his model of light assuming it made of equal perfect microscopic hard spheres, which actually move and obey the same laws of macroscopic bodies. Descartes considered analogically each sphere as a tennis ball launched by someone with a given finite speed. The ball is imagined to cross a horizontal cloth, representing the surface of separation between air and water for instance, which was so feeble and loosely woven

---

[9] Vol. 6, p. 83.

[10] Part III, 52, pp. 94–95.

[11] Vol. 6, p. 84.

[12] Vol. 6, pp. 88–89.

that the ball had the force to break it and pass [15].[13] After the impact with the cloth the vertical speed of the ball decreases causing the ball to move farther from the vertical than before the impact.

Descartes's analogy of a light-ball in motion, though fascinating presents inconsistencies. *From an empirical point of view*, as a matter of fact, light passing from lower to greater density medium bends toward the vertical, that is the angle of refraction is lower than the angle of incidence. The contrary to what expected from the ball analogy. *From a logical point of view*, there is no meaning in the assertion that light is faster or slower in different media, considered that it is transmitted instantaneously. Of course Descartes suggested answer to these difficulties. In substance he sustained that it is true that light is not associated to a true motion of particles but only to a transmission of an impulse of pressure. The motion can however be considered as virtual and virtual motion behaves much like to true motion. Moreover the virtual motion is faster in denser medium: "the more the small parts of a transparent medium are harder and stable the light pass easily" [28].[14] The contrary to what happens to the ball, so the experimental result is explained.

A reading of the *Dioptrique*, even careful but without its framing in the epistemology of the 17th century, leaves the modern reader the impression of a purely rational approach, where experience has no role and no experiment was carried out. This impression also is reinforced by repeated Descartes's claims of his derivation a priori of the laws of reflection and refraction. As for instance in the letter to Mersenne of 1st March 1638 "you should know that I demonstrated the refraction geometrically and a priori in my *Dioptrique*, and I am amazed that you still doubt it" [28].[15]

An aspect that also could point to the substantial purely a priori approach by Descartes, is the nearly complete absence in the *Dioptrique* of measurements of the indices of refraction of the various substances and thus an experimental verification of the law of refraction. This is not because Descartes was not interested at all to measurements but because he made difference between a general explanation of refraction and the justification of particular instances. The former can be given a rational explanation, the latter needs experiments whose fulfillment is possible with the help of other people and not relevant for a general theory. He discussed the problem in the *Dioptrique*. With his words: To see how different refractions should be measured and to find their values is necessary the use of experience, because they depend on the particular nature of the bodies in which they occur. It is nonetheless possible to do so reasonably certainly and easily, since all refractions are reduced to the a single measure; indeed it suffices to examine a single ray to know all the refraction over a given surface [28].[16]

A more in depth reading of the *Dioptrique*, partially modifies the impression of a purely a priori Descartes. First it must be considered that a priori according to him could have a different meaning than our post-Kantian meaning, that of no refer-

[13]pp. 514–516.
[14]Vol. 6, p. 103.
[15]Vol. 2, p. 31.
[16]Vol. 6. pp. 101–102.

ence to an empiric evidence. A priori, with a posteriori, are qualifiers of two logical approaches, from causes to effect or from effect to causes, respectively, also referred to as synthesis and analysis. In this sense a priori simply means that the explanandum is deduced from some primitive (and possible simple) assertions, of which is indifferent whether or not they are derived from or without—this is Descartes's opinion—recourse to experience. This interpretation is also validate by the quotation reported above, see Sect. 1.1.1.2, according to which the method adopted in the *Dioptrique* is that of astronomers, that starts from some hypotheses and derives their consequences. Descartes' deduction comes from the property of lights that are assumed as hypotheses, without questioning their epistemological status.

1.1.2 Pardies' Mechanics and Theory of Light

Ignace Gaston Pardies (1636–1673) was born in Pau (France), the son of an advisor at the local assembly. He entered the Society of Jesus in 1652. After his ordination he taught philosophy and mathematics at a lycée in Paris; one can thus speak of Pardies as of a mathematician with interests in natural philosophy and also in metaphysics and therefore able to increase the mixed mathematics by absorbing from philosophy. His well known scientific works are the *La statique, ou la science des forces mouvantes* of 1673 (herein after simply *La statique*), *De la connoissance des bestes* of 1672 and the *Discours du mouvement local* of 1670. His most important work is however on optics that remained unpublished. The ideas of Pardies about the nature of light were published by his confrere Pierre Ango (1640–1694) in his *L'optique divisée en trois livres* of 1682 [2].

In the *Discourse du mouvement local*, Pardies played the role of the metaphysician and criticized Descartes for his formulation of the laws of impact: "Descartes was wrong in six rules over seven" [106].[17] He then went on to describe his own rules of impact for bodies qualified as hard, which for him meant perfectly elastic. At the basis of Pardie's laws there are, first the principle of inertia, which for Pardies is not a principle of indifference to motion but rather an internal power, called impetuosity (impetuosité) [106][18]; second the Cartesian law that the motion in the collision is not lost but transferred from one body to another.

At a first sight the rules Pardies formulated appear correct to a modern reader, who can also accept their justifications. The only remark he [the modern reader] can raise is that Pardies forgot to specify that the bodies he considered were equal (with the same mass in modern term). However, the reader realizes soon that Pardies' was not a forgetfulness, but an intentional expositive (and strange) choice. In fact, in paragraph XXXI he clearly stated that the variation in speed in the impact are the same whether the bodies are equal or if they are different [106].[19]

---

[17] p. 187.

[18] p. 144.

[19] p. 171.

He had provided a justification of this occurrence in the *La statique*, asserting that:

> It is quite true that in the state we are, we find it more difficult to move a big stone than to move a little one; but there is no one who does not know that it comes from the resistance caused by the gravity of these stones. For if the great stone were not heavier than the little one, there is no doubt that we would urge it to move with the same facility [107].[20] (A.2)

In the *La statique* Pardies claimed his intention to write a complete work on mechanics, denomination which included both statics and dynamics. Pardies said that his treatise is in the footprints of Wallis' *Mechanica, sive de motu, tractatus geometricus* of 1671 [127], where was first used the term mechanics to include both statics and dynamics. Pardies considered it useful to write a new treatise because that of Wallis was too technical and was also not complete. Indeed *La statique* is interesting because his approach to mechanics is more physical than Wallis's, which was framed in the rigid schemes of the classical mixed mathematics, where mathematical demonstrations largely prevailed over the arguments of natural philosophy, used to justify the principles assumed.

The concepts of Pardies on the nature of light are summarize in the premise of the *La statique*, where it is suggested the analogy between light and sound; an analogy that had its roots in ancient time but now is treated with mathematical argumentations. Pardies's theory of light was discussed largely by Pierre Ango, in the *L'optique divisée en trois livres* of 1682. The basic hypothesis of natural philosophy is the existence of a matter infinitely more subtle than air, liquid in all its parts, named aether, which fills the whole firmament [2].[21] The sun (and the stars) is like a flame without heat. Sun contracts and expands as the flame of a candle and communicate this motion to the surrounding aether. The motion of *undulation* that results, said Ango-Pardies, is not however a transfer of matter. The reference are the waves that form when a stone in launched in still water [2][22]; this led him to consider light propagating at finite speed.

Particularly interesting is the explanation and the quantification of refraction, which is carried out under the assumption that the speed of light is the greater the lower the density of the medium, being maximum in the aether alone. It was most probably of inspiration to Huygens, which assumed a similar treatment. When light encounters a surface of separation between air and a diaphanous body, according to Pardies, waves change from spherical to ellipsoidal. Ango-Pardies explanation is not clear however, also because referring to a figure (Fig. 1.1) whose littering does not closely correspond to the description.

More clear is the explanation associated to Fig. 1.2, where a graphical algorithm to evaluate the direction of the refracted ray is suggested. Let consider two rays of light AI and *a*C, very close to each other to be considered as parallel. They encounter the surface IC of separation between two media having different density, that of the

[20] p. 262. Huygens quoted the most 'ingenious' Pardies, commenting that "we can deceive ourselves even when we assume as principles entirely probable ones" [81], p. 288.

[21] p. 7.

[22] pp. 8–17.

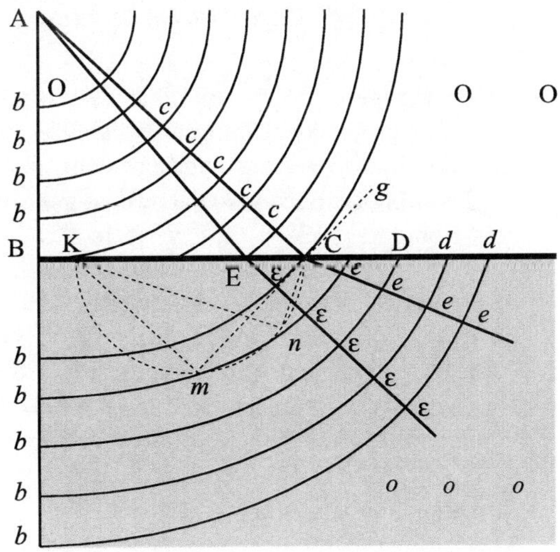

Fig. 1.1 Elliptic waves in the refraction of light according to Pardies. Redrawn from [2], p. 62

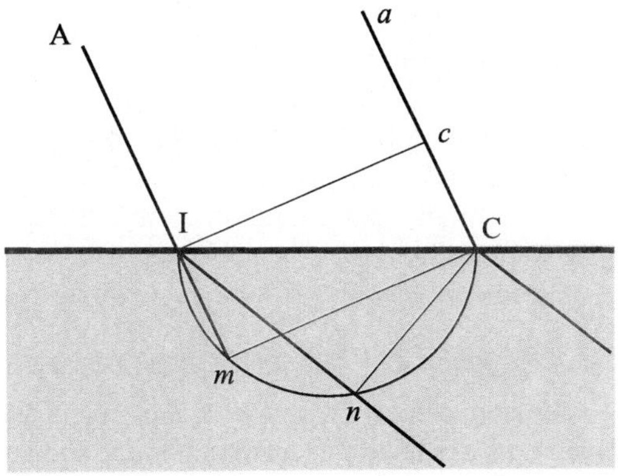

Fig. 1.2 Direction of the refracted ray. Redrawn from [2], p. 64

lower (medium 2) less than that of the upper (medium 1). Thus the speed of the light v_2 in the medium 2 is greater than the speed v_1 in the medium 1, of 1/3 for instance. The undulation in the direction of the ray AI reaches first the surface IC and continues its motion with an increased speed while the undulation in the direction of aC continues along cC. Ango-Pardies gave for granted that the ray will continue starting from the same point I but with a different direction In and that the front wave (the tangent to the undulation according to Argo-Pardies) is defined by a straight line passing through the point C. With reference to the half circumference of diameter IC, the point n which defines the direction In of the refracted ray is found by imposing that the segment Im = cC is to the segment In as v_1 is to v_2, being m the intersection of the half circumference with the extension of AI.

1.1.3 Huygens' Philosophical Mixed Mathematics

Christiaan Huygens (1629–1695) has been often considered the true heir of Galileo though a Cartesian. A common view assumes him as an eclectic scholar, a problem solver who took up yet unsolved problems and solved them without any apparent search direction. Leibniz, who in any case greatly estimated him, in a letter to Nicolas Franois Rémond (1638–1725) wrote in 1714: "he had no taste for metaphysics" [71].[23] These views could do if one considers the published great treatises only. Here the writing style is rigorous and concise, which rarely leaves space to imagine the tortuosity of elaboration surely associated to his researches. Not an uncommon trend, Newton also moved analogously though to a less extent. Because of the mathematical rigor Huygens was often referred to as a new Archimedes. He excelled in the use of mathematics in mechanical philosophy; more than Descartes who had different interests and lacked of correct impact rules, and more than Pardies, that was less good mathematicians, and even he lacked of correct dynamical rules.

A revisiting of Huygens's works occurred after the 1950s, when the so big edition of his *Oeuvres complètes*, started a sixty years before, was completed. It evidenced in correspondence and in unfinished papers that Huygens was not avulse from philosophy and metaphysics, helping to counteract the usual image of Huygens as a positivist. This is especially true for his last works, among which the so called Codex Hugeniorum 7A [81], devoted to motion and space, of which will be referred to in the following, and the *Cosmotheoros* [57], referred shortly below. In any case Huygens accepted fundamental metaphysical assumptions, those associated to mechanical philosophy.

The *Cosmotheoros* [57], Huygens' latest writing, addressed to his brother Constantijn, is a book atypical in many respects. The decision to draft this "little treatise on philosophical matters",[24] the only one of Huygens's books for which he used the cumbersome adjective *philosophical*, was due to several external motivations, among others the desire to disclose to a wider audience the cosmological consequences of the Copernican theory. The *Cosmotheoros* contained Huygens' speculations on the construction of the universe and the habitability of the planets as deduced from his own observations and those of other astronomers. The publication, though posthumous, had a remarkable success and was soon translated into several languages; much probably for the audacity of the matter treated: the existence and characteristics of the inhabitants of other worlds. Contemporaries scholars however, with the exception of Leibniz alone, judged with severity the conjectural nature of the work and felt that its editing had not increased the fame of the author [81].[25]

The *Cosmotheoros* should in any case be considered by modern historians of science worthy to be read because, besides the cosmological openings which dominate, it contains important epistemological observations. Prerequisite to Huygens' epistemological discourse is the consideration that the reasons of the Creator remain

---

[23] Vol. 3, p. 607.

[24] Letter to Leibniz, 29th may 1694, Oeuvres, t. X, 60.

[25] pp. 124–125.

impenetrable to us. With very strong accents Huygens emphasized that the reason of God is "quite other than the ours" and that the power that we assign him is usually just a confused imitation of the one we perceive in ourselves; any attempt to understand the divine attributes is therefore vain. In resuming the thesis of the infinite power of God, however, Huygens did not intend to get rid of the rules established by the mechanical philosophy which remain completely valid. Huygens made use of the omnipotence of God especially in order to reject the claims of metaphysicians and theologians to limit a priori the variety of creation, which must be established in an experimental way, and to prevent the extension of our science to hitherto unknown fields.

Huygens was anyway mainly a mathematician—very good indeed—and a supporter of the mechanical philosophy. His most important results were in mechanics. His studies on the composite pendulum, collected in the *Horologium oscillatorium* of 1673, are a milestone of dynamics. The same holds true for his studies on the impact of (elastic) bodies in the *Motu cororum ex percussione* published in 1713 and on the centrifugal force in *De vi centrifuga* of 1673. In these studies the role played by mechanical philosophy was important but not prevalent. Huygens was instead fully engaged in this philosophy in the study of the causes of gravity. His conclusions, reported in the *Discourse sur la cause de la pesanteur* of 1690, were of qualitative nature even though mathematics played an important role.

But the problem where Huygens joined the approach typical of mixed mathematics to the mechanical philosophy was that of light propagation. According to Huygens, in optics the demonstrations occur of those kinds which do not produce as great a certitude as those of geometry and which even differ much therefrom, since whereas geometers prove their propositions by fixed and incontestable principles, here the principles are verified by the conclusions to be drawn from them; the nature of these things not allowing of this being done otherwise.

> It is always possible to attain thereby to a degree of probability which very often is scarcely less than complete proof. To wit, when things which have been demonstrated by the Principles that have been assumed correspond perfectly to the phenomena which experiment has brought under observation; especially when there are a great number of them, and further, principally, when one can imagine and foresee new phenomena which ought to follow from the hypotheses which one employs, and when one finds that therein the fact corresponds to our prevision. But if all these proofs of probability are met with in that which I propose to discuss, as it seems to me they are, this ought to be a very strong confirmation of the success of my inquiry; and it must be ill if the facts are not pretty much as I represent them. I would believe then that those who love to know the Causes of things and who are able to admire the marvels of Light, will find some satisfaction in these various speculations regarding it, and in the new explanation of its famous property which is the main foundation of the construction of our eyes and of those great inventions which extend so vastly the use of them [56].[26] (A.3)

The validation of the principles at the basis of explanations (a theory in modern term) of a physical phenomena is compared to the validation of the interpretation of the characters in the decryption of an encrypted letter.

---

[26] Preface, 2nd-4th pages. English translation in [59].

> I feel that in physics there are no other demonstrations than in deciphering a letter. Here, having made assumptions about some slight conjectures, if one finds that they are verified in the following, so that according to these suppositions of letters one finds coherent words in the letter, one holds with a very great certainty that the suppositions are true, that there is need of another proof, even though it is not impossible that there are others more true [58].[27] (A.4)

1.1.3.1 The Role of Philosophy of Nature. The *Traité* de la lumière

Huygens joined the approach typical of mixed mathematics to the mechanical philosophy in a very efficient way in the study of light propagation. His mechanicistic view was for sure influenced by Descartes and at least by Hobbes (a philosopher) and Pardies (a mathematician), of whom Huygens knew the writings for sure, even if not yet published. In the following I will shortly comment the theories of Hobbes; that of Pardies and Descartes have been described in the previous section.

Thomas Hobbes (1588–1679) proposed a his own theory of light in his *Tractatus opticus* of 1644 [1]. According to him light was not a tendency to motion, but an actual motion, though very small, which is propagating instantaneously through a not specified medium. Light was generated by a luminous body which expands and swells into a greater volume, and then contracts again, continually contracting and swelling (systolem et diastolem). The motion from the luminous body is propagated to the eye through a continual pushing outwards of the contiguous parts of the medium. A characteristic and may be a drawbacks of Hobbes' theory is that the entire body of the luminous body, the sun for instance, expands and contracts together like the systole and diastole of the heart, so that all the rays of light emanate radially from the center of the sun rather than in all directions from its surface as it should be. This is exactly the sort of confusion which would arise if a theory of light were modeled too closely in an analogy with sound, as Hobbes' theory appears to be [117].[28]

Basic concepts of Hobbes's theory of light are those of *ray* and *line of light*. A ray is the path through which the motion from the luminous body is propagated through the medium. It has a thickness and so it is three dimensional. The line of light is the line from which the sides of a ray begin. With reference to Fig. 1.3 the ray is the solid ABKI, while the line of light is the line AB from which sides AI, BK begin. Any one of the lines which are derived from the line of light by a continual extension such as CD, EF, etc are lines of light as well.

Using his concepts of ray and line of light Hobbes was able to explain the refraction from air to glass, assuming that air is less resistant to motion than glass, contrary to what made by Descartes. He was conscious that the theory of expansion and contraction demanded the existence of a vacuum. At the time he wrote the *Tractatus opticus* he did not deny this possibility and thus there was no conflict. But when *De corpore* appeared in 1655, where he discussed about light again, Hobbes came to

[27] Vol. 7, p. 298.

[28] p. 149.

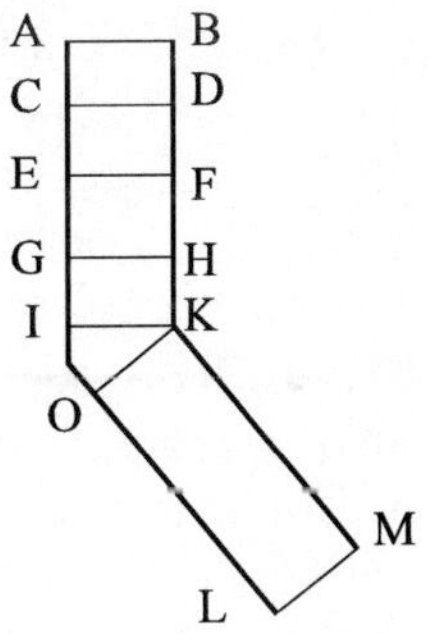

Fig. 1.3 Rays and lines of light. Redrawn from [117], p. 149

deny the existence of a vacuum, and therefore he had to reject his theory of expansion and contraction.

Huygens published his conceptions on the theory of light in the *Traité de la lumière* of 1690. Previous his published works about optics reduced to the *Dioptrica* of 1652, a treatise of geometric optics with a comprehensive theory of refraction and the configuration of lenses in telescopes. One of the reason that compelled Huygens to start a mechanist theory of light was the attempt to explain the strange behavior of the Iceland crystal, that generally exhibits two different angles of refraction and shows refraction even for vanishing angles of incidence. This behavior could not be explained with the traditional geometrical optics and for this reason its treatment was completely absent from the *Dioptrica*.

Huygens' approach to optics was not that of a canonical natural philosopher tending to explain the causes of the various phenomena exhibited by light. He restricted the range of his theory to the extent it could explain the strange refraction of the Iceland crystal and completely ignored the problems of the nature of colors as well as the phenomenon of diffraction, though his theory could explain it. Moreover he remained vague on the nature of the pulses transmitted by the particles of aether; more precisely he did not assume a periodic behavior, so that his was not a true theory of wave in modern sense but simply a pulse or vibration theory. In this section however as Huygens used the word *onde*, speaking of *wave* could be appropriate.

In Chap. 1 of the *Traité de la lumière* Huygens argued on the finite speed of light basing on astronomical observations. In particular he referred to the value obtained by the Danish astronomer Ole Rømer (1644–1710) who made use of the eclipses suffered by the 'little planets' which revolve around Jupiter and which often enter his shadow. Rømer found for the speed of light the value of 11 hundred times a hundred thousand toises (that is 110 000 000 toises) in one minute.[29]

The fact that the speed of light is finite is sufficient for Huygens to state that it propagates as spherical pulses: "Now the successive movement of light being confirmed in this way, it follows, as I have said, that it spreads by spherical waves,

[29]Because a toise is 6 feet and a (French) foot about 0.325 m, the speed suggested by Huygens corresponds to $110\,000\,000 \times 6 \times 0.325 = 214\,000\,000$ m/s, not very far from the presently accepted value of about 300 000 000 m/s.

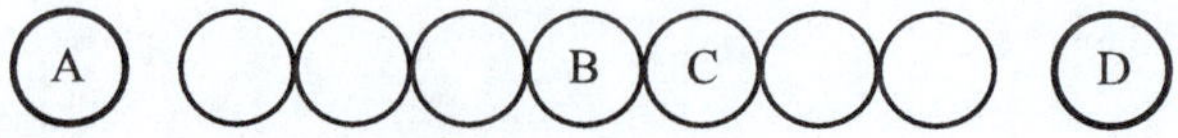

Fig. 1.4 Propagation of light in a line. Redrawn from [56], p. 16

like the movement of sound" [56].[30] Differently from sound however light does not propagate in air but in an aetherial matter. This was demonstrated very clearly, according to him, by the celebrated experiment of Torricelli, in which the tube of glass from which the quicksilver had withdrawn, remaining void of air, transmitted light just the same as when air is in it, but it did not transmit sound. For this proves that a matter different from air exists in this tube, and that this matter must have penetrated the glass or the quicksilver, either one or the other, though they are both impenetrable to air [56].[31] Whereas Pardies admitted essentially the same wave mechanism for light and sound, Huygens believed that the extremely high speed of light required a specific mechanism.

The model suggested by Huygens considered the universe completely filled with particles of the aether whose shape, for the sake of simplicity, was assumed as spherical. The luminous body communicates to these particles an impulsion that propagates through them, without any transfer of matter. Let consider, for instance, the row of equal spheres of elastic matter shown in Fig. 1.4. If against this row there are pushed from two opposite sides at the same time two similar spheres A and D, one will see each of them rebound with the same speed which it had in striking, yet the whole row will remain in its place, although the pulse has passed along its whole length twice over. And if these contrary pulses happen to meet one another at the middle sphere, B, or at some other such as C, that sphere will yield and act as a spring at both sides, and so will serve at the same instant to transmit these two movements [56].[32] If the spheres were exactly rigid the transfer of pulses would be instantaneous, as Descartes assumed, but as the spheres are elastic the transmission of the pressure from a particle to another will need a finite interval of time; the greater the lower the stiffness.

The elasticity of the spheres of aether, is explained by Huygens assuming that they, notwithstanding their smallness, were in turn composed of still smaller parts and that their springiness consists in the very rapid movement of a subtle matter which penetrates the spheres from every side and constrains their structure to assume such a disposition as to give to the subtle matter the most open and easy passage possible. In any case though one should ignore the true cause of springiness, in Huygens opinion, he still sees that there are many bodies which possess this property; and thus there is nothing strange in supposing that it exists also in little invisible bodies like the particles of aether. Moreover if one wishes to seek for any other way in which the motion of light is successively communicated, one will find none which agrees better.

[30] p. 9.

[31] pp. 10–11.

[32] p. 11.

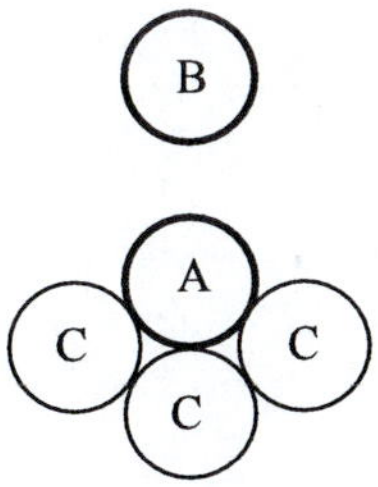

Fig. 1.5 Propagation of light in the space. Redrawn from [56], p. 14

By supposing springiness in the aetherial matter, its particles will have the property of equally rapid restitution whether they are pushed strongly or feebly and thus the propagation of light will always go on with the same speed [56].[33]

Of course, in the universe the particles of aether are not ranged in straight lines, as in Fig. 1.4, but each particle touches several others; this does not hinder them from transmitting their movement and from spreading it always forward. Indeed there is a law of impact serving for this propagation and verifiable by experiment. It says that when a sphere, such as A in Fig. 1.5, touches several other similar spheres CCC and it is struck by another sphere B in such a way as to exert an impulse against all the spheres CCC, it transmits to the spheres CCC the whole of its motion and remains motionless after the impact.

Figure 1.5 shows that any particle of aether transmits the pulse to all the particles surrounding it and thus it can be said that any particles becomes the source of (spherical) waves; what is commonly known as *principle of Huygens.*

> There is the further consideration of these waves, that each particle of matter in which a wave spreads, ought not to communicate its motion only to the next particle which is in the straight line drawn from the luminous point, but that it also imparts some of it necessarily to all the others which touch it and which oppose themselves to its movement. So it arises that around each particle there is made a wave of which that particle is the center [56].[34] (A.5)

The various spherical waves [ondes], to be considered as partial waves join in a unique (spherical) main wave. Consider Fig. 1.6 where DCEF is the spherical front wave emanating from the luminous source A, which is its center. The particle B, one of those comprised within the sphere DCEF, has originated its partial wave KCL, which touches the front wave at C at the same moment that the original wave emanating from the point A has arrived there. It is clear that it is only the part around C of the wave KCL which touches the wave DCEF, which is in the straight line drawn from AB. Similarly the other particles, such as *bb*, *dd*, etc. make their own waves. But each of these waves can be infinitely feeble only as compared with the wave DCEF, to the composition of which all others contribute by the part of their surface which is most distant from the center A.

Huygens did not define explicitly what is a ray of light, but used the term as it was well known—that is as a primitive term—coherently with his assertion that

[33]pp. 12–13.

[34]p. 17. English translation in [29].

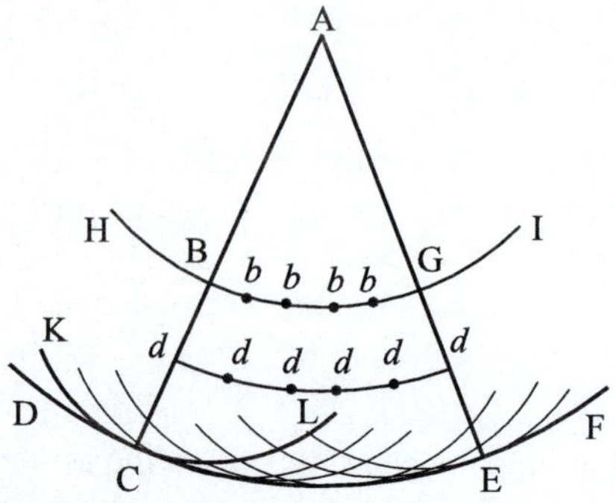

Fig. 1.6 Rectilinear rays. Redrawn from [56], p. 19

light propagates through straight lines. This stems from the assumption of spherical waves, but it is also 'proved' experimentally. Let reconsider Fig. 1.6 where BG is an opening, limited by the opaque bodies BH, GI. The wave of light which issues from the point A will always be terminated by the straight lines AC, AE; the parts of the partial waves which spread outside the space ACE being too feeble to produce light there. Huygens added upon: "Now, however small we make the opening BG, there is always the same reason causing the light there to pass between straight lines; since this opening is always large enough to contain a great number of particles of the aethereal matter, which are of an inconceivable smallness" [56][35] and concluded that each little portion of the wave necessarily advances following the straight line which comes from the luminous point. Thus then one may take the rays of light as if they were straight lines.

Notice that Huygens not only declared that light is propagating according to straight lines, but also that there is not diffraction. This last statement is quite strange for an accurate experimenter like him. Moreover he knew the work on diffraction by Grimaldi from Riccioli's *Astronomiae reformatae* and he was present when Mariotte and La Hire performed experiments at the Académie des sciences, which tended to confirm Grimaldi's results [58].[36]

One of the most marvelous property of the rays of light, according to Huygens, is that when some of them come from different or even from opposing sides, they produce their effect across one another without any hindrance. Whence also it comes about that a number of spectators may view different objects at the same time through the same opening. The mechanism of transmission of light through impact of elastic spheres allows to explain easily why the waves do not destroy nor interrupt when they cross one another [56].[37]

---

[35] p. 19.

[36] Vol. 22, p. 268.

[37] Vol. 22, p. 20.

1.1.3.2 The Role of Mathematics. Isotropic Refraction

Chapter 2 of the *Traité de la lumière* is devoted to reflection, which does not require particular attention. Chapter 4 is devoted to the refraction in non isotropic media, as the atmosphere for instance. In such a case Huygens showed that the rays are curved lines. More interesting is Chap. 3 devoted to refraction. Differently from Descartes and Newton, Huygens assumed that the speed of light was the lower the greater the density of the diaphanous medium. Lower in glass than in air for instance. This lower speed is justified by the detours imposed to the particle of aether that should move through the pores of ordinary matter: "And, moreover, one may believe that the progression of these waves ought to be a little slower in the interior of bodies, by reason of the small detours which the same particles cause. In which different speed of light I shall show the cause of refraction to consist" [56].[38]

By assuming a package of rays it results easy to show that passing from air to glass for instance, the refraction angle is lower that the incidence angle, as it is indeed. For the proof assume the setting of Fig. 1.7. Let the line AC represents a portion of a wave of light and the centre of the luminous body be supposed so distant that the rays in this portion may be considered as parallel lines. Let the piece C of the wave AC, in a certain space of time have advanced following the straight line CB which consequently cuts AC at right angles. In the same time the piece A would have come to G along the straight line AG, equal and parallel to CB and all the portion of wave AC would be at GB if the matter of the transparent body transmitted the wave as quickly as the matter of the aether. Let suppose now that the transparent body transmits this movement less quickly, by one-third, for instance. The motion then would spread from the point A, in the matter of the transparent body, for a distance equal to two-thirds of CB, making its own particular spherical wave. This wave is then represented in Fig. 1.7 by the circumference SNR, the centre of which is A and its semi-diameter is equal to two-thirds of CB. Then if one considers in order the other pieces H of the wave AC, it appears that in the same time that the piece C reaches B they will not only have reached the surface AB, but in addition, they have generated in the transparent body, from the centers K, secondary waves, represented by circumferences whose semi-diameters are equal to two-thirds of the lines KM. Now all these circumferences have for a common tangent the straight line BN; that is the same line which is drawn as a tangent from the point B to the circumference SNR. This line is the propagation (the frontwave) of the wave AC at the moment when its piece C has reached B [56, 59].[39]

The refracted rays being orthogonal to the line BN have the direction like AN, different from AD; the angle between AD and AD depending on the difference between the speed of light in the two media.

[38]p. 30.

[39]pp. 33–34.

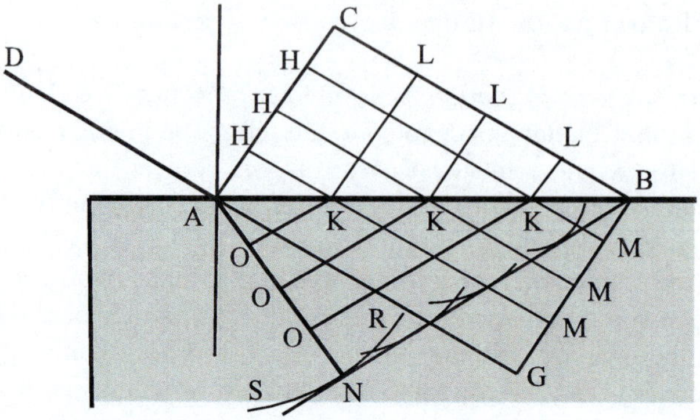

Fig. 1.7 Refraction of light. Redrawn from [56], p. 33

1.1.3.3 The Role of Mathematics. Double Refraction

As already noted, one of the reasons which pushed Huygens to develop a mechanicistic theory of the propagation of light was the desire to explain the strange behavior of the so called Iceland crystal, for which an incident ray originated two refracted rays. To explain this behavior Huygens assumed that, due the particular molecular arrangement of the crystal, two different mechanisms of propagation of light subsisted. The classical one, due to the transmission of pulses through the particles of aether which filled the pores of the crystal, the other due to pulses transmitted directly by the ordinary matter of the crystal.

> As there were two different refractions, I conceived that there were also two different emanations of waves of light, and that one could occur in the aethereal matter extending through the body of the Crystal.
> [...]
> As to the other emanation [...] I supposed it would spread indifferently both in the aethereal matter diffused throughout the crystal and in the particles of which it is composed [...]. It seemed to me that the disposition or regular arrangement of these particles could contribute to form spheroidal waves (nothing more being required for this than that the successive movement of light should spread a little more quickly in one direction than in the other) and I scarcely doubted that there were in this crystal such an arrangement of equal and similar particles, because of its figure and of its angles with their determinate and invariable measure [56].[40] (A.6)

The second way of transmission of light is anisotropic, that is it occurs with different speed in the different directions, so that the waves are not spherical but ellipsoidal. The ellipsoidal waves allow to explain easily the double refraction as well as the presence of refraction for rays incident orthogonally to the free surface of the crystal. One of the refraction is due to the classical transmission of pulses, the other to the second way of transmission.

[40] pp. 58–59.

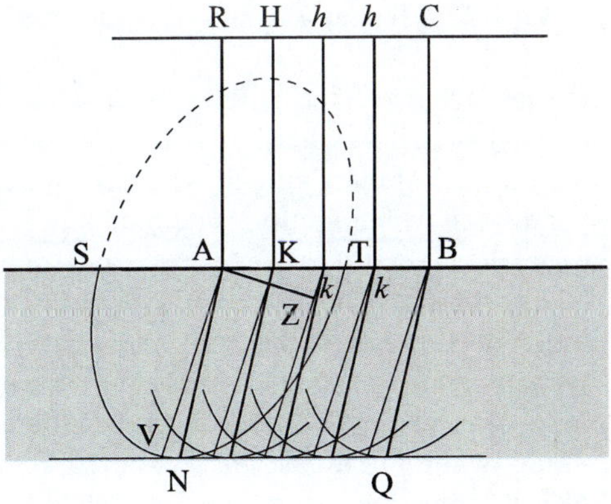

Fig. 1.8 Anisotropic refraction of light. Redrawn from [56], p. 60

With reference to Fig. 1.8 assume the straight line RC, parallel and equal to AB, to be a portion of a wave of light, in which infinite points such as RH*hh*C come to meet the surface AB at the points AK*kk*B. For the second way of transmission instead of the hemispherical partial waves which in a body of ordinary refraction would spread from each of these last points, there are semi-ellipsoidal waves. The axes of the ellipsoids are supposed to be oblique to the plane AB, as is AV, which represents the partial wave coming from the point A. Now taking a certain interval of time during which the wave SVT has spread from A, in the same time, waves similar to SVT and similarly situated occur. And the common tangent NQ of all these semi-ellipsoids would be the propagation of the wave RC which fall on AB, and would be the place where this movement occurs in much greater amount than anywhere else, being made up of arcs of an infinity of ellipsoids, the centers of which are along the line AB [56, 59].

At this point Huygens assumed to have proved that the emerging ray is not orthogonal to AB but inclined as AN, BQ are, and that the line of wave (the front wave) and the rays are not orthogonal.[41] In Huygens's reasoning there are implicit assumptions that though intuitive makes his argumentation not very stringent. The first assumption is that the bundle of rays RH*hh*C changes in another bundle, the other is that a ray maintains its identity after the refraction. That is a ray which terminates in A for instance should continue with another ray which starts from A.

Chapter 6 of the *Traité de la lumière* has an essential mathematical nature, regarding the shape to give to lenses or mirror to satisfy certain optical requisites. It is a chapter typical of any mixed mathematics treatise, where the development of a physical theory is the occasion to start a new mathematical theory. An approach that will characterize the mathematical physical treatises of the 19th century

[41] pp. 60–61. That the line connecting the center A of the ellipses with N is not orthogonal to NQ is clear from the figure. It was a known property of the ellipsoids that AN and NQ are conjugate straight lines.

1.1.3.4 The Hyper-Physics of Space and Motion

Despite Leibniz's accusation of poor sensitivity toward metaphysics, Huygens also payed attention to it when he tried to characterize motion. He acted as a mathematician with the goal of clarifying those aspects of the motion that concerned his studies of dynamics in which motion is defined in a precise way through velocity, intended as an incremental ratio between space and time.

Historians of science individuate three phases in the evolution of the concept of motion in Huygens. In the first phase (about 1650–1670) a young Huygens assumed that both rotational and translational motions were relative. After his studies, second phase, on centrifugal force Huygens attributed a character of absoluteness to rotating motion, while in the last phase, started after his reading of Newton's *Principia* in 1687, he came back to a complete relativistic vision. Regarding the conception of the space in itself the situation is less clear, even because Huygens dealt sparingly with the topic.

In the following I will refer to the concepts of motion (and space) reported by Huygens in some fragments collected in the *Codex Hugeniorum 7A*, in the version published by Gianfranco Mormino [81]. The fragments were probably written between 1687 and 1694, that is in the third phase of Huygens's reflections on motion. They present repetitions and reworking to testify Huygens efforts in the attempt to clarify his ideas; I will refer mainly to Fragment 6, which is one of the most coherent and exhaustive.

Huygens' space is neither the material space of Descartes nor the purely relational space of Leibniz. Huygens gave a positive connotation to space, which is something existing in itself; in that his position is close to that of Newton. But Huygens' space cannot be qualified with rest. Huygens asked himself: "Indeed there is neither motion nor rest if not of a substance. Thus, how could one attribute the rest to the empty space, where there is nothing? The argument is not difficult from a mathematical point of view, but from that physical or hyper-physical (*hijperphysice* in Latin)" [81].[42]

The following quotation gives one of the most exhaustive connotation of space:

> I cannot see how this space, considered in itself, without no body may be conceived at rest, since rest and motion only concern the bodies and the idea of both originated only from them. In fact, if one can say that there is a rest or a motion of space it will be of that space that is occupied or enclosed by a body, like when we say that the space of an amphora is at rest or moving together with it. But to that *infinite and empty space* [emphasis added] cannot be attributed neither the idea nor the name of motion and rest. Those who say that it is at rest do not seem to do it for any other reason than because they realize that it would be absurd to assert its motion, and therefore they thought it must necessarily be said that it is at rest, when instead it should have been thought that neither the motion nor the rest concern in any way that space [81].[43] (A.7)

[42] p. 142. In the manuscript "hijperphysice" replaced a preexisting "methaphysice", erased most probably because Huygens wanted to distance himself from the traditional metaphysics.

[43] p. 232.

To contrast the possibility of an absolute translation motion, Huygens assumed a space where there are two bodies only, A and B that move one with respect to the other. Let then one of them, A for instance, reduced to nothing; perhaps then B, which continues to exist, will no longer be in motion? asked Huygens. Certainly not, he replied, because moving is nothing but changing distance from the other. But here nothing else exists and the world has no borders or one center with respect to which B could change position. Maybe then B will be at rest? Not even, since the rest is relative to another body, respect to which the same distance and position are maintained. But does not a body have to be either at rest or in motion? Thus say scholars for whom motion and rest are something in themselves, without relation to anything else. But, according to Huygens, there is neither motion nor rest if not with respect to other bodies, B will not be neither in motion nor at rest but simply it will exist [81].[44]

In another point Huygens discussed the relative motion between free mass points in the space. Consider two bodies A and B free and at rest with each other. If one pushes A so that A and B are in relative motion, A certainly receives an impulse; but it cannot be said that it is A which moves with respect to B, because the opposite is also true. And if even A is much greater than B, the same effect is achieved either by moving one or moving the other. Then he concluded in a somewhat disconcertingly way: "although it takes more effort to move, in this way, A than B" [81].[45]

The principle of inertia is formulated by Huygens with reference to relative and not absolute motion, as Newton did instead. It is formulated as a motion with respect to other bodies. "If a body moves freely and without any obstacle with respect to some bodies at rest between them, it will travel in a straight line with respect to them and will move in a uniform motion" [81].[46] This for Huygens must be taken as an empirical principle because experience evidently proves it. It can also be justified on a rational basis; in fact it is consonant with reason that bodies put into motion one with respect to the other continue to move without deviating towards any part, if no impediment intervenes, as bodies at rest persevere in this state if it does not happen nothing else [81].[47]

More complex and less convincing is the justification of the relativity of circular motion, to which Huygens attributed the cause of the centrifugal force, demonstrated empirically. This justification affects, for example, the definition of relative rest between two mass points. According to Huygens, those bodies are at rest between them that, without being hindered by any constraint or obstacle to move away freely from one another, nevertheless maintain the same distance [81].[48] This definition of a dynamic nature of rest, serves to exclude that two bodies that move in a circular motion one around the other are at rest; in this case to keep the distance unchanged a constraint is required. Huygens also tried to argue with kinematic considerations

---

[44] p. 180.

[45] p. 182.

[46] p. 210.

[47] p. 210.

[48] p. 208.

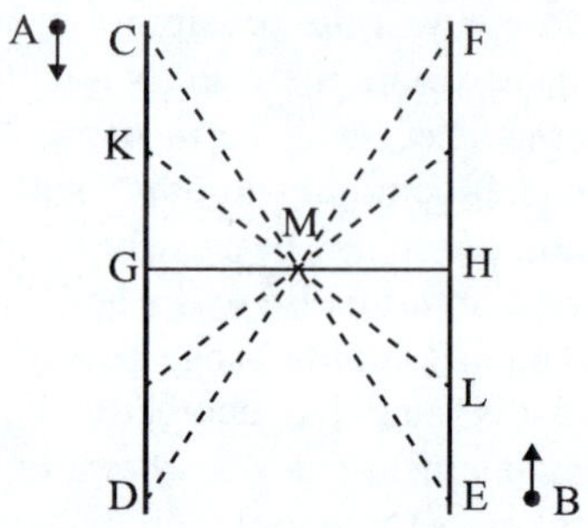

Fig. 1.9 Comparison between the relative velocity and variation of distance of two bodies. Redrawn from [81], p. 212

that in the latter case the bodies are not at rest. The arguments of Huygens, based essentially on analogical reasoning, are not very convincing and somehow presuppose what he wants to try. Before going on to examine them, it is worthwhile to report the following comment:

> For a long time I thought that in the circular motion was given the criterion of true motion, from the consideration of centrifugal force. In fact, as far as it concerns all other phenomena, it is the same thing whether a disk or a wheel that is near me moves circularly or if, being that disk at rest, I turn along its circumference. But if a stone is placed at the end, when the disk rotates, it is thrown away. For this I thought that, in this case, it [the disk] moves and really rotates, even without any reference to other bodies [81].[49] (A.8)

Then, asked Huygens, can two bodies move relatively without changing the distance? The positive answer is entrusted to two arguments of an analogical nature.

In the first argument, Huygens considered two bodies A and B that move according to lines parallel to each other in the opposite directions and equal speed, as shown in Fig. 1.9, implicitly assuming a reference against which these motions make sense.

In a first time interval A arrives in K starting from C and B arrives in L starting from E. In a second time interval of the same duration as the first, A passes from K to G and B from L to H. It is simple to demonstrate that the variation in distance between A and B is greater in the first interval than in the second. Imagining very small time intervals, one has that the variation of distance, and therefore the speed along the line joining the two bodies A and B, decreases until it nullifies when the two bodies pass through the GH line, and then continue to grow. This according to Huygens means that although the relative distance in GH does not vary, A and B still have a relative non-zero speed in the direction of the lines CD and FE. Or reversing the reasoning, if the speed of moving away two points is zero it does not mean that they are at rest.

In the second argument still reference is made to the two bodies A and B of Fig. 1.9. It is known that these two bodies are moving with respect to each other, although it is not known which of them is actually moving. Assume now a thread that by means of hooks constrains the distance between the two bodies A and B as soon as they cross the line GH. Their rectilinear motion will then turn into circular

[49] p. 236.

motion and the thread joining A and B will reveal the fact with its own tension. Huygens asked rhetorically: "Why before the two bodies met these hooks it was not known that they moved of true motion and now after they met them maybe we will know? Can one define how agitated they are in relation to that infinite and immobile space?" [81].[50]

Even this reasoning is not very convincing however; Huygens realized the difficulty and tried another reasoning:

> In free motion, given some bodies at rest [which to refer the motion], the directions and velocities along these directions are known with certainty; thanks to them one will explain the change in distance. The velocity of rotating bodies is also determined by means of those bodies [at rest]. Once those bodies are removed, it is more difficult to identify such quantities and velocities in free motions, but the circular motion of two or more bodies joined by a constraint, or parts of a single body, is recognized by the centrifugal force. Against those who maintain that this is a true movement, I affirm that it is only relative. It cannot in fact be said that the center of rotation is at rest in the world, but only in relation to other bodies [81].[51] (A.9)

1.2 Newton Philosopher, Theologian, Alchemist and Even Mathematician

In the early modern science there have been scholars who excelled both in practical and in theoretical activities. Isaac Newton (1643–1727; Gregorian calendar) was one of them and his field of interest was enormous. From mathematics to natural philosophy, to electricity, to theology, to Church history, to alchemy, to management, etc. Today, with the specialization that characterizes the modern sciences this vastness of interests would be inconceivable and if someone tried to perform researches in so a vast field, even though he were a genius like Newton or even superior, he would surely be doomed to failure. This vastness of interests was instead natural for Newton and for many of the scholars of the 18th century who should be seen not as scientists in modern sense but rather as scholars for whom it was a legitimate task to use a wide variety of materials to reconstruct the unified wisdom of the creation [78].[52]

What was exceptional in Newton is the extent to which his papers, manuscripts and books of his personal library have been preserved. Newton was a man of his times, he received a scholastic education and as many students of his time was an avid reader (and collector) of books, and like them used to summarize the results of his reading in *commonbooks*, a humanistic habit. He often relied on compiled lists of quotations in his discussions on the history of Church and on alchemy, giving the "impression of a highly erudite man without [actually] performing the impossible task of reading any source named" [47].[53]

---

[50] p. 192.

[51] p. 228.

[52] p. 138.

[53] p. 6.

Presently historians were able to consult in a easy way Newton's surviving manuscripts; they contain more or less ten millions of words which are stored in digital scans and transcriptions (see Newton project).

> We may question whether its preservation is owing to his fame or to his own meticulous cultivation of his papers, but regardless, we have in Newton's documents, now easily accessible, a remarkable insight into the world of the educated man of science and letters in late 17th century and early 18th century England. Portraying Newton as a humanist, an erudite reader of texts, and a participant in various hermeneutical communities does not detract from his value to historical narratives of the development of science. Rather, it shows both the importance of the history of scholarship to the history of science and the need to constantly evaluate the historical categories we apply to individuals from the past [47].[54]

During the whole 20th century studies on Newton concerned nearly completely the achievements in the fields of mechanics, optics and mathematics. The focus has now shifted toward what is called the *second Newton*, that is the scholar of alchemy, theology and so on; in [47, 77] an interesting survey, reviews and comments are reported on these studies. Various are the reasons of this shift; from the one hand most of Newton's papers on physics and mathematics are today well known and commented upon, so it is natural that new fields are explored, even for reasons of an academic career. It is indeed easier that works not completely known make it easier to be welcome by reviewers of journals. On the other hand, the examination of Newton's non scientific papers—this is a qualification commonly used, but it is improper because how one can say that the alchemical studies do not belong to science, for instance; Newton thought they did—would reveals a much more complex personality than it was supposed.

The variety of Newton's topics, would suggest that to understand his physics for example, one should also study, and thus to be an expert of, his theology, alchemy, etc., because all these fields of knowledge are naturally correlated in his mind. This is only partially true however. Indeed, Newton himself, I mean the individual Newton, was not a coherent author in all the fields he studied and explicitly compartmentalized his work and recognized that different subjects required discipline-specific discursive forms. So also the historian can compartmentalize his studies, only partially of course; which makes it possible to explore an aspect of Newton's though without a complete patronage of the whole, but also without completely ignoring it [60]. Of course group work would be desirable.

Newton and his contemporaries were told somehow fortunate because they found a virgin field ahead. In this regard it is worth quoting a comment by Lagrange on this fortune, reported by his biographer Jean-Baptiste Delambre (1749–1822). Lagrange considered Newton as the greatest genius that had ever existed but also "the most lucky. In his time the system of the world was still to be discovered" [22].[55]

Today no one denies Newton's greatness as a mathematician or a physicist and even as an alchemist. Many, however, criticize his approach to philosophy, seeing him as a crude empiricist, sometimes naive. Put differently, while today there is a

---

[54] p. 8.

[55] p. XX.

general agreement to treat for example Descartes, Leibniz, Hobbes and to some extent Gassendi as philosophers, there is much less agreement with Newton, as well as with Galileo and Huygens. The criticisms come from modern canonical philosophers and historians of literary education. I don't want enter the merits of the matter, but only to point out that an objective analysis shows that this image is at least questionable [123–125].

The division between philosophers and not philosophers, is something we impose on the past and profoundly anachronistic. At the time all the scholars mentioned above were treated as philosophers tout court; Locke for instance discussed with Newton on 'purely' philosophical matter and was profoundly influenced by him. Even though it was recognized that some were more specialized in one topics than in another. The reasons that today push Newton out of the list of philosophers is his apparent reluctance to deal with canonical philosophical topics in his published texts and not to write systematic treatises about metaphysics or ethics. This is certainly true, but in his published texts there are points, in particular the scholia of the *Principia* and the Queries in the *Opticks*, in which the originality and importance of his conceptions of philosophy emerge. Newton was more explicit in his letters and unpublished works.

In the following I will only try to show how the 'mathematician' Newton dealt with themes that can undoubtedly be qualified, with a modern terminology, as philosophical, in his effort to bring back wide areas of natural philosophy into mixed mathematics. I will limit myself to report his studies on motion and the nature of light and spend a few words on his theological studies, I instead have left aside his activity as a chemist and alchemist, though Newton devoted at least as much time to alchemical (and theological) studies as to his more 'scientific' ones. The process of dating his manuscripts has shown that Newton worked regularly on alchemy during his life, paralleling his 'scientific' researches; which proves that Newton's interests for alchemy was not a hobby of senility, as sometimes it is sustained [40]. Since his youth Newton had always been interested in metallurgical process. In the period 1683–1684 he carefully studied Agricola's work on metals and the transmutation of metals was his chief interest. But not for mystical reasons, rather for practical ones. For instance it would have been useful to change iron into copper, as copper mines at that time were very few and warfare and casting of cannons demanded much copper.

Some traces of Newton's alchemical studies can be found in his famous treatises also. A part from profound general connections, there is evidence of the direct influence of alchemy in the Query 31 of the *Opticks*. In particular the idea of short range forces (a somehow vitalistic conception of matter) could possibly be inspired by the Belgian iatrochemist Jean Baptiste van Helmont (1579–1644). Other influences are commented in [82]. It must be said that alchemical writings are difficult to read for a modern historian trained in chemistry, much more than the theological ones are for a modern historian trained in philosophy. Indeed the language of modern chemistry is much farther from that of alchemy than that of modern philosophy from that of theology of the 17–18 centuries. This justify the emergence of a greater number of recent studies on Newton's theology than on alchemy.

One more aspect I left aside is the role Newton played as a manager of the Mint. This job, before as Warden and then as Master, absorbed him since 1696, after he left

Cambridge for London in his fifty-four year, until his death in 1727; that is more than thirty years. The activity at Mint is substantially ignored by historians of science, considering it as a hobby, a reward given to Newton because of his fame and his belonging to the winning party of the Glorious Revolution; on some aspects on this see [120, 130]. Probably historians should devote some more effort to understand Newton's role, considering also that he took seriously it, as suggested, for instance, by a letter to John Flamsteed (1646–1719): "I do not love to be printed on every occasion, much less to be dunned and teased by foreigners about mathematical things, or to be thought by our own people to be trifling away my time about them, when I am about the King's business".[56]

When in 1661 Newton began the study of natural philosophy at Trinity College, Cambridge, Aristotelianism, broad meaning, was still the central system of thought in the educational system [21]. One textbook on natural philosophy was Johannes Magirus's *Physiologia peripatetica*. Magirus dealt with the full sweep of topics proper to physics of the time: the principles of natural things, place, vacuum, motion, time; the planets, fixed stars, eclipses; the elements, primary, secondary and occult qualities, mixed bodies; meteors, comets, tides, winds; metals, minerals, plants, spirits, man, zoophytes; the soul, the senses, dreams, the intellect, the will. This was the broad agenda for natural philosophy throughout Newton's lifetime, unimpaired in his case by a possible inclination toward the Stoic view [42].[57] One more text at Newton's disposal was *Axiomata philosophica sub titulis XX comprehensa* of 1645 by Daniel Sthal (1589–1654), more philosophical than Magirus'. From copies pertaining to Newton's private library it is possible to reconstruct his reading basing on annotations and corrections. In [36][58] it is stressed the relevance of the treatise on logic by Samuel Smith (1587–1620), *Aditus ad logicam* of 1613, as Newton's source for his conception of analysis and synthesis in natural philosophy.

Newton was involved in metaphysical studies in his youth, even though he is often charged to have an anti-metaphysical attitude. One of his more canonical philosophical text, the unpublished manuscript *De gravitatione* was concerned with metaphysics. Of uncertain dating it deals with God and his management of Creation, doctrines of substance, the nature of mind and body and their interaction and union. In the Scholium generale of the *Principia*, added to the second editions, which for about sixty percent is concerned with theological writings, one finds the famous passage on God, Lord over all, which has a high metaphysical and theological content:

> The supreme God is an eternal, infinite, and absolutely perfect being [...]. He is eternal and infinite, omnipotent and omniscient, that is, he endures from eternity to eternity, and he is present from infinity to infinity; he rules all things, and he knows all things that happen or can happen. He is not eternity and infinity, but eternal and infinite; he is not duration and space, but he endures and is present. He endures always and is present everywhere, and by existing always and everywhere he constitutes duration and space. Since each and every particle of space is *always*, and each and every indivisible moment of duration is *everywhere*, certainly the maker and lord of all things will not be *never* or *nowhere* [...]. God is one and the same

[56]Quoted from [120], p. 217.

[57]p. 423.

[58]pp. 33–36.

God always and everywhere. He is omnipresent not only virtually but also *substantially* [...]. We know him only by his properties and attributes and by the wisest and best construction of things and their final causes we admire him because of his perfections [...]. This concludes the discussion of God, and to treat of God from phenomena *is certainly a part of natural philosophy* [emphasis added] [90].[59] (A.10)

With its natural philosophy, metaphysics and theological apologetics, the Scholium generale offers an important sample of the interaction between these fields. An interesting analysis of the theological implications of the scholium can be found in [121] where its different drafts and related private writings are studied. In particular it is shown that Newton was much more explicit about his antitrinitarian view in private writings than in official writings, because the denial of Trinity was prohibited by law.

Newton was a dualist and a libertarian [42],[60] a choice whose defense required a use of metaphysics. There is abundant textual evidence of Newton's belief in the motive powers of the will [42].[61] Interesting the contrasting view on the role of will between Newton and Leibniz is referred to in a review of the *Commercium epistolicum*, published in the Philosophical Transaction of 1715 (Gregorian calendar), which also gives some hints of Newton's mechanical philosophy.

It must be allowed that these two Gentlemen differ very much in Philosophy. The one proceeds upon the Evidence arising from Experiments and Phaenomena, and stops where such Evidence is wanting; the other is taken up with Hypotheses, and propounds them, not to be examined by Experiments, but to be believed without Examination. The one for want of Experiments to decide the Question, doth not affirm whether the Cause of Gravity be Mechanical or not Mechanical: the other that it is a perpetual Miracle if it be not Mechanical. The one (by way of Enquiry) attributes it to the Power of the Creator that the least Particles of Matter are hard: the other attributes the Hardness of Matter to conspiring Motions, and calls it a perpetual Miracle if the Cause of this Hardness be other than Mechanical. The one doth not affirm that animal Motion in man is purely mechanical: the other teaches that it is purely mechanical, the Soul or Mind never acting upon the body so as to alter or influence its Motions [110].[62]

But metaphysics, for Newton, was not just about divinity and will. It dealt with the definition of the fundamental concepts of the philosophy of nature, in particular the concepts of space, time and motion.

1.2.1 Space, Time and Motion

Newton elaborated concepts of space and time having in mind both his needs as a mathematician and those as a philosopher of nature. As a mathematician he needed a structure to support his laws of motion; as a philosopher he felt the need to give a sense of reality to his concepts.

---

[59] Scholium generale, pp. 528–529. English translation in [103].

[60] p. 437.

[61] p. 438.

[62] p. 224.

In the modern approach of Newtonian mechanics[63] the introduction of the concepts of space and time is relatively simple and not problematic; the two concepts are unified in one: space-time.[I] That the universe be absolute and infinite does not create any embarrassment in a prevalently mathematical theory, as mechanics is considered today, simply because here there is no reference to ontological aspects. Only definitions are concerned and as such perfectly legal and thus only discussion about usefulness are meaningful.

Newton did not proceed in this way because he was part of a community equipped with a mathematical apparatus less inclusive than the current one. Newton was a mathematician rooted in the tradition of mixed mathematicians. They need propositions that are mathematical in nature but must be extracted from the real world and require elaboration that had to take place within the natural philosophy of the time.

Newton's main problem was the definition of motion, and the concepts of space and time had to be introduced in such a way as to explain it. On motion the canonical philosophers of nature had long disputed; mathematicians a little less. A first idea of motion the mathematicians had in geometry in which motion was the basis for the definition of some fundamental geometric entities, such as the spiral for instance. But it was a motion in which space and especially time were purely ideal. Very interesting attempts to define the motion of bodies, essentially of mass points, occurred in the Renaissance with Tartaglia and Benedetti. But it was only with Galileo that motion entered forcefully into mixed mathematics. In particular, to time in which motion is developed, a physical reality character was attributed and measurement criteria were also given, with the pulse beats, the water clock, exploiting the isochronism of the pendulum. Space did not appear problematic because it was the space that surrounds us, in which the position of bodies is uncritically referred to the earth's surface. Galileo faced the problem to establish whether the reference he was considering was fixed or mobile. And solved the difficulties involved introducing what is today called the Galilean relativity principle, which he expressed intuitively and with little precision with the image of an observer performing experiments inside a cabin of a ship, with the windows darkened so that he cannot notice if the ship is in motion or at rest. The observer according to Galilei would observe the same phenomena in both cases.

By introducing a cosmological context the concept of space became much more challenging. Descartes faced the problem by providing a solution that appears to us to be twisted, and it appeared so to many of his contemporaries, probably also because of the Church's opposition to considering the earth in motion in space. Descartes, to save his planetary vortex theory in which the earth is dragged with circular motion, introduced the definition of philosophical motion, distinct from vulgar motion, in which he could somehow say that the motion, in a philosophical sense, of the earth, with respect to the layers of fluid that immediately surrounds it, was at rest [43].[64]

---

[63] Newtonian mechanics is a term used in modern times almost as a synonym for classical mechanics. Thus it is not Newton's mechanics as developed in the *Principia*.

[64] pp. 156–196.

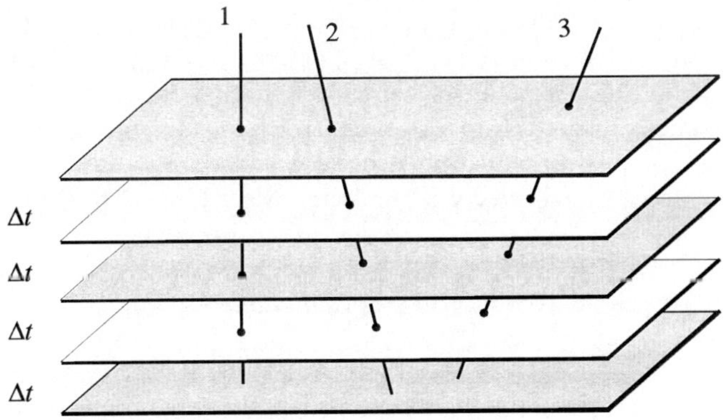

Fig. 1.10 Galilean space time structure

Newton was very critical of this definition and discussed it at length. One of Newton's problems was to give a rational foundation to the principle of inertia, whose validity is professed on at a cosmological level. How to say that a body without interactions from the outside moves in a uniform straight line? For Newton the simplest solution, in reality the only one he proposed, is to think of an absolute space, a space that exists in itself, and although infinite, it allows in some way to provide a position and define a motion, which are absolute. Newton's choice is not the only one possible. A modern scholar of the fundamentals of mechanics has clear that Newton asked more than it was structurally necessary. The laws of his mechanics require a weaker structure than absolute space-time, they only need one in which systems in uniform motion with respect to one another are indistinguishable; in other words it is sufficient to assume a *Galilean space-time* structure.[II]

A geometrical representation of the Galilean space-time can be given with reference to Fig. 1.10, where for the sake of simplicity (and possibility) of representation a two-dimensional space—and thus a three-dimensional space-time—is considered. The planes, which represent the space of contemporary events are replicated at each time interval Δt. Lines 1, 2, 3 represent rectilinear trajectories of a mass point that moves in the space time; the intersections with the planes give the position of this mass point in space at different instant of time, a position that is assumed endowed with individuality even if the positions in space are not countable. If the space time is Galilean, one cannot say which of the three trajectories is the one, or even too there is one, that represents rest. From the figure it would seem that 1 represents rest, but this is only due to the impossibility of giving a geometrical representation of motion of a mass point in the boundary less space-time; indeed without having introduced an observer or a reference frame, nothing can be said. If as a time axis (understood in a general sense as one of the three axes necessary to define the position of the mass point in space-time) instead of choosing 1, one chooses 2, or 3, then 2 or 3, represent points at rest.

It is not clear how much Newton in his choice was influenced by technical reasons, or by philosophy of nature, or metaphysics, or even theology. As a mathematician there was nothing more natural to assimilate the physical space to the space of Euclidean geometry. Moreover in his time there were numerous discussions on the reality of space among canonical philosophers, including Pierre Gassendi (1592–1655) and Newton's friend and colleague Henry More (1614–1687) [50].

1.2.1.1 A Metaphysical View. The *De gravitatione*

The *De gravitatione* is a text commonly classified as metaphysical; according to Howard Stein "this fragment deserves to be considered one of the most interesting metaphysical disquisitions of the seventeenth century" [122].[65] It is written in Latin but some English translations exist [99, 100, 104, 105].

The text is of controversial dating. Since the 1960s, following work by A. Rubert Hall and Mary Boas Hall [99], it was assumed to be composed around 1760s, or a few years later. But in 1991 it was suggested by Betty Jo Dobbs it was written in early 1685, as part of the preparation of the *Principia* [32].[66] This position is now prevalent [111] even though not shared by all [10, 54]. A correct dating of the *De gravitatione*, that is establishing if it was an early work or a mature one, is important to evaluate the evolution of Newton's thought on force, space and time. Luckily for the present book this point is not fundamental.

The *De gravitatione* is mainly a critique to the Cartesian concepts of space and motion (the indirect famous controversy with Leibniz is later); it also refers to topics with a theological background, which have been the subject of harsh criticism by modern science historians. Interesting considerations on the concept of force are also included.

The first part of the *De gravitatione* is a summary of Descartes's doctrine about the nature of motion, carefully supplied with references to the *Principia philosophiae* with quotations of passages in which Descartes himself contradicts his own position and a series of arguments demonstrating the utter incoherence of Descartes's conceptions as a foundation for the physical theory of motion. After this summary Newton's own conception of space is presented, followed by the introduction of the concept of force. No room is instead left for the discussion of the nature of time.

The documents starts and ends, in a puzzling way, as an unfinished hydrostatic treatise. It begins by stating that the two sciences that deals with gravity and equilibrium of fluids and solids in fluids belong to mixed mathematics, in which the principles are extracted from the phenomena and finishes with a discussion on elastic fluids which lasts for several pages. In between these hydraulic 'digressions' there is the metaphysical content.

At first definitions concerning place, body and motion are introduced:

1. Place is part of space which things fill evenly.

---

[65] p. 28.

[66] p. 141.

2. Body is that which fills place.
3. Rest is remaining in the same place.
4. Motion is change of place.

Newton paused at length on these definitions, pointing out against Descartes that there is a difference between place and body. Then he introduced his conceptions-definitions of extension and space. Extension is neither substance, nor accident, nor else nothing at all. But it has its own manner of existing which is proper to it. It is not substance: on the one hand, because it is not absolute in itself, but is "as it were an emanative effect of God [tanquam Dei effectus emanativus]; on the other hand, because it is not endowed with the proper affections that denote substance, namely actions, such as thoughts in the mind and motions in body" [104].[67]

1. In all directions, space can be distinguished into parts whose common boundaries we usually call surfaces; and these surfaces can be distinguished in all directions into parts whose common boundaries we usually call lines; and again these lines can be distinguished in all directions into parts which we call points.
2. Space is extended infinitely in all directions. For we cannot imagine any limit anywhere without at the same time imagining that there is space beyond it.
3. The parts of space are motionless.
4. Space is an affection of being just as being [Spatium est entis quatenus ens affectio]. *No being exists or can exist which is not related to space in some way* [emphasis added]. God is everywhere, created minds are somewhere and body is in the space that it occupies; and whatever is neither everywhere nor anywhere does not exist. Hence it follows that space is an emanative effect of the first existing being, for if any being whatsoever is posited, space is posited.[68]
5. The positions, distances and local motions of bodies are to be referred to the parts of space. And this appears from the properties of space enumerated as 1 and 4 above, and will be more manifest if one conceives that there are vacuities scattered between the particles, or if he pays heed to what formerly said about motion. *To this it may be further added that in space there is no force of any kind that might impede, assist, or in any way change the motions of bodies* [emphasis added].
6. Lastly, space is eternal in duration and immutable in nature because it is the emanative effect of an eternal and immutable being. If ever space had not existed, God at that time would have been nowhere; and hence he either created space later (where he was not present himself), or else, which is no less repugnant to reason, he created his own ubiquity [104].[69]

Two things should be underlined in these definitions. First, although Newton said (point 4) that space is "as it were an emanative effect of God", this passage explicitly does not derive space from theology. Space is "an emanative effect of the first-existent thing", which, according to Newton's theology, is indeed God; but the reasoning holds

[67]p. 21.

[68]This translation is nearly verbatim the same as that referred to in [122], p. 32.

[69]pp. 22–26.

good even though God is not considered at all, because what does matter is that "posit any thing [not necessarily God], space is posited" [122].[70] Second, space is declared to be inert (point 5). This constitutes the empirical foundation of the metaphysics of space. It is drawn from astronomical observations of the motion of the planets and comets and from the experiments on pendulum, carefully designed, reported in Book II of the *Principia* which showed that if there is an aether its resistance should be either zero or entirely insensible [aut nulla erit aut plane insensibilis] [87].[71] The requisite of inertness is essential to exclude space from the list of substances and authorized Newton to declare that there is void in the open space [104].[72]

After the definition of space that of body follows. For Newton, however, the introduction of bodies is more difficult, for they only exist by divine will. He declared himself reluctant to say positively what the nature of bodies is, but he would rather describe a certain kind of being similar in every way to bodies (whose concept is given at the moment assumed as primitive), but not necessarily a body, and whose creation one cannot deny to be within the power of God, so that we can hardly say that it is not body [104].[73]

Thus, said Newton, suppose that there are empty spaces scattered through the world, one of which, defined by certain limits, happens by divine power to be impervious to bodies, and by hypothesis it is manifest that this would resist the motions of bodies and perhaps reflect them, and assume all the properties of a corporeal particle, except that it will be regarded as motionless. If we should suppose that impenetrability is not always maintained in the same part of space but can be transferred here and there according to certain laws, yet so that the quantity and shape of that impenetrable space are not changed, there will be no property of body which it does not possess [104].[74] In the same way, if several spaces of this kind should be impervious to bodies and to each other, they would all sustain the vicissitudes of corpuscles and exhibit the same phenomena. And so if all of this world were constituted out of these beings, it would hardly seem to be inhabited differently. And hence these beings will either be bodies, or very similar to bodies. One can thus define bodies as determined quantities of extension which omnipresent God endows with certain conditions. These conditions are:

1. That they be mobile and therefore one did not say that they are numerical parts of space which are absolutely immobile, but only definite quantities which may be transferred from space to space.
2. That two quantities of this kind cannot coincide anywhere, that is, that they may be impenetrable, and hence that oppositions obstruct their mutual motions and they are reflected in accord with certain laws.

---

[70] p. 32.

[71] p. 353.

[72] p. 34.

[73] p. 27.

[74] p. 28.

3. That they can excite various perceptions of the senses and the imagination in created minds, and conversely be moved by them, which is not surprising since the description of their origin is founded on this [104].[75]

1.2.1.2 A More Mathematical View. The *Principia*

The *De gravitatione* is characterized by a metaphysical view including the ultimate ontological status of space and its relation to God. The *Principia*, by contrast, has a more restricted domain of entities, appropriate to mixed mathematics. The treatise saw three editions in Newton's life time: in 1687, in 1713 and in 1726; all of them in Latin [87, 89, 90]. The first English translation was by Andrew Motte (1696–1734) in 1729 [93], which was revised by Florian Cajori (1859–1930) and published posthumous in 1934 [97]. Two modern important editions exist, one, without an English translation, due to Koyré and Ierome Bernard Cohen with the assistance of Anne Withman of 1972 [101], a starting point for any serious research on the *Principia*, with critical notes commenting variants and Newton's annotations; and another edition completed with the English translation due to Cohen and Withman of 1999 [103]. In the following all the translations are drawn from this last edition.

Next to the absolute space (that is the space that has a reality in itself), "in its own nature, without regard to anything external, remains always similar and immovable", in the *Principia* there is the relative space, which is "is some movable dimension or measure of the absolute spaces; which our senses determine by its position to bodies; and which is vulgarly taken for immovable space" [90].[76] At the end of the scholium in which space and time are introduced, Newton in controversy with Descartes said that the measure of space must not be confused with the space itself, otherwise there is the risk of doing violence to the sacred scriptures:

> And if the meaning of words is to he determined by their use, then by the names time, space, place and motion, their measures are properly to be understood; and the expression will be unusual, and purely mathematical, if the measured quantities themselves are meant. Upon which account, they do strain the sacred writings, who there interpret those words for the measured quantities. Nor do those less defile the purity of mathematical and philosophical truths, who confound real quantities themselves with their relations and vulgar measures [90].[77] (A.11)

An important innovation of the *Principia* with respect to the *De gravitatione* is the introduction of time. Next to the "absolute, true, and *mathematical time* [emphasis added], of itself, and from its own nature flows equably without regard to anything external, and by another name is called duration", there is the "relative, apparent, and common time" which is "some sensible and external (whether accurate or unequable)

[75] pp. 28–29.

[76] p. 6.

[77] p. 11. English translation in [103].

measure of duration by the means of motion, which is commonly used instead of true time; such as an hour, a day, a month, a year" [90].[78]

The meaning of the statement according to which time flows equably seems to presuppose a substratum with respect to which the flow takes place. This cannot be the case and the phrase "flows equably" refers not to the ontology of time but rather to its structure. This should allow to say that it is meaningful to ask of any two events how much time elapses between their occurrence [39].[79]

The concept of time is more elusive than that of space and on it thousands of books and articles have been written by authors of any education. Since ancient times the idea of absolute space had a substantial consensus on the part of both philosophers and mathematicians, while that of time had less. For example, in the *De rerum natura*, Lucretius gave a substantially relativistic definition of time: "Even time exists not of itself; but sense reads out of things what happened long ago, what presses now, and what shall follow after: No man, we must admit, feels time itself, disjoined from motion and repose of things" [72].[80] And Aristotle, in the *Physica*, besides associating motion with time gave a subjective connotation of it, wondering if its existence is connected to the human soul [4].[81]

The absolute concept of time started to affirm as the instruments of its measure were perfected. It is natural to think that they always measure the same thing and that this thing exists in itself. Said with Newton, it is natural that time is something absolute and that it always flows in the same way. Time became not only absolute but also mathematical with Galileo, who among the first, took time as a physical magnitude that intervenes in the formulation of the laws of nature. Certainly there were difficulties in the measurement of time, for which one had to resort to phenomena observable with the senses, for example motion. Time was thus measured with some motions which could be considered uniform, such as the flow of water and the rotation of celestial bodies. It could happen that these motions, with the introduction of new physical theories and new measuring instruments, were downgraded to non-uniform, not always with obvious reason. Physicists, however, were convinced that they could carry out increasingly more consistent (absolute) time measurements. Of course with the theory of relativity everything has changed.

Historians are left with a number of problems regarding Newton's concepts. These include questions about Newton's early ideas about space and time and their relation to his atomist ideas; questions about the role of his theory of fluxions for fostering his ideas about time in physics and of course the relationship between the concepts figuring in his physics and those in his metaphysics [68].[82]

When Newton wrote on absolute time in the *Principia*, along with its correlate, absolute space, he seemed to assume them as something selfevident. One indication that supports this possibility is that the discussion on the matter is free of caveats.

---

[78]p. 6.

[79]p. 8.

[80]Translated into English by Leonard WE, I, 459, 461.

[81]IV, 10, 218.

[82]p. 18.

But not a few scholars disagreed. Leibniz did not contrasted directly Newton, but exposed his ideas in the famous correspondence with Samuel Clarke (1675–1729); his ideas are summarized below[83]:

> As for my own opinion, I have said more than once, that I hold space to be something merely [purement] relative, as time is; that I hold it to be an order of coexistences, as time is an order of successions. For space denotes, in terms of possibility, an order of things which exist at the same time, considered as existing together; without enquiring into their manner of existing [17].[84]

However Leibniz with his conception of relational space offered only criticisms and not an organic alternative to be used as basis for the foundation of mechanics. For example there was no reaction to Newton's bucket experiment.

The criticisms toward the concepts of absolute space and time introduced by Newton were also taken up by modern philosophers and (some) scientists. Howard Stein was among the first to argue that the modern critique of these concepts is misdirected and confuses Newton's ontological conceptions with those that are actually definitions. According to Stein Newton did not try to answer the metaphysical question if space and time are actually absolute or not; on the contrary, he did not even take for granted that such a question was well-posed. His primary aim, instead, was to define absolute space, absolute time and absolute motion for applying the concepts and to reveal the roles that they play in solving the problems of mechanics. The corresponding concepts defined by his contemporaries, as purely relative notions, were for any mechanical purpose quite useless [30].[85]

1.2.2 The Concept of Force

That Newton 'helped' to spread the term force in natural philosophy is a shared opinion. What however he intended with this term has been the object of heated debates, probably not yet concluded, to which in the following I will give my contribution. To exemplify the nature of contention it is enough to cite the opinions of who are among the most influential interpreters of Newton: Richard Westfall and Ierome Bernard Cohen. Westfall states that Newton rejected the prevailing mechanical philosophy by insisting that force must be endowed with fundamental ontology [129].[86] Cohen, on his side, contends that never Newton even addressed the question of the existence of (true) forces [20].[87]

If Newton's concept of force at his time was the subject of discussion (by philosophers), its use was (almost) immediately unquestioned (by mathematicians). With

[83]The correspondence of Clarke and Leibniz has been the object of countless papers; for a modern interesting comment see [6].

[84]III, 4, p. 57.

[85]p. 17.

[86]p. 377.

[87]p. 346.

the idea of providing a measure of force by means of the variation of velocity, or more precisely by means of acceleration, and therefore of transforming it into a mathematical magnitude, thanks to the use of Calculus Newton was able to present a very efficient tool in the *Principia* that allowed the immediate solution of some problems of the mechanics of the times and laid the foundation for the solution of all the others.

The concept of force has evolved in Newton, especially in the youth period. In the following I report his ideas in the essentially definitive phase. First as he expresses them in the *De gravitatione* and then in the *Principia* and in the Queries of the *Opticks*.

1.2.2.1 Definition of Force (and Mass) in the *De gravitatione*

The *De gravitatione* besides discussion on space and time left room for a discussion about forces. They were introduced with the following main definition:

> Definition 5. Force is the *causal principle of motion and rest* [emphasis added]. And it is either an external one that generates, destroys, or otherwise changes *impressed motion* [emphasis added] in some body, or it is an internal principle by which existing motion or rest is conserved in a body, and by which any being endeavors to continue in its state and opposes resistance [104].[88]

Notice that here Newton assumed the existence of two kinds of force, the external and the internal. The latter is named *inertia* in definition 8, reported below.

After Definition 5, other definitions follow which give some characterization of force:

Definition 6. *Conatus* [endeavor] is resisted force, or force in so far as it is resisted.
Definition 7. *Impetus* is force in so far as it is impressed on a thing.
Definition 8. *Inertia* is the inner force of a body, lest its state should be easily changed by an external exciting force.
Definition 9. *Pressure* is the endeavor [conatus] of contiguous parts to penetrate into each other's dimensions. For if they could penetrate [each other] the pressure would cease. And pressure is only between contiguous parts, which in turn press upon others contiguous to them, until the pressure is transferred to the most remote parts of any body, whether hard, soft, or fluid. And upon this action is based the communication of motion by means of a point or surface of contact.
Definition 10. *Gravity* is the force in a body impelling it to descend. Here, however, by descent is not only meant a motion towards the center of the earth, but also towards any point or region, or even from any point.
Definition 11. The *intension* of any of the above mentioned powers is the degree of its quality.
Definition 12. Its *extension* is the quantity of *space* [emphasis added] or time in which it operates.
Definition 13. Its absolute quantity is the product of its intension and its extension.

---

[88] p. 36.

Definition 14. *Velocity* is the intension of motion, slowness is remission.
Definition 15. Bodies are denser when their inertia is more intense and rarer when it is more remiss.

From the last definition it seems that density is a definite (or secondary) concept and inertia a primitive one, which would be an approach unusual and contrary to what is usually attributed to Newton. A reading of the comment to this definition shows however that things are not so clear and density is also definite in a geometrical way, as the volume of the matter, once pores are ignored: "so that one may consider inertia to be remitted by the increase of the pores and intensified by their diminution, as though the pores, which offer no inertial resistance to change, and whose mixtures with the truly corporeal parts give raise to all the various degrees of inertia, bear some ratio to the parts" [104].[89]

May be it is only an improper way to express his ideas, but a duality in the conception of mass as a geometrical magnitude and a dynamical one (inertia) also occurs in the *Principia*. According to Cotes these two definitions are incompatible. In 1712 he wrote to Newton, commenting the Proposition 6, Corollary 3 of Book III:

> Let us suppose two globes A & B of equal magnitudes to be perfectly fill'd with matter without any interstices of void Space; I would ask the question whether it be impossible that God should give different vires inertia to these Globes [...]. Therefore when You define or assume the quantity of Matter to be proportionable to its Vis Inertia, You must not at the same time define or assume it to be proportionable to ye space which it may perfectly fill without any void interstices; unless you hold it impossible for the 2 Globes A & B to have different Vires Inertia. Now in the 3rd Corollary I think You do in effect assume both these things at once [96].[90]

Newton was reticent to accept Cotes' conclusions. His commitment to the homogeneity of matter and the essential, determinate proportion between extension and inertia was difficult to overcome [10].[91]

Newton assumed moreover that bodies though made of particles can be considered as continua. To the purpose one can suppose its parts to be infinitely divided and dispersed everywhere throughout the pores, so that in the whole composite body there is not the least particle of extension without an absolutely perfect mixture of infinitely divided parts and pores. "Certainly such reasoning is suitable for contemplation by mathematicians; or if you prefer the manner of the peripatetics: things seem to be captured differently in physics" [104].[92]

---

[89] p. 38.
[90] Letter of Cotes to Newton 16th February 1712. pp. 65–66.
[91] p. 19.
[92] p. 38.

1.2.2.2 Definition of Force in the *Principia*

Newton reached a more mature position in the *Principia*, where he separated clearly internal and external forces at the ontological level. The internal force is qualified as inherent (insita):

> Definition III. Inherent force [vis insita] of matter is the power of resisting by which every body, so far as it is able, perseveres in its state either of resting or of moving uniformly straight forward [90].[93] (A.12)

This is Newton's comment to the previous definition:

> This force is always proportional to the body and *does not differ in any way from the inertia of the mass except in the manner in which it is conceived* [emphasis added]. Because of the inertia of matter, every body is only with difficulty put out of its state either of resting or of moving. Consequently, inherent force may also be called by the very significant name of *force of inertia* [emphasis added]. Moreover, a body exerts this force only during a change of its state, caused by another force impressed upon it, and this exercise of force is, depending on the viewpoint, both resistance and impetus: resistance insofar as the body, in order to maintain its state, strives against the impressed force, and impetus insofar as the same body, yielding only with difficulty to the force of a resisting obstacle, endeavors to change the state of that obstacle. Resistance is commonly attributed to resting bodies and impetus to moving bodies; but motion and rest, in the popular sense of the terms, are distinguished from each other only by point of view, and bodies commonly regarded as being at rest are not always truly at rest [90].[94] (A.13)

There are dark sides and many interpretations of the concept of the inherent force reported in the literature. According to Ierome Bernard Cohen "Def. 3 is in many ways, the most puzzling of all the definitions in the *Principia*" [103].[95]

Because the vis insita opposes to the exhaustion of motion, or said in another way, contributes to maintain the motion, it looks like the medieval impetus, as introduced by Buridan. The assimilation is not however entirely satisfactory: firstly Newton's force of inertia is a substantial property of the bodies; it acts both if the body is at rest and in motion, unlike the impetus which is defined only for a body which moves. In the second place the vis insita tendency to keep a body in its state of uniform rectilinear motion, did not exist in the theory of impetus, which was alien to the concept of direction and which also justified the uniform circular motion. Whatever the interpretation is accepted the adoption of the word used by Newton to indicate the force of inertia represents a concession to the pre-Galilean mechanics.

A very thorough analysis on the concept of vis insita is reported by Westfall in a still actual book [129], which reconstructs its evolution from the earliest times. According to Westfall, Newton gradually changed his thought passing from a conception of vis insita as internal force to a concept of vis insita as inertia. At the same time gradually he introduced the concept of force as an external cause of change of motion. The two concepts of force are incompatible, the latter rejects inevitably the former, since

---

[93] p. 2. English translation in [103].

[94] p. 2. English translation in [103].

[95] p. 96.

one conceives the force as the cause of the perdurance of motion the other as the cause of its variation; Newton will never be able to fully carry out the separation of the two concepts, and here and there, also in the *Principia*, tracks of the vis insita conceived as internal force remained. Ernan McMullin concentrated on the role Newton's attributed to matter and challenges the synonymy between *vis viva* and *inertia* suggested by Newton and generally accepted by historians [79]. For a textual reconstruction of Definition 3 see [9].

The external forces, that is the causes of variation of motion, are characterized by their action on bodies; different kind of forces can give raise to the same action, or effect:

> Definition IV. Impressed force [vis impressa] is the action exerted on a body to change its state either of resting or of moving uniformly straight forward [90].[96] (A.14)

The vis impressa, that is the action that determines the change of motion of a body, differs from the vis insita for two aspects of ontological type: vis insita is an universal attribute of matter, not further reducing, it is permanent and always responsible of the preservation of motion; vis impressa (intended as action) has instead a transient nature and vanishes when the force (intended as cause) has finished its work.

> This [impressed] force consists solely in the action and does not remain in a body after the action has ceased. For a body perseveres in any new state solely by the force of inertia. Moreover, there are various sources of impressed force, such as percussion, pressure, or centripetal force [90].[97] (A.15)

Similarly to what happened for vis viva, many opinions have been expressed about the meaning of vis impressa; especially for the attribute *impressed* before force—an attribute also found in the definition of force in the *De gravitatione*: *impressed* motion, see Sect. 1.2.2.1—and Newton has been also accused to be incoherent and imprecise [33, 34, 108]. Actually this is not the case. Newton took much care in the use of words and gave a very precise meanings to the terms he used to indicate 'force'.

Impressed forces are not forces intended as causes of motion; the latter may have different nature, as Newton specified: percussion, pressure and gravity, etc. Of none of them can be assigned a measure because they are causes. Impressed forces instead are all of the same nature; they have lost most (all?) of their ontology and can be given a measure according to the second law of motion with the variation of quantity of motion (or in modern term by mass times acceleration). Sometimes Newton seems to confuse the measure of the effect with the impressed force itself; that is the action of the force is not only measured by the effect but is identified with it, similarly to d'Alembert.

Focusing on actions rather than on hypostatized forces, Newton could deal with attraction and centripetal forces in a similar way as with pressure and percussion, paying attention to mathematical aspects only. At the beginning of section XI of the

[96]p. 2. English translation in [103].

[97]p. 2. English translation in [103].

first book, he, referring to gravity, clearly said that he only was interested in these aspects of forces:

> Up to this point, I have been setting forth the motions of bodies attracted toward an immovable center, such as, however, hardly exists in the natural world [...]. For this reason I now go on to set forth the motion of bodies that attract one another, considering centripetal forces as attractions, although perhaps – if we speak in the language of physics – they might more truly be called impulses. *For here we are concerned with mathematics; and therefore, putting aside any debates concerning physics, we are using familiar language so as to be more easily understood by mathematical readers* [emphasis added] [90].[98] (A.16)

While in the *Principia* there is no definition of the force of percussion and pressure, whose grasping is given for granted, the definition of centripetal force (vis centripeta)—the term was coined by Newton himself to contrast the well established concept of centrifugal force [vis centrifuga]—is instead highly developed, the only full exemplification of an impressed force. This for what concerns both mathematical and physical interpretations. Newton's approach is partly explained by the fact the treatise had as one of its main purpose the study of the orbital motions, partly because this force is the one which has a greater difficulty to be introduced.

> Definition V. Centripetal force is the force by which bodies are drawn from all sides, are impelled, or in any way tend, toward some point as to a center [90].[99] (A.17)

An interesting comment follows Definition V:

> One force of this kind is gravity, by which bodies tend toward the center of the earth; another is magnetic force, by which iron seeks a lodestone; and yet another is that force, whatever it may be, by which the planets are continually drawn back from rectilinear motions and compelled to revolve in curved lines [90].[100] (A.18)

The centripetal force is a cause (a physical entity) which can be seen under different points of view and given different measurement or impression [14].[101]

It is worth quoting one more comment to the definition of centripetal forces where for the first time in the history of mechanics weight is introduced in a completely modern way. The weight of a body decreases moving away from the earth because the amount of accelerating force decreases.

> An example is weight, which is greater in a larger body and less in a smaller body; and in one and the same body is greater near the earth and less out in the heavens. This quantity is the centripetency, or propensity toward a center, of the whole body, and (so to speak) its weight, and it may always be known from the force opposite and equal to it, which can prevent the body from falling [90].[102] (A.19)

---

[98]p. 160. English translation in [103].

[99]p. 3. English translation in [103]. It is worth noting that today the definition of the centripetal force is something different from the Newtonian one. The centripetal force that acts on a mass point in motion along a curved trajectory is the component of the force in the direction normal to the trajectory. Only if the trajectory is circular today's and Newtonian definition coincide; the centripetal force of Newton is now qualified just as central force.

[100]pp. 3–4. English translation in [103].

[101]pp. 249–250.

[102]p. 5. English translation in [103].

Newton drawn a clear distinction between the words attraction and gravity. This is not seen clearly by browsing the *Principa*, where the words having as their roots either *gravitas* or *attractiones* are spread nearly uniformly. But if one distinguishes between harmful and technical use he finds that things are different. This at least is the opinion of I Bernard Cohen, for whom the *index verborum* of the *Principia*, would record about 300 instances of the noun attraction (or other nouns with the same root), of which more than ninety percent in the first two books. The noun *gravitatis*, with its variants, is never used in the first two books it would find its natural place in the third book [19].[103]

When considering a single body Newton used the word centripetal force, but when he considered more than one body he used the word attraction, even if he did not want to give any physical meaning to the words, he took into account that in the case of more bodies there is no a single center to refer to, so the centripetal attribute would be ambiguous. If it were not for the third book, having a more physical character, Newton's arguments on the purely mathematical significance of the forces impressed would have had a complete plausibility. When Newton said that he only needed the mathematical expression of the centripetal forces he officially distanced himself from a possible charge of having introduced occult forces or forces at a distance: the word *attraction* is not associated with a physical meaning (as instead is for gravity), it is only a manner of speaking, it only has a purely mathematically meaning.

In the final scholium of section XI of the first book Newton paused on the nature of the centripetal force:

> I use the word *attraction* here in a general sense for any endeavor whatever of bodies to approach one another, whether that endeavor occurs as a result of the action of the bodies either drawn toward one another or acting on one another by means of spirits emitted or whether it arises from the action of aether or of air or of any medium whatsoever—whether corporeal or incorporeal—in any way impelling toward one another the bodies floating therein [90].[104] (A.20)

1.2.2.3 Gravity and Action at a Distance

In the previous section attention has been focused on the mathematical interpretations of forces, continuously advocated by Newton; here some considerations are referred to about the causes of impressed forces [67]. Quite interesting from this point of view is the reading of the *De mundi systemate*, translated into English as *A treatise of the system of the world*, written around 1685 as a draft of Book III of the *Principia* for people with little expertise in mathematics, using the traditional discursive language of natural philosophy [91, 92].

A first step passing from mathematics to physics (and *vice versa*) is to assume forces as proximate causes, thus endowing them with a minimal ontology. Certainly, Newton suggested, this is only a partial explanation, but it is a good approach to

---

[103] pp. 82–83.

[104] p. 6. English translation in [103].

science, otherwise: "no phenomenon could be rightly explained by its cause, unless the cause of this cause and the cause of prior cause were to be delivered and so successively continuously as long as the primary cause were to arrived in" [36].[105] A further step is to look for the causes of the proximate causes. Newton discussed problems related to this step, for gravity in particular, and advanced physical explanations, spread here and there in the *Principia*, in the Queries of the *Opticks*, in his letters and unpublished papers.

He declared quite strongly not to believe in an action at a distance but rather in some mediation. However he was not taken seriously by his contemporaries, philosophers and mathematicians. And he is not believed by modern mechanicians too, who mostly give for granted and accepted an unexplained action at a distance both at microscopic and macroscopic levels.

Indeed Newton was not always consistent in the few points where he discussed the cause of gravity, and many different opinions were expressed since the issue of the *Principia* in 1687. Recently the interpretation of action at a distance and gravity by Newton has become of renewed interest by historians [35–37, 54, 63, 64, 66, 116], I must confess without reaching a shared conclusion.

The possibilities that are explored are:

1. Mechanical agent; the action at a distance is explained with the action by contact of a material aether.
2. Attribution given by God to the crude matter at the moment of Creation. That is gravity is substantial to matter.
3. Non mechanical but material agent, due to the action of particles of aether endowed with force at distance in the short range.
4. Non mechanical and non material mysterious agent, endowed with not well specified properties.
5. The continuous intervention of God to produce the power of attraction in the otherwise inert matter.

The possibility (1) is explicitly excluded by a mature Newton, as documented by the following quotation, at the end of the *Principia*, in the Scholium generale:

> Thus far I have explained the phenomena of the heavens and of our sea by the force of gravity, but I have not yet assigned a cause to gravity. Indeed, this force arises from some cause that penetrates as far as the centers of the sun and planets without any diminution of its power to act, and that acts not in proportion to the quantity of the surfaces of the particles on which it acts (*as mechanical causes are wont to do* [emphasis added]) but in proportion to the quantity of solid matter, and whose action is extended everywhere to immense distances, always decreasing as the squares of the distances [90].[106] (A.21)

Newton excluded mechanical causes (that is impact of particles) in the explanation of gravity, because they used to act in proportion to surface and not volume, as gravity do, and also because heavens are substantially void of matter. However he was not fully explicit in the point and, in any case, his assertion is not correct because Euler

[105] p. 23.

[106] p. 530. English translation in [103].

will show later that mechanical causes can explain the dependence of gravity with volume; see Sect. 3.4.1.3.

Newton excluded possibility (2) also, because for him it would limit God's will and paved way for atheism. To remember that in his preface to the second edition of the *Principia*, Roger Cotes (1682–1716) wanted to introduce gravity as a substantial characteristic of matter, but Newton modified this thesis, presenting it as a primary quality (an idea probably not too different). Newton position is clearly expressed in a famous letter to Richard Bentley of 1693 (Gregorian calendar) where he clearly denied the possibility of a direct [robust] action at a distance.

> *Tis inconceivable, that inanimate brute Matter, should (without ye mediation of something else which is not material), operate upon & affect other matter without mutual contact* [emphasis added]; as it must if gravitation in the sense of Epicurus, be essential & inherent in it. And this is one reason why I desired you not to ascribe innate gravity to me. That gravity should be innate inherent & essential to matter so yt one body may act upon another at a distance through a vacuum without the mediation of any thing else & by & through which their action and force may be conveyed from one to another is to me such an absurdity that I beleive no man who has in philosophical matters any competent faculty of thinking can ever fall into it. Gravity must be caused by an agent acting constantly according to certain laws, but whether this agent be material or immaterial is a question I left to ye consideration of my readers [98].[107]

The possibility of a non mechanical but material agent, point (3), is suggested by the Queries of the *Opticks*, in particular Query 21 and Query 31. Here Newton, seems to assume that force at a distance existed at a microscopic level. He thought that the physical reality was made up by atomic hard particles which exchange forces at a distance of different nature, a mechanicistic vision all considered. The difference between Newton and classical mechanical philosophers is his decision to accept the ultimate inscrutability of nature, due to the presence of mysterious forces.

In Query 31 Newton first asked whether or not forces at a distance among particles exist; later he declared his interest in the proximate causes, leaving open the question of more remote causes.

> Have not the small Particles of Bodies certain Powers, Virtues or Forces, by which they act at a distance, not only upon the Rays of Light for reflecting, refracting and inflecting them, but also upon one another for producing a great part of the Phaenomena of Nature? For it's well known that Bodies act one upon another by the Attractions of Gravity, Magnetism and Electricity; and these Instances shew the Tenor and Course of Nature, and make it not improbable but that there may be more attractive Powers than these. For Nature is very consonant and conformable to herself. How these Attractions may be perform'd, I do not here consider [94].[108]

The two other points (4 and 5) presuppose an immaterial agent. The former point still maintains the possibility of a natural (if not mechanical) explanation, the latter requires a continuous intervention of God and is out of the sensibility of a modern reader, but also possibly of many Newton's contemporaries.

---

[107] Vol. 3, pp. 253–254.

[108] pp. 350–351.

Interest of Newton for immaterial agents is rooted in Neoplatonist British philosophy, in his alchemical studies [31] and in his interest on ancient science with his commitment on a *prisca sapientia* which would have been lost in the years [36, 78]. In some drafts for propositions from IV to IX of the *Principia*, composed in the 1690s for a second edition of this treatise, many references to the thought of Graeco-Roman antiquity are reported. These writings are today known as *Classical scholia* [16]. Here motion and gravity are attributed to God, gods and some spirits, all immaterial agents.

Quite interesting is the following quotation from the *Classica scholia*:

> Up to this point I have explained the properties of gravity. I have not made the slightest consideration about its cause. However I would like to relate what the ancients thought about this. Quite apparently the heaven are nearly free of bodies, but nevertheless filled everywhere with a certain infinite spiritus, which they call God. The bodies, however, move around freely in this *spiritus*, as a consequence of its forces and natural efficiency they are constantly thrust toward each other, more or less *strongly* in accordance with the harmonic ratio of the distances and gravity consists in this impact. Some differentiated this spiritus from the highest God and called it the soul of the world [16].[109] (A.22)

A cross reference to the action of spirits appears at the end of the *Principia* (eds. 1713 and 1726):

> A few things could now be added concerning a certain very subtle spirit pervading gross bodies and lying hidden in them; by its force and actions, the particles of bodies attract one another at very small distances and cohere when they become contiguous; and electrical [i.e., electrified] bodies act at greater distances, repelling as well as attracting neighboring corpuscles; and light is emitted, reflected, refracted, inflected [...]. But these things cannot be explained in a few words; furthermore, there is not a sufficient number of experiments to determine and demonstrate accurately the laws governing the actions of this spirit. [90].[110] A.23)

Notice however the absence of any reference to gravity and the admission of the difficult to explain the action of the spirit.

Andrew Janiak has suggested an interesting, even though 'curious', in my opinion, explanation of gravity which is out of the list presented above. Taking for granted that Newton did not believe both in an action at a distance and in a mechanical explanation, Janiak calls for the presence of God in the world, which for Newton necessarily exists and is always and everywhere. If one assumes the action of God as the cause of gravity, this action is surely non-mechanical, but it is also not at a distance because God is close to each body [63].[111]

---

[109] p. 45. Gregory ms. 247, f. 14^v. English translation in [36].

[110] p. 530.

[111] pp. 39–40.

1.2.3 *Theory of Light*

Optics, in ancient times, was among the mixed mathematics that than more the others was studied with strong reference to natural philosophy: nature of light, physiology of the eye and explanation of perceptions. With music, optics made frequent recourse to contrived experiments, some of which had a rhetorical value only, as those related to the law of reflection, others like those concerning refraction, presented quantitative aspects also.

Newton fitted with this tradition, by integrating it with that of physico mathematica of the second half of the 17th century, where the mathematical strictness was introduced in the experimental philosophy, with the use of an akin deductive approach using the classification of propositions in principles and theorems. Here theorems were not always proved with the argumentation of logic, but were based on experimental results also.

1.2.3.1 Composition of Light and Nature of Colors

The climax of a long way of Newton's studies on the nature of light and colors, whose beginning is to be found in the 1760s, is represented by the *Opticks* first edited in English in 1704. The standard view see it, besides other Newton's works on light, as the other side of the *Principia*. Though the label of empiric foundation is maintained to both of them, the latter would be based in the simple observation, the former in devised experimentation. The latter deeply rooted in geometry, the former in natural philosophy. Of course there is something true in this view, but the difference of the two texts is not due to Newton only. He moved, as any mathematician of the time, in the mainstream of mixed mathematics where, optics, astronomy, music and mechanics were each dealt with in different ways.

First researches of Newton on the nature of light should be found among the entries headed *Quaestiones quaedam philosophicae*, the name given to a set of notes Newton kept for himself during his earlier years in Cambridge, where two experiments on colors were reported. They seem to date from late 1665 or early 1666 [48]. Studies of Newton on light however saw their official start with the letter of Newton to Henry Oldenburg, the secretary of the Royal society, in 1672 [84].[112] The letter,

[112]Attention should payed in reading the date of all English letters and published documents of the 17th and the first half of 18th century. In the published Newton's correspondence the letter to Oldenburg it is reported 6 February 1671/2, with an apparent ambiguities in the year [98], vol. 1, p. 92. This is due to the fact that in 1751 only England adopted the Gregorian calendar (leaving the Julian calendar), already in use since 1582 in the Continent and at the time the Julian calendar was ten days less than the Gregorian calendar. Moreover the beginning of the year was different in the two calendars. Thus between 1st January (Continental new year in the Gregorian calendar) and 25 march (English new year in the Julian calendar) the designation of the English year was one year (and ten days) less than in the Continent. In the end 6th February 1671, the date of Newton's letter in English calendar, designates 16th February 1672 in the continental calendar. If not differently specified in the following the dates of Gregorian calendar is adopted. With the notable exception:

a paper indeed, was read at the society meeting and published shortly after in the Philosophical Transactions of the Royal Society of London containing in the title the expression *New theory about colours and light* (Herein after simply *New theory*), with which the paper is currently known [84].

The *New theory* presented the main topics of Newton's theory of light; in particular the assertion that light is compounded of colored rays which can be separated by refraction. The paper describes the famous *experimentum crucis*, that according to Newton should prove the compounded nature of light. This is made without any reference to illustrations. Let consider, said Newton, two small tables, and placed one of them close behind the prism at the window, so that the light might pass through a small hole made in it [in one of the small table] for that purpose and fall on the other small table which is placed at about 12 feet (~3.5 m) distance, having first made a small hole in it also, so that some of the incident light to pass through. Another prism is placed behind this second table so that the light projected through both the tables might pass through that also and is again refracted before it arrived at the wall. This done, the first prism is turned slowly about its axis so much as to make the several parts of the image cast on the second table, successively pass through the hole in it, so that one might observe to what places on the wall the second prism would refract them. It can be seen, by the variation of those places, that the light tending to that end of the image towards which the refraction of the first prism was made, did in the second prism suffer a refraction considerably greater then the light tending to the other end. "And so the true cause of the length of that image [the image refracted by one prism only] was detected to be no other then that light consists of rays differently refrangible which without any respect to a difference in their incidence were according to their degrees of refrangibility transmitted towards divers parts of the wall" [84].[113]

Figure 1.11 shows a drawing by Newton himself of the period when the *New theory* was redacted; it illustrates the experimental set of the *experimentum crucis*. The drawing exhibits a writing in Latin which reads: "Nec variat lux fracta colorem", that is light does not change color when refracted. This is what a modern reader expected the *experimentum crucis* should reach. But Newton supposed he had not stringent arguments to explicitly assert the link between refrangibility and colors, and in his published letter the Latin sentence is not present.

He had been more explicit in his early works, in particular in a manuscript known as *Of colours*, of uncertain dating, to which is conventionally associate the date of 1666. This manuscript contains a list of 64 experiments, those connected to the *experimentun crucis* being identified with numbers 44 and 45.

> 44. Refracting the Rays through a Prisme into a darke rome And holding another Prisme about 5 or 6 yards from the former to refract the rays againe I found ffirst that the blew rays did suffer a greater Refraction by the second Prisme then the Red ones.

---

all the papers of the Philosophical Transactions of the Royal Society of London will be identified using the calendar in use in England (be it Julian or Gregorian) at the time of issue.

[113] p. 3079.

Fig. 1.11 Experimental set of the *experimentum crucis* [98] (vol.1, after p. 106.)

> 45. And secondly that the purely Red rays refracted by the second Prisme made noe other colours but Red & the purely blew ones noe other colours but blew ones [83].

A long debate about the role and the experimental consistency of the *experimentum crucis*, and more in general on the theory of colors, followed, which as well known troubled Newton very much [65, 114, 119, 126, 130].

Worth noting a radical criticism of Newton's experiment brought forward more than a century after the publication of the *New theory*. The author was Johann Wolfgang Goethe (1749–1832) with his *Zur Farbenlehre* of 1810 [44, 45]. His criticisms, in some cases well motivated and interesting, affected not only the *experimentum crucis*, but all the theory of light and also the result of the experiments. They were also the criticisms of a philosopher of nature toward the new science, rooted in the mixed mathematics. Goethe was not taken very seriously by the scientists of the time even because he was known as a poet (actually he was a polymath with good scientific training). His chief supporters have been philosophers, artists, and physicians; in the scientific community there has been occasional support from researchers studying the physiology and psychology of color perception, especially since the later 19th century [115].

Newton's experiment was not easy to be replied, because of the lack of precision with which it was described. It can be said that the English scholars accepted it more easily than the continental ones. The two tables below report the results of some replications of the *experimentum crucis*.

In the first series of experiments the failure of the Jesuits is probably associated with the nature of the prisms they used, so imperfect as to prevent from observing anything Newton had observed. The failure reported by Edme Mariotte (1620?–1684)

Table 1.2 Replications of Newton's experimentum crucis before *Opticks*. Success is defined as being able to observe unequal refrangibility and color immutability. Adapted from [65], p. 44

Year	Person	Place	Successful	Witnesses
1666	Newton	Cambridge	Yes	Private
1670–71	Newton	Cambridge	No	Students
1672	Gregory	Edinburgh	Yes	Private
1672	Flamsteed	London	Yes	Private
1672	Hooke	London	Yes	–
1674–76	Jesuits	Lige	No	Jesuits
1676	Royal society	London	Yes	R.S. members
1676	Lucas	Lige	No	Jesuits
1681	Mariotte	France	No	Private

Table 1.3 Replications of Newton's experimentum crucis after *Opticks*. In Poleni's case, we only know that he replicated some experiments. Galiani, wrote of observing color immutability [65], p. 44

Year	Person	Place	Successful	Witnesses
1707–14	Whiston	London	Yes	Lecture course
1707	Poleni	Venice	–	Private
1707–08	Galiani	Rome	–	Private
1710	Bernoulli	Basel	Yes	Private
1714	Galiani	Rome	Yes	Public
1716	De Marian	Beziers	Yes	Private
1720's	Rizzetti	Venice	No	Private, witnesses

can be explained by the different arrangement of the prisms he prepared in the lack of a precise description by Newton [119].[114] Mariotte, then considered a French leading experimental scientist, published *De la nature des couleurs* in 1681 [76]. Here he granted that many experiments agreed with Newton's theory, but the *experimentum crucis* did not. This was his response on the point: "By this experiment it is evident that the same portion of light got different colors because of the different modifications, and that the ingenious hypothesis of Mr. Newton should not be accepted" [76].[115]

According to [65], the difference of the results between Tables 1.2 and 1.3 depends not only on the difficulty of replicating the experiment but also on sociological grounds. Newton since 1672, the year of *New theory*, though well known in England was instead a perfect stranger on the Continent. After the publication of the *Opticks* he was then a very famous and respected character throughout the western world;

[114] p. 79.

[115] p. 211.

his word, that is, had greater prestige than when he was younger, and it was easier for everyone to accept his conclusions.

Newton's *New theory* can be considered as a summary, but non only, of his *Lectiones opticae* delivered at Trinity College during the years 1670–1672. Newton intended to publish his lessons shortly after his letter to Oldenburg, but because of the controversies that followed he decided to give up. The topics of the *Lectiones opticae* were taken up in the *Opticks*, a twenty years later. Newton began the first book in 1687 or so and did not complete the last book and its Queries until 1703, shortly before it appeared in press in the early 1704. Three editions of *Opticks* were published before Newton's death. The English edition of 1704, a Latin edition in 1706, a second English edition in 1718 and a third English edition in 1721. A fourth English edition was published shortly after Newton's death, in 1730, based on Newton's changes.

The treatise, organized, as then classic in physico mathematica, in definitions, axioms, propositions is made of three books. Newton's methodology is declared at the very beginning of the Book I:

> My Design in this Book is not to explain the Properties of Light by Hypotheses, but to propose and prove them by Reason and Experiments. In order to which I shall premise the following Definitions and Axioms [94].[116]

The purpose of Newton was to rule out modification theories of light; theories which treated white sun light as a basic homogeneous entity which can take different colors after interacting with matter, but essentially stays the same. Thus, devices like prisms only modify the sunlight.

Immediately after the declaration of the methodology eight definitions started, among them that of ray is to be cited:

> Def.1. By the Rays of Light I understand its least Parts, and those as well Successive in the same Lines as Contemporary in several Lines. For it is manifest that Light consists of Parts both Successive and Contemporary; because in the same place you may stop that which comes one moment, and let pass that which comes presently after; and in the same time you may stop it in any one place, and let it pass in any other. For that part of Light which is stopt cannot be the same with that which is let pass. The least Light or part of Light, which may be stopt alone without the rest of the Light, or propagated alone, or do or suffer any thing alone which the rest of the Light doth not or suffers not, I call a Ray of Light [94].[117]

So the ray of light, at least according to the definition, ceases to be a purely geometric entity, a continuous line, to assume a physical connotation, a row of particles. Newton adopted the analogy with matter theory: the ray is a least part, just as an atom is the least part of matter. Newton's definition is sufficiently vague, most probably deliberately, to raise the doubt that the small particle of light may be for instance simply a portion of space which is active [51],[118] or a geometrical entity, an element of volume. In practice, it remained however fundamental the geometrical idea of

[116] p. 2.

[117] p. 4.

[118] p. 94.

line along which the minimal particles of light propagate, which reconnects to the classical definition of ray in the geometrical optics. However, differently than in classical geometrical optics, Newton's ray may be a curved line, when its particles are deflected by forces due to the interaction of matter, as it occurred in the case of diffraction (or inflection according Newton's nomenclature).

Also def. VII and VIII deserve to be quoted.

> Def. VII. The Light whose Rays are all alike Refrangible, I call Simple, Homogeneal and Similar; and that whose Rays are some more Refrangible than others, I call Compound, Heterogenal and Dissimilar. The former Light I call Homogeneal, not because I would affirm it so in all respects; but because the Rays which agree in Refrangibility, agree at least in all those their other Properties which I consider in the following Discourse.
> Def. VIII. The Colours of Homogeneal Lights, I call Primary, Homogeneal and Simple; and those of Heterogeneal Lights, Heterogeneal and Compound. For these are always compounded of the colours of Homogeneal Lights; as will appear in the following Discourse [94].[119]

The definitions, are not simply syntactic as in the modern axiomatic theories. They are rather real definitions and presuppose the reality of what is defined. For example it is presupposed that a ray may be compound.

To definitions eight axioms follow. They have an empirical character, that is they are not evident in themselves. What to a modern may seem an abuse of language, to name a proposition true but not evident in itself as an *axiom*, was indeed a quite common habit in the experimental philosophy of the 18th century. The eight axioms, a part some isolated points, are in any case propositions shared by any scholar of opticks, and in this sense the analogy with axioms of geometry is not strange even for a modern reader.

Axiom 5 concerns the law of refraction; it is formulated in an apparently surprising way. That is, it is presented as a law that applies only approximately: "The Sine of Incidence is either accurately or very nearly in a given Ratio to the Sine of Refraction" [94].[120] Newton's long explanation of the statement of this axiom makes it clear how one can talk about approximate validity. The various colors of light have different indices of refraction; however the difference is so little that it needs seldom be considered and the mean value can represent the refraction as a unique phenomenon. Note however that Newton in the explanation gave for granted what is still to be proved: different colors have different index of refraction. Axiom 5 is followed by two axioms related to flux of homogenous light.

Next to axioms are propositions, classified as in Euclid's *Elements* as theorems and problems. Proposition 1 asserts that "lights which differ in colour, differ also in degrees of refrangibility" [94].[121] This is proved by showing that it is in agreement with experimental results, or deduced from phenomena. There are two different proofs based on two different experimental situations. Below I only refer the one Newton presented as experiment 1.

---

[119] p. 4.

[120] p. 5.

[121] p. 16.

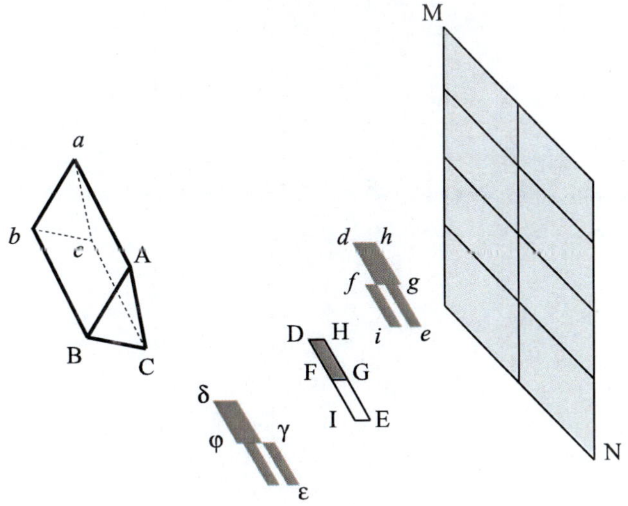

Fig. 1.12 Different refraction of lights. Redrawn from [94], Book I, Par I, Table II, Fig. 11

The slice of paper DE of Fig. 1.12, was distinguished into two equal parts DG and FE. One of them was painted with a red color and the other with a blew. This paper was viewed through a prism of solid glass, whose two sides through which the light passed to the eye were plane and well polished and formed an angle of about 60^0 (that is the triangular base of the prism is equilateral), named the angle of the prism. The two strips appear split as in $dfge$ or $\delta\gamma\varphi\epsilon$ depending on the rotation of the cylinder about its axis. According to Newton this depends on the different index of refraction of red and blew lights.

Newton, a very prudent experimenter, specified that from these experiments it does not follows, that all the light of the blue is more refrangible than all the light of the red; for both lights are mixed of rays differently refrangible, so that in the red there are some rays not less refrangible than those of the blue, and in the blue there are some rays not more refrangible than those of the red. But these rays, in proportion to the whole light, are but few, and serve to diminish the event of the experiment, but are not able to destroy it [94].[122]

Result of Proposition 1, but the same holds true for most of the others, is not proved from the axioms and definitions of the theory, as proposition of geometry are. However axioms and definition made possible the interpretation Newton gave. The most important concept that allow to understand Newton's interpretation is that of compound ray, only in this way rays with different refrangibility can be conceived. A modern reader would say that the experiment is compatible with the *hypothesis* (a word Newton was reluctant to use) that the rays of light have different indices of refraction. Newton instead saw the result as an empirical inductive proof.

[122] p. 21.

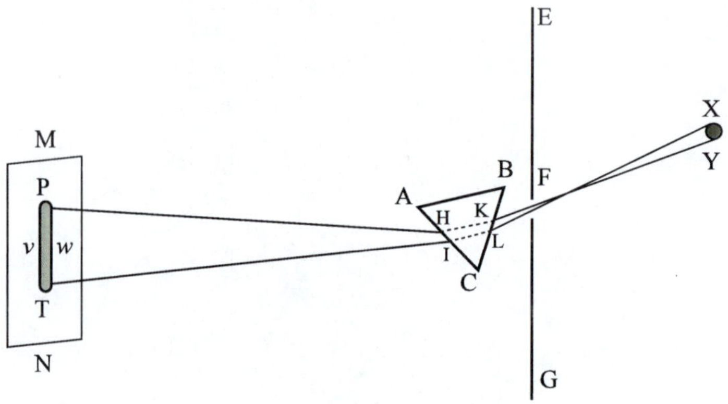

Fig. 1.13 Dispersion of sun light [94], Book I, Par I, Table III, Fig. 13

But the most important proposition that contains the essence of the theory of light and colors is Proposition 2, which states that sunlight, that is light par excellence, is a mixture of colors: "The Light of the Sun consists of Rays differently Refrangible" [94].[123]

It was proved by various experiments; of them below are reported with some detail only the first (Newton's Experiment 3) and the fourth (Experiment 6). Experiment 3 is very simple and reproduces with greater critical spirit experiments already carried out by other scholars, among them Descartes and Boyle, but introducing precise measurements. In Fig. 1.13, the scheme of the experiment is shown.

Let XY be the sunlight passing through a small hole F in the window EG; it is refracted by the prism ABC which projects the light coming out of it onto the screen MN. Newton observed that rotating the prism around the prism axis in a certain way the image moves first upward then downward. He decided to keep the prism still in the intermediate position, that of transition of the motion of the image from one verse to another. Newton noted that since the index of refraction passing from air to glass through the surface BC of the prism is equal to the inverse of the index of refraction passing from the glass to the air though the surface A, the angles between incoming and outgoing rays remain unchanged and thus the image of PT should be circular as the XY hole is. This did not happen however; and besides the already known rainbow effect, Newton highlighted a phenomenon that had never been observed: the elongation of the image, *with the red part in T and the violet part in P*. This can (only?) be explained if the light is composed of rays with different refractive index.

Figure 1.14a shows the elongated image of light. They are seen as the superposition of many circles (in the figure there are only six for illustrating purpose). Each circle represents the image there would be if light were monochromatic. Colors are pure only at the extremity P and T. In Fig. 1.14b it is reported the situation in case the circles (the images of the holes) were smaller. Here the circles are more distanced and

---

[123] p. 21.

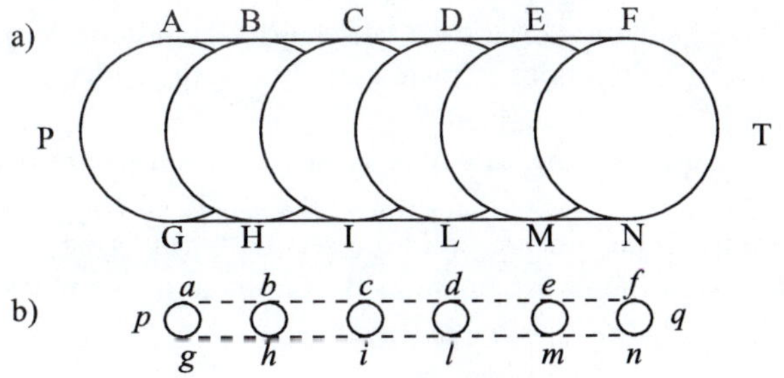

Fig. 1.14 The elongate image [94], Book I, Par I, Fig. 23

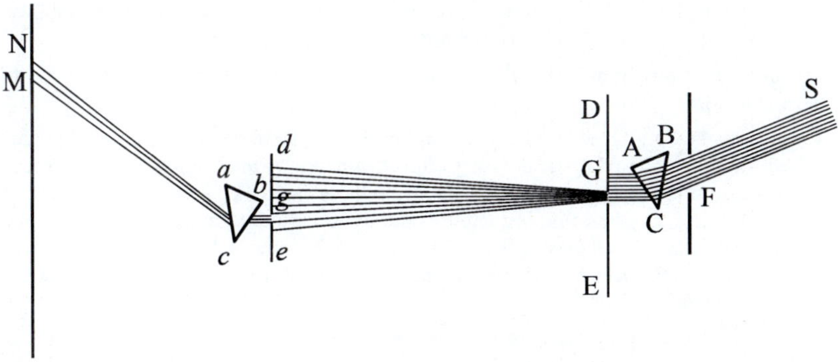

Fig. 1.15 *Experimentum crucis*. Redrawn from [94], Book I, Par I, Table IV, Fig. 18

the colors less mixed: "Now he that shall thus consider It, will easily understand that the Mixture is diminished in the same Proportion with the Diameters of the Circles. If the Diameters of the Circles whilst their Centers remain the same, be made three times less than before, the Mixture will be also three times less; if ten times less, the Mixture will be ten times less, and so of other Proportions" [94] (Fig. 1.14).[124]

Experiment 6 is what in the *New theory* was referred to as *experimentum crucis*, a locution not found in the *Opticks*, where the experiment is simply indicated with a number. In Fig. 1.15 the hole F lets a beam of light pass which is refracted by the prism ABC on the screen where a hole G is made. The light passing through G meets another screen and through another hole reaches the prism *abc* from which it is refracted on the screen NM. During the execution of the experiment the first prism ABC is rotated around its horizontal axis (that is an axis orthogonal to the plane of the figure) and the position of the image on the screen NM is observed. The result of the screen is so summarized: "in that common Incidence some of the Rays were more refracted and others less. And those were more refracted in this Prism which by

[124] p. 56.

a greater Refraction in the first Prism were more turned out of the way, and therefore for their constancy of being more refracted are deservedly called more refrangible" [94].[125]

Note that commenting on the second refraction Newton speaks only of the luminous image but does not say explicitly whether it is made up of monochromatic light or not. This aspect will be considered in the second part of Book I, where the following propositions/problems are proved/solved that associate index of refractions to colors.

1. The Phaenomena of Colours in refracted or reflected Light are not caused by new Modifications of the Light variously impress'd, according to the various Terminations of the Light and Shadow.
2. All homogeneal Light has its proper Colour answering to its Degree of Refrangibility, and that Colour cannot be changed by Reflexions and Refractions.
3. To define the Refrangibility of the several sorts of homogeneal Light answering to the several Colours.
4. Colours may be produced by Composition which shall be like to the Colours of homogeneal Light as to the Appearance of Colour, but not as to the Immutability of Colour and Constitution of Light. And those Colours by how much they are more compounded by so much are they less full and intense, and by too much Composition they may be diluted and weaken'd till they cease, and the Mixture becomes white or grey. There may be also Colours produced by Composition, which are not fully like any of the Colours of homogeneal Light.
5. Whiteness and all grey Colours between white and black, may be compounded of Colours, and the whiteness of the Sun's Light is compounded of all the primary Colours mix'd in a due Proportion [94].[126]

Newton used the term *colour spectrum* for the image of the refracted sun light, and although the it appears to be continuous, with no distinct boundaries between the various colors, he chose to divide it into seven *primary colors*: red, orange, yellow, green, blue, indigo, and violet. Newton chose the number seven because of the ancient Greek belief that seven is a mystical number and seven were the note of the musical scale of his time, which allowed him to make analogies between light and music.

1.2.3.2 Fits of Easy Reflection and Transmission

Book II of *Opticks*, part I, describes the appearance of colored rings, now known as *Newton rings*, both from reflected and refracted rays, exhibited in thin transparent films being illuminated with monochromatic light. The phenomenon is today explained by the interference of light waves, that is the superimposing of trains of waves so that when their crests coincide, the light brightens; but when trough and crest meet, the light is destroyed; light waves being reflected from both top and bottom surfaces of the film. Rings appear also with white light, but in these case the phenomenon is much more complex because of the interaction of the various colors.

[125] p. 39.

[126] pp. 99–117.

Newton found that rings also appeared in thick plates and he discussed the fact in Book II, Part IV.

Newton knew about the rings, exhibited in thin transparent plates enlightened with white light, under the influence of Hooke's *Micrographia* [55];[127] Hooke's observations have been already recorded in the *Of colours*. Newton studied long the problem; here reference is made mainly to what he wrote in the *Opticks*; for more information see [112, 118, 131].

In his experiments with rings Newton constructed a film of air of continuously varying thickness, enclosed between the plane surface of a planoconvex lens and the curved surface of a double convex lens supporting the first, as shown in Fig. 1.16c with monochromatic light coming from above. The occurrence of rings with this arrangement of lenses is described in the Observation 15 of Book II.

> These Rings were not of various Colours like those made in the open Air, but appeared all over of that prismatick Colour only with which they were illuminated. [...]. *And from thence the origin of these Rings is manifest. namely that the Air between the Glasses, according to its various thickness is disposed in some places to reflect and in others to transmit the Light of any one Colour (as you may see represented in the fourth Figure [Fig. 1.16d]) and in the same place to reflect that of one Colour where it transmits that of another* [emphasis added] [94].[128]

The explanation given in italic in the above quotation is exactly the same at that referred to in the *Discourse concerning light and colours*, a Newton's writing which could be dated around 1675 [85].

To justify the phenomenon of rings, both in thin and thick plates, Newton, as usual, separated the mathematical theory from physical hypothesis. The former, concerns the proximate causes only and according to him can be deduced from the experiments, the latter concern the *causae primae*, and in the *Opticks* they have only a didactical role. The proximate cause of the appearance of the colored rings was attributed by Newton to *fits of easy transmission*, or simply *fits*, which are first introduced in Proposition XII of Book II, part II:

> Every Ray of Light in its passage through any refracting Surface is put into a certain transient Constitution or State, which in the progress of the Ray returns at equal Intervals, and disposes the Ray at every return to be easily transmitted through the next refracting Surface, and between the returns to be easily reflected by it [94].[129]

and then specified with a definition:

> Definition. The returns of the disposition of any Ray to be reflected I will call its Fits of easy Reflexion, and those of its disposition to be transmitted its Fits of easy Transmission, and the space it passes between every return and the next return, the Interval of its Fits [94].[130]

Newton's choice of the term *fits* [paroxysms], quite strange for a modern reader, is drawn from contemporary medical language. A fit is one of recurrent attacks of a

[127] pp. 47–67.

[128] pp. 186–187.

[129] p. 252.

[130] p. 256.

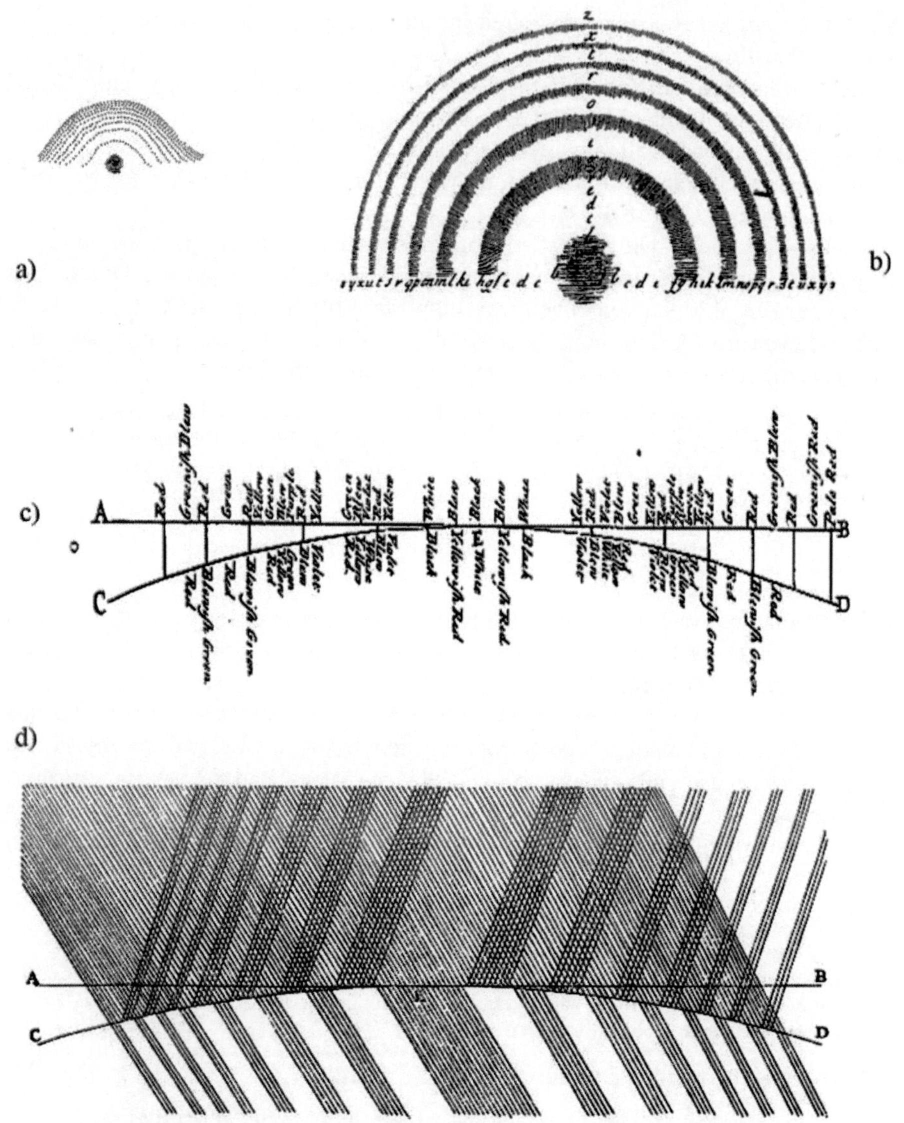

Fig. 1.16 Rings. Redrawn from [94], Book II, Tab I

periodic ailment, in particular, of ague or intermittent fever (today know as malaria). The fits of ague have a number of features in common with fits of easy reflection and transmission that make it an apt term. Fits alternate between two opposite phases, cold and hot, which are alternate bodily states and not alternations of some bodily substance such as blood. Fits are periodic and, depending on the type of ague, have periods of return of one, two, or three days [117].[131]

Of a certain interest in the definition of fits is the introduction of the concept of the interval (or length) of fits. Analysis of previous Newton's work and also what is written in the *Opticks* suggests that this quantity could be associated to the wave length of some vibratory motion.

Newton arrived at the theory of fits after *causae primae* were looked for, in a unsatisfactory way indeed. In the previously referred *Discourse concerning light and colours* Newton explained the origin of fits with reference to air. In the *An hypothesis explaining the property of light discoursed in my severall papers*, more or less of the same period, air was replaced by aether. And an analogy with a stone that falls in still water was assumed; just "as stones thrown into water do in its Surface; and that these vibrations are propagated every way into both the rarer & denser Mediums, as the vibrations of Air which cause Sound are from a Stroke, but yet continue Strongest where they began, & alternately contract & dilate the aether in that Physicali Superficies" [98].[132]

In any case Newton maintained that light was not vibration of the aether, but it has a different nature and its propagation is simply associated to aether vibration. Around 1687 in a projected section of the *Opticks* Newton introduced an explanation or the rings without recourse to the vibration of the medium (air or aether) [118].[133] Instead he called for the agitation of the particles of refracting and reflection bodies, assuming a periodic motion:

> Prop. 12 The motion excited by a ray of light in its passage through any refracting surface is reciprocal & by its reciprocations doth alternately increase & decrease the reflecting power of the surface.
>
> [...]
>
> The proper argument for ye truth of this Proposition is the alternate reflections & transmissions of light succeeding one another in a thin transparent plate accordingly as the thickness of the plate encreaseth in an arithmetical progression [85].[134]

An idea which is presented an hypothesis in Book II, Proposition 13 of the *Opticks*, which read nearly verbatim as: So the rays of light, by impinging on any refracting or reflecting surface, excite vibrations in the refracting or reflecting medium, and by exciting them agitate the solid parts of the body, causing it to grow warm or hot. The vibrations thus excited are propagated in the refracting or reflecting medium, much after the manner that vibrations are propagated in the air for sound. They move faster than the rays so as to overtake them; so that when a ray is in that part of the vibration

---

[131] p. 180.

[132] Vol. 1, p. 374.

[133] p. 172.

[134] Quoted from [118], p. 173.

which conspires with its motion, it easily breaks through a refracting surface, but when it is in the contrary part of the vibration which impedes its motion, it is easily reflected. By consequence, every ray is successively disposed to be easily reflected, or easily transmitted, by every vibration which overtakes it [94].[135]

Newton however ended his explanation by distancing from it "but whether this hypothesis be true or false I do not here consider. I content myself with the bare discovery, that the rays of light are by some causes or other alternatively disposed to be reflected or refracted for many vicissitudes" [94].[136]

Newton's theory of fits has too many obscurities to be considered successful from a modern point of view. It is remarkable anyway in the sense that when a purely corpuscular view proved inadequate to explain the periodicity of an optical phenomenon, waves were introduced, for the first time in the history of physics, to cooperate with the light particles [112].[137]

Nothing is said about the actual size of the light particles referred to in the definition of ray, although violet rays being more refrangible are considered to be made by smaller particles than those red rays. The length of the fits is calculated for the border between yellow and orange to be the 1/89000th part of an inch (1/35039 cm).[138]

1.2.4 Theological Writings. The Treatise on Apocalypse

In recent years historians have tried to give some explanations of why one of the world's greatest scientists should have spent so much time thinking and writing about religious matters. Even though paradoxically one can reverse the question: "why did one of the greatest anti-Trinitarian theologians of the 17th century take time off to write works on natural science, like the *Principia mathematica*?" [49].[139]

Indeed though provocative the question has some reason to be raised. Newton wrote on religion and theology from his college days down to the end of his life. Almost half of the pages that he physically wrote, most still unpublished, deal with explicating the Bible, interpreting it, and developing a theory of scriptural and natural revelation. It must be said, however, that Newton was not alone. In England, other great mathematicians such as John Wallis and Isaac Barrow had strong interests in theology. According to [41] this phenomenon was not confined to English puritanism, but also to the Continent and it is a characteristic of the 17th century.

> Galileo and Descartes, Leibniz and Newton, Hobbes and Vico were either not clergymen at all or did not acquire an advanced degree in divinity. They were not professional theologians, and yet they treated theological issues at length. Their theology was secular also in the sense

[135]pp. 255–256.

[136]p. 255.

[137]p. 126.

[138]The length of fits represent actually a half wavelength; modern estimation of the yellow wave length is of about 1/20000 cm, not very different from Newtons' value.

[139]p. 81.

> that it was oriented toward the world, *ad seculum*. The new sciences and scholarship, they believed, made the traditional modes of theologizing obsolete; a good many professional theologians agreed with them about that. Never before or after were science, philosophy, and theology seen as almost one and the same occupation. True, secular theologians seldom composed systematic theological treatises for the use of theological faculties; some of them, mainly the Catholic, pretended to abstain from issues of sacred doctrine; but they dealt with most classical theological issues–God, the Trinity, spirits, demons, salvation, the Eucharist. Their discussions constituted theology inasmuch as they were not confined to the few truths that the "natural light" of reason can establish unaided by revelation–God's existence perhaps, or the immortality of the soul [41].[140]

There was a melting of language between scientific and theological propositions, so that physical principles were expressed in theological terms and *viceversa*. Newton lived in an era where religion was still at the basis of society. It is thus not strange he was a religious man, even though a heretic one. His religious ideas (mainly God omnipotence and omnipresence) influenced all his life and also his natural philosophy. The same his temperament, his political conceptions and the fortuitous events of his life, did. These are difficult aspects to analyze and are out of the scope of the present book. Here interest is focused to see how his studies on theology influenced his studies on natural philosophy.

In his printed books, in particular in the scholia of the *Principia* and in the Queries of the *Opticks* Newton claimed that arguing on God from the occurrence of experimental phenomena belongs to natural philosophy. But he also (post 1710) stated that "religion and philosophy are to preserved distinct. We are not to introduce divine revelations into philosophy, nor philosophical opinions into religion [...]. That Religion & polity or the laws of God & the laws of man are to be kept distinct. We are not make the commandements of men a part of the laws of God" [88].

Newton's religious manuscripts have always enjoyed scant attention from science historians. No inventory of Newton's theological manuscripts has been published, for what I know. Nor any of them was published in the 18th century; with the exception of some fragments as *Observations upon the prophecies of Daniel and the Apocalypse of St. John*, published in 1733 [95]; its last reprinting was in 1922 [49].[141] The text had a good editorial success and was translated into various languages. Most of the early fundamentalist interpreters, who saw the American and French Revolutions as fulfillments of prophecies in this text, used Newton both as a source and as a theorist explaining why exact predictions often failed. Newton's *postfacto* method of interpretation allowed for reconsideration and restudying of prophecies when prediction failed.

Recently it has been edited and published a study on Newton's *Treatise on apocalypse* [102], basing on a manuscript conserved in the Jewish National and University Library of Jerusalem [102], known as Newton MS I, to be dated most probably in the first 1670s [102],[142] and thus predating the *Principia*. A its reading suggests that

---

[140] p. 3.

[141] p. 91.

[142] p. XLIII.

Newton gave to the holy scriptures a value of knowledge not very different from the scientific one and the same axiomatic approach of the *Principia* was used there [74, 75].

This could indicate that Newton's way of conceiving the philosophy of nature, his philosophizing rules were first developed in his theological writings than in his treatises on natural philosophy. If this is true it would be necessary to re-direct the studies on Newton's methodology by seriously reading theological writings, also because they are not the result of a schizophrenic malaise or of a love for an encyclopedic knowledge that explores different branches of knowledge, opening and closing one drawer at a time.

In the *Treatise on apocalypse* Newton put the interpretation of the Scripture and that of nature on the same ground; both are attributable to God and in both cases the truth is unique and the method to achieve it the same. According to Newton the language of the Scripture must be univocal and homogeneous because it refers to realities that are simple and harmonious. Confusion does not belong to the Scripture:

> Truth is ever to be found in simplicity, & not in ye multiplicity & confusion of things. As ye world, wch to ye naked eye exhibits the greatest variety of objects, appears very simple in its internall constitution when surveyed by a philosophic understanding, & so much ye simpler by how much the better it is understood, so it is in this vision [102].[143]

Newton addressed the study of the Apocalypse with the tools that are most familiar to him; those of mathematics. There are definitions and propositions and general rules of interpretation (16) that enable the reader to know when an interpretation is genuine [102].[144] One of the fundamental rule, typical of mathematics, stated not to use synonyms. "To assign but one meaning to one place of scripture, unless it be by way of conjecture [For a man cannot be obliged to believe more meanings of a place the one]" [102].[145]

In [23] it is suggested that the relationship of mathematics to the analysis of prophecies is not that close. Newton, with other contemporary exegetes, considered mathematics and prophecy as quite different matter and declared that prophetical interpretation was not a demonstrative science, differently from mathematics [102].[146] A notable exception was represented by More and Cambridge Platonists who thought that mathematics was able to reveal the truth of prophecies.

To testify the role of the Apocalypse in the development of a general methodology for scientific inquires, the following Table 1.4 compares the famous Regulae philosophandi of the *Principia* with the Rules for interpreting the words and language in Scripture of[147] the *Treatise on Apocalypse*.

Notice that the Apocalypse rule 2, which in the table is split into part I and part II, contains both rule I and IV of the *Principia*.

[143] p. 28.

[144] p. 18.

[145] p. 20.

[146] pp. 34, 36.

[147] pp. 398–399.

Table 1.4 Comparison between the rules from *Treatise on Apocalypse* and *Principia*. Adapted from [74], pp. 398–399

Treatise on Apocalypse	*Principia*
2 [part I]. To assigne but one meaning to one place of scripture. 3. To keep as close as may be to the same sense of words	Regula I (1687) Causas rerum naturalium non plures admitti debere, quam quae et verae sint & earum phaenomenis explicandis sufficiant
1. To observe diligently the consent of Scripture. 8. To choose those constructions which reduce contemporary visions to the greatest harmony of their parts. 9. To choose those constructions which reduce things to the greatest simplicity	Comment to Regula I Natura enim simplex est & rerum causis superfluis non luxuriat
Rules 4, 6, 7, 10, 12, 14, 15	Regula II (1687) Ideoque effectuum naturalium eiusdem generis eaedem assignandae sunt causae, quatenus fieri potest
5. To acquiesce in that sense of any portion of Scripture as the true one which results most freely & naturally from ye use & propriety of ye Language & tenor of the context in that & all other places of Scripture to that sense. 11. To acquiesce in that construction of the Apocalyps as the true one which results most naturally & freely from the characters imprinted [...] for insinuating their connexion	Regula III (1713) Qualitates corporum quae intendi & remitti nequeunt, quaeque corporibus omnibus competunt in quibus experimenta instituere licet, pro qualitatibus corporum universorum habendae sunt
2 [part II]. If two meanings seem equally probable he is obliged to believe no more then in general ye one of them is genuine untill he meet with some motive to prefer one side	Regula IV (1726) In philosophia experimentalis, propositiones ex phaenomenis per inductionem collectae, non obstantibus contrariis hypothesibus, pro veris aut accurate aut quamproxime haberi debent, donec alia occurrerint phaenomena, per quae aut accuratiores reddantur aut exceptionibus obnoxiae

The comparison between the two columns suggests that either the regulae philosophandi of the *Principia* have their source in the rules of the *Apocalypse* or that both of them have the same source, possibly a not theological one. Mamiani suggests that the two possibilities are not disjoined and that the common source should be searched, besides the *Discours de la méthode* of Descartes, in the studies of logic of Newton while an undergraduate at Cambridge. In particular in the *Logicae artis compendium* by Robert Sanderson [113] of 1618, an author Newton studied when at Cambridge. In [23][148] this derivation is considered as possible but not necessary. This difference of opinion testifies to the difficulty in interpreting Newton's complex interaction between theology and natural philosophy and the need for further studies; for some of them see [46, 61–63, 74, 98, 121].

[148] p. 237.

1.2.5 A New Form of Mechanicism. The Queries

In a previous section I commented upon the first three books of the *Opticks* that concern geometric and physical properties of light in continuity with the optics understood as a mixed mathematical science. But *Opticks* became famous not only and perhaps not primarily for the theory of light reported there, but also for the *Queries* put at the end of Book III. The content and number of the Queries, changed over the various editions, starting from sixteen of the first edition, concerning exclusively the nature of light, to the thirty-one of the third edition. They are not only about optics; other branches of physics are touched, indicating the topics that for Newton were the most important to be investigated in the immediate future [3].

The queries will be commented for their methodological aspects in Sect. 1.2.6.2; here I dwell instead on the aspects of natural philosophy properly said. In the queries Newton laid the foundations of a new mechanical philosophy, a philosophy in which the phenomena of nature are not explained using the laws of kinematics (impact included) but using the laws of dynamics. The world is seen as a set of corpuscles (hard) that exchange forces at a distance. On the nature of these forces Newton was ambiguous. The prevailing position was that they might be understood in a mathematical sense, that is, they simply expressed the interaction of bodies whose effects are measurable for example through the accelerations of reciprocal motion. He considered as possible that the true causes of these forces could eventually be explained, but for the moment this explanation did not exist.

If the ontology of interaction forces was the object of discussion among the new philosophers of nature, what was not in question was the extreme fertility of this new mechanicism which made it possible to explain all the physical and chemical phenomena of nature. This mechanicism was accepted by almost all the scientists since 1750. It must be said that it was not accepted only for his fertility but also because proposed by a philosopher of nature who had by then become an authority of absolute greatness and that among other things with his *Principia* had succeeded in explaining the motions of the bodies of the Heaven.

In addition to some general considerations on Newtonian 'mechanical' philosophy, I will focus on an important element of it, the idea of aether, a name and a concept that has its origins in ancient Greece but that with Newton took on a new identity and also an aura of mystery, because even though Newton dedicated many years and writings to the aether, he hardly ever made it enter into his most important treatises. Because of space reasons I will only relate to the ideas that Newton expressed in his final version of *Opticks*, the 1730 edition. This is enough considering the aim of this book.

1.2.5.1 Matter and Short Range Forces

The Newtonian mechanical philosophy is exposed in the Query 31 of the *Opticks*, the one added in 1718. At the beginning Newton gave for granted the corpuscular nature of matter. He asked rhetorically: "Have not the small particles of bodies certain powers, virtues, or forces, by which they act at a distance, not only upon the rays of light for reflecting, refracting, and inflecting them, but also upon one another for producing a great part of the phaenomena of nature?" [94]:[149] And continued, by noticing that it is well known, that bodies act one upon another by the attractions of gravity, magnetism and electricity; and these instances make it not improbable that there may be more attractive powers than these. How these attractions may be performed, said Newton, are not considered. What is called attraction may be performed by impulse, or by some other means. It only signifies in general any 'force' by which bodies tend towards one another, whatsoever be the cause. The attractions of gravity, magnetism, and electricity, reach to very sensible distances, and so have been observed by vulgar eyes, but there may be others which reach to so small distances as hitherto escaped observation; "and perhaps electrical Attraction may reach to such small distances, even without being excited by Friction" [94].[150]

As an example of action of forces other than the known one, Newton referred to the phenomenon of capillarity, by quoting results obtained by Francis Hauksbee (1660–1713), the first to study the phenomenon in a systematics way in the 1700s, who measured, among other situations, the rise of water between two glass plates versus their distance [80]: "Now by some Experiments of this kind (made by Mr. Hauksbee), it has been found that the Attraction [...] within the same quantity of attracting Surface, is reciprocally as the distance between the Glasses. And therefore where the distance is exceeding small, the Attraction must be exceeding great" [94].[151] It must be said that Hauksbee was cited less that it was due considered that he carried out many experiments on capillarity very well done, published in the *Philosophical Transactions* of the Royal Society in the years 1711–1713, after having published his results in the *Physico-mechanical experiments on various subjects* of 1709 [53].

Newton could conclude that there are therefore agents in nature able to make the particles of bodies stick together by short range forces; they allow Newton to present a his own idea on the constitution of matter, which is similar to that suggested by Boyle. The smallest particles of matter may cohere by the strongest attractions and compose bigger particles of 'weaker virtue' [attractive force]; and many of these may cohere and compose bigger particles whose virtue is still weaker, and so on for some successions, "until the Progression end in the biggest Particles on which the Operations in Chymistry, and the Colours of natural Bodies depend, and which by cohering compose Bodies of a sensible Magnitude" [94].[152]

The different bodies may be solid, liquid or aeriform depending on the arrangement of elementary particles. But there are not attractive force only. Indeed as in algebra, where positive quantities vanish and cease, negative ones begin, so in mechanics, where attraction ceases, there a repulsive force ought to succeed. And that there is such a repulsive virtue, seems to follow, for instance, said Newton, from the reflex-

---

[149] Q. 31, p. 350.

[150] Q. 31, p. 351.

[151] Q. 31, pp. 368–369.

[152] Q. 31, p. 370.

ions and inflexions [diffractions] of the rays of light. Indeed the rays are repelled by bodies in both these cases, without the immediate contact of the reflecting or inflecting body. It seems also to follow from the nature of air and vapor. Here "vast contraction and expansion seems unintelligible, by feigning the particle of air to be springy and ramous, or rolled up like hoops, or by any other means than a repulsive power" [94].[153] Here the spring of air—and also of aether discussed later—is attributed, and this is an original position, by Newton to repulsive forces acting at a distance.

Newton advocated the simplicity of nature which is "very conformable to her self" [94],[154] performing all the great motions of the heavenly bodies by the attraction of gravity which intercedes those bodies by some other attractive and repelling powers or active principles. And the presence of active principles is necessary to justify the motion in the world. Indeed the *vis inertiae* is a passive principle by which bodies persist in their motion or rest, but by this principle alone there never could have been any motion in the world. Some other principles are necessary for putting bodies into motion; but also they are necessary for conserving this motion, because there are causes that decrease the quantity of motion of bodies and it is very certain that there is not always the same quantity of motion in the world [94].[155] Among the causes Newton considered responsible of the decrease of motion there are non-elastic impacts and frictions. Thus, he can conclude, active principles are needed, such as are the cause of gravity, by which planets and comets keep their motions in their orbs (and the cause of fermentation, by which the heart and blood of animals are kept in perpetual motion and heat etc.) [94].[156]

After these preliminaries Newton specified his mechanical philosophy with the following statement, which though very well known is worthy to be quoted one more once:

> All these things being consider'd, it seems probable to me, that God in the Beginning form'd Matter in solid, massy, hard, impenetrable, moveable Particles, of such Sizes and Figures, and with such other Properties, and in such Proportion to Space, as most conduced to the End for which he form'd them; and that these primitive Particles being Solids, are incomparably harder than any porous Bodies compounded of them; even so very hard, as never to wear or break in pieces; no ordinary Power being able to divide what God himself made one in the first Creation [94].[157]

While the particles remain entire, according to Newton, they may compose bodies of one and the same nature and texture in all ages. And therefore, assuming that nature may be lasting, the changes of corporeal things are to be placed only in the various separations and new associations and motions of these permanent particles. Compound bodies being apt to break, not in the midst of solid particles, but where those particles are laid together and only touch in a few points.

---

[153] Q. 31, p. 371.

[154] Q. 31, p. 372.

[155] Q. 31, p. 373.

[156] Q. 31, p. 375.

[157] Q. 31, pp. 375–376.

Newton continued by reasserting that these particles have not only a *vis inertiae*, accompanied with such passive laws of motion as naturally result from that force, but also that they, as the macroscopic bodies, are moved by certain active principles. "These principles I consider, not as occult qualities, supposed to result from the specific forms of things, but as general laws of nature, by which the things themselves are formed; their truth appearing to us by phenomena, though their causes be not yet discovered" [94].[158] According to Newton, to derive two or three general principles of motion from phenomena and afterwards to tell how the properties and actions of all corporeal things follow from those manifest principles, would be a very great step in philosophy, though the causes of those principles were not yet discovered.

1.2.5.2 Mechanical and Non-mechanical Aether

The idea of aether was always present in Newton's thought but he discussed it mainly privately. In the Queries of the *Opticks* the concept found a public audience however. The idea changed a lot in time; from a pseudo Cartesian concept of a fluid made of inert particles in the writings before 1680s to an elastic medium made of corpuscles endowed with active principles (forces at a distance) [109].

The aether was discussed in depth to explain gravity in two letters, one of 1675 to Oldenburg with the title *The hypothesis explaining the properties of light discovered in my severall papers* [98],[159] the other in a letter to Boyle of 1679 [86]. In the letter to Oldenburg Newton assumed a kinematic reasoning, according to which material bodies were carried toward the surface of the earth by the circulation of the aether which: "For nature is a perpetuall circulatory worker, generating fluids out of solids, and solids out of fluids, fixed things out of volatile, & volatile out of fixed, subtile out of gross, & gross out of subtile, Some things to ascend & make the upper terrestriall juices, Rivers and the Atmosphere; & by consequence others to descend for a Requitall to the former [...]. And that the vast aethereall Spaces between us, & the stars are for a sufficient repository for this food of the Sunn & Planets" [98].[160]

In the letter to Boyle Newton started by arguing that the aether pervades all gross bodies, but yet so as to stand rarer in their pores than in free spaces, and so much the rarer as their pores are less. The rarer aether within bodies and the denser out them, is not terminated in a mathematical superficies but grows gradually into one another: the external aether beginning to grow rarer and the internal to grow denser at some little distance from the superficies of the body, and running through all intermediate degrees of density in the intermediate spaces. This property of aether explain the refraction of light and cohesion of bodies [86].[161]

In the previous description of the properties of the aether it is not fully clear what Newton meant with rarer and denser. Apparently, he spook of only one type of matter

[158] Q. 31, pp. 376–377.
[159] Vol. 1, pp. 362–392.
[160] Vol. 1, p. 366.
[161] p. 62v.

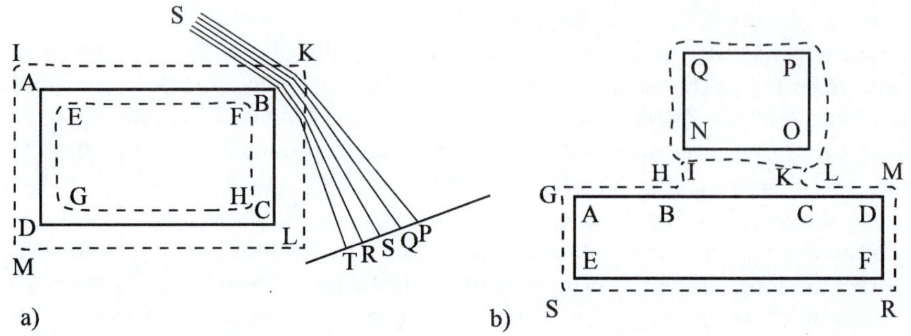

Fig. 1.17 Diffraction (a) and cohesion (b) [86]

that can be found more or less thickened. This will be clarified later in the letter, when he formulated the explanation of gravity.

For what the explanation of the diffraction is concerned, let consider the dense body ABCD of Fig. 1.17a, either opaque or transparent, EFGH the outside of the uniform aether which is within it, IKLM the inside of the uniform aether which is outside it. Conceive the aether which is between EFGH and IKML to pass through all intermediate degrees of density between that of the two uniform kinds of aether on either side. This being supposed, the rays of the sun SB, SK, which pass by the edge of this body between B and K, ought in their passage through the unequally dense aether there, to receive a ply from the denser aether which is on that side toward K and thereby to be scattered through the space PQRST, as by experience they are found to be.

Newton's aether can explain cohesion. When two bodies approach one another and come so close as to make the aether between them start to rarefy, they begin to have a reluctance from being brought nearer together and endeavor to recede from one another. This reluctance and endeavor will increase as they come nearer together because thereby they cause the interjacent aether to rarefy more and more. But at length, when they come so close that the excess of pressure of the external aether which surrounds the bodies is so great as to overcome the reluctance of the bodies from being brought together, said Newton, that excess of pressure drives them with violence together and make them adhere strongly to one another.

For instance in Fig. 1.17b when the bodies ED and NP are so close, the spaces of the aether graduated rarity begin to reach to one another and meet in the line IK. The aether between the two bodies suffer much rarefaction and the endeavor which the aether between them has to return to its former natural state of condensation will cause the bodies to have an endeavor of receding from one another. But if the bodies come nearer together so as to make the aether in the mid-way-line IK grow rarer than the surrounding aether, there will arise from the excess of density of the surrounding aether a compression of the bodies towards one another: which when by the nearer approach of the bodies it becomes so great as to overcome the aforesaid endeavor the bodies have to recede from one another, they will then go towards one another and adhere together (Newton reason is not fully sound here. How can the pressure of the surrounding aether overcame the repulsion of the two bodies which increases with the decrease of their distance?).

Newton assumed that the variation in density of the aether could explain gravity also with a play of the pressure around common bodies. Differently than in the circulatory explanation of the letter to Oldenburg, in the letter to Boyle the role of aether had a cosmological character, explaining the gravity of the whole universe. For this end suppose aether to consist of parts differing from one another in subtlety by indefinite degrees. Thus in the pores of bodies there is less of the grosser aether in proportion to the finer then in open spaces, and consequently in the great body of the earth there is much less of the grosser aether in proportion to the finer then in the regions of the air. Imagine now any body suspended in the air or lying on the earth; the aether being by hypothesis grosser in the pores which are in the upper parts of the body then in those which are in its lower parts, and this grosser aether being less apt to be lodged in those pores then the finer aether below, it will endeavor to get out and give way to the finer aether below, which cannot be without the bodies descending to make room above for it to go out into. And thus gravity [86].[162] A similar explanation of the cause of gravity will be referred to by Leonhard Euler in Sect. 3.4.1.3 some years later; where the aether was however an elastic continuous medium.

In the Queries of *Opticks* Newton introduced the idea of aether, in dubitative form as usual. The presence of a medium other than air, is supposed in Query 18 on the basis of experimental measurements of temperature in vacuo and in plenum: "And is not this Medium exceedingly more rare and subtile than the Air, and exceedingly more elastic and active? And doth it not readily pervade all Bodies? And is it not (by its elastick force) expanded through all the Heavens?" [94].[163]

Query 20. The very word aether is introduced in Query 20, where it is asked: does the *aetherial medium* in passing out of water, glass, crystal and other compact and dense bodies into empty spaces, grow denser and denser by degrees and the gradual condensation of this medium extend to some distance from the bodies, and thereby cause the diffraction of the rays of light, which pass by the edges of dense bodies, at some distance from the bodies? A reasoning in agreement with that developed in the letter to Boyle.

Query 21. By means of the properties of this aether Newton suggested an explanation of gravity in Query 21 which is apparently similar to that referred to in the letter to Boyle of 1679. But the aether of Query 21, at a scrutiny, reveals to be very different from that introduced in the letter to Boyle. *Here* the aether is a 'living' matter, formed by particles aimed by repulsive forces which give it the property of elasticity. And the difference in density is due to a greater or lesser concentration of equal particles; *there* the aether was an inert matter and the difference in density was due to grains of different thickness.

---

[162]p. 65r.

[163]Q. 18, p. 324.

Newton started by asking: "Is not this Medium much rarer within the dense Bodies of the Sun, Stars, Planets and Comets, than in the empty celestial Spaces between them? And in passing from them to great distances, doth it not grow denser and denser perpetually, and thereby cause the gravity of those great Bodies towards one another, and of their parts towards the Bodies; every Body endeavoring to go from the denser parts of the Medium towards the rarer?" [94].[164] For if this medium be rarer within the sun's body than at its surface and rarer there than at the orb of Saturn, and though this increase of density may at great distances be exceeding slow, it may suffice to impel bodies from the denser parts of the medium towards the rarer, with all that power which we call gravity.

Query 21 ends with the suggestion that the particles of aether repel each other with a short range force at a distance:

> And so if any one should suppose that *Aether* (like our Air) may contain Particles which endeavour to recede from one another (for I do not know what this *Aether* is) and that its Particles are exceedingly smaller than those of Air, or even than those of Light: The exceeding smallness of its Particles may contribute to the greatness of the force by which those Particles may recede from one another, and thereby make that Medium exceedingly more rare and elastick than Air [94].[165]

Query 22. In Query 22 Newton discussed if his aether does disturb the motion of planets. He asked: "May not planets and comets, and all gross bodies, perform their motions more freely, and with less resistance in this aethereal medium than in any fluid, which fills all space adequately without leaving any pores, and by consequence is much denser than quick-silver or gold? and may not its resistance be so small, as to be inconsiderable?" [94].[166] The answer is yes. Newton did not specify the way he reached his results, but surely basing on the studies of Book II of the *Principia*, according to which the speed of propagation of a pulse in a medium is proportional to the square root of the ratio between elastic force (that is in modern term stiffness) and density and because the very high value of the speed of light, Newton concluded that the density of the aether should be 700 000 less than that of air (and 600 000 000 of water), and consequently because in the *Principia* (Book II) it is proved that the resistance encountered by a body moving in a medium is proportional to its density (and the square of speed), the resistance encountered by a planet moving in the aether should be 700 000 times less than that it encountered in air, and thus negligible.

Query 28. In Queries 28 and 29 Newton discussed the nature of light and the role of aether. In Query 28 he wrote for instance: "Are not all Hypotheses [the vibration theory] erroneous, in which Light is supposed to consist in Pressure or Motion, propagated through a fluid Medium? For in all these Hypotheses the Phaenomena of Light have been hitherto explain'd by supposing that they arise from new Modifications of the Rays; which is an erroneous Supposition" [94].[167]

---

[164] Q. 21, p. 325.

[165] Q. 21, p. 326.

[166] Q. 22, p. 327.

[167] Q. 28, p. 336.

And as an example of the defect of the vibration theory he considered the explanation of fits which is hard to accept unless one might suppose that there are in the world two aethereal vibrating media, and that the vibrations of one of them constitute light and the vibrations of the other which are swifter constitute fits. "But how two Aethers can be diffused through all Space, one of which acts upon the other, and by consequence is re-acted upon, without retarding, shattering, dispersing and confounding one another Motions, is inconceivable" [94].[168]

Query 29. In Query 29 Newton presented his projectile theory, in a dubitative form but giving it a high probability, specially because the linear propagation of light. "Are not Rays of Light very small Bodies emitted from shining Substances? For such Bodies will pass through uniform Mediums in right Lines without bending into the Shadow, which is the Nature of the Rays of Light" [94].[169]

1.2.6 Newton's Methodology

1.2.6.1 Regulae Philosophandi

Toward the end of his *Principia*, at the beginning of Book III, Newton appended his famous *Regulae Philosophandi*. In the final version, as reported in the third edition of 1726, they are:

> Rule 1. *No more causes of natural things should be admitted than are both true and sufficient to explain their phenomena.* As the philosophers say: nature does nothing in vain, and more causes are in vain when fewer suffice. For nature is simple and does not indulge in the luxury of superfluous causes.
> Rule 2. *Therefore, the causes assigned to natural effects of the same kind must be, so far as possible, the same.* Examples are the [...] the falling of stones in Europe and America.
> Rule 3. *Those qualities of bodies that cannot be intended and remitted [i.e., qualities that cannot be increased and diminished] and that belong to all bodies on which experiments can be made should be taken as qualities of all bodies universally.*
> Rule 4. *In experimental philosophy, propositions gathered from phenomena by induction should be considered either exactly or very nearly true notwithstanding any contrary hypotheses, until yet other phenomena make such propositions either more exact or liable to exceptions.* This rule should be followed so that arguments based on induction may not be nullified by hypotheses [90].[170]

Rivers of words have been spent about these rules and little can be added. The content, the denomination and the position of the *Regulae philosophandi* varied a

[168] Q. 28, p. 339.

[169] Q. 29, p. 345.

[170] pp. 387–389. English translation in [103].

lot with the various editions of the *Principia*, proving the attention Newton devoted to them. In [11, 38] it is documented, both in Newton's published and unpublished works, the development of the rules from their early inception to the final form under which they appeared in 1726.

Rules 1 and 2 are rules of causal simplicity and uniformity of nature, a position which was accepted by many scholars of the 17th and 18th centuries, but that for Newton was also suggested by his conception of God, who has created a world so that man could discover its laws, at least partially [11].[171] Rule 1 says that the cause should be both sufficient (the phenomenon should follow from the cause) and also necessary (the cause should be implied by the phenomenon). The cause has to be single because the same phenomenon could not be explained by more than one true cause. Rule 1 contrasts probabilistic epistemologies, which allows for various explicative hypotheses without one could express preference for one of them. Rule 2 is more or less a corollary of rule 1, and represents a sort of Ockham's razor.

Rule 3 makes induction the appropriate method of studying physics. In [8, 36, 38] it is suggested that the rule is not an endorsement of induction only, but also of a something more sophisticated inference, which belongs to what is known as *transductive inference*, a term introduced in logic by Vladimir Vapnik in the 1990s, to indicate a reasoning from observed, specific (training) cases to specific (test) cases. In contrast, induction is reasoning from observed training cases to general rules, which are then applied to the test cases. With the rule 3, Newton declared that the properties detected experimentally on all bodies, that is, bodies at a macroscopic level, can also be extended to unobservable bodies, or atoms. A transductive inference would have been used in the proof of the Proposition 70, Theorem 30 [90],[172] in which it is proved that a corpuscle placed inside a spherical surface is acted by gravity forces in any direction. In this case the property of gravity and the law of the inverse square of distance, which is undoubtedly universal for all observable bodies (by induction) is also attributed to bodies of infinitesimal dimension, which belong to a different category of bodies and therefore their property could not be inferred according to the classical criteria of induction.

Rule 4 is the one currently most discussed by historians and philosophers of science. It concerns the so called Newtonian method, or Newtonian style, of investigation for *deducing* propositions, or laws, of physics from phenomena. It is an alternative to the approaches put forward by some scholars of the 17th and 18th century, based on a probabilistic epistemology, for which one never could reach truth in physics. What it could be make for them was only to propose, based on a priori knowledge, more or less probable hypotheses and verify their validity with experimental observations. Newton was against this approach because it could explain the empirical evidence of a phenomenon by proposing many alternative hypotheses, without a criterion of choice all being able to explain the phenomenon in object. The only way, according to Newton, to prevent this indeterminacy, is to avoid the formu-

[171] p. 17.

[172] p. 189.

lation of hypotheses a priori, and formulate them (that thus are no longer hypotheses but experimental laws) starting from experimental data.

In an unpublished manuscript, when elaborating on the meaning of rule 4, Newton wrote this clarifying passage:

> For if arguments from hypotheses would be admitted against inductions, inductive arguments, on which the whole of experimental philosophy is based, could always be overturned by contrary hypotheses. If a certain proposition collected by induction should be not sufficiently accurate, it ought be corrected, not by hypotheses but by phenomena of nature that are to be more widely and accurately observed [38].[173]

Newton proposed thus a gradual process of acquisition of the correctness of a law, by correcting it—if necessary—when new data is available. To a modern his approach does not seem different from the hypothetical deductive one of scholars who supported a probabilistic epistemology, but for Newton this was not the case. Mostly because for him a theory deduced from the phenomena could be improved by a better knowledge of the phenomena, but never overturned. While working by hypotheses, either the ascertainment of new data leads to a radical change of the explanatory picture, or it does not change anything because no element intervene to prefer one theory from another. It is not here the case to discuss if an inductive approach is possible in physics—this is a debated matter—but only to argue that Newton was convinced of the fact independently of that according some historians and philosophers of science, he unconsciously pursued a hypothetical deductive approach.

The substantial breaking of Newton's empiricism with Descartes's rationalism has possibly part of its roots in the different role they attributed to God in establishing the laws of nature. Both of them supposed that these laws were imposed by God to the world making it regular. Descartes thought they, at least the basic ones or the principles, were impressed in man's soul and thus a simple reflection would suffice to discover them, without any recourse to experiments [52]. Newton, with the English, thought that the world was a contingent creation of God and only the observation discovers its laws; and possibly the first causes (that is the principles) are very difficult to discover.

As the rules were placed at the beginning of the third book of the *Principia*, entitled *De systemate mundi*, at the end of the deductive part of the treatise, this means that for Newton they did not have the same nature of axioms or definitions. The regulae are architectural in nature and relate to the way in which the principles of natural philosophy properly intended should be determined. They are thus methodological criteria and as such contain wide margins of arbitrariness and indefiniteness.

Sometimes, certainly exaggerating, it is said that the regulae represent Newton's fundamental legacy to modern science; they were explicitly considered as a model by many scholars, even in the 19th century, such as for instance William Whewell (1794–1866) and Charles Darwin (1809–1882) [11].[174] The situation is actually not so clear. Meanwhile, the rules were presented by Newton in a non-definitive form only, as will be clarified later, in the first edition of the *Principia* of 1687, when he

---

[173]p. 157.

[174]p. 1.

had by then already substantially developed his mechanics and optics. They therefore represent an explicit epistemology, that is, an epistemology that Newton wearing the clothes of the philosopher considered as the correct one, but that when he was wearing the clothes of the physicists he did not apparently consider. In fact, Newton referred to his rules explicitly only in very few points of the *Principia*. Referring to the third edition of this treatise he used Rule 3 and Rule 4 only once (and only in the third book), respectively while defending the Proposition III.5 "For the cause of the centripetal force which retains the moon in its orbit will extend itself to all the planet, by Rule I, II and IV" [90][175]; and III.6 "This is the quality of all bodies within the reach of our experiments; and therefore (by Rule III) to be affirmed of all bodies whatsoever" [90].[176] Rule 2 is cited in support of Proposition III.5 "For the circumjovial planets about Jupiter [...] and the circumsolar planets, about the sun, are appearances of the same sort with the revolution of the moon about the earth; and therefore by rule II must be owing to the same sort of causes" [90].[177] And both Rule 1 and Rule 2 are cited in support of proposition III.4 ("And therefore, by rule I and II, the force by which the moon is retained in its orbits is that very same force which is commonly called gravity" [90],[178] and III.5 "Therefore since both these forces, that is, the gravity of heavy bodies and the centripetal forces of their moons, with respect, they will (by Rule I and II) one and the same causes" [90],[179]

Moreover the rules don't seem particularly original. They represent a synthesis of medieval theories of knowledge and an explicitness of the approach followed by the mixed mathematicians since ancient Greece. In Sect. 1.2.4 their possible derivation from Newton's theology is considered; here I limit to note that even Isaac Barrow (1630–1677), Newton's mentor, had expressed similar ideas, framing them in an Aristotelian epistemology.

> From which it appears that according to Aristotle, the principles of all science depend wholly upon the testimony of the senses and particular experiments [...]. But where any proposition is found agreeable to constant experience, especially where it seems not to be conversant about the accidents of things, but pertains to their principal properties and intimate constitution, it will at least be most safe and prudent to yield a ready assent to it. For as we are justly accused of a rash temerity, by suffering ourselves to be so much as once deceived by our faith, so we are guilty of the greatest imprudence, if we shew the least distrust, and do not yield our stedfast assent and obstinately adhere, when we still find our expectations answered as accurately as possible (quam accuratissime), after a thousand researches; and especially when we have the constant agreement of nature to conérm our assent, and the immutable wisdom of the first cause forming all things according to simple ideas, and directing them to certain ends: which consideration alone is almost sufficient to make us look upon any proposition confirmed with frequent experiments, as universally true (universaliter vera), and not suspect that nature is inconstant and the great author of the universe unlike himself. [...] As Aristotle observes and confirms by a most appropriate instance [7].[180]

[175] p. 399.
[176] p. 399.
[177] p. 399.
[178] pp. 397–398.
[179] p. 398.
[180] pp. 73–74.

The label *Regulae philosophandi* appeared only from the second edition of the *Principia*. In the first edition the rules went under the name of *Hypotheses*, for a total of nine propositions, in the second there were three rules and in the third four, listed above.

According to Alexandre Koyré the (logical) disorder present in the first draft of the rules must be sought in the haste with which the *Principia* were drawn up. The commonly accepted motivation for the change of name from hypotheses to rules is attributed to Newton's desire to clearly distinguish his way of philosophizing from Descartes's, avoiding the use of hypotheses not justified by experience. It must be said, however, that in the 17th and 18th centuries the use of the terms principle, axiom, theorem law, hypothesis, contained large margins of ambiguity. In particular, there was not made always difference between purely empirical propositions and arguments derived with geometric deductions [70].[181]

Newton had a fifth hypothesis to publish, which is found in his manuscripts.

> Whatever is not derived from things themselves, whether by the external senses or by the sensation of internal thoughts, is to be taken for a hypothesis. Thus I sense that I am thinking, which could not happen unless at the same time I were to sense that I am. But I do not sense that any idea whatever may be innate. And I do not take for a phenomenon only that which is made known to us by the five external senses, but also that which we contemplate in our minds when thinking; such as, I am, I believe, I understand, I remember, I think, I wish, I am unwilling, I am thirsty, I am hungry, I rejoice, I suffer, etc. And those things which neither can be demonstrated from the phenomenon nor follow from it by the argument of induction, I hold as hypotheses [70].[182] (A.23)

As it can be seen, this rule has a more pronounced metaphysical character and is clearly anti-Cartesian. The reasons why Newton did not publish it may be various. Partly in order not to accentuate the anticartesian polemic, partly because the rule is too speculative and not suited to a treatise on mixed mathematics like the *Principia*.

1.2.6.2 Hypotheses Non Fingo

Newton is famous for his empiricist statement *Hypotheses non fingo* appearing in the Scholium generale of the *Principia* [87].[183] Disputes have also been raised for its translation into English, without however helping to clarify the problem [18]. Motte, the first editor of the *Principia* in English, translates: "I frame no hypotheses". However it seems that neither Newton, nor the Newtonians such as Richard Bentley, Henry Pemberton, or Colin Maclaurin gave an official approval to this translation [18].[184] Samuel Clarke translated: "Hypoteses I make not". Roger Cotes, the editor of the second edition of the *Principia* left the sentence in Latin. Recently Koyré suggested to use "feign" for *fingere*. For him with "hypotheses non fingo" Newton meant "I feign no hypotheses" [69].

---

[181] pp. 261–272.

[182] p. 272.

[183] p. 530.

[184] p. 379.

In any case, more than the translation of the verb fingere, the interest should be addressed on the name "hypotheses".

Newton on several occasions distinguished between *hypothesis*, and *true doctrine*, or *theory*, a term this latter used more rarely. For example *theory* is found in the title of *A new theory of light and colors* of 1672 [84], but in the body of the text only *doctrine* is used. There is no doubt that the hypotheses Newton did not approve were those of mechanical philosophy, in particular in the version given by Descartes, in which the invention of an elegant mechanism based on the interaction of various corpuscles, that is an hypothesis, was more important than it empirical verification.

Newton used *hypothesis* in the *Principia* in a very few instances (40 in 1726 edition). In most cases the term is used referring to Descartes and other scholars; only in three cases the term was given an apparently technical relevance. In Book II, in the study of the motion of fluids:

> HYPOTHESIS.
> The resistance arising from the want of lubricity in the parts of a fluid, is, caeteris paribus, proportional to the velocity with which the parts of the fluid are separated from each other [90].[185](A.24)

and two times in Book III:

> HYPOTHESIS I.
> *The center of the system of the world is at rest [Centrum systemati mundani quiescere].* No one doubts this, although some argue that the earth, others that the sun, is at rest in the center of the system. Let us see what follows from this hypothesis [90].[186](A.25)

> HYPOTHESIS II.
> If the ring discussed above were to be carried alone in the orbit of the earth about the sun with an annual motion (supposing that all the rest of the earth were removed from it), and if this ring revolved at the same time with a daily motion about its axis, inclined to the plane of the ecliptic at an angle of 23° degrees, then the motion of the equinoctial points would be the same whether that ring were fluid or consisted of rigid and solid matter [90].[187](A.26)

In the first case, Book II, the hypothesis has a similar role of the hypothesis of astronomer; the second case the hypothesis has a metaphysical nature and consequently cannot be verified. More complex is to classify the third hypothesis; to a modern it seems to be verifiable simply inside mathematics, by calculation. But for Newton this was not probably the case.

If in the *Principia* the role of hypotheses in Newton appears only in few, but meaningful points, it is instead more important in his optical writings and in the letters in which he defended his theory of colors from the disputes of Leibniz, Huygens, Hooke, Pardies, and others. In the *A new theory of light and colours*, Newton distinguished quite well between his doctrine, derived for him from experiments, and a hypothetical explanation about the nature of light. The doctrine is summarized in the following points:

---

[185]p. 374.

[186]p. 408. English translation in [103].

[187]p. 476. English translation in [103].

Prop 1. Colours are not *qualifications of light* derived from refractions or reflections of naturall bodies as 'tis generally beleived, but originall & connate properties, which in diverse rays are divers.

Prop 2. To the same degree of refrangibility ever belongs the same colour, & to the same colour ever belongs the same degree of refrangibility.

Prop 3. The species of colour & degree of refrangibility proper to any particular sort of rays, is not mutable by refraction.

Prop 4. Yet seeming transmutations of colours may be made where there is any mixture of divers sorts of rayes

Prop 5. There are therefore two sorts of colours. The one originall & simple; the other compounded of these.

Prop 6. The same colours in *specie* with these primary ones may be also produced by composition.

Prop 7. [Whiteness is] ever compounded and to its composition are requisite all the aforesaid primary colours mixed in a due proportion.

Prop 8. Whiteness is generated if there is a due proportion of the ingredients; But if any one predominate, the light must incline to that colour.

Prop 9. Since those [rays] which differ in colour proportionally differ in refrangibility, *they* by their unequall refractions must be severed and dispersed into an oblong form in an orderly succession from the least refracted scarlet to the most refracted violet.

Prop 10. Why the colours of the *rainbow* appear in falling drops of rain is also from hence evident. For those drops refract the rays.

Prop 11. Coloured bodies appear in one of colour in one position and another colour in another position because they are apt to reflect one sort of light and transmit another.

Prop 12. Namely that though they were severally transparent enough yet both together became opake. For if one transmitted only red, and the other only blew, no rays could pass through both.

Prop 13. The colours of all naturall bodies have no other origin then this, that they are variously qualified to reflect one sort of light in greater plenty then another [84].[188]

The hypothetical explanation is summarized in a few lines:

> These thinges being so, it can be no longer disputed whether there be colours in the dark, nor whether they be the qualities of the objects wee see, no nor *perhaps* [emphasis added] whether light be a body [84].[189]

Newton was clearer in his replies to Hooke's criticism, particularly in his letter to Oldenburg of 11 June 1672. Newton stated that Hooke did not interpret his hypothesis correctly be expressing them in these words: "*But grant his first supposition that light is a body, and that as many colours or degrees thereof as there may be so many bodies*

[188]pp. 3081–3084.

[189]p. 3085.

there may be, all which compounded together would make white" [98].[190] Indeed it is true that Newton argued the corporeity of light, but he did it without any absolute positiveness, as the word *perhaps*, in the above quotation, intimates, and make it at most but a very plausible consequence of the doctrine and not a fundamental supposition. And if it is true that the properties of light can also be explained by Hooke's theory, they "were in some measure capable of being explicated non onely by that, but by many other Mechanicall Hypotheses" [98].[191]

Newton, ironically apologized, he did not take Hooke's theory up, since he did not think necessary to explicate his doctrine by any Hypothesis at all. For "I can as easily conceive that ye severall parts of a shining body may emit rays of differing colours & other qualities, of all wch light is constituted, as that the severall parts of a false or uneven string, or of unevenly agitated water in a Brook or Cataract, or ye severall Pipes of an Organ inspired all at once, or all ye variety of sounding bodies in ye world together, should produce sounds of severall tones, & propagate them through ye Air confusedly intermixed" [98].[192]

The letter to Oldenburg ends with a section entitled "That the science of colours is most properly a Mathematicall Science" [98],[193] which predates what will be written in the *Opticks*, and clarify Newton conceptions about mixed mathematics.

> I said indeed that the *science of colours was mathematical & certain as any part of optiques*; but who knows not that optiques & many other mathematical sciences depends as well on physical principles as on mathematical principles. And the absolute certainty of a science cannot exceed the certainty of its principles [98].[194]

Also interesting is the reply to Huygens, in a letter to Oldenburg dated 3 April 1673, which supported the need for a mechanicistic explanation, even limited to two colors.

> But to examin how colours may be thus explained Hypothetically is besides my purpose. I never intended to show wherein consists the nature and difference of colours, but onely to show that *de facto* they are originall & immutable qualities of the rays wch exhibit them, & to leave it to others to explicate by Mechanicall Hypotheses the nature & difference of those qualities; wch I take to be no very difficult matter. But I would not be understood as if their difference consisted in the different refrangibility of those rays. For that different refrangibility conduces to their production no otherwise then by separating the rays whose qualities they are. Whence it is that the same rays exhibit the same colours when separated by any other meanes; as by their different reflexibility; a quality not yet discoursed of [98].[195]

For Newton however, hypotheses are not necessarily useless if constructed with common sense. In [128][196] it is suggested that hypotheses in Newton had two functions:

1. To illustrate the theory.

[190] Vol.1, p. 173.
[191] Vol.1, p. 174.
[192] Vol.1, p. 177.
[193] Vol.1, p. 187.
[194] Vol.1, p. 187.
[195] Vol. 1, pp. 264–265.
[196] p. 71.

2. To suggest experiments.

The first function of hypotheses was declared by Newton in the paper entitled *An hypothesis explaining the property of light discoursed in my severall papers*, sent to Oldenburg in December 1675, with a cover letter, and one more paper, *The discourse of observations*, which is accessible in the *Opticks*.

> I had formerly purposed never to write any Hypotheses of light & colours, fearing it might be a means to ingage me in vain disputes: but I hope a declar'd resolution to answer nothing that looks like a controversy (unles possibly at my own time upon some other by occasion) may defend me from yt fear. And therefore considering that such an Hypothesis would much illustrate ye papers I promis'd to send you, & having a little time this week to spare: I have not scrupled to describe one so far as I could on a sudden recollect my thoughts about it, not concerning my self whether it shall be thought probable or improbable so it do but render ye papers I send you, and others sent formerly, more intelligible [98].[197]

The second function of hypotheses can be found in some queries Newton began to report in the correspondence. For example in the aforementioned letters to Oldenburg of 11 June 1672, Newton introduced three Queries.

Q_1 Whether the unequal refractions made without respect to any inequality of incidence, be caused by the different refrangibility of several rays, or by the splitting breaking or dissipating the same ray into diverging parts.
Q_2 Whether there be more then two sorts of colours.
Q_3 Whether whitenesse be a mixture of all colours [98].[198]

A modern reader would refer to these queries as working hypotheses. Newton did not adopt this term; for two reasons: (a) it is not a question of verifying the correctness of a mechanism made up of corpuscles, but rather the occurrence of certain phenomena. That is, the queries have an empirical character. (b) Newton had already given an answer to the queries in his works. In particular, Q_1 is resolved by the *experimentum crucis*. Q_2 from the fact that in the light there are more than two indices of refraction and Q_3 from the fact that the colors appear refracting the white light and this can be reconstructed by the colored rays.

A greater articulation of queries, can be found in the so-called *Queries paper*, a letter to Oldenburg of 6 July 1672.

> 1. Whether rays that are alike incident on ye same Medium have unequall refractions, & how great are the inequalities of their refractions at any incidence?
> 2. What is ye law according to wch each ray is more or lesse refracted, whether it be yt the same ray is ever refracted according to the same ratio of the sines of incidence & refraction; & divers rays, according to divers ratios; Or that the refraction of each ray is greater or lesse without any certain rule? That is, whether each ray have a certain degree of refrangibility according to wch its refraction is performed, or is refracted without that regularity?

[197] Vol. 1, p. 361.
[198] Vol. 1, p. 178.

3. Whether rays wch are indued with particular degrees of refrangibility, when they are by any meanes separated, have particular colours constantly belonging to them: viz, the least refrangible, scarlet; the most refrangible, deep violet; the middle, Sea-green; & others, other colours? And on the contrary? fraction?
4. Whether the colour of any sort of rays apart may be changed by refraction?
5. Whether colours by coalescing do really change one another to produce a new colour, or produce it by mixing onely?
6. Whether a due mixture of rays, indued with all variety of colours, produces light perfectly like that of the Sun, & wch hath all the same properties & exhibits the same Phaenomena?
7. Whether there be any other colours produced by refractions then such, as ought to result from the colours belonging to the diversly refrangible rays by their being separated or mixed by that refraction [98]?[199]

Queries will find their natural place, and a technical meaning, in the *Opticks*, at the end of the Book 3. It was easy to see that Queries were not just Cartesian hypotheses under a different name; they were rather empirical questions that were to be resolved by experiments. While the early queries, those preceding the *Opticks*, seemed to be tied to a specific experimental program and theoretical points of the theory of light, the queries in the *Opticks* explored a broader range of ideas.

Though the specific functions were different, there were at least two general similarities between the early queries and the queries of the *Opticks*. First, the former tended to take the form, "whether it is the case that p?", while the latter took the form, "is it not the case that p?". Even though the latter might be a slightly stronger form of indirect assertion, they both function in the same way. Second, they shared a general experimental outlook, concerned with leading the discussion towards an empirical solution.

It has been argued that many of the queries that appeared in the first edition of the *Opticks* look like contributions to an experimental natural history [3].[200] In the whole however, despite the speculative content and concern with the nature of light, the Queries were experimental research programs. Some of them contained lots of discussion of observation and experiment, others, little or none. Moreover, the experimental discussion was, for the most part, sketchy and qualitative [128].[201]

1.3 Quotations

A.1 La mesme force qui peut lever un poids, par exemple, de cent, livres a la hauteur de deux pieds, en peut aussy lever un de 200 livres, a la hauteur d'un pied, ou un de 400 a la hauteur d'un demi pied, & ainsy des autres.

---

[199] Vol. 1, pp. 209–210.

[200] p. 265.

[201] pp. 123–124.

A.2 Il est bien vrai que dans l'état où nous sommes, nous avons plus de peine remuer une grosse pierre, qu'à en remuer une petite; mais il n'y a personne qui ne sache que cela vient de la resistance que cause la pesanteur de ces pierre. Car si la grande pierre n'étoit pas plus pesante que la petite, il n'y a point de doute que nous la poussions mouvoir avec la m me facilité.

A.3 Il est possible toutefois d'y arriver a un degré de vraisemblance qui bien souvent ne cede guere à une evidence entiere. Sçavoir lors que les choses, qu'on a demontrées par ces Principes supposez, se raportent parfaitement aux phenomenes que l'experience a fait remarquer; sur tout quand il y en a grand nombre, & encore principalement quand on se forme & prevoit des phenomenes nouveaux, qui doivent suivre des hypotheses qu'on employe, & qu'on trouve qu'on cela l'effet repond a nostre attente. Que si toutes ccs preuves de la vraisemblance se rencontrent dans ce que je me suis proposé de traiter, comme il me semble qu'elles sont, ce doit etre une bien grande confirmation du succês de ma recherche, & il se peut malaisement que les choses ne soientpeu pres comme je les represente. Je veux donc croire que ceux qui aiment a connoitre les causes, & qui sçavent admirer la merveille de la Lumiere, trouveront quelque satisfaction dans ces diverses speculations qui la regardent, & dans la nouvelle explication de son insigne proprieté, qui fait le principal fondement de la construction de nos yeux, & de ces grandes inventions qui en étendent si fort l'usage.

A.4 Je respons que dans les choses de physique il n'y a pas d'autres demonstrations que dans le déchiffrement d'une lettre. Ou ayant fait des suppositions sur quelques légères conjectures, si l'on trouve qu'elles se vérifient en suivre, the sorte que suivant ces suppositions de lettres on trouve des paroles bien suivies dans la lettre, on tient d'une certitude trs grande que les suppositions sont vraies, quoy qu'il n'y ait pas autrement de demonstration, et qu'il ne soit pas impossible qu'on n'est poisse y avoir d'autres plus véritable.

A.5 Il y a encore considerer dans l'émanation de ces ondes, que chaque particule de la matiere, dans laquelle une onde s'etend, ne doit pas communiquer son mouvement seulement Ia particule prochaine, qui est dans la ligne droite tirée du point lumineux; rnais qu'elle en donne aussi necessairement toutes les autres qui la touchent, & qui s'opposent a son mouvement. De sorte qu'il faut qu'autour de chaque particule il se fasse une onde dont certe particule soit le centre.

A.6 Comme il y avoit deux refractions differentes, je coçnus qu'il avoit aussi deux differentes emanations d'ondes de lumiere, & que l'une se pouvoit faire dans la matiere ètherée repandue dans le corps du cristal.

[...] Qant l'autre emanation [qui devoir produire la refraction irreguliere, je voulus essaier ce que seroient des ondes Elliptiques, ou pour mieux dire spherodes; lesquelles [...] je supposay qu'elles s'entendoient indifferemment, tant dans la matiere ètherée repandue dans le crstal, que dans les particules dont il est composé; suivant la derniere maniere dont j'ay explique la transparence. Il me sembloit que la disposition, ou arrangement regulier de ces particules, pouvoit contribuer a former les ondes spheroides, (n'estant requis pour cela si non que le mouvement successif de la lumiere s'etendit un peu plus viste en un sens qu'en l'autre) & je ne doutay presque point qu'il n'y eust dans ce cristal un tel arrangement de particules égales & semblables, cause de sa figure & de ses angles d'une mesure certaine & invariable.

A.7 Hoc spatium ita solum absque ullo corpore consideratum, quomodo quiescere intelligi possit non video. Cum quies et motus non sint nisi corporum, et utriusque idea ab his solis exorta sit. Nam si spatij quies aut motus esse aliquis dici potest, illius spatij erunt, quod a corpore occupatur, vel quod a corpore includitur, ut sis amphorae spatium una cum amphora quiescere aut moveri dicamus. At spatio illi infinito et inani neque motus neque quietis idea aut appellatio convenit. Qui vero quiescere ipsum statuunt, non alia ratione id facere videntur, quam quod animadvertunt absurdum esse si moveri dicatur, unde m necessaria quiescere dicendum putarunt. Cum potius cogitare debuerint nec motum nec quietem ad spatium illud omnino pertinere.

A.8 Diu putavi in circulari motu haberi veri motus χριτηριον, ex vi centrifuga. Etenim ad caeteras quidem apparentias idem fit sive orbis aut rota quaepiam c me juxta adstante circumrotetur, sive stante orbe illo ego per ambitum ejus circumferar, sed si lapis ad circumferentiam ponatur, projicietur circumeunte orbe, ex quo vere tunc et nulla ad aliud relatione eum moveri et circum gyrari judicari existimabam. Sed is effectus hoc tantummodo declarat impressione in circumferentiam facta partes rotae motu relativo ad se invicem in partes diversas impulsas fuisse.

A.9 In motu libero praesentibus corporibus inter se quiescentibus certo cognoscantur directiones et in his celeritates per quas mutatio distantiae explicetur et horum opera etiam circulantium celeritas defmiturw. Illis sublatis corporibus, difficilius hoc cognoscitur in liberis sed motus circularis duorum vel plurium vinculo conjunctorum, vel partium unius corporis, deprehenditur ex vi centrifuga. contra eos qui verum motum hunc esse volunt. dico non esse nisi respectivum. non enim potes dicere centrum circulationis quiescere in mundo, sed etiam respective tantum ad alia corpora.

A.10 Deus summus est ens aeternum, infinitum, absolute perfectum [...]. Aeternus est & infinitus, omnipotens & omnisciens, id est, durat ab aeterno in aeternum, & adest ab infinito in infinitum: omnia regit; & omnia cognoscit, quae fiunt aut fieri possunt. Non est aeternitas & infinitas, sed aeternus & infinitus; non est duratio & spatium, sed dura & adest. Durat semper, & adest ubique, & existendo semper & ubique, durationem & spatium constituit. Cum unaquaeque spatii particula sit *semper*, & unumquodque durationis indivisibile momentum *ubique*, certe rerum omnium fabricator ac dominus non erit *nunquam, nusquam*. [...] Deus est unus & idem deu semper & ubique. Omnipraesens est non per *virtutem* solam, sed etiam per *substantiam* [...] Hunc cognoscimus solummodo per proprietates ejus & attributa, & per sapientissimas & optimas rerum structuras & causas finales & admiramur ob perfectiones [...]. Et haec de deo, de quo unique ex phaenomenis disserere, ad philosophiam naturalem pertinet.

A.11 At si ex usu definiend sunt verborum significationes; per nomina illa temporis, spatii, loci & motus proprie intelligend erunt h mensur sensibiles; & sermo erit insolens & pure mathematicus, si quantitates mensurat hic intelligantur. Proinde vim inferunt sacris literis, qui voces hasce de quantitatibus mensuratis ibi interpretantur. Neque minus contaminant mathesin & philosophiam, qui quantitates veras cum ipsarum relationibus & vulgaribus mensuris confundunt.

A.12 Definitio III. Materiae vis insita est potentia resistendi, qua corpus unumquodque, quantum in se est, perseverat in statu suo vel quiescendi vel movendi uniformiter in directum.

A.13 Haec semper proportionalis est suo corpori, neque differt quicquam ab inertia mass, nisi in modo concipiendi. Per inertiam materi fit, ut corpus omne de statu suo vel quiescendi vel movendi difficulter deturbetur. Unde etiam vis insita nomine significantissimo vis Inerti dici possit. Exercet vero corpus hanc vim solummodo in mutatione status sui per vim aliam in se impressam facta; estque exercitium illud sub diverso respectu & resistentia & impetus: Resistentia, quatenus corpus ad conservandum statum suum reluctatur vi impress; impetus, quatenus corpus idem, vi resistentis obstaculi difficulter cedendo, conatur statum obstaculi illius mutare. Vulgus resistentiam quiescentibus & impetum moventibus tribuit: sed motus & quies, uti vulgo concipiuntur, respectu solo distinguuntur ab invicem; neque semper vere quiescunt quae vulgo tanquam quiescentia spectantur.

A.14 Definitio IV. Vis impressa est actio in corpus exercita, ad mutandum ejus statum vel quiescendi vel movendi uniformiter in directum.

A.15 Consistit hc vis in actione sola, neque post actionem permanet in corpore. Perseverat enim corpus in statu omni novo per solam vim inerti. Est autem vis impressa diversarum originum, ut ex ictu, ex pressione, ex vi centripeta.

A.16 Hactenus exposui motus corporum attractorum ad centrum immobile, quale tamen vix extat in rerum natura [...]. Qua de causa jam pergo motum exponere corporum se mutuo trahentium, considerando vires centripetas tanquam attractiones, quamvis fortasse, si physice loquamur, verius dicantur impulsus. In mathematicis enim jam versamur; & propterea, missis disputationibus physicis, familiari utimur sermone, quo possimus a lectoribus mathematicis facilius intelligi.

A.17 Definitio V. Vis centripeta est, qua corpora versus punctum aliquod, tanquam ad centrum, undique trahuntur, impelluntur, vel utcunque tendunt.

A.18 Hujus generis est gravitas, qua corpora tendunt ad centrum terr; vis magnetica, qua ferrum petit magnetem; & vis illa, qucunque sit, qua planet perpetuo retrahuntur a motibus rectilineis, & in lineis curvis revolvi coguntur.

A.19 Uti pondus majus in majore corpore, minus in minore; & in corpore eodem majus prope terram, minus in coelis. Haec quantitas est corporis totius centripetentia seu propensio in centrum, & (ut ita dicam) pondus; & innotescit semper per vim ipsi contrariam & aequalem, qua descensus corporis impediri potest.

A.20 Vocem attractionis hic generaliter usurpo pro corporum conatu quocunque accedendi ad invicem: sive conatus iste fiat ab actione corporum, vel se mutuo petentium, vel per spiritus emissos se invicem agitantium; sive is ab actione theris, aut aris, mediive cujuscunque seu corporei seu incorporei oriatur corpora innatantia in se invicem utcunque impellentis.

A.21 Hactenus phaenomena caelorum & maris nostri per vim gravitatis exposui, sed causam gravitatis nondum assignavi Oritur utique haec vis a causa aliqua, quae penetrat ad usque centra solis & planetarum, sine virtutis diminutione; quaeque agit non pro quantitate *superficierum* particularum, in quas agit (ut solent causae mechanicae) sed pro quantitate materiae *solidae*; & cujus actio in immensas distantias undique extenditur, decrescendo semper in duplicata ratione distantiarum.

A.22 Hactenus proprietates gravitatis explicui. Causas ejus minime expendo. Dicam tam en quid Veteres hac de re senserint. ⟨nimirum spiritum quendam per caelos⟩ Nempe caelos esse corporis prope vacuos (?) sed spiritu tamen quodam infinito quem Deum nuncupant undique impleri: (?) corpora autem in spiritu illo libere moveri ejus vi et virtute ⟨corpora⟩ naturali ad invicem (impelli) perpetuo impelli, idque magis vel minus pro ratione harmonica distantiarum, & in hoc [impulsu] gravitatem consistere. Hunc spiritum aliqui a Deo summa distinxerunt & animam mundi vocarunt.

A.23 Adjicere jam liceret nonnulla de spiritu quodam subtilissimo corpora crassa pervadente,& in iisdem latente; cujus vi & actionibus particulae corporum ad minimas distantias se mutuo attrahunt, & contiguae factae cohaerent; & corpora electrica agunt ad distantias majores, tam repellendo quam attrahendo corpuscula vicina; & lux emittitur, reflectitur, refringitur, inflectitur [...]. Sed haec paucis exponi non possunt; neque adest: sufficiens copia experimentorum, quibus leges actionum. hujus spiritus accurate determinari & monstrari debent.

A.24 Reg. V. Pro hypothesibus habenda sunt quaecunque ex rebus ipsis vel per sensus externos, vel per sensationem cogitationum internarum non derivantur. Sentio utique quod Ego cogitem, id quod fieri nequiret nisi simul sentirem quod ego sim. Sed non sentio quod Idea aliqua sit innata. Et pro Phaenomenis habeo non solum quae per sensus quinque externos nobis innotescunt, sed etiam quae in mentibus nostris intuemur cogitando: Ut quod, Ego sum, ego credo, doleo, etc. Et quae ex phaenomenis nec demonstrando nec per argumentum inductionis consequuntur, pro Hypothesibus habeo.

A.25 HYPOTHESIS. Resistentiam, quae oritur ex defectu lubricitatis partium fluidi, caeteris paribus, proportionalem esse velocitati, qua partes fluidi separantur ab invicem.

A.26 HYPOTHESIS I.
Centrum systemati mundani quiescere.

A.27 HYPOTHESIS II.
Si annulus praedictus terra omni reliqua sublata, solus in orbe terree, motu annuo circa solem ferretur, & interea circa axem suum, ad planum eclipticae in angulo graduum $23^{1/2}$ inclinatum, motu diurno revolveretur: idem foret motus punctorum aequinoctialium, sive annulus iste fluidus esset, sive is ex materia rigida & firma constaret.

Notes

[I]This is a possible definition:

1. The space time is a four dimensional affine space A^4, named *universe*. The points of the universe are called *world points* or *events*. The parallel displacements of the universe A^4 constitute a vector space R^4.
2. Time is a linear mapping $\varphi : R^4 \rightarrow R$ from the vector space of parallel displacements of the universe to the 'real time' axis. The kernel of the mapping φ is a three-dimensional linear subspace R^3, named space of the contemporary events.
3. The space R^3 is endowed with a metric structure which makes it a three dimensional Euclidean space E^3 [5], p. 5.

[II]The space-time A^4 is named a Galilean space-time if it is invariant with respect to the Galilean transformations that assuming for the sake simplicity A^4 as $R \times R^3$, are defined as follows:

Uniform motion with velocity **v**

$$g_1(t, \mathbf{x}) = (t, \mathbf{x} + \mathbf{v}t) \quad \forall t \in R; \mathbf{x} \in R^3$$

Translation of the origin of time (s) *and space* $(\mathbf{s})$

$$g_2(t, \mathbf{x}) = (t + s, \mathbf{x} + \mathbf{s}) \quad \forall t \in R; \mathbf{x} \in R^3$$

Rotation by means of an orthogonal matrix **G**

$$g_3(t, \mathbf{x}) = (t, \mathbf{G}\mathbf{x}) \quad \forall t \in R; \mathbf{x} \in R^3$$

The invariance with respect to the first transformation states that it is not possible to distinguish a space from another if they move of translatory uniform motion one with respect to the other. The invariance with respect to the second and third transformations says that the distinction neither occur for a simple translation (homogeneity of space) nor for a rotation (isotropy of space).

References

1. Alessio F (1963) Thomas Hobbes: Tractatus opticus. First integral edition. Riv Crit Stor Della Filos 18(2):147–228
2. Ango P (1682) L'optique divise en trois livres: Où l'on démontre d'une manière aisée tout ce qui regarde; La propagation et les proprietez de la lumiere; La vision; La figure et la disposition des verres qui servent la perfectionner. Michallet, Paris
3. Anstey PR (2004) The methodological origins of Newton's queries. Stud Hist Philos Sci Part 35(2):247–269
4. Aristotle (2018) Physica. The internet classical archive. Translated into English by Hardie RP, Gaye RK

5. Arnold VI (1989) Mathematical methods of classical mechanics. Springer, New York
6. Ballard KE (2018) Leibniz's theory of space and time. J Hist Ideas 21(1):49–65
7. Barrow I (1734) The usefulness of mathematical learning explained and demonstrated. Translated into English by Kirkby J. Austen, London
8. Belkind O (2017) On Newtonian induction. Philos Sci 84(4):677–697
9. Bertoloni Meli D (2006) Inherent and centrifugal forces in Newton. Arch Hist Exact Sci 60(3):319–335
10. Biener Z (2017) De gravitatione reconsidered: the changing significance of empirical evidence for Newton's metaphysics of space. J Hist Philos 55(4):583–608
11. Biener Z (2018) Newton's regulae philosophandi. In: Smeenk C, Schliesser E (eds) Oxford handbook for Isaac Newton. Oxford University Press, Oxford, pp 1–23
12. Capecchi D (2012) History of virtual work laws. Birchäuser, Milan
13. Capecchi D (2014) Attempts by Descartes and Roberval to evaluate the centre of oscillation of compound pendulums. Early Sci Med 19(3):211–235
14. Capecchi D (2014) The problem of motion of bodies. Springer, Cham
15. Capecchi D (2018) The path to post-Galilean epistemology. Springer, Cham
16. Casini P (1984) Newton: the classical scholia. Hist Sci 22(1):1–46
17. Clarke S (1717) A collection of papers which passed between the late learned Mr. Leibnitz and Dr. Clarke. Knapton, London
18. Cohen IB (1962) The first English version of Newton's hypotheses non fingo. Isis 53(3):379–388
19. Cohen IB (1980) The Newtonian revolution. Cambridge University Press, Cambridge
20. Cohen IB (1992) The review of the first edition of Newton's Principia in the Acta Eruditorum, with notes on the other reviews. In: Harman PM, Shapiro AE (eds) The investigation of difficult things. Cambridge University Press, Cambridge, pp 323–354
21. Costello WT (1958) The scholastic curriculum at early seventeenth century Cambridge. Harvard University Press, Cambridge
22. Delambre JBJ (1812) Notice sur la vie et les oeuvres de M. le Comte J.L. Lagrange. In: Serret JA, [Darboux G] (1867–1892) (ed) Oeuvres de Lagrange (14 vols). Gauthier-Villars, Paris, pp I–LI
23. Delgado-Moreira R (2006) Newton's treatise on revelation: the use of a mathematical discourse. Hist Res 79(204):224–246
24. Descartes R (1637) La dioptrique. In: Descartes R (1668) Discours sur la methode, plus la dioptrique et les meteores. Girard, Paris
25. Descartes R (1644) Principia philosophiae. Ludovicum Elzevirium, Amsterdam
26. Descartes R (1650) Musicae compendium. Ackersdijck and Zijll, Utrecht
27. Descartes R (1664) Le monde de Mr. Descartes ou le traité de la lumière et des autres. Girard, Paris
28. Descartes R (1964) Oeuvres de Descartes; nouvelle édition complètes (1896–1913) (11 vols). In: Adam C, Tannery P (eds). Vrin, Paris
29. Dijksterhuis FJ (2004) Lenses and waves. Christiaan Huygens and the mathematical science of optics in the seventeenth century. Kluwer, New York
30. Disalle R (2016) Newton's philosophical analysis of space and time. In: Iliffe R, Smith G (eds) The Cambridge companion to Newton. Cambridge University Press, Cambridge, pp 34–60
31. Dobbs BJT (1988) Newton's alchemy and his "active principles" of gravitation. In: Scheurer P, Debrock G (eds) Newton's scientific and philosophical legacy. Kluwer, Dordrecht, pp 55–80
32. Dobbs BJT (1991) The Janus faces of genius: the role of alchemy in Newton's thought. Cambridge University Press, Cambridge
33. Dolby R (1966) A note on Dijksterhuis' criticism of Newton's axiomatization of mechanics. Isis 57(1):108–115
34. Dolby R (1996) 'F = ma' and the Newtonian revolution: an exit from religion through religion. Hist Sci 34(3):303–346
35. Ducheyne S (2011) Newton on action at a distance and the cause of gravity. Stud Hist Philos Sci Part A 42(1):154–159

36. Ducheyne S (2012) The main business of natural philosophy. Isaac Newton's natural-philosophical methodology. Springer, Dordrecht
37. Ducheyne S (2014) Newton on action at a distance. J Hist Philos 52(4):675–701
38. Ducheyne S (2015) An editorial history of Newton's regulae philosophandi. Etudios Filos 51:143–164
39. Earman J (1989) World enough and space-time. MIT, Cambridge
40. Figala K (2002) Newton's alchemy. In: Cohen IB, Smith GE (eds) The Cambridge companion to Newton. Cambridge University Press, Cambridge, pp 370–386
41. Funkenstein A (1986) Theology and the scientific imagination from the middle ages to the seventeenth century. Princeton University Press, Princeton
42. Gabbey A (2016) Newton, active powers, and the mechanical theory. In: Iliffe R, Smith G (eds) The Cambridge companion to Newton. Cambridge University Press, Cambridge, pp 421–453
43. Garber D (1992) Descartes' metaphysical physics. The University of Chicago Press, Chicago
44. Goethe JW (1810) Zur Farbenlehre. 3 vols. Cotta'schen Buchhandlung, Tübingen
45. Goethe JW (1840) Goethe's theory of colours. Translated into English by Eastlake CL. Murray, London
46. Greenham P (2017) Clarifying divine discourse in early modern science: divinity, physico-theology, and divine metaphysics in Isaac Newton's chymistry. Seventeenth Century 32(2):191–215
47. Greenham P (2017) Isaac Newton, scholar: an exceptional example of normal erudition. Hist Compass 15(6):e12389
48. Guerlac H (1983) Can we date Newton's early optical experiments? Isis 74(1):74–80
49. Hall RA (1988) Newton's biblical theology and his theological physics. In: Scheurer PB, De Brock G (eds) Newton's scientific and philosophical legacy. Kluwer, Dordrecht, pp 81–98
50. Hall RA (1992) Newton and the absolute. In: Harman PM, Shapiro AE (eds) The investigation of difficult things. Cambridge University Press, Cambridge, pp 261–286
51. Hall RA (1993) All was light. Oxford University Press, Oxford
52. Harrison P (2013) Laws of nature in seventeenth-century England. In: Watkins E (ed) The divine order, the human order, and the order of nature: historical perspectives. Oxford Univeristy Press, New York, pp 127–148
53. Hauksbee F (1709) Physico-mechanical experiments on various subjects containing an account of several surprizing phenomena touching light and electricity, producible on the attrition of bodies: with many other remarkable appearances, not before observ'd: together with the explanations of all the machines, (the figures of which are curiously engrav'd on copper) and other apparatus us'd in making the experiments. Brugis, London
54. Henry J (2011) Gravity and De gravitatione: the development of Newton's ideas on action at a distance. Stud Hist Philos Sci Part A 42(1):11–27
55. Hooke R (1665) Micrographia. Martyn & Allestry, London
56. Huygens C (1690) Traité de la lumiere, où sont expliquées les causes de ce qui luy arrive dans la reflexion, & dans la refraction. Et particulierement dans l'etrange refraction du cristal d'Islande. Vander, Leiden
57. Huygens C (1698) ΚΟΣΜΟΘΕΟΡΟΣ, sive de terris coelestibus, earumque ornatu, conjectura. Moetjens, The Hague
58. Huygens C (1888–1950) Oeuvres complètes de Christiaan Huygens (22 vols). Nijhoff, The Hague
59. Huygens C (1911) Treatise on light. Translated into English by Thompson SP. MacMillan and Co, London
60. Iliffe R (2004) Abstract considerations: disciplines and the incoherence of Newton's natural philosophy. Stud Hist Philos Sci Part A 35(3):427–454
61. Iliffe R (2016) The religion of Isaac Newton. In: Iliffe R, Smith G (eds) The Cambridge companion to Newton. Cambridge University Press, Cambridge, pp 485–523
62. Iliffe R (2017) Priest of nature: the religious worlds of Isaac Newton. Oxford University Press, New York

63. Janiak A (2008) Newton as a philosopher. Cambridge University Press, New York
64. Janiak A (2010) Substance and action in Descartes and Newton. The Monist 93(4):657–677
65. Kansichik P (2009) Newton's experimentum crucis from a constructivist point of view. PhD thesis, Humbolt University of Berlin
66. Kochiras H (2009) Gravity and Newton's substance counting problem. Stud Hist Philos Sci Part A 40(3):276–280
67. Kochiras H (2013) Causal language and the structure of force in Newton's system of the world. HOPOS: J Int Soc Hist Philos Sci 3(2):210–235
68. Kochiras H (2016) Newton's absolute time. In: Gerhardt K (ed) Time and tense. Philosophia. Cambridge University Press, Munich, pp 169–195
69. Koyré A (1956) L'hypothse et l'expéerience chez Newton. Bull Soc Fr Philos 50(2):59–79
70. Koyré A (1965) Newtonian studies. Champan & Hall, London
71. Leibniz GW (1685–1690) Die philosophischen Schriften. In: Gerhardt KI (ed) (7 vols). Weidman, Berlin
72. Lucretius Carus T (1942) De rerum natura: Libri sex. Translated into English by Leonard WE and Smith SE. The University of Wisconsin Press, Madison
73. Mach E (1919) The science of mechanics: a critical and historical account of its development. Translated into English by McCormack TJ. Open Court, Chicago
74. Mamiani M (2002) Newton on prophecy and apocalypse. In: Cohen IB, Smith GE (eds) The Cambridge companion to Newton. Cambridge University Press, Cambridge, pp 387–408
75. Mamiani M (2004) To twist the meaning: Newton's regulae philosophandi revisited. In: Buchwald J, Cohen I (eds) Isaac Newton's natural philosophy. MIT, Cambridge, pp 3–14
76. Mariotte E (1681) De la nature des couleurs. Michallet, Paris
77. Mazzotti M (2007) The two Newtons and beyond. Br J Hist Sci 40(1)
78. McGuire J, Rattansi P (1966) Newton and the pipes of Pan. Notes Rec R Soc Lond 21(2):108–143
79. McMullin E (1978) Newton on matter and activity. University of Notre Dame, Notre Dame
80. Millington EC (1947) Studies in capillarity and cohesion in the eighteenth century. Ann Sci 5(4):352–369
81. Mormino G (1993) Penetralia motus. La fondazione relativistica della meccanica in Christiaan Huygens con l'edizione del Codex Hugeniourum 7A. La Nuova Italia, Florence
82. Newman W (2002) The background to Newton's chemistry. In: Cohen IB, Smith GE (eds) The Cambridge companion to Newton. Cambridge University Press, Cambridge, pp 358–369
83. Newton I (1666) Of colours, MS Add. 3975, pp 1–22. The Newton project. http://www.newtonproject.sussex.ac.uk
84. Newton I (1671) A letter of Mr. Isaac Newton, professor of the mathematicks in the university of Cambridge; containing his new theory about light and colors. Philos Trans R Soc Lond 6(80):3075–3087
85. Newton I (1675) Discourse concerning light and colous, MS Add. 3970.3, ff. 501r–517
86. Newton I (1679) Letter from Newton to Robert Boyle, dated 28 February 1678/9, MS Add. 9597/2/18/62-65. The Newton project. http://www.newtonproject.sussex.ac.uk
87. Newton I (1687) Philosophiae naturalis principia mathematica. Jussu Societatis Regiae ac Typis Josephi Streater, London
88. Newton I (1710) Statements on religion, Keynes Ms. 6, King's College, Cambridge, UK. The Newton project. http://www.newtonproject.ox.ac.uk/view/texts/normalized/THEM00006
89. Newton I (1713) Philosophiae naturalis principia mathematica, 2nd edn. Crownfield, Cambridge
90. Newton I (1726) Philosophia naturalis principia mathematica, 3rd edn. Innys, London
91. Newton I (1728) A treatise of the system of the world. Fairam, London
92. Newton I (1728) De mundi systemate: Liber Isaaci Newtoni. Tonson & Osborn & Longman, London
93. Newton I (1729) Isaac Newton's Principia. Translated into English by Motte A. Adee, New York

94. Newton I (1730) Opticks: Or, a treatise of the reflections, refractions, inflections and colours. Innys, London
95. Newton I (1733) Observations upon the prophecies of Daniel and the apocalypse of St. John. Darbin and Browne, London
96. Newton I (1850) Correspondence of Isaac Newton and professor Cotes. In: Edleston J (ed). Parker, London
97. Newton I (1934) Sir Isaac Newton's mathematical principles of natural philosophy and his system of the world. Revised by Cajori Florian. University of California Press, Berkley
98. Newton I (1959–1960) The correspondence of Isaac Newton. In: Turnbull HW (ed) (2 vols). Cambridge University Press, Cambridge
99. Newton I (1962) De gravitatione et aequipondio fluidorum. In: Hall A, Hall M (eds) Unpublished scientific papers of Isaac Newton. Cambridge University Press, Cambridge, pp 89–156
100. Newton I (1962) De gravitatione et aequipondio fluidorum. Translated from Latin into English by Allan B. http://williambarclayallen.com/translationsDe-Gravitatione-et-Aequipondio-Fluidorum-translation.pdf
101. Newton I (1972) Isaac Newton's Philosophiae naturalis principia mathematica. Assembled and edited by Koyré A and Cohen IB (assisted by Withman A). Harward University Press, Cambridge
102. Newton I (1994) Trattato sull'apocalisse. In: Mamiani M (ed). Bollati-Boringheri, Turin
103. Newton I (1999) The Principia. Mathematical principles of natural philosophy. Translated into English by Cohen IB, Withman A (assisted by Budenz J). University of California Press, Oakland
104. Newton I (2004) De gravitatione. In: Janiak A (ed) Isaac Newton philosophical writings. Cambridge University Press, Cambridge, pp 12–39
105. Newton I (Unknown) De gravitatione et aequipondio fluidorum (selected passages; translation and interpolated commentary by Howard Stein). Available for download at http://strangebeautiful.com/other-texts/newton-de-grav-stein-trans.pdf
106. Pardies IG (1725) Discourse du mouvement local. In: Pardies I (ed) Oeuvres du R. P. Ignace Gaston Pardies, Bruyset, Lyon, pp 133–183
107. Pardies IG (1725) La statique, ou la science des forces mouvantes. In: Pardies I (ed) Oeuvres du R. P. Ignace Gaston Pardies, Bruyset, Lyon, pp 211–322
108. Pourciau B (2006) Newton's interpretation of Newton's second law. Arch Hist Exact Sci 60(2):157–207
109. Rosenfeld L (1969) Newton's views on aether and gravitation. Arch Hist Exact Sci 6(1):29–37
110. Royal Society of London (1715) An account of the book entituled commercium epistolicum collinii & aliorum, de analysi promota. Philos Trans R Soc Lond 29(342):173–224
111. Ruffner JA (2012) Newton's de gravitatione: a review and reassessment. Arch Hist Exact Sci 66(3):241–264
112. Sakkopoulos S (1988) Newton's theory of fits of easy reflection and transmission. Eur J Phys 9(2):123–126
113. Sanderson R (1965) Logicae artis compendium. In: Ashworth EJ (ed). CLUEB, Bologna
114. Schaffer S (1989) Glass works: Newtons prisms and the use of experiment. In: Gooding D, Schaffer S, Pinch T (ed) The use of experiment: studies in the natural sciences. Cambridge University Press, Cambridge, pp 67–104
115. Sepper DL (1988) Goethe contra Newton. Polemics and the project for a new science of color. Cambridge University Press, Cambridge
116. Sfectu N (2019) Isaac Newton on the action at a distance and gravity: with or without God? multiMedia Publishing. https://www.setthings.com/en/e-books/isaac-newton-on-the-action-at-a-distance-in-gravity-with-or-without-god/
117. Shapiro AE (1973) Kinematic optics. 'A study of the wave theory of light in the seventeenth century'. Arch Hist Exact Sci 11(2/3):134–266
118. Shapiro AE (1993) Fits, passions, and paroxysms: physics, method and chemistry and Newton's theories of coloured bodies and fits of easy reflection. Cambridge University Press, Cambridge

119. Shapiro AE (1996) The gradual acceptance of Newton's theory of light and color, 1672–1727. Perspect Sci 4:59–140
120. Shirras GF, JH C (1945) Sir Isaac Newton and the currency. Econ J 55(218/219):217–241
121. Snobelen SD (2001) "God of Gods, and Lord of Lords": the theology of Isaac Newton's general scholium to the Principia. Osiris 16:169–208
122. Stein H (1955) On metaphysics and method in Newton. http://strangebeautiful.com/other-minds.html#stein
123. Stein H (1977) Some philosophical prehistory of general relativity. In: Glymour C, Stachel J, Earman J (eds) Foundations of space-time theories. Minnesota studies in the philosophy of science, vol 8. University of Minnesota, Minneapolis, pp 8–49
124. Stein H (1990) The great Hugenius and the incomparable Mr. Newton. In: Bricker P, Hughes R (eds) Philosophical perspectives on Newtonian science. MIT, Cambridge, pp 17–47
125. Stein H (1993) On philosophy and natural philosophy in the seventeenth century. Midwest Stud Philos 18:177–201
126. Takuwa Y (2013) The historical transformation of Newton's experimentum crucis: pursuit of the demonstration of color immutability. Hist Sci 23(2):113–140
127. Wallis J (1671) Mechanica, sive de motu, tractatus geometricus. Godbib, London
128. Walsh KE (2014) Newton's epistemic triad. PhD thesis, University of Otago, New Zeland
129. Westfall R (1971) Force in Newton's physics. The science of dynamics in the seventeenth century. Neal Watson Academic Publications, New York
130. Westfall R (1980) Never at rest. A biography of Isaac Newton. Cambridge University Press, Cambridge
131. Westfall RS (1964) Isaac Newton's coloured circles twixt two contiguous glasses. Arch Hist Exact Sci 2(3):181–196

Chapter 2
The Birth of Physics as an Academic Discipline

Abstract The chapter deals with the way mathematicians were successful in replacing canonical philosophers nearly completely in the study of natural philosophy, both in research and academic contexts and how they invented an academic discipline that was called simply physics, concerned only with the study of inanimate matter, excluding alchemy. The new conception of physics for at least the whole of the 18th century still continued to be called natural philosophy, and even maintained some of the characteristics of old physics. Following the spread of mechanical and experimental philosophies in the European universities and colleges, the theoretical explanations of natural philosophy were accompanied by experiments, mainly concerning mechanics, hydraulics, pneumatics, electricity. Later, especially in France, teaching began to be supported by mathematics. The complex relationship between experimental and mechanical philosophies (and the heuristic role of theories) is also addressed. In principle, experimental philosophy did not require the knowledge of mechanical philosophy. The latter, however, was helpful because it suggested explanatory models and made it possible to make predictions, which if sometimes proved to be false were, however, a starting point. For this reason many experimental philosophers supported mechanical philosophy.

2.1 Mechanical Philosophy

In the first half of the 17th century a new form of philosophy of nature emerged, which became gradually dominant: the mechanical philosophy. It had at its basis a very simple theory of causation. Final causes were generally not considered, a part from their appearance as preambles of metaphysical nature. Formal and material causes changed nature, with the former that assumed the meaning of geometrical configuration and the latter which referred to a unique kind of matter assumed divided in particles of different size and shape. There remained efficient causes. All changes in the world was considered due to the collision of particles that moved in plenum or in vacuum with varying velocities.

The term mechanical philosophy is today often used as a synonymous of mechanicism. Notice that in the English literature instead of mechanicism it is often used

D. Capecchi, *Epistemology and Natural Philosophy in the 18th Century*,
History of Mechanism and Machine Science 39,
https://doi.org/10.1007/978-3-030-52852-2_2

the term mechanism; its use is however avoided here because its polysemic nature; indeed it also may mean contrivance. In the following the two terms *mechanical philosophy* and *mechanicism* are considered to be distinct. For mechanicism it is here stipulated to mean the theory for which any phenomenon should occur and should be explained by means of mechanical causes, or more generally without a will. This should be considered both in a methodological sense, for which the explanations should be looked for by means of the laws of mechanics only and in a ontological sense, for which the reality of nature is made of entities endowed with qualities, named *primary qualities*, as extension, shape and motion.

Mechanicism (and mechanical) derive from the Greek term μηχανικῄ, the science of machines. It associated nature with a great machine, a clock. An old concept that in the 17th century replaced the animistic Renaissance idea of nature as a big animal. A concept that eliminated psychology from physics, by replacing it with mechanics, the world of efficient causes of material kind, where all is explained by means of body, motions (and forces). By and large Pomponazzi's philosophy which avoided a substantial intervention of intelligences in the material world, can be classified as mechanicistic. But the fundamental move to establish mechanicism was due to Kepler that in the first years of the 17th century replaced the soul of planets with natural forces. Very well known is his letter to Johan Hans Georg Herwart von Hohenburg (1553–1622) of 16th February 1605: "My aim is to say that the machinery of the heavens is not like a divine animal but like a clock (and anyone who believes a clock has a soul gives the work the honor due to its maker) and that in it almost all the variety of motions is from one very simple magnetic force acting on bodies, as in the clock all motions are from a very simple weight" [70].[1] The passage from macrocosmo to microcosmo was immediate; already in the1630s Descartes proposed analogies of all natural bodies, the human body included, with machines.

The term mechanical philosophy has usually a restrict meaning, as clear from its definition at the beginning of the section. Besides avoiding the recourse to occult qualities and limiting to apply the laws of mechanics, it assumes also that matter has a corpuscular nature and all phenomena in the heaven or in the earth must be explained in terms of size, shape and motion of such corpuscles. Thus a machine is seen at a microscopic level.

Broadly it can be said that nearly all the mathematical practitioners embraced mechanicism (and many of them mechanical philosophy also) and nearly all more or less canonical philosophers embraced mechanical philosophy with the aim to replace the whole of old natural philosophy. This is a simplified view, as tracing a clear division between mathematicians and philosophers is difficult. Using modern categories one could say that on the one hand there were canonical philosophers who were mainly devoted to what are today considered as philosophical problems, and that besides philosophy of nature dealt also with metaphysics, ethics and logic. On the other hand, there were mathematical practitioners who, even though have carried out in-depth philosophical studies, were implied in sectors that today can be

[1] vol. 15, p. 146.

classified as scientific. There were also many characters that did not fall into any of these classifications and there were some that belonged to both categories.

Among the promoters of mechanical philosophy were prominent canonical philosophers, such as Henry More, René Descartes, Thomas Hobbes, Pierre Gassendi, Baruch Spinoza, Nicolas Malebranche, etc. In particular More was among the first to introduce the term mechanical philosophy with a technical meaning; he used two times the expression *mechanick philosophy* and once *mechanical philosophy* in the preface of his *Immortality of the soul* [84] of 1659 [1, 11],[2] even though the term mechanism was used also before at least in the English literature [11].[3]

But the person that mostly contributed to diffuse the term and the program of a strict mechanical philosophy was Robert Boyle, who usually is not considered as a philosopher, or at least is scarcely studied by modern philosophers. He was a promoter of an irenic approach toward mechanical philosophy, based on the recognition of 'essential' properties of matter that are relevant from a practical point of view. He aimed to avoid discussions on the possibility of vacuum—which he personally believed as possible—and the infinite divisibility of matter. So various corpuscular conceptions such as those of Descartes and Gassendi could be reconciled.

Boyle discussed the meaning of the term mechanical philosophy, in *The origin of forms and qualities according to the corpuscular philosophy* of 1666 [27] for instance. He considered the expression *corpuscular philosophy*—a his own denomination—appearing in the title, and *mechanical philosophy*, appearing in the body of the text, as synonymous. And this was also the feeling of his contemporaries. In this context it is interesting to note that the editor of *The philosophical work of the honourable Robert Boyle*, printed in 1725, changed slightly the title of the referred text, by replacing *corpuscular* with *mechanical*, to give *The origin of forms and qualities according to the mechanical philosophy* [27].[4] It was clear to Boyle that the two expressions, corpuscular and mechanical, had two different meanings; one that referred to the constitution of matter (corpuscularism), the other to the laws that regulate its motion (mechanicism), but assumed that the two meanings coalesce.

According to a restrict meaning of the term mechanical philosophy, that proposed by Boyle, neither Galileo, who applied the laws of mechanics but was little interested in the explanation in terms of corpuscles, nor Newton who equipped his corpuscles with action at a distance, were mechanical philosophers. Nor Beeckman, who had a complex conception of corpuscles on which matter is based, but used mechanics to study their motion [61].[5] Nor philosophers and physicians who dealt with chemical processes and had a corpuscular conception of matter, as Sennert for example, but they made no important use of the laws of mechanics.

The mechanical philosophy spread rapidly near scholars who had not received a thorough training in philosophy, or that if they had, at universities or religious

---

[2] p. 82; p. 12. note 2.

[3] pp. 80–81.

[4] p. 197.

[5] p. 74.

colleges, they were interested more in aspects related to mathematics (broad meaning), experimentation, medicine or technology. For them the mechanical philosophy was easier to understand, both because it was actually less nuanced than the traditional natural philosophies, in which metaphysical and theological discourses are difficult to follow and because it was carried out by philosophers who had a similar background to them, with some contaminations of the mathematical approach.

Most of the followers of mechanical philosophy adopted an expository style that took the rhetorical form of mathematicians as a model. After all, mechanical philosophy and mathematics were closely related; the very notion of corpuscles, their shape and configuration refer to geometry; the concept of motion also, after the studies of the Renaissance, refers to geometry. Even when were no explicit formulations of algebraic equations or geometric theorems, there was however the stringent language of mathematics, with the effort to limit synonymy and polysemy, with conclusions that derived from assumptions clearly specified in advance. However it must be said that the lexicon presented a great instability and therefore if the use of synonyms tended to be avoided within a treatise of a given author, it was not so of treatises of different authors. This also concerns the naming of fundamental concepts. In a fairly large case study the use of numerous synonyms for the modern terms is reported: force (11), particle (10), law (6) velocity (4) [90].[6]

It is true that the explanations of the mechanical philosophy were purely hypothetical because they have at their basis unobservable entities, the corpuscles. But they lend themselves in the form of geometric (modern meaning) models that allowed not only to explain but also to predict new phenomena. The congruence between experiments and theory, if there was any, allowed to attribute some truth value to the theory; if the verification was not successful, it nevertheless provided useful information to adjust the model. An important example of the heuristic power of the mechanical philosophy is provided by Huygens' optical studies reported in the previous chapter. Another example is provided by the explanation of the electrical phenomena which will be referred to in a later chapter.

The mechanical philosophy of the 18th century was essentially hegemonic and taught in universities and colleges, it must be said alongside elements of Aristotelianism that still resisted especially in schools of religious inspiration. However, it began to assume a different form from the mechanical philosophy of the previous century. Thanks to the influence of Newton and the alchemical school, the interaction between the corpuscles was no longer reduced to the impact. Even though in a not very explicit way, Newton in the Query 31 of the *Opticks*, nearly completely devoted to chemistry, made reference to the presence of forces of attraction and repulsion among the particles, what allowed the new mechanical philosophy with a greater heuristic power.

[6]pp. 95–96.

2.2 Experimental Philosophy

In the second half of the 17th century, after the death of Galileo, besides the *traditional* speculative approach to natural philosophy represented by the mechanical philosophy a *new* approach was born in Europe, which gave great relevance to empiric observations and contrived experiments. Some of the protagonist of this approach referred to it as the *experimental philosophy* and experimental philosophy has quite recently became a historiographic category that received a great deal of attention. To distinguish it from a modern movement known similarly as experimental philosophy or x-philosophy, the 17th century approach is sometimes named early modern experimental philosophy.

The term early modern experimental philosophy can be used in a rather broad sense, to indicate the prevalent use of experience, especially contrived experiments, in the study of nature; from this point of view the name philosophy could be replaced by the modern term science and experimental philosophy become experimental science. By using this broader meaning the term early experimental philosophy does not define a significant historiographic category. Basically reference is about the history of modern science from Renaissance to today.

Sometimes the term is used however in a narrower sense to indicate an approach in which no use at all is made of predefined theories; and even the declared goal of providing theories with an inductive approach is seen only as a very remote arrival point, to be left to posterity. The goal is to accumulate as many experimental results as possible. Some historians, believe that the term experimental philosophy should be taken according to this narrow meaning, as a typically English phenomenon carried on by the fellows of the Royal society of London. For example, this is the position of Peter Dear [46], who believes that this type of philosophy is characterized by a historical narrative, that is, a narrative where the reference to the experiments is carried out according to a historical approach, reporting in faithful way the results, without generalizations. In such a case one can speak of Baconian natural history.

The use of the restricted meaning defines a historiographic category that can be used and is used especially by the English writers [5, 6]. The period of interest is constituted by the second half of the 17th century, from the foundation of the Accademia del Cimento to the affirmation of Newton's approach, which was linked to the tradition of mixed mathematics. Its exhaustion appears determined by the substantial sterility of an experimental research disconnected from a speculative analysis.

In the following the term early experimental philosophy is considered in its broader meaning and restricted to that part dealing with natural philosophy only, as classically considered in the Aristotelian tradition, even though later on in the 18th century the approach of experimental philosophy was extended also to moral problems [6]. For the sake of simplicity early modern experimental philosophy will be referred to simply as experimental philosophy.

In the second half of 17th century the term experimental philosophy spread in England to indicate an approach to natural philosophy opposed to speculative philosophy, or *armchair philosophy*. Boyle composed a work entitled *Of the usefulness of speculative & experimental philosophy to one another*, in the1660, though it is no longer extant [3].[7] Hooke in his *Micrographia* of 1665 distinguished between experimental philosophy and "philosophy of discourse and disputation" [65].[8] Speculative philosophy was the study of natural phenomena basing on some a prior assumptions or hypotheses without a recourse to systematic observations or experiments. Experimental philosophy involved instead gathering of experimental data that was supposed to be made independently of any pre-constituted assumption. The experimental data might be the basis for the individuation of regularities or laws. Before this dichotomy appeared, natural philosophy had been considered to be only speculative.

The promoters of this new approach to natural philosophy were not canonical philosophers; rather they were mathematicians, physicists, lawyers, naturalists, chemists, architects, technicians, etc. The speculative philosophy that they fought was not only that of the schools but also the modern mechanical philosophy as carried out by very famous canonical philosophers such as Descartes, Hobbes and to some extent by Gassendi, scarcely interested in experiments.

The origins of the process to favor experimentation were varied and still object of discussion. The theoretical elaborations of an experimental philosophy can be rooted in the Aristotelian philosophy. In the Middle Ages and Renaissance, there were approaches not purely theoretical, or speculative, to the study of nature. Roger Bacon (1214c-1294), for instance, is generally considered as the promoter of a discipline called *experimental science* (scientia experimentalis), which should give a mathematical description of natural phenomena, promote technological applications and prognosticate the future on the basis of astrological knowledge. It is not clear if the writing of Roger Bacon were known in the 16th and 17th centuries. There was however evidence of at least an indirect knowledge. Indeed, a text widely read in late 16th century England took inspiration from Roger Bacon's scientia experimentalis. This was the English scholar John Dee (1527–1608)'s *Mathematical praeface* to the English translation of Euclid's *Elements*, first published in 1570 [5].[9]

Mathematicians and engineers, had carried out a their own projects, that were influenced only in part by the work of canonical philosophers. Science (modern meaning) had its own life. Mathematicians and engineers had sometimes a deep knowledge of natural philosophy; of it they chose freely enough the theoretical approaches that were more congenial to them without getting to a systematic elaboration work.

There were social and political reasons that brought to give more attention to facts instead than theories. Theories could pronounce on important aspects of nature, such for example cosmology, and could easily get in conflict with social ideologies, especially religious ones. This was partly one of the reasons that influenced Italian scientists (Academia del cimento and Jesuits). Another reason in between the epistemological and sociological, was given by the coexistence of different conceptions of philosophy of nature at odds with each other, flourished to justify new scientific dis-

[7] p. 218.

[8] Preface. Not numbered pages, third page.

[9] p. 19, free on line version.

coveries. With the birth of scientific associations such as the Accademia del cimento, the Académie des sciences de Paris, the Royal society of London, the best way to hold together scholars of different philosophical backgrounds was to rely on raw facts. Indeed to the experimental activity was recognized a higher epistemological status with respect to speculations based on more or less sophisticated hypotheses.

Of course promoting an experimental philosophy in the 17th century was a speculative move; the character usually called for as the major theoretician of experimental philosophy is Francis Bacon (1561–1626). Also Niccolò Cabeo and the young John Locke are sometime named.

Bacon saw natural philosophy divided into speculative and operative. The speculative component comprehended physics, metaphysics and natural history. The operative component has a less defined subdivision that evolved in time; it comprehended magic and mechanics but not only. Bacon discussed natural history in some works published during his lifetime, such as: *Advancement of learning* of 1605; *Novum organum* of 1620; *Historia naturalis et experimentalis* of 1622 and the *De augmentis scientiarum* of 1623. His natural history differed from the traditional (classificatory) natural history. It was made of collections of facts and was an undertaking of very great size and requires great labour and expense, involving many people in its execution and also comprehended the results of contrived experiments [2].[10]

Bacon's natural history belonged to the speculative side of natural philosophy but interacted with its operative side and thus considered also aspects from mechanics and magic. Some considerations on Bacon's conceptions on natural histories as well as science and natural philosophy can be found in [2, 4, 5]. Figure 2.1 shows the division of natural philosophy as reported in Bacon's *Advancement of learning* of 1605. The main division of natural philosophy is between speculative and operative. The former is concerned with the acquisition of causes, the latter with the production of effects. In the *Advancement learning* of 1605 Bacon considered a subdivision of mathematics into pure mathematics and mixed mathematics. Pure mathematics are two, geometry and arithmetic, the one handling continuous quantities and the other discrete quantities. Mixed mathematics have for subject some axioms or parts of natural philosophy: "For many parts of Nature can neither be invented with sufficient subtlety, nor demonstrated with sufficient perspicuity, nor accommodated unto use with sufficient dexterity, without the aid and intervening of the mathematics, of which sort are perspective, music, astronomy, cosmography, architecture, engineery, *and divers others* [emphasis added]" [7].[11]

The evaluation of the influence attributed to Bacon by historians has been largely motivated by the evaluation of Bacon himself. For historians and scientists of the 19th century, when inductivism held as an account of the success of science, it was natural to consider that Bacon's rules about scientific knowledge were applied directly by the Royal society and many experimental philosophers of the Continent as well. With the development of the hypothetic-deductive epistemologies of the 20th century, Bacon's role as a philosopher of science receded and his influence was seen differently. In

---

[10] pp. 70–71.

[11] pp. 124–125.

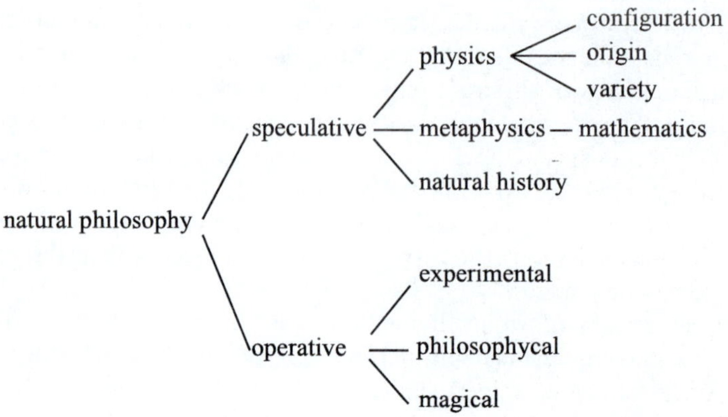

Fig. 2.1 Classification of natural philosophy in Francis Bacon. Redrawn from [4], p. 19

particular Bacon was seen as a promoter of a set of general commitments rather than of a strict research program [76].[12] Even though English writers still devote to Bacon a plenty of attention.

Reference to Bacon as to the theoretician of the experimental philosophy, I believe also depends on the difficulty that exists in modern philosophy to consider as a philosopher who does not respect modern standards to be defined a philosopher. There is indeed a tendency to consider philosophy not so much as a form of knowledge that aims to answer fundamental questions about the world and man, but rather as an academic discipline carried out by those who belong to a particular professional category that has self-assigned the philosopher's label over the last few centuries. From this point of view Bacon can be considered a philosopher, though a particular one. Instead, many experimental philosophers, Boyle included, are to be considered at most as scientists (using a term that did not exist before the 19th century) and more often simple practitioners. An if one wants to look for a promoter of a branch of philosophy, the experimental philosophy in this case, he looks for a 'true' philosophers, thus Bacon.

In the 17th and 18th centuries things were seen differently. Much of the natural philosophy scholars, regardless of the approach followed, prized the label of philosopher and were recognized as such even by those who today are considered 'true' philosophers. With some exceptions. For example, Leibniz and Huygens criticized Boyle, the champion of the experimental philosophy, for his lack of interest in speculation. In a letter to Leibniz of 1692, Huygens wrote:

> Mr. Boyle is dead, as you will probably already know. He seems pretty strange that he has not founded anything [any theory] on so many experiences of which his books make full; but the thing is difficult, and I have never believed him capable of a great application necessary to establish probable principles [hypotheses] [69].[13] (B.1)

[12] p. 3.

[13] vol. 10, p. 239.

Today in the face of the development of biology and computer science, the leading sectors of modern research in natural philosophy and mathematics, where what is commonly classified as science is scarcely distinguishable from technology, even canonical philosophers begin to recognize a cognitive value to practical activity. This cognitive value should also be recognized in retrospective to the protagonists of the so called early modern experimental philosophy.

2.2.1 *The Experimental Philosophy of the Accademia del Cimento*

After Galileo's death his pupils and admirers tended to present the maestro as the founder of a method of inquire of the material world strongly based on experiments, the still mysterious experimental method. Vincenzo Viviani (1622–1703) in particular to keep alive the memory of his teacher, was involved in the Bologna edition of Galileo's works in 1656 [54] and devoted much time to the patient and systematic collection of documents, testimonies and letters of Galileo with a generous grant from de' Medici. That allowed Antonio Favaro, at the turn of the 20th century, to complete the national opera in the style of completeness that certainly would have pleased Viviani.

In 1654 Viviani wrote a lucky *Racconto istorico della vita del sig. Galileo Galilei* [55],[14] a biography to be appended to Galileo's works published posthumously in 1717. Here he referred to numerous experiences, among which the famous ones on the synchronism of the oscillations of the pendulum and the fall of a heavy body from the leaning Pisa tower.

> In this while with the sagacity of his genius he invented that simple and adjusted time measurement by means of the pendulum, not yet known, taking the opportunity to observe it from the motion of a lamp, while he was one day in the Cathedral of Pisa; and making very precise experiences, he ascertained the equality of its vibrations, and by then thought to adapt it to the use in medicine for the measurement of the frequency of the wrists, with amazement and delight of the doctors of those times and that today we practice vulgarly: of which invention he then gained various experiences and measures of times and motions, and he was the first to apply it to the celestial observations, with incredible purchase in astronomy and geography
>
> [...]
>
> At this same time, it seems to him that to investigate the natural effects one necessarily requires a true knowledge of the nature of motion, given that philosophical and vulgar axiom *Ignorato motu ignoratur natura*, thus he gave to the contemplation of that. Then, with great dismay of all philosophers, by means of experience and with solid proofs and discourses, many conclusions of the same Aristotle on the matter of motion were revealed as falsehood, since then held for very clear and indubitable; as, among others, that the speeds of the mobiles of the same matter, but unequally heavy, moving for a given means, do not retain the proportion of their gravity assigned to them by Aristotle, on the contrary all moved with equal speed, demonstrating this with repeated experiments from the height of the Campanile

[14]vol. 19, pp. 599–632.

of Pisa in the presence of the other teachers and philosophers, and the whole assembly of students [55].[15] (B.2)

In another biography, the *Vita di Galileo*, Niccolò Gerardini (1604–1678) in about 1653–1654, after an ample discussion of the activity of Galileo as experimenter, wrote:

He possessed a little amount of books, and his study depended on continuous observations, deducing the subject of philosophizing from all the things he saw, heard or touched; and he said that the book in which one has to study was that of nature, which is open to all people [55].[16] (B.3)

This perspective of Galileo experimenter seems to be not faithful however. If it is true that Galileo considered very relevant contrived experiments, he was not so involved in strict experimentation as Viviani and his friend suggested. In particular there are stringent historical reasons to assert that Galileo neither made his observation of the oscillations of the lamp in the dome of Florence nor he left heavy body to fall from Pisa tower [102]. In Galileo's manuscripts and letters there are important reference to experiments, as documented by historians such as Naylor, Clavelin, Segre, Drake, and Settle, but in his official writings he made reference nearly only to astronomical observations.

A question then raises: Assuming that his biography of Galileo, as typical of the Renaissance, was more an hagiography, why Viviani depicted Galileo as a pure experimentalist? And why pure experimentation was so largely evaluated in the 1650s much before English experimental philosophers established? Historians have not given yet, for what I know, an answer to such questions. Is it possible what appears to us as a feeble experimentalism was seen by Galileo's pupils a fundamental break with the traditional approach to natural philosophy?

In Florence, even before the death of Galileo, experimental activity spread encouraged by the Grand Duke Ferdinando II de' Medici and his brother Leopoldo. This activity culminated in founding in 1657 the Accademia del cimento. It never had a statute and its birth is associated with the first meeting of a group of scholars on 18th June 1657. Apart from the Grand Duke and his brother, that company ranked: Vincenzo Viviani (1622–1703), Giovanni Alfonso Borelli (1608–1679), Carlo Rinaldini (1615–1698), Alessandro Marsili (1601–1670), Candido Del Buono (1618–1676), Paolo del Buono (1625–1659), Antonio Oliva (1624?-1691), Lorenzo Malagotti (1637–1712), Francesco Redi (1626–1697), Carlo Dati (1619–1676), Alessandro Segni (1633–1697) [20].[17] Correspondents were: Michelangelo Ricci, Giovanni Domenico Cassini, Geminiano Montanari, Donato Rossetti, Ottavio Falconieri, Niels Steensen, Jean de Thévenot, HonoréFabri [80].[18]

The Accademia del cimento purposes are declared in the preface to the readers in the *Saggi di naturali esperienze* (herein *Saggi*, see Fig. 2.2) [80], the only publi-

---

[15]vol. 19, p. 603, 606.

[16]vol. 19, pp. 646.

[17]p. 10.

[18]p. 82.

cation of the academy. In it there were reports of experimental activities on various natural phenomena using a refined and very numerous instrumentation. A main role of the Accademia del cimento was to experience and narrate the results of the experiments, with the development of a particular language immediate and flexible. The importance the academicians attributed to the instrumentation is documented by its abundance and its extensive descriptions in the *Saggi*, as if only observations with the help of instruments were worthy of being part of science. To give an idea, the collection of Leopoldo alone contained 1282 glass instruments [10] [19] The equipment had a non-trivial cost, which was supported by de' Medici. It could be justified only by the collective nature of the research and would not make sense for an isolated scientist, even of the caliber of Galileo.

The academy was the first modern society whose members worked together in an unique collective project. For example, the Accademia dei Lincei, founded much earlier, in 1603, although had among its members some high level scientists, including Galileo, had the main function of promoting the publication of the works that its members carried on individually. The training of the academicians was varied, including in addition to mathematicians—some members of the academy had an excellent mathematics education, for example Borelli and Viviani—also physicians and naturalists, and the natural philosophy ideas they professed were different. To make possible a collective undertaking, the academicians were required to limit as much as possible any interpretation of data through theories, maintaining an objective reading [41].[20]

In the following excerpt, taken from the *Saggi*, after having praised mathematics, its limits in application to the natural sciences are stressed:

> This is what the Mind attempts in the search of Nature; wherefore we must Confess, we have no better means then Geometry, which at first Essay hits the Truth, and frees at once from all doubts, and wearying Researches. And indeed she leads into the way of Philosophical Speculations, but at last leaves us; not that Geometry has not a large Field to expatiate in, and Travels not over all Natures Works; as they all submit to those Mathematick Laws, by which the Eternal Decree freely Rules, and Commands them; but because we hitherto are unable to follow her in so long, and wide a Path onely a few steps. Nowhere we may not trust our selves to go farther, we can relye on nothing with greater Assurance than the faith of Experience, which (like one that having several loose and scattered Gems, endeavours to fix each in its proper Collet) by Adapting the Effects to the Causes; and again the Causes to the Effects if not at first Essay, as Geometry yet at last succeeds so happily, that by frequent *trying and rejecting* [emphasis added] she hits the mark [80].[21] (B.4)

Using a Baconian terminology one could say that the goal of the academicians was to carry on natural history researches. It must be said however that a direct influence of Bacon on the empirical choices of the academicians is practically absent and it

[19] p. 135.

[20] pp. 440–449.

[21] p. 6. Translation in [113]. Notice the number of pages starts again from 1 after the preface of Magalotti.

is in fact clearly documented that most of the academicians had not even read the works of Bacon [10].[22]

The academy motto, trying and rejecting (*provando e riprovando*), that was also accepted by the Royal society, is justified both by the fact that a single experiment can be conducted incorrectly, and therefore not able to provide certain data, and by the fact that some events were not observed with due attention, mainly because no one knew what he had to concentrate in; by repeating experience one see new things. According to the academicians, one ought to proceed with much circumspection, lest too great a reliance and trust in experience, turn us out of the way and impose upon us; since it sometimes falls out, that before the clear truth appears to us, when the first more open veil of deceit are taken off, we discover some cheating appearances that indeed have some likeness, and resemblance of truth. These are the imperfect lineaments that are seen through the last coverings that more nearly veil the lovely face of truth; through the fine web whereof she some-times seems so plain and lively, that some might conclude, she was nakedly discovered [80, 113].[23]

One of the purpose of the natural histories of the academy was to verify the assertions of natural philosophy that had became commonplaces, such as for instance: nature abhors a vacuum. But also important experiences made by others European experimenters were repeated. Facts were the only authority recognized. To Leopoldo de' Medici is attributed the will to contrast authority, because the reputation of great authors proved too often hurtful to the studious, who through too much confidence and veneration of their names, fear to call in question what is delivered upon their authority; wherefore its is worthy to confront with the most accurate and sensible experiments, the force of their assertions [80].[24] In a 'democratic' way, as the academy verified the experiments carried out by others, it allowed others to check its own; and the record of the experiences reported in the *Saggi* also had this purpose with the wishes for a free communication to different 'meetings' scattered for the most distinguished and substantial regions of Europe [80].[25]

Luciano Boschiero suggests that the above description is deliberately artificial and not very responsive to the actual functioning of the academy [20]. Even though the official publication, the *Saggi*, declares very clearly a purely experimental activity, without discussing the principles and conclusions of natural philosophy, the unpublished texts and correspondence would show that the academicians also debated among themselves vividly on the interpretation to give to the experiments, and many of them used the experiments to verify their own theories [20]. According to Boschiero, the official account of a purely experimental activity would have been dictated by the lords of Florence, Ferdinando and Leopoldo. Since the Renaissance, de' Medici had gained interest about natural philosophy, mathematics and engineering as a means to increase their prestige in Italy and Europe. After Galileo's death they promoted an experimental activity stressing that this was carried out in the footprints

[22] p. 140.

[23] pp. 6–7. To the reader.

[24] pp. 7–8.

[25] p. 8.

of the great Galileo and gave a strong support to Vincenzo Viviani to collect works and news about Galileo which could enforce such point of view. The Accademia del cimento was founded to pursue this objective and its achievements had to be shown externally by means of well prepared publications. De' Medici strongly influenced the way to expose the results of the academy by stressing the experimental activity. However even though Boschiero's reasoning seems to be stringent it does not explain why in the 1650s' there was the idea that the experimental activity could be considered as the most interesting one in the study of natural philosophy

The activity of the Accademia del cimento is documented, as already noticed, by a single publication, *Saggi* of 1667, issued in a year that coincided with the closing of the academy itself. The *Saggi* presented a summary of experimental works over the course of a decade. The book, lavishly illustrated, collected a considerable editorial success. In 1684 the first English translation appeared under the title *Essayes of natural experiments made in the Academie del Cimento* [113] by Richard Waller (d. 1715) on the recommendation of the Royal society of London. In 1731, the Dutch scientist Pieter van Musschenbroek (1692–1731) prepared a Latin translation [85]. Among the later editions, very important is the one edited by Vincenzio Antinori in 1841 [80], where together with the original text also some appendixes relating to experiences, not reported in the *Saggi*, but documented in the archive of the academy, can be found.

Experiences dealt with various problems, some related to Galileo's researches, on mechanics, others concerning subjects only by very short time object of 'scientific' investigation, such as heat, electricity and magnetism, just explored by Galileo. In all cases the experiences contained qualitative flanked by quantitative descriptions. It should be said, however, that in most cases the numerical values of the measurements performed are not reported. Indeed numbers appearing in the various experiences are very few and generally referred to the description of the instrumentation; they are normally reported in literal form (that is 'a thousand' instead of '1000'). The exception is a long series of tables that gave the temperature of water in a freezing process [80].[26]

The reliability of the reported results, not being documented by numerical values that could facilitate comparison to people who wanted to reiterate the experience, was entrusted to the prestigious of the academy, to its sponsor, Leopoldo, besides, in some cases, to the call of similar experiences. Gassendi for instance is mentioned in several places. There are not, at least I have not seen them, references to the presence of distinguished witnesses. A rhetorical form of validation that instead was widely used at the Royal society and by many natural philosophers of the second half of the 17th century.

The activity of the glorious academy ceased in 1667 in a quite inglorious way, for several causes. Most notable was the abandonment by important members as Borelli, Oliva and Rinaldini and then the appointment as Cardinal of Leopoldo de' Medici, who had been the academy engine, resulting in disengagement considered the new heavy and delicate commitments to be undertaken. Alongside these imme-

---

[26] pp. 95–104.

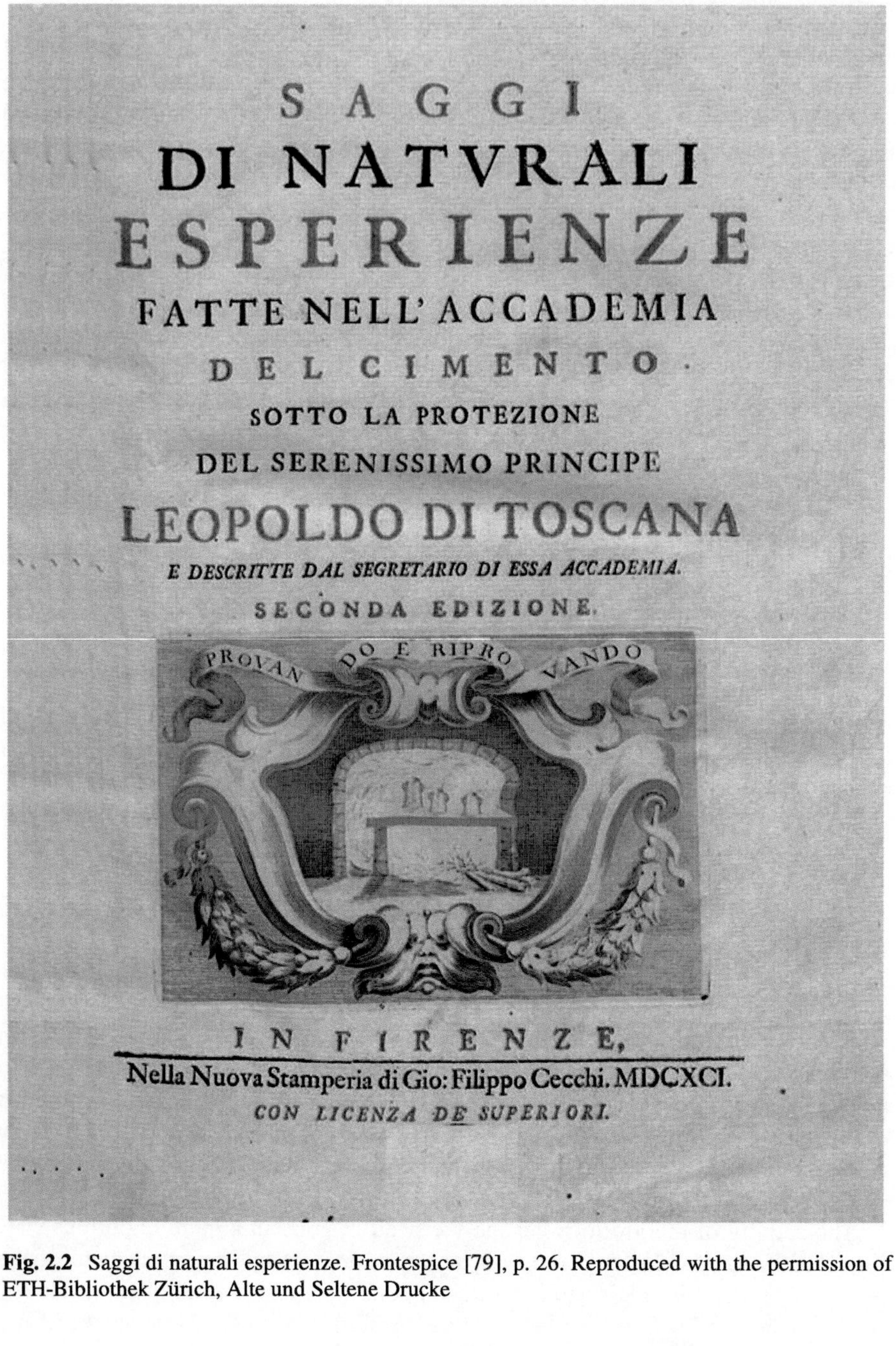
SAGGI
DI NATVRALI
ESPERIENZE
FATTE NELL' ACCADEMIA
DEL CIMENTO.
SOTTO LA PROTEZIONE
DEL SERENISSIMO PRINCIPE
LEOPOLDO DI TOSCANA
E DESCRITTE DAL SEGRETARIO DI ESSA ACCADEMIA.
SECONDA EDIZIONE.

PROVANDO E RIPROVANDO

IN FIRENZE,
Nella Nuova Stamperia di Gio: Filippo Cecchi. MDCXCI.
CON LICENZA DE' SUPERIORI.

Fig. 2.2 Saggi di naturali esperienze. Frontespice [79], p. 26. Reproduced with the permission of ETH-Bibliothek Zürich, Alte und Seltene Drucke diate causes, however, there was probably a most important hidden cause, inherent the very structure of the academy, that is the choice of the form of anonymous publication, probably the first in history, and the commitment not to formulate theories [10]. Anonymity frustrated the ambitions of individuals, especially the most talented, ambitious, and among them certainly figured Borelli. The banning of scientific theories made partly sterile the experimentation, among other things preventing the creation of new and more interesting experiments, the need for which could only be conceived within a theory. The other European academies will adopt a different attitude. For example, the Royal society (see below) allowed and encouraged publications by individual members, which in addition to report their contributions to the experiments carried out at the academy, could also interpret them with their own categories of physics and mathematics.

2.2.2 The Natural Histories of the Royal Society of London

The Royal society of London founded in 1660 was in many respects the heir of the Academia del cimento. No coincidence that Robert Southwell (1635–1702), who will be chairman of this society from 1690, was a *protégé* of Viviani from whom he learned the methods and organization of the Accademia del cimento.

There are many works that relate to the Royal society and its foundation [12, 15, 16, 76, 107]; here there is no space and perhaps there is no need of an in depth analysis of the society by studying original sources; thus for many considerations I relay on published studies. The society saw its origin in a meeting of various characters (twelve), more or less famous and more or less well versed in the sciences. As it appears from the journal book, on the 28th of November 1660, the lord viscount Brouncker, Mr. Boyle, Mr. Bruce, Sir Robert Moray, Sir Paul Neile, Dr. Wilkins, Dr. Goddard, Dr. Petty, Mr. Balle, Mr. Rooke, Mr. Wren, and Mr. Hill, after the lecture of Mr. Wren at the Gresham College, withdrew, for mutual conversation, into Mr. Rooke's apartment, where, amongst other matters discoursed of, something was offered about a design of founding a college for the promoting of physico-mathematical experimental learning [12].[27]

A 1663 statute of the Royal society provided instructions for separating facts from their interpretations, giving preferences to facts. And if any fellow shall think to suggest any conjecture, concerning the causes of the phenomena in such experiments, the same shall be done apart; and so entered into the register of the society. Hypotheses thus should not be avoided, simply they should be formulated on when a very great collection of facts was available. This is what the statute of 1663 required:

> The secretary shall jointly draw up the Report of the matter of fact, in every such Experiment or Observation; or if any difference shall happen between them in their apprehensions there about, the same shall be related in the Report. In all Reports of Experiments to be brought into the Society, the matter of fact shall be barely stated, without any prefaces, apologies, or

[27] vol. 1, p. 3.

> rhetorical flourishes; and entered so in the Register-book, by order of the Society. And if any Fellow shall think fit to suggest any conjecture, concerning the causes of the phaenomena in such Experiments, the same shall be done apart; and so entered into the Register-book, if the Society shall order the entry thereof [99].[28]

Someone gave a restricted interpretation, for example Robert Moray (1608?–1673):

> In the mean time this Society will not own any Hypothesis, systeme, or doctrine of the principles of Naturall philosophy, proposed or maintained by any Philosopher Auncient or Moderne. And till there be a sufficient collection made, of Experiments, Histories, and observations, there are no debates to be held at the weekely meetings of the Society, concerning any Hypothesis or principle of philosophy, nor any discourses made for explicating any phenomena, except by speciall appointment of the Society, or allowance of the president: But the time of the Assemblyes is to be employed, in proposing and making Experiments, discoursing of the trueth, manner, grounds & use therof; Reading & discoursing upon Letters, reports, and other papers concerning philosophicall & mechanicall matters; Viewing and discoursing of curiosities of Nature and Art; and doing such other things as the Councel, or the president alone shall appoint [67].[29]

Another example of this attitude was furnished by John Evelyn's (1620–1706) *Sylva, or a discourse of forest-trees, and the propagation of timber in his majesties dominions* of 1664, the first work published by the Royal society [76].[30] The majority, Boyle included, saw an empirical experimental basis for all theories and the rejection of any hypothesis non clearly grounded upon experimental evidence. Other still, like John Wallis (1616–1703) and Hooke, allowed room for mathematically derived theories.

When Robert Hooke (1635–1703) published his *Micrographia* in 1665 the Royal society had to question his use of hypothesis and interpretations, pressing him to reply:

> After my Addressè to our Great Founderè and Patron, I could not but think my self oblig'd, in consideration of those many Ingagements you have laid upon me, to offer these my poor Labourse̊ to thisMOST ILLUSTRIOUS ASSEMBLY. YOU have been pleas'd formerly to accept of these rude Draughts. I have since added to them some Descriptions, and some Conjecturesè of my own. And therefore, together with YOUR Acceptance, *I must also beg YOUR pardon* [emphasis added]. The Rules YOU have prescrib'd YOUR selves in YOUR Philosophical Progress do seem the best that have ever yet been practis'd [65].[31]

Below a summary of how a matter of fact should be established, according to Thomas Spratt (1635–1713) the author of *History of the Royal society of London, for the improving of natural knowledge* of 1667 [107]. After the experimenters have performed the trial, said Spratt, they brought all the history of its process back again to the test. Then came the second great work of the experimenters; which was to *judge and resolve upon the matter of fact*. In this part of their employment, they used to take an exact view of the repetition of the whole course of the experiment; and observed

[28] pp. 289-290.

[29] p. 173.

[30] pp. 34–69.

[31] Letter to the Royal society, just before the preface.

all the chances and the regularities of the proceeding; what nature does willingly and what constrained; what with its own power, what by the succors of art; what in a constant mode and what with some kind of extravagance; industriously marking all the various shapes into which it turns itself and by how many secret passages it at last obtains its end. Never giving it over, *till the whole company has been fully satisfied of the certainty and constancy; or, on the other side, of the absolute impossibility of the effect*. This critical and reiterated scrutiny of those things, which are the plain objects of their eyes must put out of all reasonable disputes, the reality of those operations, which the society shall positively determine to have succeeded [107].[32]

Though the fellows of the Royal society, at the early phase of its institution and long after, differed as to the manner in which experiments and observations could be best used, all of them were, or at least declared to be, convinced that the improvement of natural knowledge could be achieved by following methodological reform of knowledge and the establishment of collective (and preferably state supported) institutionalization of knowledge [76].[33]The Royal society was proposing a pedagogical role believing that scientific knowledge should be shared and that it should somehow become a collective enterprise. And certainly for people who were not introduced to the study of science it was more simple and interesting to read reports of experiments, rather than explanation of theories based on a mathematical approach, not always conclusive.

The idea that knowledge should be based primarily, exclusively according to some, on experimental evidence, had different origins, as already discussed in previous sections. In England however the influence of Francis Bacon's philosophy had most probably a major role. Indeed Bacon's writings on scientific method began to be rediscovered in the 1640s, just some years before the foundation of the Royal society [76][34] and were carefully read by some of the Royal society founders. Moreover the religious contrasts and bloody wars that crossed Britain from the beheading of Charles I, the Government of Oliver Cromwell and the Restoration of Charles II were still alive. The founder of the Royal society wanted to avoid a climate similar to that; certainly less bloody but equally pernicious—giving room for animated and not resolvable discussions, among the supporters of different scientific theories.

How much Bacon influenced the single characters is difficult to say and not yet sufficiently studied; it is a matter of fact however that officially to most of the fellows of the Royal society Bacon was of inspiration. Because of the variety of interest and approaches of the various members of the society some historians had concluded that Baconianism served to give a convenient public image, glossing over internal disagreements. That is the Baconianism served to the Royal society more as a nominal than a real construct: "It is even questionable whether the Royal society had 'a philosophy' which extended beyond immediate apologetic purposes. Behind their unified front of Baconianism, which was readily adopted as a defensive mechanism against critics, lay diverse philosophical outlooks, which betray many

[32]p. 99.

[33]p. 17.

[34]p. 15.

other influences than Bacon and provided the basis for considerable philosophical controversy in the pre-Newtonian period" [114].[35] Some other historians however assumed Bacon's influence much more profound; see for instance [15].

The heterogeneous nature of early English Baconianism has been used as an incentive by many historians to classify the fellows into two separate camps: serious scientists on the one side and amateurish dabblers on the other; moreover opinions are divided on whether the serious scientists stand with or against Bacon. According to [76][36] this is an oversimplification of the question and even the role of the amateurs should be analyzed with much more attention

A characteristic of the Royal society was its openness to the outside; both with respect to the English society and to the scholars of the Continent. Particularly important, at least initially, were the contacts with Italy, which had seen the birth of the first scientific institution of experimental character, the Accademia del cimento, and with France, where in 1666 a similar institution was founded, the Académie des sciences de Paris. After the demise of the Accademia del cimento, Italian natural philosophers looked at the Royal society for inspiration; partially seeing it as the heir of the Accademia. Not many Italian were fellows of the society however, but in the early decades of its foundation the number of correspondent was great; among them there were Marcello Malpighi and Vincenzo Viviani who later became fellow respectively in1669 and 1661. Correspondents were from many part of Italy, but especially from Bologna. Relations with the Académie des sciences de Paris and French scientists varied greatly over the years.

2.2.2.1 The Journal of the Royal Society

One of the ways the Royal society advertised its activities was through the regular publication of a magazine, the Philosophical Transactions of the Royal Society of London (herein after Philosophical Transactions), established by the (permanent) secretary Oldenburg in 1665. The Philosophical Transactions should be printed the first Monday of every month, if there was sufficient matter for it [15].[37]

The first number of the Philosophical Transactions contained accounts of inventions and discoveries derived partly from Oldenburg's own knowledge, partly from accounts read to the Royal society, partly from letters and partly from printed sources. By the second number, the journal was settling into what became an its pattern: extracts of letters, English and foreign—the latter translated into English unless in Latin—and, at the end, one or more book reviews.

Despite current belief in experiments as one of the foundations of science, only a small part of the volumes of the Philosophical Transactions examined up to the 19th century were devoted to reporting on contrived experiments. Both in terms of the percentage of total articles and percentage of pages, experimental articles

[35]p. 123.

[36]p. 5.

[37]p. 59.

Table 2.1 Contents of the Philosophical Transactions of the Royal Society of London, 1720–79, arranged by decades. Drawn from [105], p. 37

Years	NHist	MMath	Med	ExNP	Anat	Antiq	SpNP	PMath	Misc	Total
1720s	86	80	76	38	40	2	11	12	13	358
1730s	109	94	60	43	29	9	12	6	1	363
1740s	157	84	124	60	58	39	10	8	17	557
1750s	268	82	97	51	24	38	17	11	3	591
1760s	162	172	59	31	17	27	22	9	6	507
1770s	183	94	36	55	28	11	14	19	10	450
Total	965	606	452	278	196	126	86	65	50	2826
%	34	21	16	10	7	4	3	2	2	99

Abbreviations: NHist: natural histories (natural sciences); MMath: mixed mathematics; Med: medicine; ExNP: experimental natural philosophy; Anat: anatomy and physiology of animals and plants; Antiq: antiquities; SpNP: Speculative natural philosophy; PMath: pure mathematics; Misc: miscellanea

accounted for only 5–20% of each volume up to volume 80. Only in volume 90, opening the 19th century, did the percentages rise substantially to 39% of the articles and 38% of the pages. Experiments were only one of many types of information to be transmitted among those interested in science. Most articles and pages were devoted to observations and reports of natural events, ranging from earthquakes, through astronomical sightings, anatomical dissections, and microscopical observations; to accounts of technological and medical advances, and travelogues of journeys to China and Japan or an interview with the prodigy Wolfgang Amadeus Mozart [9].[38]

Table 2.1 shows the distributions of the memoirs of the Philosophical Transactions in the various field of natural philosophy in the 18th century.

Memoirs concentrated on natural sciences, mixed mathematics, medicine, experimental natural philosophy, and anatomy. Very occasionally members contributed papers on pure mathematics, speculative natural philosophy, and antiquities. The prevalence of natural sciences or medicine is hardly surprising, as the Royal society was heavily populated with country gentlemen and physicians.

Table 2.2 shows the internal distributions of the mixed mathematics papers. As one can see accounts of astronomical observations are largely prevailing; yet numerous are papers about mechanics.

In the first volumes of the *Philosophical Transactions*, some of the experiments were simply cookbook recipes for creating marvelous effects or effects of practical use, such as the instructions for coloring marble. However by volume 20 some experiments had clear hypothesis-testing functions. Experiments were recognized as events designed with specific claims about nature in mind. In volume 25, for example, Francis Hauksbee (1660–1713) wrote: “Since the greatest satisfaction and demon-

[38]p. 65.

Table 2.2 Mixed mathematics in the of the Royal Society of London, 1720–79, arranged by decades. Drawn from [105], p. 38

Years	Astronomy	Geography	Mechanics	Miscellaneous	Total
1720s	32	38	6	4	80
1730s	23	45	19	7	94
1740s	40	19	14	12	85
1750s	30	18	15	19	82
1760s	106	36	11	19	172
1770s	33	35	8	18	94
Total	264	191	73	79	607
%	43	31	12	13	99

stration that can be given for the credit of any hypothesis is, that the experiments, made to prove the same, agree with it in all respects, without force" [62].[39]

The comparison between the papers published in the Philosophical Transactions and the treatises of natural philosophy of the same period, even of those inspired by mechanicism, makes it evident the great change that is intervening in the study of nature. Essentially all subjects of natural philosophy are treated; using modern categories: physics, chemistry, natural sciences, medicine. However, the approach is not that of the canonical philosophers; there is a lack of attention to metaphysics and to the construction of systems, although the rigor of the treatment is often sufficiently high even for today's standards. As far as the study of the inanimate world is concerned, at the beginning mainly the writings of mechanics, optics and astronomy were presented, carried forward by mathematicians who followed the approach of mixed mathematics but who did not disdain philosophy of nature; since the 18th century, electricity and magnetism began to receive a great deal of attention.

The form of communication, a memoir of a few pages, instead of a long treatise, also contributed to modify the study of the philosophy of nature. Given their relative shortness, the memoirs did not allow a systematic treatment of the whole philosophy of nature; therefore, they dealt with very specific subjects beginning to outline a certain form of specialization. Scholars with a mathematical background wrote about astronomy, mechanics, optics, electricity, thermology and magnetism, that is, topics today classified as physics and chemistry. Other scholars, especially amateurs, instead provided reports on journeys, quirks of medicine, animals, stones. Arguments that are classifiable as natural sciences and medicine.

The results of the experiments were reported with increasing precision as a debate among the authors was established. They saw forced to pay more attention and the making of increasingly precise measurements. As experiments became something more than a private affair between the researcher and his colleagues, there was increasing likelihood that the events described were real successes. While at first the report was little more than a summary of the information reporting that the fact

[39] p. 2415.

had happened, in the following period the detailed reports of experiments became the point of making the experiment replication possible, assuming a force of historical report, usually referred in first person. The presence of illustrious witnesses was often referred to.

2.3 Mechanical Philosophy, Experimental Philosophy and Mixed Mathematics

Mechanical philosophy and experimental philosophy are two very useful historiographic categories, which have the advantage of using terms and concepts also used in the 17th and 18th centuries. But today historians tend to use them in a much more exclusive way than the scholars whom they make history of.

The use of different terms lends to considering them as representative of two uncorrelated activities. In reality things are more complex and a scholar could be both a supporter of mechanical philosophy and an experimentalist. In the following, for simplicity, I will talk of two philosophies, even if the mechanical, in its corpularistic form, represents a true form of philosophy of nature whereas the experimental may represent, in fact, more an approach (to natural philosophy), which can be mechanicistic or not.

As seen in the previous sections, the two philosophies can be considered in a broad sense or in a narrow sense. The mechanical philosophy considered in a broad sense, more properly referable only as mechanicism, faces the study of nature with the exclusive use of the laws of mechanics. Strictly understood, corpularism also is assumed. Experimental philosophy understood in the strict sense refers to the study of nature in which the empirical aspect derived from devised experiments cannot be ignored. Understood in a broad sense it indicates the compilation of natural histories, with or without any attempt to verify hypotheses or deduce laws by induction.

Table 2.3 illustrates how experimental and mechanical philosophy can be combined. The first row of the table, in which use is made of mechanical theory and experiment, represents the activity of scholars who are generally qualified as mixed

Table 2.3 Interaction between mechanical and experimental philosophy

	Mechanical philosophy		Experimental philosophy	
	Broad	Strict	Broad	Strict
Mixed mathematics	+	–	–	+
Philosophy	–	+	+	–
Emergent sciences	–	+	–	+
Mathematics	+	–	–	–

mathematicians. Among them Galileo and Newton. The second row represents the scholars in whom the mechanical philosophy (corpularistic) theory does not actually interact with experiments. This is the case, for example, of Descartes who was interested in natural stories and also referred to Bacon:

> If someone with this mood would undertake to write a history of celestial appearances, according to the method of Verulamius [Francis Bacon], and, without putting in reasons or hypotheses, he described heavens exactly as they appear, what position each fixed star in respect of its neighbors, what difference, of size, of color, of clarity, or to be more or less sparkling, and if that responds to what the ancient astronomers wrote, and noted the difference he finds (because I have no doubt that stars change little their position even they can be considered as fixed) [...] this would be a work that would be useful to people, much more than it could appear, and relieve me of much pain [50].[40] (B.5)

According to Descartes the experiment should not be used to test hypotheses, but simply to highlight what are the phenomena that occur in our world that is regulated by necessary mechanical laws, but which is contingent because it depends on the way God has set its initial conditions, according to his will. After revealed these phenomena, a mechanical explanation is provided, which always exists, although not unique, because it is not always possible to solve the contingent condition of the world.

The third row of Table 2.3 refers to the approach in which there is a very intense experimentation connected to a corpuscular conception of matter. This is the case of Boyle and of all the new sciences that study electricity, magnetism, thermology, chemical reactions which today are often called Baconian sciences (after Kuhn). The fourth row sees mechanics as a purely rational science. A typical representative of this category is d'Alembert. But also Descartes and Euler can be considered, with the necessary clarifications.

The combination expressed in the first row, that of mixed mathematics, is the one that had the greatest development, at least in the 18th century. Part of the phenomena studied with the approach of the third row, after the experiments have succeeded in clarifying and quantifying them, is gradually brought back into the ground of mixed mathematics. Biology and natural sciences remained outside for the time being. The second row concerns an approach that today is no longer considered scientific but is relegated to natural philosophy.

Experimental and mechanical philosophy together have played a fundamental role in the development of mixed mathematics. The factual, contingent truths, empirically revealed, are foundations of mixed mathematics. Thus, experimental philosophy played a crucial role in their development. In the past, the empirical basis, with the exception of astronomy, was founded on simple observations of the regularities of nature that did not require the use of laboratory experiments, with some exceptions for optics. Since the Renaissance and in particular with Galileo, daily observation has been replaced by contrived experiments. New phenomena were observed, from which with an inductive approach, experimental laws were derived that could be put

---

[40] vol. 1, pp. 251–252.

at the basis of mixed mathematics, or individual experiences that could be used to verify theories were recorded.

For instance, accurate measurements were taken of the fall of bodies, of the period of oscillation of simple and compound pendulums, of the speed of sound propagation, of the breaking and deformability of bodies. In biology the use of the microscope allowed a greater understanding of vital phenomena, advancing hypotheses and verifying them. The practices of alchemy were made more rigorous, stripped of their mystical content, to give rise to modern chemistry. Magnetism and electricity, phenomena already known in the ancient world, were subjected to an intense study. The invention of the thermometer opened the possibility of a quantitative study of thermal phenomena.

The astronomical observations carried out with the use of the telescope or in any case with more accurate optical equipments than those used in antiquity, made it possible to study precisely the motion of the planets, to discover irregularities and satellites, to visualize new stars. At the beginning of the 18th century astronomy was with optics, and more than optics, the field of natural philosophy that had underwent profound changes. Already with Kepler mathematicians had begun to regain possession after so long, after Ptolemy indeed, of a discipline that had become the prerogative of natural philosophers, a discipline to which one can refer as physical astronomy. Galileo played an important role. He acted as a pure experimental philosopher, limiting himself to an observational work, with interpretations that essentially had the task of giving a geometric description of the phenomena. For example, in the case of the Medicean satellites he limited to say that the phenomena he observed were explained by hypothesizing bodies revolving around Jupiter, from the observation of the phases of Venus he deduced that this planet must rotate around the sun and not the earth. For the spots of the moon and the sun he limited to geometric interpretations, perhaps with something more for sunspots. Alfonso Borelli resumed the work of Kepler and carried out a mechanical study of the motion of the planets. To do this he felt into the shoes of the natural philosopher and provided explanations in terms of efficient mechanical causes. But he did not stop at purely qualitative aspects. If perhaps his speculations of natural philosophy could not be impeccable he took them as a starting point for the application of the laws of mechanics, arriving at a 'satisfactory' enough explanation of the elliptical shape of the orbits of the planets around the sun.

Mechanical philosophy could become, and in fact became, an important approach to the development of mixed mathematics, especially for the so-called Baconian sciences. It allowed to provide interpretative keys to the experimental results. For example, Huygens, as already discussed in Chap. 1, thanks to the corpuscular hypothesis of medium transmitting the light, built a geometrical mechanical model that served as an explanation of the phenomena. He could thus justify for example that the angle of refraction is smaller than the angle of incidence passing from a less dense medium to a denser medium. Or he could explain the phenomenon of double refraction. The model could be used not only for the explanation of known phenomena, but also, at least partially, for the prediction of new ones. Corpularism helped to explain the motion of the planets of the solar system. A typical example of the application of corpularism to cosmology is provided by the *Principia philosophiae* of Descartes.

2.4 Robert Boyle, an Experimental and Mechanical Philosopher

Robert Boyle (1627–1691) is generally considered a 'scientist' who carried out important experiments in hydrostatics and pneumatics (well known is Boyle-Mariotte's law on gas compressibility) and especially in chemistry. Very often indeed Boyle is framed as a (great) chemist. Thomas Kuhn considers that he made no substantial innovation in chemistry [74], William Newman partially contrasted this point of view [87] and Marie Boas considers Boyle to be a physicist [13].

An important aspect of modern studies on Boyle is the attempt to consider his thought as a whole rather then looking at only his more known achievements. As for Newton indeed for him too the label 'scientist' is very restrictive. Boyle saw himself, and was seen by his contemporaries, as a (new) philosopher of nature; he was a promoter of the experimental philosophy and with Descartes, Gassendi and Hobbes one of the greatest supporter of mechanical philosophy of the 17th century. He wrote philosophical treatises on mechanical philosophy, specifically on the qualities, the most important of which was *The origine of formes and qualities* of 1666 [34].[41] He also wrote fundamental texts on experimental philosophy among which *The sceptical chymist* [21] and *The christian virtuoso* [26] and many important reports of experiments on hydrostatics, pneumatics, chemistry etc. Like Newton he was deeply involved in theology, to the point that in *The great historical, geographical, genealogical and poetical dictionary* of 1701 edited by Jeremy Collier (1650–1726), more emphasis was given to his role as a lay theologian than as a natural philosopher. For some bibliographical references on Boyle see the still influential [66][42] and the website *Robert Boyle project* [35]. Very important is also a new edition of Boyle's works [34].

Boyle lived in a time when natural philosophy was in strong identity crisis. The old canonical philosophy of the schools was attacked from all sides and many scholars presented new points of view on the nature of things discordant with each other. He maintained that the only way to introduce a new effective philosophy of nature, whose principles could be shared by many, was to found it on experiments seen as source of incontestable *matters of fact*. Apart from the circumstance that some (canonical) philosophers, among whom Hobbes, did not accept a philosophy founded on experiments, Boyle had to face the problem to define what precisely facts were. A problem whose difficulty is perfectly clear to modern epistemologists, who commonly believes that facts cannot be separated from assumed theories.

---

[41] vol. 5, pp. 281–291.

[42] pp. 215–226.

The refusal of a speculative approach to the study of nature led to a natural rapprochement towards those who had always followed an operative view to the understanding and manipulation of nature, that is technicians and craftsmen. Many of whom were somehow, more or less consciously, the heirs of the alchemical and magic culture. But the relationship with artisans could not be one-way. It was also necessary for scholars to interact with them so that they would become good artisans, thus allowing a virtuous loop to be established. This required that the propagation of knowledge occurred in an understandable way, which had first of all to avoid the use of the language of the learned: Latin. The artisan should have to become a *virtuoso*, a term with which Boyle, and people of his entourage, did not mean so much a willing dilettante but rather those who understood and cultivated the experimental philosophy [33].[43] The virtuoso should not have been interested only in the philosophy of nature but also in addressing the God of Christians, that is, he had to become a virtuoso Christian, a Christian experimental philosopher.

According to Boyle, the experimental philosopher has a great advantage over the scholastic. For in the peripatetic schools, where things are wont to be ascribed to certain substantial forms and real qualities (the former of which are acknowledged to be very abstruse and mysterious things and the latter are confessedly occult). The accounts of nature's works may be easily given in a few words, that are general enough to be applicable to almost all occasions. But these do neither oblige a man to deeper searches into the structure of things and consequently are very insufficient to disclose the exquisite wisdom. To be told, that an eye is the organ of sight, continued Boyle, and that this is performed by that faculty of the mind which from its function is called visual, will give a man but a sorry account of the instruments and manner of vision itself. Different is the situation for an experimental philosopher who takes it necessary to sustain the pains to dissect the eyes of animals and accordingly to have a view of the contrivance of the organ. He being profoundly skilled in anatomy and optics, by their help takes asunder the several coats, humors, and muscles, of which that exquisite dioptrical instrument consists, and having separately considered the figure, size, consistence, texture, diaphaneity, or opacity, situation, and connections of each of them, and their coaptation in the whole eye, shall discover, by the help of the laws of optics, how admirably this little organ is fitted to receive the incident beams of light, and dispose them in the best manner possible [33].[44]

Boyle was a theorist of experimental philosophy, despite the contrast toward the speculative philosophy that is attributed to him and that he officially attributed to himself. His vision of experimental philosophy concerned both epistemological and sociological aspects (see below). In both of them one could see a Baconian influence, but if it was the case such an influence was possibly indirect as "Bacon did not play a particularly prominent role in Boyle's early natural philosophical writings" [68].[45]

[43]vol. 5, p. 513.

[44]vol. 5, pp. 516–517.

[45]p. 7.

2.4.1 Mathematics and Experimental Philosophy

Although the mathematical culture of Boyle was for sure not negligible [103],[46] he used mathematics very sparingly; at least in his experimental reports. There are two main reasons for this.

On the one hand as Boyle wanted to spread scientific knowledge at the widest possible level of society, mathematics, which required a thorough study and thus was mastered only by an elite, certainly put a limit to the spread of writings that used it. He, for example, criticized Mersenne because, in his opinion, affecting brevity, had made himself obscure; so what he wrote could scarcely be understood, but by mathematicians [103].[47] But Boyle did not intend to lower the level of natural philosophical knowledge; he hoped that people being properly educated could work to increase knowledge. He believed that the scientific knowledge should be public and therefore the scholars had to disclose it with simple language and mainly had to carry out researches that could have a public utility. This sociological face was somehow reinforced by his puritanical ideology, which also contained democratic demands. It was also reinforced by the profound change of English society in the 17th century due to strong economic development followed by the opening of extra-European markets.

On the other hand he had objection at an epistemological and ontological level; according to Boyle the book of nature is not written in mathematical language, but in a less rigid and precise language. He saw in the use of mathematics in natural philosophy a certain degree of immorality, that is a certain form of arrogance that pretends to idealize and have a control on the variety God has diffused in the world he created. In particular, he thought it was difficult to assign invariant properties to the various material components in the world, as mathematicians should do. For example what is named gold, one of the most pure metals, it is not always the same metal. Not only because one cannot obtain a pure product, but because the texture and compactness and the specific weight and the mechanical properties that may be found in several samples can vary, even though to a small extent [31]. Thus fluctuations inevitably associated with experiments are not only due to imperfection of measuring instruments or presence of accidental impediments, such as friction, and other impurities, but are structural.

In any case, Boyle commended mathematics, especially pure mathematics, as a general form of culture and training for the mind. For him mathematics may bring help to the minds of men, to whatever study they apply and consequently to the minds of the students of natural philosophy. Mathematical disciplines make men accurate and very attentive; they much improve reason, by accustoming the mind to deduce successive consequences and judge of them without easily acquiescing in any thing but demonstration. Moreover the operations of symbolical arithmetics (or the modern algebra) seem to afford men one of the clearest exercises of reason, nothing being there to be performed without strict and watchful ratiocination, and the whole

[46] p. 26.

[47] p. 43.

method and progress of that appear at once upon the paper, when the operation is finished [33].[48]

In describing his experiments Boyle used numerical values; integer numbers to quantify the various operations; real numbers to represent the measurements of various magnitudes involved. He also made a limited use of mathematics to express the regularities or laws. A most famous example is the measurement of the contraction of the volume of air versus the increase the pressure (modern term) to which it is subjected. In [103][49] it is suggested that in this case Boyle was influenced by his assistant Hooke more oriented toward mathematics to describe laws of nature.

One of the aim of experimental philosopher, and Boyle was such, was to show the compatibility between experimental findings and the principles of mechanical philosophy. "Boyle's was a program for the interpretation of nature rather than the interpretation itself. In fact Boyle never attempts to determine what is the texture or the mixture of particular elements or compounds" [103].[50] He did not need to offer specific mathematical accounts of particular bodies or events in the invisible realm of corpuscles, because he who did so would risk of subjecting the visible to the invisible, the readily intelligible and conceivable to the less intelligible and the esoteric, the concrete to the abstract [103][51]

The most important role mathematics played was indirect. Boyle, like many other experimental philosophers of the period, employed the way of reasoning typical of mathematicians in which every proposition must be derived from previous assumptions, without resorting to tricks of rhetoric—and a limited use of synonyms and homonyms. The assumptions should be defined precisely and only based on experimental observations. In *The origine of formes and qualities*, Boyle contrasted the old scholastic philosophy of nature, which dealt with forms and qualities, and where the language to explain generation, corruption and alteration was usually so obscure, tangled and unsatisfactory. Here, said Boyle, discussions of these subjects consisted so much more of logical and metaphysical notions and hair-splitting than of observations and reasonings about the real world, and it was difficult for a reader of average intelligence to understand what they meant and equally difficult for any intelligent and unprejudiced reader to accept what they taught [33].[52]

In some important works, those that made him famous in the 18th century, Boyle wrote explicitly as a mixed mathematician. This is the case of his studies on the air compressibility, of which his *New experiments physico-mechanical touching the air* of 1660 [25] is an exemplary representative, and of the researches of the statics of fluids referred toon the *Hydrostatical paradoxes* of 1666 [22]. These works are judged by some modern historians of science almost a form of evasion of Boyle from

[48]vol. 3, p. 426.

[49]p. 35.

[50]p. 40.

[51]p. 41.

[52]vol. 1, p. 4.

his real job of mechanical philosopher [15].[53] But in my opinion they represent the other side of Boyle.

To illustrate the most advanced use Boyle made of mathematics I refer below some considerations whose main purpose was to prove that the air had spring besides weight. That is that air exercises forces that do not depend only on the weight of the atmospheric open sea that dominates us but also on its tendency to expand, as suggested by the experiences of Torricelli in 1644.

Regarding the weight of the air Boyle measured and found it 938 times less dense than water, a much more accurate results—from the point of view of modern standard—than the known ones [33].[54] The elasticity of air was afforded mainly on qualitative basis, drawing from corpuscular conceptions. Boyle's hypothesis assumed air made of small corpuscles, elastic in themselves, like little springs, as may be resembled by a fleece of wool, that transfer their elasticity to the whole air. This explanation of air elasticity may seem strange to a modern accustomed to the idea of an air made of small particles: the pressure is due to impact of these particles, which move very fast, against walls of a vessel that contains the air. This view of pressure, however, emerged only in the 18th century when in 1738 Daniel Bernoulli published his *Hydrodynamica*.

To signal however that Boyle in a late work, *General history of air*, published posthumously, presented an account not very different from Bernoulli's. Here after having discussed about the constitution of air and having introduced the elastic corpuscles resembling a fleece of wood, he spoke about the existence of other particles responsible of elasticity because their motion due to heat:

> And I will allow you to suspect, that there may be sometimes mingled with the particles, that are springy, upon the newly mentioned account, some others, that owe their elasticity, not so much to their structure, as their motion, which variously brandishing them, and whirling them about, may make them beat off the neighbouring particles, and thereby promote an expansive endeavour in the air, whereof they are parts [30].[55]

Boyle's air then is a heterogeneous substance, composed of at least two sorts of elastic particles; some produce elasticity because of their shape and some other because of their motion under the influence of external agitation due to heat.

Boyle referred his work about the compressibility of air especially in the *New experiments physico-mechanical touching the spring of the air, and its effects* [33],[56] which saw three editions, 1660, 1662, 1682. In the second and third editions Boyle reported two 'additions', *A defence of the doctrine touching the spring and weight of the air* [33][57] to reply the criticisms of Francis Line (1595–1675) and a discussion on Hobbes' ideas, *An examen of Mr. T. Hobbes his Dialogus physicus de natura aeris* [33].[58]

---

[53] p. 112.

[54] vol. 1, Experiment 36, p. 86.

[55] p. 615.

[56] vol.1, pp. 1–185.

[57] vol.1, pp. 118–185.

[58] vol.1, pp. 186–242.

The *New experiments physico-mechanical touching the spring of the air, and its effects* referred to many experiences (43), all of them quite interesting. Boyle experimentation, conducted with great skill, was both narration and explanation. Though his explanations were based on proximate causes (see below) and qualitative, Boyle distanced himself from Aristotelian philosophers (and even from Cartesian ones). Occult qualities did not appear; or rather they were represented by forces, pressures and weights, which were hidden because they were not referred to their first causes, the interactions of the particles, but justified experimentally.

Experiments were performed using the air-pump Boyle had built with Hooke's help; for a description of this machine see for instance [104].[59] The first two experiments used lamb bladders closed at the extremities and containing a little amount of air. By placing these bladders into the receiver of the air-pump and letting the air out, one could see their swelling up to the burst, even without the resort to a very strong vacuum. A modern reader is a little disconcerted by the naivety and ingenuity together with which the experiments were conducted and reported. They were not carried out blindly, however, but had the precise objective of demonstrating the hypothesis that air has elasticity, or rather that it increases or decreases the volume according to the forces that urge the containers in which air is contained. There is no reference to quantitative aspects, neither about some values of the experiment results, nor about the size of the objects used. For instance, by referring to lambs blasters, he said they are were small enough but never he specified, for example, their size or their thickness. About the air-pump it is only said that a little or a lot of air was expelled, but not exactly how much. In this sense, Boyle's narrative was not unlike to that found in the *Saggi* of the Accademia del cimento. At the end of his experiments Boyle believed he had proved that the air was elastic.

Once ascertained the elasticity of air, Boyle intended to measure it in some way. For instance in the Experiment 6 it is a matter of verifying the expansion of an air bubble created in a tiny glass tube sealed on one side and filled with water. He complained, and here his rhetoric clearly intervenes, also an unfortunate fact that happened, that is that the chosen glass tube had broken and resulted shorter than desired. The experiment however was carried out the same. The experience was then repeated with a longer tube that, said Boyle, was at hand available and used it for this reason even if it was not completely suitable because too wide. The experiments revealed that the air expands over 100 times the initial volume when the water that fills the tube was sucked out to make the vacuum. In this experience there were also some measures, the initial volume and final volume of air which were measured in quite ingenious way. In a case, said Boyle, a little emphatically, their ratio was of one to 152. A modern reader could criticize Boyle by saying that his quantitative determinations make no sense because the expansion of the air is limitless and the final volume can be a multiple great pleasure of the initial one. But Boyle still considered hypothetical the expandability of the air to think that it could be infinite [25].[60]

[59]pp. 26–30.

[60]pp. 30–31.

Boyle some time later resumed the experiments on air elasticity, reported on the second edition of the *New experiments physico-mechanical touching the spring of the air, and its effects* of 1662. The one most known that makes Boyle famous is contained in *A defense of the doctrine touching the spring and weight of the air*, where he exposed for the first time the now well known law of Boyle -Mariotte, for which the pressure in a gas and the volume occupied by it are inversely proportional, for a fixed temperature.

The approach was different with respect to that of older experiments, partly because the starting point is different. Once the elasticity of the air has been established, Boyle could concentrate on precise measuring. The designed test was quite simple and did not require the use of the vacuum machine. It involved compressing the air contained in a thin tube with the weight of a column of quicksilver and measuring the relationship between the length of the section of tube in which there was air and that in which there was quicksilver. On the basis of his measurements Boyle was able to formulate the following simple mathematical law: "the pressures and expansions to be in reciprocal proportion" [24],[61] but it is unclear whether as a result of induction from the experimental measurements or as a priori hypothesis. Notice that Boyle had not the modern concept of pressure, intended as force per unit of surface; his pressure is simply a force. So most probably Boyle would not have understood the modern formulation of his law.

Boyle gave his law a contingency character, or rather perhaps considered it interesting from a practical point of view, but attributed to it no scientific value in the strict sense, because there were no guarantees on its actual truth, not only for all gases but neither for different portions of air because a substance with varying composition. Mariotte, who performed experiment on air compression more or less in the same period of Boyle—and whose nome was associated to that of Boyle in the so called law of Boyle-Mariotte—had a different conception; he believed that mathematics could capture the actual behavior of the phenomena and that it could be described by relatively simple mathematical laws [81]. If the mathematical laws did not observe exactly experimental, it depends only on the imperfections of matter and the experimental errors (difficult to avoid) [41].[62]

2.4.2 *Hypotheses and Matters of Fact*

Boyle's epistemology was empiric in the sense that it was founded on matters of fact, recognized by observations and experimentations. It left however much space to theoretical elaborations, or hypotheses, about the causes at play. This theoretical activity was not sufficiently noticed by contemporaries. For instance, when Boyle died, Leibniz and Huygens while deploring his loss, in mutual correspondence declared that

[61] p. 58.

[62] pp. 478–484.

he had wasted his talent in only performing experiments [69].[63] A quite ungenerous and false appreciation.

For Boyle a matter of fact once verified was true; an hypothesis was simply a possible explanation of a fact. However the division was not so sharp. For Boyle the identification of causes was a process that involved steps ranging from the closest to the more remote causes; in such a process a low level hypothesis (a proximate cause) might be assumed as a matter of fact if carefully verified by contrived experiments. An example that may help to clarify how Boyle dealt with hypotheses and matters of fact is the different ways in which he treated the spring of air in different periods of his researches. The spring (that is elasticity) of the air is a tendency to expand or to contract in dependence of the external constraints to which it is subjected with a pressure that opposes to these constraints. There is difference between a fluid, such as for example water, in which there is a pressure dependent on its weight but there is no trend to expand (or springiness)–at least not so obviously–and a gas, for example air, which owes its pressure to the weight of the surrounding atmosphere but tends to expand. In the first experiment of his *New experiments physico-mechanical touching the air* Boyle declared that the existence of the spring of the air can be assumed as a reasonable hypothetical cause to explain a lot of phenomena. Using his words: "I thought it not superfluous, nor unseasonable in the recital of this first of them, to insinuate that notion by which it seems likely that most, if not all of them, will prove explicable [...]. That there is a Spring, or Elastical power in the Air we live in [...]. That our Air either consists of, or at least abounds with, parts of such a nature; that in case they be bent or compress'd by the weight of the incumbent part of the Atmosphere, or by any other Body, they do endeavour, as much as in them lieth, to free themselves from that pressure, by bearing against the contiguous Bodies that keep them bent" [25].[64] In some later experiments he instead gave for granted the existence of the spring of air, as proved by experiments and as "from now on acknowledged by the most eminent modern naturalists" [25]:[65] in substance as a matter of fact.

Boyle's goal was to provide an intelligible explanation or a hypothesis for the various phenomena. But for some of them the causes are unknown and thus they are inexplicable to human beings; the category of inexplicable phenomena is quite large. For instance why the body fall, how the cohesion of the smallest particles works, how human soul can move bodies, how human memory operates. Besides things that are inexplicable there are others that are mysterious and incomprehensible, such as space and time and anything requiring the concept of infinity [115].[66]

Boyle thought that though many phenomena were inexplicable in themselves for the human being, excellent hypotheses (see below) could be formulated of them which are intelligible. He assumed that the same phenomenon could be explained by different hypothesis and suggested criterions of choice, by classifying hypotheses as

[63] vol. 10, p. 239; pp. 228–229.

[64] Experiment 1, p. 12.

[65] Experiment 20, p. 71.

[66] pp. 153–157.

good or excellent. Hypotheses to be accepted should be grounded in the phenomena; the formation of premature or purely speculative hypotheses should be regarded as a most serious error of natural philosophy.

In the following the requisites of good and excellent hypotheses are reported, expressed by Boyle in a mnemonic form:

1. To frame a good Hypothesis, one must see First, that it clearly Intelligible be.
2. Next that it nought assume, nor do suppose That flatly dos any known Truth oppose.
3. Thirdly, that with itself it do consist So that no One part, th'other do resist.
4. Fourthly, Fit and sufficient it should be, T'explain all the Phenomena that we Upon good grounds, may unto It refer: Or those at least, that do the Chief appear.
5. Fifthly the Framer carefully must see That with the Rest, it do at least agree, And contradict no known Phenomena Of th' Universe, or any Natural Law.
6. Sixthly, An Hypothesis to be Excellent, Must not beg a praecarious Assent; But be built on Foundations Competent.
7. Next of all good, the Simplest it must be: At least from all that is superfluous, free.
8. Eighthly, It should the only be, that may The given Phaenomena & so wel display.
9. Ninthly, It should inable us to foreshow The' Events that will, from welmade Tryals flow [36].[67]

A good hypothesis must be intelligible and must not contain anything manifestly impossible or false. It must explain the phenomenon under study and not in contradiction with other known phenomena. An excellent hypothesis must be good and in addition based on sufficient evidence, it must be the simplest among all the good hypothesis that explain the phenomena and lastly, it should have a predictive power.

An example of a hypothesis which is neither excellent nor good, is the hypothesis of the existence of substantial forms of the schoolmen. For Boyle this hypothesis should be rejected on the ground it was unintelligible and superfluous [29].[68] The corpuscular hypothesis is instead an excellent hypothesis as it provides accounts of most phenomena which are easily understandable.

Although Boyle provided criteria for choosing hypotheses, he did not provide a truth criterion. It was possible for him that an hypothesis was excellent, but nevertheless it could not be declared true with certainty, as there was no guarantee that a not good hypothesis be false. The reason for which Boyle avoided any statement about the truthiness of a hypothesis should be searched in his theology, as already suggested. Boyle believed that God had created the world according his infinite understanding and will and maintained the power to change it at pleasure when he desired it. Human beings were created later and independently of the world, with limits in their understanding. This voluntaristic view of God is very different from Descartes's . For him too God was free to create the world with fully freedom; but after that God was bound by his immutability to make arbitrary change, which guarantees the possibility of a certain knowledge to men.

---

[67]vol. 36, fol. 57v. Transcribed in [115], p. 167.

[68]pp. 117–188.

Boyle thought than human being was capable in the uncovering of nature secrets, but there were limits that God in his infinite wisdom has seen fit to impose on human understanding [36, 115].[69] Most of intellectual weakness will disappear in the afterlife.

> In heaven our faculties shall not only be gratified with suitable and acceptable objects, but shall be heightened and enlarged, and consequently our capacities of happiness as well increased as filled.
>
> [...]
>
> Our then enlarged capacities will enable us, even in objects which were not altogether unknown to us before, to perceive things formerly undiscerned, and derive thence both new and greater satisfactions and delights [34, 115].[70]

In the second part of the *Christian Virtuoso* Boyle made clear that not only in the afterlife there will be an understanding of theological mysteries, but knowledge of of the world by natural philosophers (but not for common men?) will be increased as well: "For, at least, in the great renovation of the world, and the future state of things, those corporeal creatures, that will then, be knowable, notwithstanding such a change, as the universe will have been subject to, shall probably be known best by those, that have here made their best use of their former knowledge" [34].[71]

2.4.3 *Corpuscular Philosophy and Chemistry. Physical Chemistry*

Because of his experimental work was largely on what today is classified as chemistry, Boyle is often labeled as a chemists. But he was seen by his contemporaries more as a natural philosopher than an alchemist or a *chymist* [13, 14].[72] An interesting distinction between a chemist and a mechanical philosopher as seen in the 18th century can be appreciated by the following quotations, the former due to Bernard de Fontenelle, the latter by Boyle himself.

> Chemistry, by means of visible operations, resolves the body into certain gross and palpable principles, salts, sulfur, &c. But Physics, by delicate speculations, acts on these principles, as chemistry has done on bodies; she herself resolves them into even more simple principles, into small bodies and figures of infinite variety. This is the main difference between physics and chemistry [53].[73] (B.6)

> To be short, those I reason with, do concerning blackness what the chymists are wont also to do concerning other qualities; namely, to content themselves to tell us, in what ingredient of a mixt body, the quality enquired after does reside, instead of explicating the nature of it, which (to borrow a comparison from their own laboratories) is much as if in an inquiry after the cause of salivation, they should think it enough to tell us, that the several kinds of

---

[69] p. 188; vol. 8, f. 187r.

[70] vol. 1, Seraphic love, p. 283; p. 210.

[71] vol. 6. Christian virtuoso. Second part, p. 776.

[72] pp. 496–497; p. 91.

[73] p. 54.

> precipitates of gold and mercury, as likewise of quicksilver and silver (for I know the make and use of such precipitates also) do salivate upon the account of the mercury, which though disguised abounds in them; whereas the difficulty is as much to know upon what account mercury itself, rather than other bodies, has that power of working by salivation [28].[74]

There are reason to consider Boyle as the first physical chemist of history, a title he deserves by reason of his attempts to apply physical methods to chemistry and to use the corpuscular hypothesis to elucidate chemical as well as physical phenomena [13].[75]

Some Boyle's *chemical* theories had profound influence upon later chemists (and physicists); for example his hypothesis of the material nature of fire, shared for instance by Boerhaave, and his explanation of calcination of metals in terms of combination of fire particles with calcined matter. When the idea of fire as corporeal was combined with the current view that heat was caused by the motion of the particles of matter the result was a 18th century theory that heat is associated to the component particles agitated by fire, an all-pervasive, material substance. Later another particulate fluid, phlogiston, was substituted for fire in the explanation of calcination, and also this concept was inspired by Boyle [87].[76] The account of the historical genesis of Boyle's ideas and in particular understanding how much he drew from his predecessors is not at stake here, however; it enough to comment what he wrote in his papers.

Boyle was the champion of mechanicism and corpularism. His approach to corpuscular philosophy, albeit inevitably based on some metaphysical assumptions, had a strong empirical character. Differently from Descartes who grounded his view on rational and indisputable (for him) assumptions, Boyle assumed the existence of corpuscles as a hypothesis, a well founded one, but a hypothesis that could in principle be reviewed.

To Boyle experience showed that matter was composed of particles. The corporeal substances were formed by *minima naturalia* (Boyle's nomenclature), which have not the meaning of indivisible elements in an absolute sense; they were, however, indivisible in fact, in the sense that known chemical and physical operations failed to decompose them [23].[77] Boyle still did not rule out that there were elementary particles of undifferentiated matter, atoms in Democritean (or Epicurean) sense, whose combinations give rise to atoms in Boyle sense (that is minima naturalia), characterized by peculiar qualities determined by the texture of the component corpuscles.[78]

Out of the minima naturalia were formed "primitive concretions or cluster", which although capable to being decomposed into minima naturalia usually act as indissoluble components in chemical reactions and thus can be considered as the seeds or immediate principles of common matter [23].[79] Boyle spook of *prima mista* as

[74]p. 724.

[75]p. 497.

[76]p. 65.

[77]p. 71.

[78]Atoms is a word Boyle used in a not technical way to mean irreducible corpuscles.

[79]pp. 71–72.

the simplest assembly of minima naturalia, which continues toward more and more complex aggregates called *compounded* and *de-compounded*. Where this last peculiar term, de-compounded, actually means super-compounded [87].[80] The minima are never directly exemplified in nature, but the prima mixta play the role of the elementary atoms or molecules of various naturally occurring bodies (gold, silver, mercury, sulphur, etc.). In many reactions change of quality is associated either with a rearrangement of the prima mixta (mercury to mercury oxide and vice versa) or with the secondary union of the prima mixta of two relatively elementary substances (synthesis and analysis of the mercury sulphides) [74].[81]

In *The origine of formes and qualities* [23], Boyle stated that two principles are at the foundation of material world: undifferentiated matter and motion: "I agree with the generality of philosophers so far, as to allow, that there is one *catholick or universal matter* [emphasis added] common to all bodies, by which I mean a substance extended, divisible and impenetrable" [23].[82] By adding that because this matter all has the same intrinsic nature, the qualitative variation we see in bodies must arise from something other than the matter they consist of. "And since one does not see how matter could change if all the parts that it is or could be divided into were perpetually at rest among it follows that the universal matter can sort itself out into a variety of natural bodies only if it has motion in some or all its distinguishable parts" [23].[83]

Differently from the Greek, Boyle called for the will of God for the existence of corpuscles, with a sentence that parallels the famous one by Newton in the Query 31 of the *Opticks*:

> But for (most of) the other phaenomena of nature, methinks we may, without absurdity, conceive, that God in the scripture it is affirmed, *That all his works are known to him from the beginning*, having resolved, before the creation, to make such a world as this of ours, did divide (at least if he did not create it incoherent) that matter, which he had provided, into an innumerable multitude of very variously figured corpuscles, and both connected these particles into such textures or particular bodies, and placed them in such situations, and put them into such motions, that by the assistance of his ordinary preserving concourse, the phaenomena, which he intended should appear in the universe, must as orderly follow, and be exhibited by the bodies necessarily acting according to those impressions or laws, though they understood them not at all, as if each of these creatures had a design of self-preservation, and were furnished with knowledge and industry to prosecute it [32].[84]

Boyle was not completely clear about the divisibility of matter. He thought that some instruments, more appropriate than the commonly used fire, could break up prima mixta (and even minima naturalia?) to obtain their components [86]. This chance allowed him, for instance, to consider as possible the transmutation of metals; indeed Boyle was deeply involved in archetypal alchemical approaches as the transmutation of metals in gold and the philosopher's stone, that was not such a mystical activity

[80] pp. 66–67.

[81] p. 25.

[82] p. 3.

[83] p. 3.

[84] p. 39.

at it can be supposed as his atoms were not the solid and impenetrable units of Democritus or Epicurus, but structured composites made up of smaller particles of catholic matter, there was every reason to imagine that a sufficiently powerful chemical agent could be able to penetrate and break the particles of a metal into their components and recompose them in a new metal. A procedure which is assumed to be possible today, though not with chemical agents, but through physical ones (bombing with elementary particles).

Strictly connected with Boyle's opinion about the divisibility of matter, is his definition of element, reported below as given in *The sceptical chymist*:

> And, to prevent mistakes, I must, advertize You, that I now mean by Elements, as those Chymists that speak plainest do by their Principles, certain Primitive and Simple, or perfectly unmingled bodies; which not being made of any other bodies, or of one another, are the Ingredients of which all those call'd perfectly mixt Bodies are immediately compounded, and into which they are ultimately resolved: now whether there be any one such body to be constantly met with in all, and each, of those that are said to be Elemented bodies, is a thing I now question [21].[85]

A definition considered by someone as very modern, by someone else as not particular new, as already given in Aristotle's *De caelo* [87].[86] Indeed the definition is not modern, because as stated in the last rows of the previous quotation, a body to be an element should enter in the composition of all bodies. Which is not true for the modern definition, where any body may be composed by an arbitrary number of elements. Moreover the definition should not be considered as Aristotelean, because from this Boyle arrived to the experimental evidence that all the presumed elements, fire, air, water, earth for the Peripatetics and sulfur, salt, and mercury for the Paracelsians, are not such. According to Boyle the only element, if one want to consider it, is the catholic matter, while no consideration is given as possible candidate to elements to the *prima mixta*, that to a modern could appear as natural candidates.

Boyle devoted much of his scientific work to explain phenomena of chemical nature, referring to his corpuscular theory. While from an ontological point of view, in the weak sense in reality, Boyle considered that any phenomena could be explained by recourse to corpuscles and motion, by an epistemological point of view he considered as problematic, and in some cases even impossible, this approach.

> For although such explications be the most satisfactory to the understanding, wherein it is shewn, how the effect is produced by their more primitive and catholick affections of matter, namely, bulk, shape and motion; yet are not these explications to be despised, wherein particular effects are deduced from the more obvious and familiar qualities or states of bodies, such as heat, cold, weight, fluidity, hardness, fermentation, &c. though these themselves do probably depend upon those three universal ones formerly named [33].[87]

And in his studies on 'chemistry' Boyle often left out any argumentation based on corpuscles and motion and spoke *about chymical* qualities', such as fixity, easiness to precipitate, volatility, ability to undergo amalgamation with quicksilver, and so on

[85]p. 350.

[86]p. 64.

[87]vol. 1, p. 308.

[86].[88] Even in fields today part of mechanics, such as hydrostatics and pneumatics, Boyle often avoided the recourse to corpuscular components, preferring *physical qualities* such as forces and pressures considered as proximate causes, as illustrated in the study of the spring of air. In such a case Boyle could behave as a mixed mathematician and did so giving fundamental contributions. He however—this is the opinion of most historians—maintained as fundamental an empirical and experimental analysis and qualified his study as physical-mechanics, preferring this term to that of physico-mathematica, used to indicate the approach in the wake of mixed mathematics, that was then spreading.

2.5 Newtonian Philosophy

If it is true that Newton was not the isolated genius it is asked to believe and that physics and mechanics (besides mathematics) of the 18th century were not his exclusive creation it is neverthless true that he was seen by contemporaries as a very important natural philosopher and mathematician. Gradually since 1750 s a myth grew around Newton who became a reference for all the scholars of the western world for any matter concerning physics (and chemistry), more or less as Aristotle was in the previous centuries, so that one can say that Newtonianism replaced Aristotelianism in the schools. "Newton was most and foremost an *emblem* of a new era [...] With time, the historical Newton receded into the background, overshadowed by the very legacy he helped create" [112].[89] This occurred notwithstanding Newton's writing were known only partially. Indeed by the middle of the 18th century, a part from the *Principia* that was mainly a treatise of mixed mathematics, a physicist or a chemist could find elements of the Newtonian philosophy only in the *Opticks* and in the *Queries* in the *Principia*. He could read in Thomas Birch's *History of the Royal society* the early statement on Newton's *The hypothesis explaining the properties of light discovered in my severall papers* of 1675, while in Boyle's *Works* he could find the letter written to him by Newton in 1679 concerning the properties of the aether (see Sect. 1.2.5.2). Such a reader would be puzzled by the Scholium generale to Book III of the *Principia*, in the third edition of 1726.

Within the limits of the influence that an individual can have in the development of a community, it indeed should be said that the role of the individual Isaac Newton was great. But this gave no reason of the myth of Newtonianism, for which much of what was known before him and much of what was discovered after him was attributed to the individual Newton.

There are very many texts on the history of science of general character where 'Newtonianism' and its affirmation first in England and then in the Continent are discussed in depth. This point of view is however not very interesting for the present book and will therefore not be considered; or rather only some aspects of it will be

---

[88] p. 67.

[89] pp. 21–22.

commented. Questions has been raised as to whether the ground on which Newtonianism developed was already fertile and that rather than conquering it, Newtonianism established itself because it had already entered into current practice independently of Newton, following the same path Newton had followed. For instance s' Gravesande and Musschenbroek are usually labelled as strict Newtonians. They met Newton in England and popularized the theory of gravitation and the heterogeneity of light. But they accepted also the Leibizian theory of vis viva and recent researches have shown that they drew on methodological sources different from Newton's [51].[90]

The moral to be drawn is that our understanding of Newton's legacy in the 18th century will not be advanced by producing taxonomies of different kinds of 'Newtonianism'. Rather, advance will result from studying the specific contexts in which Newton and his work were mobilized and from paying equal attention to both similarities and dissimilarities between Newton's work and that of 18th century scholars. Differently put, we will be able to further our understanding of Newton's legacy once we realize that the label 'Newtonianism' has misled more than enlightened [51].[91]

One realizes the fame Newton gained in the 18th century if he looks for example at the entry *Newtonianisme* of the *Encyclopédie*, which certifies a definition already introduced in the English dictionary *Lexicon Thecnicum*, at least since the edition of 1736 [42].[92]

> Newtonianism, or Newtonian Philosophy, is the theory of the mechanism of the universe, and particularly of the motions of the heavenly bodies, their laws and their properties, as this has been taught by Mr. Newton. See Philosophy.
>
> The term Newtonian philosophy has been variously applied, and from this, several ideas of the word have arisen.
>
> Some authors understand by it the corpuscular philosophy, as reformed and corrected by the discoveries with which Mr. Newton has enriched it. It is in this sense that Mr. Gravesande calls his elements of physics an *Introductio ad philosophiam Newtonianam*. In this sense, the Newtonian philosophy is no other than the new philosophy, different from the Cartesian and Peripatetic philosophies, and from the ancient corpuscular philosophies.
>
> Others mean by Newtonian philosophy the method which Mr. Newton follows in his philosophy, i.e. the method which consists in deducing his reasoning and his conclusions directly from phenomena, without any previous hypothesis; starting from simple principles; deducing the basic laws of nature from a small number of selected phenomena; and then in using those laws to explain other things.
>
> As others understand Newtonian philosophy, it considers physical bodies mathematically, and applies geometry and mechanics to solve [questions about] phenomena. Taken in this sense, Newtonian philosophy is no other than mechanical and mathematical philosophy.
>
> Others mean, by Newtonian philosophy, that part of physics which Mr. Newton has handled, extended and explained in his book of the Principia.
>
> And still others, finally, understand by Newtonian philosophy the new principles which Mr. Newton has brought into philosophy, the new system he has founded on these principles, and the new explanations of phenomena that he has deduced from them [52].[93] (B.7)

---

[90] p. 110.

[91] p. 125.

[92] p. 180.

[93] Article Newtonianisme. English translation by Terry Stancliffe.

The definition of the *Encyclopédie* was reiterated by d'Alembert, Condorcet, de Lalande et als. [42].[94]

In substance there were two main meanings of Newtonianism in the 18th century, one connected with the *Principia* and another with *Opticks* and *Queries*. The former will be examined in the next chapter where the developments of mechanics in the Continent is discussed. Of the latter, that more properly is linked to the experimental philosophy, more precisely to that part of it that will later be called physics, is given an outline below to then resume the subject in Chap. 1.

In the following I will use the label Newtonianism, but with a very weak meaning, intending with this term the ideas for which forces are mathematical entities rather than physical ones, a law $f = ma$ is adopted, matter is corpuscular and each particle is endowed with attractive or repulsive force acting at a (short) distance, vacuum is accepted and light has a prevalently corpuscular nature; moreover experiments plays a major role. Considering that the appreciation of Newton's ideas was different in the different fields of a scientific society that had already started a process of specialization, below I distinguish between 'chemists' and 'physicists', intending with these labels respectively, the scholars that were mainly inspired by alchemy and those inspired mainly by mechanical and experimental philosophy.

2.5.1 *Influence of Newtonianism on Physicists*

About physicists only two exemplary cases are considered: John Theophilus Desaguliers in England and Georges-Louis Leclerc, Comte de Buffon in the Continent.

2.5.1.1 John Theophilus Desaguliers

In England one of the most faithful interpreter of the new physics was John Theophilus Desaguliers (1683–1744), a French-born British natural philosopher, mathematician, clergyman, engineer and freemason whose father has been exiled as a Huguenot by the French government. He attended lectures by John Keill, who used innovative demonstrations to illustrate difficult concepts of Newtonian natural philosophy and obtained a master's degree in 1712. He soon became most successful in delivering public lectures in experimental philosophy, offering them in English, French or Latin. By the time of his death he had given over 140 courses of some 20 lectures each, on mechanics, hydrostatics, pneumatics, optics and astronomy. He kept his lectures up to date, published notes for his auditors, designed his own apparatus, including a renowned planetarium to demonstrate the solar system and a machine to explain tidal motion. In 1714 Isaac Newton, then president of the Royal society, invited Desaguliers to replace Francis Hauksbee as demonstrator at the society's weekly meetings; he was soon thereafter made a fellow of the Royal society

[94] p. 180.

itself. Desaguliers applied his knowledge to practical situations. His interest in steam engines and hydraulic engineering made him an expertise in ventilation. He devised a more efficient fireplace which was used in the House of Lords and also invented the blowing wheel which removed stale air from the House of Commons for many years [47].

Desaguliers was eager to publicize rather than to publish his lectures; eventually in 1734 five long lectures and many additional notes were published as the *Course of experimental philosophy*, which saw a second volume in 1744. The first volume was devoted wholly to theoretical and practical mechanics, including both a simple treatment of Newton's system of the world and a description of Mr. Allen's railroad at Bath. Desaguliers attributed the ten-year delay before the appearance of the second volume to his desire to improve the treatment of machines, especially waterwheels. In it he added seven more lectures discussing impact and elasticity, vis viva and momentum, heat, hydrostatics and hydraulics, pneumatics, meteorology and more machines. This volume is even more concerned with applied science and engineering than the first and entitles Desaguliers to be considered a forerunner of the more advanced knowledge of machinery that characterized the Industrial Revolution [60].

By referring to Newton's *Queries*, in the *Annotations upon the eleventh lecture* of the second volume, Desaguliers said that the questions raised by Newton in the *Opticks* must be solved positively upon close examination and that only the "incomparable philosopher's modesty made him propose those things by way of queries" [49].[95] And this, according to Desaguliers, was not only his opinion, but also that of the "reverend and learned Dr. Stephen Hales". This was a move shared with many post-Newtonian scholars indeed.

According to Desaguliers there are two main kinds of attraction in nature, that is gravity and cohesion. Another type not so strong as cohesion but stronger than gravity exists. Its proportion in removal of bodies attracting is nearly as the cube—this is also Newton opinion—of distance: "This is the magnetical attraction" [49].[96] But also, according to Desaguliers there are repulsive powers in nature "and very often the same bodies attract one another at a certain distance, and under some circumstances do repel one another at different distances" [49],[97] and all the phenomena of nature, such for instance the elasticity can be reduced to these powers.

Here Desaguliers is more direct than Newton for the dependence of force between two corpuscles with distance; Newton only alluded to a change of forces from attractive to repulsive as it happens in algebra, without specifying which parameter was changing; Desaguliers specified the parameter: the distance between corpuscles. A more clear statement of the variation of forces between two particles can be found in the *Compendious system of natural philosophy* by John Rowning—the dates of publication of the four parts of this book are confusing, but part 1 seems to have appeared first in 1735 and part 2 in 1736 [97] [98]—an author now unknown but well renowned at

[95] vol. 2, p. 403.

[96] vol. 1, p. 16.

[97] vol. 1, p. 17.

[98] p. 66.

his own time. His treatise, reissued seven times, was used at Cambridge and Oxford, at the College of William and Mary in Virginia, at many dissenting academies and by John Wesley as a text for his itinerant preachers. It was also mentioned in the correspondence of people as various like John Adams, William Beckford and Joseph Priestley [101]. Below as Rowning characterized the forces between particles, in the section of his treatise concerning fluids:

> Further, since it has been proved that if the parts of fluids are placed just beyond their natural distances from each other, they will approach and run together; and if placed further asunder still, will repel each other; it follows, upon the foregoing supposition that each particle of a fluid must be surrounded with three spheres of attraction and repulsion one within another: the innermost of which is a sphere of repulsion, which keeps them from approaching into contact; the next a sphere of attraction diffused around this of repulsion, and beginning where this ends, by which the particles are disposed to run together into drops; the outermost of all, a sphere of repulsion whereby they repel each other, when removed out of that attraction [98].[99]

The 'supposition' of the three spheres of attraction and repulsion allow to explain as a fluid can be converted into a solid and *viceversa*. If the action of the first sphere is destroyed by cold, the particles of fluids must necessarily be brought into closer contact with the forces of the second spheres and by that means constitute an harder body than before. An inverse mechanism acts in passing from solid to liquid.

The theory proposed by Rowing, of force alternating from repulsion, attraction, repulsion and again attraction (due to gravity), is close to the famous theory proposed by Boscovich in his *Philosophiae naturalis theoria redacta ad unicam legem virium in natura existentium* of 1758, discussed below in Chap. 4. For what I know Rowning's and Boscovich's findings are independent of each other.

2.5.1.2 Georges-Louis Leclerc, Comte de Buffon

In France the acceptance of the Newtonian approach was a little slower; but in the end it becomes robust, even if the attention was more towards the *Principia* than *Opticks*. An exception is constituted by Georges-Louis Leclerc, Comte de Buffon (1707–1788).

Because of the enormous editorial, but non only, success of his *Histoire naturelle*, published in 36 volumes between 1749–1789, Buffon is today esteemed as a naturalist, like the Swedish Carl Linnaeus (1707–1778). Indeed, as most his contemporaries Buffon's interests ranged over various subject matters, among which physics and mathematics (in this last subject he was encouraged by Gabriel Kramer). And mathematical probability was one of his most interesting work, the *Mémoire sur le jeu the franc-carrau*, presented in 1733 before the Académie des sciences de Paris and favorably commented by Fontanelle in the memoirs of the academy of the same year. Here he showed his mastery of Calculus. The work was however published only in 1777, as part of the *Essai d'arithmetique morale*, added to the fourth supplement

[99] Part. II, pp. 5–6.

to the *Histoire naturelle, généraleet particuliére* [38][100] and for this reason remained buried until Morgan Crofton (1826–1915) discovered it in 1869, with great surprise and appreciation [92].[101]

In the memoir, Buffon compared the different kind of 'truth', a problem very popular in the period, differentiating among mathematical, physical, moral truth. According to Buffon as physical truth is concerned, there is no demonstration, no evidence whatsoever as is instead the case of mathematics. A physical truth, necessarily based upon observation or experiment, is only probable. And Buffon tried to evaluate the degree of probability, with an interesting though scarcely convincing procedure. He measured the reliability of the physical truth with the probability that the sun would rise tomorrow: "If one wants to reduce here the seniority of the world and of our experience to six thousand years, the sun has risen for us only 2 million 190 thousand times—the days of six thousand years—and as to date back to the second day that it rose, the probabilities to rise the next day increase, as the sequence $1, 2, 4, 8, 16, 32, 64$ [...] or 2^{n-1} (where n is equal to 2 190 000). One will have, I say, $2^{n-1} = 2^{2\,189\,999}$; this already is such a prodigious number that we ourselves cannot form an idea, and it is by this reason that one must look at the physical certainty as composed from an immensity of probabilities" [38].[102]

For the probability of the moral truth, Buffon started with the idea that the most important event for a man is his own death. Now, the tables of mortality show that the probability for a man at the age of fifty to die within the following 24 hours is a little less than one out of 10 000. But an average healthy man is not afraid of dying because he does not believe he will die the next day. Accordingly, any event whose probability is equal or inferior to one out of 10 000 is of no concern for us [38].[103] Notice that the probability of moral truth is much lower the probability of the physical truth.

Notwithstanding his confidence with mathematics, at least in his youth, Buffon did not believe it could be very useful in physics, where very complex phenomena are dealt with. He expressed his pessimism in the *Premier discourse* of 1749 which opened the first volume of the *Histoire naturelle*. Here he first expressed the impossibility to evaluate the cause of physical phenomena. According to Buffon, suppose that after having determined the facts through repeated observations and having established new truths through precise experiments, one wished to search for the causes, or reasons, for these occurrences. He finds himself suddenly baffled, reduced to trying to deduce effects from more general effects and obliged to admit that causes are and always will be unknown to him, because senses, themselves being the effects of causes of which one has no knowledge, can give ideas only of effects and never of causes. Thus one must be content to call cause a general effect, and

---

[100] Supplement, tome 4, pp. 46–148.

[101] p. 32.

[102] Supplement, tome 4, p. 53. Notice that Buffon measured here probability differently from us; it may be greater than unity. In modern language Buffon would say that the probability that sun does not rise tomorrow is $1 : 2^{2\,189\,999}$, a negligible value indeed.

[103] Supplement, tome 4, p. 56.58.

must forego hope of knowing anything beyond that. "These general effects are for us the true laws of nature" [38, 78].[104]

According to Buffon, the union of mathematics and physics can be accomplished only for a very small number of subjects. In order for this to take place it is necessary that the phenomena to explain are susceptible to being considered in an abstract manner and that their nature be stripped of almost all physical qualities. But mathematics is inapplicable to the extent that such subjects are not simple abstractions. "The most beautiful and felicitous use to which this method has ever been applied is to the system of the world" [38, 78].[105] For Buffon there are very few other subjects in physics in which the abstract sciences can be applied so advantageously.

When the problems are too complicated to allow the application of calculation and measurement, as is almost always the case in natural history and physics, according to Buffon, the true method of guiding one's mind is to make observations, to gather these together and from them to make new observations in sufficient numbers to ensure the truth of the main phenomena. Mathematics should be used only to estimate the probabilities of the consequences that can be drawn from observed facts. Above all, it is necessary to try to generalize these facts and to distinguish well those that are essential from those that are only ancillary to the subject in question. It is therefore necessary to link these facts together by analogy, to confirm or destroy certain equivocal points by experiments, to form one's own explanations based on the combination of all the connections and to present them in the most natural order [38].[106]

The true goal of experimental physics is to experiment with all the things that cannot be measured by mathematics, all the effects of which one does not yet know the causes and all properties whose circumstances are not known. This only can lead new discoveries, whereas the demonstration of mathematical effects will never show anything except what already known [38].[107]

> But this abuse is as nothing in comparison with the inconveniences into which one stumbles when one wishes to apply geometry and arithmetic to quite complicated subjects of physics, to objects whose properties we know too little about to allow us to measure them. One is obliged in all such cases to make suppositions which are always contrary to nature, to strip the subject of most of its qualities, and to make of it an abstract entity which has no resemblance to the actual being. And after long reasoning and calculation on the connections and the properties of this abstract entity, and after having arrived at a conclusion equally abstract, when it appears that something real has been found, and the ideal result is transferred back upon the real subject. This process produces an infinity of false consequences and errors [38].[108] (B.8)

Buffon assumed a corpuscular structure of matter, admitting the existence of vacuum. For him, at a microscopic level, infinitesimal corpuscles attract each other with a force

[104]vol. 1, p. 57; p. 175.

[105]vol. 1, p. 58; p. 176.

[106]vol. 1, p. 62.

[107]vol. 1, p. 60.

[108]vol. 1, Premier discours. De la manière d'tudier & de traiter l'Histoire naturelle, pp. 60–61. Translation in [78].

whose law varies with the inverse of the square of their distance as it happens for gravitation. This is because he maintained that nature must always act in the same way. For not infinitesimal corpuscles the inverse square law, as is the case for gravitation also, strictly applies only to spherical shapes. For corpuscles of other shape it is no longer valid and becomes a function of their shape and distance measured from some characteristic points. This variation of the law did not disturb Buffon, because it could be obtained, at least in principle, by imagining a corpuscle as formed by infinitesimal spheres for which the inverse square law holds true. If the force between two aggregates of corpuscles is referred to the distance of their centers of gravity, a law different from the inverse of the square is in general obtained, but if the aggregates are very small it differs only slightly from the inverse square.

The chemist Louis-Bernard Guyton de Morveau (1737–1816), a friend and follower of Buffon shared his views on this subject. He attempted, as an example of the effect of shape, to calculate the force between two tetrahedra, each composed of an array of close-packed spheres. Assuming that each sphere of one tetrahedron is attracted by any sphere of the other tetrahedron with the law of the inverse square, the whole force referred to the centers of gravity of the two tetrahedra clearly do not follow the inverse square [97].[109]

Buffon definitive statement on the subject is to be found in the preface to the volume XIII of his *Histoire naturelle* that deals with a wide range of animals. There he wrote:

> All matter is attracted to itself in the inverse ratio of the squares of the distance, and this general law does not seem to vary in particular attractions, except by reason of the shape of the constituent particles of each substance, since this shape enters as a factor into the [evaluation of the] distance [38].[110] (B.9)

According to Buffon all the powers of Nature with which we are acquainted to, may be reduced to two primitive forces; the one which causes weight and that which produces heat, that is expansion and attraction [39].[111] With his words: it is sufficient that the forces of attraction and expansion are two general, real, and fixed effects, for us to receive them for causes of particular ones; and impulsion is one of these effects, which we must not look upon as a general cause, known and demonstrated by our senses, since we have proved that this force of impulsion cannot exist nor act, but by the means of attraction, which does not fall upon our senses. The first reduction being made, it might be perhaps possible to adduce a second, and to bring back the power even of expansion to that of attraction, insomuch that all the forces of matter would depend solely on a primitive one. Now cannot we conceive that this attraction changes into repulsion every time that bodies approach near enough to rub together, or strike one against the other? Impenetrability, which we must not regard as a force, but as an essential resistance to matter, not permitting two bodies to occupy the same place, what must happen when two molecules, which attract the

---

[109]p. 38.

[110]vol. 13, p. XIII.

[111]vol. 10, p. 27.

more powerfully as they approach nearer, suddenly strike against each other? Does not then this invincible resistance of impenetrability, become an active force, which, in the contact, drives the bodies with so much velocity, as they had acquired at the moment they touched? And from hence the expansive will not be a particular force opposed to the attractive one, but an effect derived therefrom [39].[112]

2.5.2 *Influence of Newtonianism on Chemists*

Evaluating the contribution of Newton to chemistry (or alchemy, or chymistry, terms that at the time were not clearly differentiated) is a more difficult task. If it may be true that Newton made no fundamental discovery in chemistry, it is equally true that his ideas on matter constitution were considered useful and in agreement with their topics by most chemists. These ideas can be resumed in four points. Matter is no longer a mysterious subject, or at least it was such at an ontological level, but there was a method of measuring matter: to weigh it, and the balance was a familiar instrument to chemists. The corpuscular nature of light helped chemists to reason about the influence of light in chemical reactions. Matter was made by corpuscles which may attract or repel each other at microscopic level. These force were not mysterious metaphysical beings, they were simply experimental ascertainment of tendency to motion. Indeed the traditional mechanicism, either Descartes's or Gassendi's and even Boyle's could not be helpful for chemists.

In the following I will consider only a few meaningful chemists, those who also influenced the physics of the 18th century in electricity, magnetism, thermology: the English Stephen Hales (1677–1761) and the Dutch Herman Boherhaave (1668–1738). I will refer very shortly about the German chemist and physician Georg Ernst Stahl (1659–1734); this also because Stahl's link with Newton is weak. His conception of matter constituted with Newton's the two alternative pillars of the theory of matter of the 18th century [73].[113]

Stahl was with Boherhaave one of the reference chemist of the first half of the 18th century. He used the works of Johann Joachim Becher (1635–1682) to come up with explanations of chemical phenomena. The main theory that Stahl got from Becher was the theory of phlogiston. This theory did not have any experimental basis before Stahl worked with metals and various other substances in order to separate phlogiston from them. Stahl proposed that metals were made of calx, or ash, and phlogiston and that once a metal is heated, the phlogiston leaves only the calx within the substance. Phlogiston provided an explanation of various chemical phenomena and encouraged the chemists of the time to rationally work with the theory to explore more of the subject. This theory was later replaced by Antoine-Laurent Lavoisier's theory of oxidation.

---

[112] vol. 10, pp. 30–31.

[113] p. 331.

Sthal like many chemists of his period was very diffident toward classical mechanicism. For instance, when one speaks of salt, he stated, and says that it is composed by water and by one of the two types of earths, he gives a real and clear idea of what he intends for salt. And from this he will be sure that when obtaining a salt from a whichever body, he will find that is made up of earth and water. To the contrary when one says that a salt is made of sharp particles, longer than wide, and he is asked to look for this salt, he certainly could neither find nor discover it [83].[114]

2.5.2.1 Stephen Hales

An author very appreciated and named by Desaguliers who also influenced Benjamin Franklin was Stephen Hales (1677–1761), an English 'clergyman' who made major contributions in botany, pneumatic chemistry and physiology and was soon recognized as a leading English scientist during the second third of the 18th century. He received a good scientific and mathematical education while in Cambridge and Newton was there as a professor. In 1718 Hales was elected a fellow of the Royal society. His published writings are very few and well represented in the edition of 1733 of the *Statical essays*, containing in vol. 1 the *Vegetable staticks*, a revised version a his edition of 1727 [58] and in vol. 2 the *Haemastaticks* [59]. In the following I will briefly present some meaningful aspects of *Vegetable staticks* only, which were also well summarized by Desaguliers in the *Philosophical Transactions* of 1727 [48].

Hales stated he was using the statical method of enquiring, that is the examination of the amount of fluids, and solids dissolved into fluids, an animal daily takes in and with what force and different rapidities those fluids are carried about in their proper channels etc. This now (and also then) obsolete use of the word *statikcs*, to mean weighting, comes from Nicholaus Cusanus and his *Idiota de staticis experimentis* of which an English translation is available [43].[115]

For Hales, science was more than the avocation of a country minister: it was a natural extension of his religious life. If he was a devote of the corpuscular world view and held that the living organism was a self-regulating machine, which was in no way incompatible with his faith. For him, as for many other *physical theologians*, nature testified the wisdom, power, and goodness of the all-wise Creator. Hales derived from Newton the fundamental concepts discussed in the *Queries*: matter is particulate and the particles are subject to very special laws of attraction and repulsion.

The *Vegetable staticks*, the most known Hale's treatises, is one of the first work on biology where an extended use of mathematics is made [95]; a mathematics, that though Hales was skilled enough in the matter, was kept at an elementary level, mainly consisting in algebraic manipulations. To certify his belief in a quantified science, Hales opened his treatise with the words:

> And since we are assured that the all wise Creator has observed the mod exact proportions, of *number, weight and measure*, in the make of all things; the most likely way therefore, to get

[114] p. 102–103.

[115] pp. 605–624.

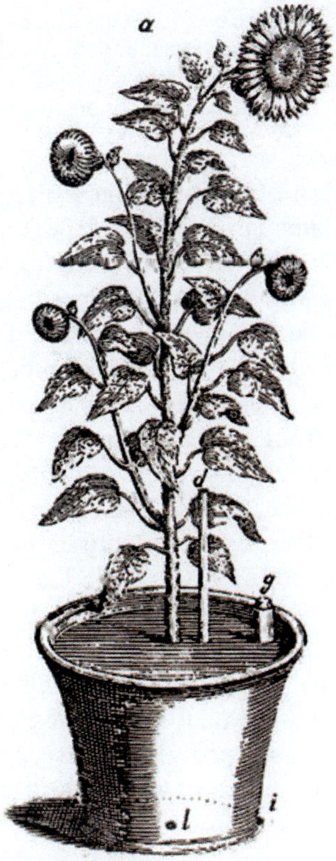

Fig. 2.3 Watering of the sun-flower. Redrawn from [58], p. 28, Fig. 1

> any insight into the nature of those parts of the creation, which come within our observation, must in all reason be to number, weigh and measure [58].[116]

A declaration which was outdated for the 18th century and echoes for instance what Luca Pacioli wrote in the *Summa de arithmetica, geometria, proportioni et proportionalita* of 1494 [41].[117] The reference to the Old Testament for a quantified science, not infrequent however, instead to Archimedes, derived from Hales's theological attitude.

As an example of the way Hales used mathematics, I summarize below his experiment I of the *Vegetable staticks*, carried out with the purpose to measure the quantity of water imbibed and perspired by the sun-flower illustrated in Fig. 2.3. In this experiment Hales was in the need to evaluate the surfaces of the leaves of the plant. To avoid the counting of all the leaves he recurred to a sampling procedure described below.

> I cut off all the leaves of this plant, and laid them in five several parcels, according to their several sizes, and then measured the surface of a leaf of each parcel, by laying over it a large

[116] p. 1.

[117] p. 85.

> lattice made with threads, in which the little squares were 1/4 of an inch each; by numbering of which I had the surface of the leaves in square inches, which multiplied by the number of the leaves in the corresponding parcels, gave me the area of all the leaves; by which means I found the surface of the whole plant, above ground, to be equal to 5616 square inches, or 39 square feet [58].[118]

It is of particular interest for the present book the role Hales attributed to air. I do not intend to discuss in depth the matter, in particular the difficulty of speaking about air in a period where there was no notion about the nature and composition of gases, in particular the difference between CO_2 and O_2, discovered by Lavoisier. I only want to underline the influence of Newton and the final idea Halles reached of air, which had a role in the development in the theory of heath and electricity.

According to Newton, in their free state the particles of air exert upon each other strong repulsive forces, which accounts for the air elasticity. Yet this elasticity is not an immutable property, for he had remarked that "true permanent air arises by fermentation or heat, from those bodies which the chymists call fixed, whose particles adhere by a strong attraction" [58].[119] When air enters into dense bodies and becomes fixed, its elasticity is lost because strong attractive forces overcome the forces of repulsion between its particles [57]. For Newton the particles of fluids which do not cohere strongly are of such a smallness as render them most susceptible of agitations and are most easily separated and rarified into vapor. In the language of the chemists, they are volatile, rarifying with heat, and condensing with cold. But those which are grosser, and so less susceptible of agitation or cohere by a stronger attraction, are not separated without a stronger heat, or perhaps not without fermentation [88].[120]

Hales found after many experiments that permanent air could be obtained by the action of fermentation to free the air incorporated in the substances of vegetables. This air is permanent because continue to persist, is elastic and dilated as common air. In the end Hales could conclude that ordinary matter contains particles of a special kind of substance, referred to as air, that under particular circumstances (for instance by heating and fermenting) can be released as an elastic fluid. This substance should take the place of mercury or spirit as a fifth element.

> Since then air is found so manifestly to abound in almost all natural bodies; since we find it so operative and active a principle in every chymical operation, since its constituent parts are of so durable a nature [...], may we not with good reason adopt this now fixt, now volatile *Proteus* among the chymical principles, and that a very active one, as well as acid sulphur; notwithstanding it has hitherto been overlooked and rejected by Chymists as no way intitled to that denomination [58].[121]

For Hales what is commonly called air, that is the atmosphere, is a "Chaos, consisting not only of elastick, but also of unelastick air particles, which in great plenty float in it, as well as the sulphureous, saline, water and earth particles, which are in no

[118]pp. 5–6.

[119]p. 165. Quoted from Newton's Query 31.

[120]Query 31. p. 372.

[121]p. 316.

ways capable of being thrown off into a permanently elastic state, like those particles which constitute true permanent air" [58].[122]

2.5.2.2 Herman Boerhaave

Herman Boherhaave (1668–1738), an older colleague of s' Gravesande and Musschenbroek, is now virtually unknown but in his own day had a great reputation. "One of the greatest teachers of all time" and "perhaps the most celebrated physician that ever existed, if we except Hippocrates" [111].[123] Destined to study theology he devoted himself to medicine and chemistry. He should also have some confidence with mathematics, because at the beginning of his career he augmented the income by giving lessons of mathematics. He was created a foreign member of the Académie des sciences de Paris in 1728 and a fellow of the Royal society in 1730 [77].

Boherhaave is often portrayed as a disciple of Newton and an adherent of Newtonian science. Moreover he would have had a great role in the diffusion of Newtonianism in The Netherlands and in the rest of the Continent. Actually the situation is much more complicated. In 1715 Boherhaave in his rectorial oration *Sermo academicus de comparando certo in physicis* addressed the question of the certainty in physics and endorsed Newton's approach, but also distinguished himself from him. In [51][124] it is suggested that reference to Newton in his *Sermo academicus* was instrumental to attach Descartes. Indeed references to Newton in Boherhaave's writings, apart from some orations, are scanty and a modern reader could hardly recognize a direct and important role of Newton. The impression of the contemporaries was however different and reference to Boherhaave as a strict Newtonian appeared consistent.

Even though he was mainly famous as a physicians with a medical system based on mechanics, Boherhaave most important contribution to science was in chemistry. He introduced here exact quantification methods by measuring temperatures and using the precise balance made by Fahrenheit.

When a spurious edition of his chemical lectures was published in 1724 under the title *Institutiones et experimenta chemiae*, he felt impelled to publish in 1732 (followed soon by a second edition in 1733) his master piece *Elementa chemiae* [17], which was later translated into English and French and remained the authoritative chemical manual for decades. Boherhaave's treatise is remarkable for its clarity, its systematic presentation not very common in the chemical treatises of the period. There is a very substantial treatment of heat which influenced Black's work on specific and latent heat ad also influenced Lavoisier's theory of caloric [71].[125]

There were two early English translation of Boherhaave's treatise, one due to Timoty Dallowe of 1735 [44] and another due to Peter Shaw (1694–1763) of 1741 [19]. In the following I will refer to Shaw's translation only, which is quite faithful to

---

[122] p. 315.

[123] p. 212, 209.

[124] p. 112.

[125] p. 37.

the original. It is filled with footnotes; too much may be however. In them the editor presented his view on Boherhaave as a strict Newtonian, expressing his personal understanding but altering the appreciation of Boherhaave proper ideas. Regarding the reference to Newton, it must be said that in the original Latin text they occurred in 11 instances in the two volumes; in the English translation there were 69 instances (footnotes included).

Probably the only part in which Newton ideas of the force at a distance reported in the *Queries* are made explicit and partially accepted is in the discussion of the *menstrua*, or solvents. Here Boherhaave declared that a purely mechanical action of the solvent is not acceptable, but also active principles are needed. Boherhaave defined a mestrum as follows: "The term is a barbarous term; and denotes a body, which, when artificially applied to another, divides it subtly, so that the particles of the solvent remain thoroughly intermixed among those of the solvend". The reason why this solvent was called a menstruum, is "because the chemists, in its application to the solvend, first used a moderate fire, for a philosophical month, or forty days" [19].[126] But, said Boherhaave, it rarely happens that any menstruum exerts all its dissolving power mechanically. And hence, "Sir Isaac Newton, in his researches, has found reason, from observation, to add other necessary causes" [19].[127] They are the actions at a distance of attraction and repulsion.

This views of a Newtonian chemistry, supported for instance in [42][128] and [83] was contrasted in [72], where it is assumed than Newton's scheme, based on a limited number of forces possibly describable by mathematical laws, was too rigid for Boherhaave, who like Boyle, adopted the idea of seminal principles or *thread of the warp*, that is corpuscles endowed by God with a very large variety of plastic powers. These principles "are entwined and woven together, so as to form the foundation and support for each single body existing, growing, moving, maintaining itself and propagating itself by fruitful generation. You realize that I allude to the seeds of things" [18, 72].[129] According to [72] the greater richness attributed to nature by Boherhaave should be due to his Dutch Calvinism that more that Newton's Puritanism made Good free to do all what he wanted.

The part of the *Elementa chemiae* devoted to the characterization of matter appears at first as a traditional treatment of natural philosophy of the 17th century, with the introduction of the four elements: fire, air, water and earth, not uncommon in chemical writings. However, a more careful reading reveals its peculiarity. Boherhaave's is an experimental chemistry treatise, and although the quantitative aspects are not relevant, it takes as its model, at least officially, Newton's approach to experimentalism.

Particularly important for the development of physical theories of the 18th century is his conception of fire, which for Boherhaave was a substance, a very thin fluid,

[126] p. 489.

[127] p. 511.

[128] p. 224.

[129] pp. 22, 178.

present in all bodies. The English translation of Boherhaave's *Elementa chemiae* started the chapter devoted to fire, with comments by the translator Peter Shaw:

> The great and fundamental difference in respect of the nature of fire, is, whether it be originally such, form'd thus by the Creator himself at the beginning of things or whether it be mechanically producible from other bodies, by inducing some alteration in the particles thereof. Among the modern writers, Homberg, Boherhaave, the younger Lemery and s'Gravesande maintain the former: the latter is chiefly supported by the English authors [19].[130]

Boherhaave at the beginning did not pronounce on the nature of fire, because "by introducing hypotheses a priori one may fall into error" and such a caution can never be more necessary, according to him, than on the present occasion.

Boherhaave distinguished between vulgar and pure fire. Vulgar fire, as supported by combustible matter is very different from pure fire, both in nature and effects. Many errors have arose among the chemists, for want of distinguishing, with sufficient accuracy, between these two kinds of things, which are known by the common name of fire [19].[131]

After a long list of experiments, Boerhaave removed concerns about the nature of fire, giving some caracterization of it, after having stated that it is a body:

1. From a careful consideration of what has been above laid down, we may perhaps be enabled to assert divers things concerning the nature of fire. First then it appears, that true elementary fire is corporeal, since under the name corporeal is included any thing geometrically measurable by three lines, drawn perpendicularly to each other from the same centre; or, as we more usually express it, an extended surface.
2. But whether fire have also that further property, which some of the greatest men of the present age hold inseparable from all bodies, viz. weight or gravity, in proportion to its solidity, does not so certainly appear from every way the confederation of the whole history of fire.
3. The particles of fire, which have already been shewn to be corporeal, appear further to be the smallest of all the bodies yet known: for if they be corporeal, they must necessarily be exceedingly subtile, as they readily penetrate all, even the densest bodies, and pervading the thickest parts thereof, shew themselves present in every assignable part thereof.
4. The small particles which constitute the ultimate elements of fire, appear to be the most solid of all bodies.
5. These corporeal, solid, subtile particles appear perfectly Smooth, even, and polished on their Surfaces.
6. From the whole history of fire we may infer its absolute simplicity, by which we mean that condition of a body, whereby each particle of it retains the same nature which is observed in the whole.
7. The sixth property of fire is its mobility, which is so great, that we are almost certain it never absolutely rests in any place [19].[132]

---

[130] vol. 1, p. 206.

[131] vol. 1, p. 298.

[132] vol. 1, pp. 357–364.

A modern reader cannot probably see much novelty in Boerhaave's description of the properties of fire. It appears as a restatement of ancient theories, Aristotelian, Epicurean. The novelty is seen in the fact that Boerhaave derived the properties of fire not from metaphysical reasonings but with reference to a lengthy series of accurate experiments. A novelty can also be found in the assertion that fire pervades all the substances, it is weightless and extremely mobile. However no explicit reference is made to the property of elasticity, a property that Newton associate to the aether and Hales to the air.

Boherhaave's conception of fire, flanked to that of air due to Hales, more or less of the same period, offered the scholars of the 18th century two new fluids pervading all bodies, which was the occasion for the 'invention' of other fluids such as caloric, phlogiston, electric fluid and so on. These fluids represent an ontological basis, or a reification, of the various kinds of force at distance. In [63][133] this process is seen as the passage from homogeneous (one kind of object) explanation of the phenomena of nature, to inhomogeneous explanations (more objects).

2.6 The Treatises of Experimental Physics

At the beginning of the 18th century *physics* or *physica* was more or less synonymous of *natural philosophy*, even though a little bit of ambiguity remained, because *Physics* was the title of Aristotle's treatise which dealt only with one aspect of natural philosophy, that is that of the inanimate world. Below what d'Alembert wrote in the *Encyclopédie*:

> PHYSICS. This science, sometimes also called natural philosophy, is the science of the properties of natural bodies, of their phenomena and their effects, as of their different affections, movements, &c. See &c. See Philosophy & Nature. This word comes from the Greek φύσιζ, nature [52].[134] (B.10)

The interplay between the two terms, physics and natural philosophy continued for the whole 19th century, when the content of the matter they indicated had already largely changed with respect to previous centuries.

The spreading of the term physics with a quite modern meaning started when the subjects belonging to natural sciences (modern term) were excluded from courses on natural philosophy in the universities and colleges. That is when the distinction between the two faces of natural philosophy, general physics and particular physics were stressed. General physics was since then simply named physics. It excluded topics related to natural science, though maintained some parts as electricity, magnetism, thermology, chemistry. In any case still at the turn of the 19th century, neither the name physics nor its nature of a well integrated scientific discipline had a clear status.

[133] pp. 65–70.

[134] Article *Physique*.

There were two main reasons that led to the breaking up of the unity of natural philosophy. First the spreading of the mechanical philosophy, second that of the experimental philosophy. Mechanical philosophy, especially in the form given to it by Descartes, had replaced the traditional Aristotelian natural philosophy in the schools, though not everywhere or even completely, but the teaching still remained traditional, based on written texts. The spreading of experimental philosophy modified the form of teaching: not only books but and mainly experiments. The discipline that used the experimental approach to mechanics, optics, electricity, magnetism, thermology and so on was usually referred to as *experimental physics*.

The first professor who used experiments in teaching, in a coherent and organized way was the Cartesian scholar Jaques Rohault (1618–1672), who gave private lessons about mechanical philosophy all around France [82, 106]. He held weekly Wednesday conferences from 1659 until his dead, by illustrating and discussing experiments and, probably more than any other contributed to establish Cartesianism in France. The treatise where Rohault exposed his main ideas on natural philosophy is the *Traitéde physique* of 1671 [93], but the English translation with Newtonian comments added by Samuel Clarke, *System of natural philosophy* of 1723, is much more known [94]. The treatise was based on Cartesian philosophy, but avoided any metaphysical reference. Differently from Descartes, Rohault held that experiments had a theoretical role, hinting an empirical approach to knowledge. The *Traitéde physique*, as well as the *System of natural philosophy* devoted little space to living beings.

The role of experiments in the *Traité de physique* had various faces. Often experiments are used to validate a theory or to test compering theories, sometimes to confirm well known result, sometimes to find new results. Note however that in his experiments Rohault never contradicted Descartes. This is very evident where he experienced about the impact of bodies. Here he referred to only cases for which the Cartesian laws of motion were conform to experience, avoiding thus experiments that would have contradicted some 'absurds' Descartes's rules [94].[135]

An example of a quite trivial experiment is concerned with the explanation of the functioning of a syringe [96].[136] It is obvious, wrote Rohault, that when the end of a syringe is open, the piston can be drawn back with a circular motion of surrounding air. This is obvious because of the general principle that in a full world, all motion is circular. But what if the end is closed? Either (1) the syringe has pores, and the motion will take place, or (2) the syringe does not have pores and there will be no motion. But the experiment says that motion does take place, thus there must be pores in the glass of the syringe. The experiment intervenes not to allow for a choice between alternative theories. Instead, the general principles of Cartesian theory being given, the experiment is simply about a choice between two different instances, equally possible, in our contingent world [93].[137] In another experiment Rohault reproduced

[135] vol. 1, pp. 48–53.

[136] pp. 55–56.

[137] Part I, pp. 73–74.

what already done by Huygens to prove that rotatory motion of the subtle particles surrounding the earth can produce the gravity [94].[138]

An experiment which had not only a didactic value as the two referred above explored a scarcely known phenomenon, concerned what today is known as capillarity. Rohault considered two glass plates very close to each other, immersed in a vessel filled with water. The water between the two plates raised above the level of the water in the vessel, the more the lesser the distance between the plates:

> 81. If two plain Bodies which the Water will wet, such as two pieces of clean Glass, be put very near one another, and dipped a little way into a Vessel of Water; (I) the Air which moves from one Side of the Vessel to the other, in order to get over the Obstacle that lies in its èway, ought rather to pass over the Top of the two Glasses, èthan to descend into that streight Place, which isè between them: So that the Water is not so much pressed èhere as it is in other Places, where the Air can go without bending its so much, and so it ought to rise to a considerable Height above the Level of the Water contained in the Vessel; and thus we see by Experience that it does. 82. And there is no doubt but that the Water would rise still higher, if the two Pieces of Glass Were closed on both Sides, for by that means almost all the Air which moves cross, without bending its Course, would beè hindred from entering in. Or, which is the same Thing, we may take a very small Glass Tube open at both Ends, and dip it in the Water, for then the Air cannot enter in by the Sides: so that the Water must rise very high in such fort of Tubes, if they be very slender: And indeed I have made the Water rise a Foot high in a GlassTube so small, èthat one could scarce get a Horse-hair into it. 83. However, we must not conclude from hence that it ought to rife on without End in these small Tubes; for it is easy to see, that the Water must top, when the Weight of that which is risen, tends downwards with greater Force than the Pressure of the external Air has to thrust it up [94].[139]

The experience is close to one carried out by Hauksbee, published in the Philosophical Transactions in the years 1711–1713 and in the *Physico-mechanical experiments on various subjects* of 1709, already referred to in Sect. 1.2.5.1.

In England John Keill (1671–1721) delivered a course on Newtonian physics using experiments at the university of Oxford from 1694 until 1709 [37].[140] But the most influential scholar to spread experimental activity in the universities was Willem Jacob's Gravesande (1688–1742), professor of mathematics and astronomy at the university of Leiden, starting from 1717, with Newton still active. He published in Latin the treatise *Physices elementa mathematica, experimentis confimiata. Sive, introductio ad philosophiam Newtoninanam* of 1720–1721 (herein after *Physices elementa mathematica, experimentis confimiata*). This text was translated into English by Desaguliers, as the *Mathematical elements of natural philosophy confirmed by experiments, or an introduction to Sir Isaac Newton's philosophy* [100] in the same years of the publication of 's Gravesande's treatise.[141]

---

[138] vol. 2, p. 94.

[139] vol. 1, p. 148.

[140] p. 63.

[141] Although the title page gives the date as 1720, the book had in fact already appeared in 1719; even the English translation of Desaguliers had appeared in December of that year [108], footnote 17.

’s Gravesande’s masterpiece—both in Latin and in English—went through a complex evolution. The first edition written specifically for students, was composed of four books that dealt, respectively, with the body in general and the movement of solids, with fluids, with light, and with celestial mechanics. In the second edition (1725), mathematical proofs were added. Here as s’ Gravesande presented the second edition: “When I first intended to write these Elements, my Design was that my Auditors shou’d be able to re’collect, with ease such things as they had heard more largely explained and demonstrated [...] But that the second Edition might be likewise of service to such of my Readers as were better acquainted with Mathematics, I annex’d the mathematical Demonstrations of all such Propositions, in the scholia to those chapters, in which they are mention’d” [100].[142] From the third edition of 1742 on, the treatise acquired its final structure: the books were now six and embraced the totality of natural phenomena. The first book was devoted to the body in general and its properties (extension, solidity, divisibility and mobility), as well as to several specific physical issues (among which balance of forces, gravity, pressure). The second book dealt with the inner forces and collision of bodies, the third discussed fluids, the fourth air and fire, the fifth light, the sixth the system of the world. Table 2.4 shows the entries of ’s Gravesande-Desaguilers’ treatise in its final version, into two volumes.

If a modern reader leafs through’s Gravesande’s treatise, he does not immediately think that it is a text from the early 18th century; in many respects it could be taken for a modern elementary textbook of physics, in particular in optics and mechanics. Going forward, in particular starting from Chap. 1 of Book II (more or 1/3 of the first volume), the reader realizes that things are a little different; the main difficulty encountered is the way in which’s Gravesande treated force and its measure. Before Book II, the modern reader thinks he is faced with a text largely inspired by Newton’s ideas; here the use of force is not precise; the prevailing term are power and pressure. From Book II onwards, the impression changes and a certain bewilderment intervenes. Said in modern terms while in the first chapters the action of a force seems to be given by the product the force by time, subsequently the action is defined as the product of the force by displacement (that is a work in modern term), which leads to the measure of the action of force as the product of mass by the square of speed (mv^2). Thus the readers sees that in the mechanics of ’s Gravesande, notwithstanding the label *Newton’s philosophy* in the title, there is a strong injection of the ideas of Leibniz and Johann Bernoulli [112].

The following quotation from the second edition of *Physices elementa mathematica, experimentis confimiata*, helps to clarify’s Gravesande’s view:

> ‘Altho’ in many Things relating to the fore-mentioned Theories, I differ in my Opinion from Sir Isaac Newton, yet I made no scruple to keep the title of an Introduction to the Newtonian Philosophy, and to prefix it to the second Edition [...]. He only, who in Physics reasons from Phenomena, rejecting all feign’d Hypotheses, and pursues this Method

[142] p. X.

Table 2.4 Table of contents of *Mathematical elements of natural philosophy confirmed by experiments, or an introduction to Sir Isaac Newton's philosophy* [100], vol. 1, pp. LVII–LXIII

Vol. I		
Book I	Part I	Of body in general
Book I	Part II	Of the actions of powers
Book I	Part III	Concerning motions, chang'd by the action of powers
Book II	Part I	Of innate forces
Book II	Part II	Of the simple congress of bodies, direct and oblique
Book II	Part III	Of compounded congress or collision
Book II	PartI V	Of the laws of elasticity
Book III	Part I	Of the gravity and pressure of fluids
Book III	Part II	Of the motion of fluids
Book III	Part III	Of the actions and resistances of fluids in motion
Book III	Part IV	Of bodies mov'd in fluids
Vol. II		
Book IV	Part I	Of air and other elastick fluids
Book IV	Part II	Of fire
Book V	Part I	Of the motion and inflexion of light
Book V	Part II	(No title)
Book V	Part III	Of the reflection of light
Book V	Part IV	Of opacity and colours
Book VI	Part I	Of the system of the world
Book VI	Part II	The physical causes of the celestial motions

> inviolably to the best of his Power, endeavours to follow the Steps of Sir Isaac Newton, and very justly declares that he is a NEWTONIAN Philosopher [100].[143]

's Gravesande tried to justify his measure of forces with a series of experiments, some of the most interesting concern the measurement of the effect produced by bodies of spherical shape dropped on a layer of clay. The effect, at least in the 1742 edition, is measured by the volume of the imprinting left in the clay. Experiments show that this effect is proportional to the product of the weight of the sphere by the height from which it is dropped and therefore to mv^2; for the axiom action = effect = force, force should thus be measured by the square of speed [100].[144]

The *Physices elementa mathematica, experimentis confimiata* was replaced only after many years by Pieter Musschenbroek's (1692–1761) *Introductio ad philosophia naturalem*, posthumously published in 1762, that besides topics of mechanics and optics, introduced electricity and magnetism also.

Figure 2.4 shows the apparatus's Gravesande prepared to prove that the motion of a heavy body, given a horizontal initial velocity, is a parabola. The description of

[143] p. XI

[144] pp. 197–208.

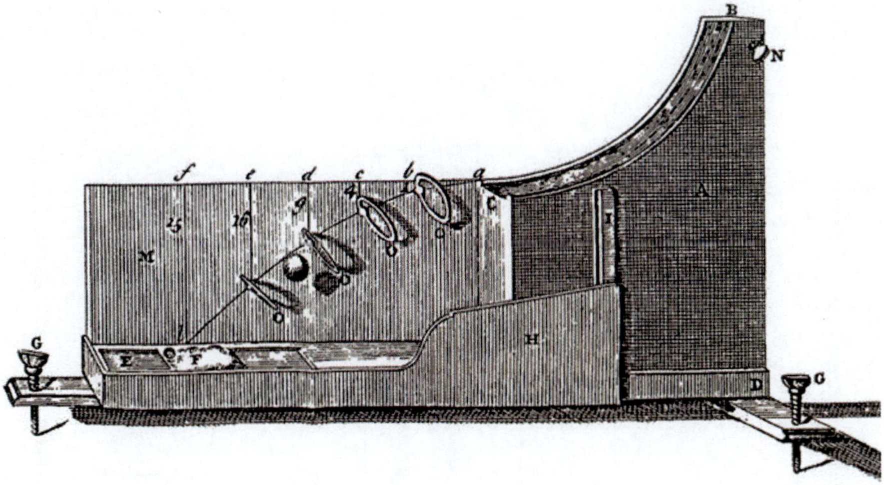

Fig. 2.4 Parabolic motion of a heavy body [100], vol. 1, p. 128, plate XIX

the experience, compared with those referred to by Galileo a century before [40], is much more precise and more simple and convincing.

> The Ball is let down from B, having roll'd to C, is there horizontally projected, and falls at F, and in the mean time passes thro' the Rings O, O, O, O. What has been said of the Curve run thro' by a Body horizontally projected, belongs also to any Projection whatever. Let a Body be projected along AE; and let *ab, bc, cd, de* be equal; the Body will pass along the Curve AFGHI so, that the vertical Lines BF, CG, DH, EI, will be to one another, as 1, 4, 9 and 16; in which Case also the Curve is call'd a *Parabola* [100].[145]

In France the tradition of experimental physics was pursed by Jean Antoine Nollet (1700–1779), on the footprints of 's Gravesande and Desaguiler. Nollet wrote a successful six volume treatise, the *Leçons de physique* of 1743–1748 [89], concerned with experiments on mechanics, electricity and so on. Nollet had the great credit of substituting experimental physics to the speculative Cartesian physics in France. But his disaffected and uncritical neglecting of mathematics, common however to many experimental physicists of the time, contributed to lead physics into a dead end from which it will only come out at the end of the 18th century.

Below the description of a simple experiment about mechanics, carried on with a quite complex apparatus, shown in Fig. 2.5, to prove that in vacuum all bodies fall with the same speed, independently of their weight. Let consider a frame which contains a glass tube which has a length of six feet (just less than 2 m) and a diameter two and half inches, wider and open at two extremes, AB, as shown in the Fig. 2 of Fig. 2.6. A copper plate is attached to the top by means of a ring to which is fastened the hatch of a wheel formed by six spokes which turns vertically as shown in the Figs. 3 and 4. Before placing the piece of Fig. 4 on the glass pipe, it is necessary

[145] vol. 1, p. 124.

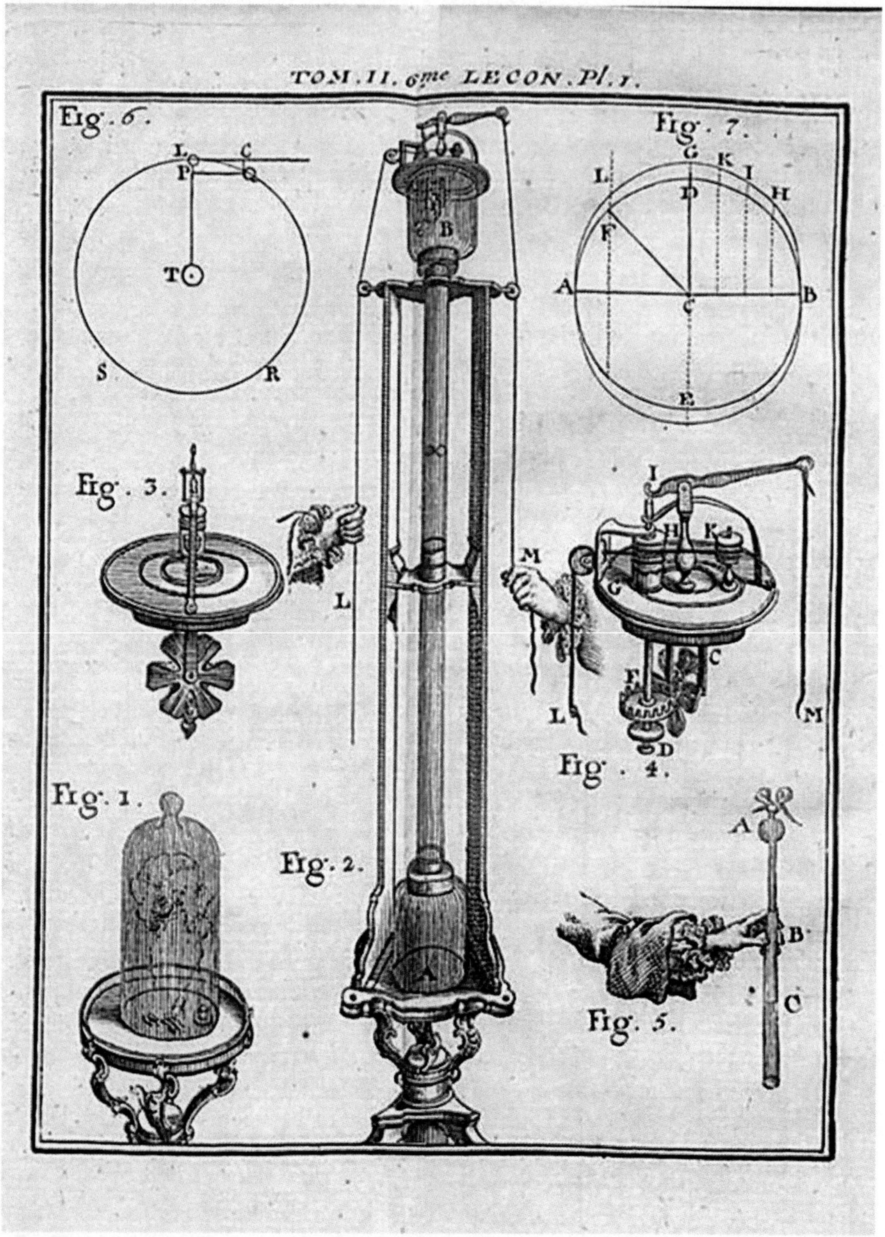

Fig. 2.5 Fall of heavy bodies in the vacuum [89], tome II, p. 154, leçon VI, plate1. Reproduced with the permission of ETH-Bibliothek Zürich, Alte und Seltene Drucke

to be careful by putting on each radius pairs of small bodies with volumes that are nearly similar, but different by weight. For example, a piece of lead and a feather may be placed in the first pair; in the second a piece of copper and a small sheet of paper, etc. When the air in the pipe has been thinned as much as possible with the pump, by pulling the rope L and M wheel F two bodies are made to drop down. Having done its work, it is passed another stroke. The two bodies of each couple fall at the same time and there is no visible difference in the duration of their fall. But if one repeats, added Nollet, the experiment, leaving the tube full of air in its natural state, those who have the most weight fall faster, the slowness is more sensitive as its mass is less. Thus wood falls more slowly than iron, but its slowness is not so great as that of paper and feather [89].[146] The experimental apparatus prepared by Nollet lended itself to perform experiments, at least six falls of couples of heavy body with different weights, in an operationally simple way. It was thus particularly suitable for salons presentations.

Courses in experimental physics, possibly still held under the title of natural philosophy, gradually replaced the traditional ones, both Aristotelian and Cartesian, in universities and colleges. In the newly established universities, and these were many in the 18th century England, the new courses in experimental philosophy were taught by specialists, who had more a mathematical training, broad meaning, instead by philosophers of Aristotelian mould, even though the skill required to understand such courses was not very heigh. Mathematics still remained the cinderella in the undergraduate curriculum in the universities. At Oxford, David Gregory (1661–1708), and after him some professors of astronomy, provided injection of mathematics, but they had only a few auditors [37].[147] Still in 1750 in Cambridge, Newton's own university, hardly a student had mathematical skill to demonstrate the propositions of the *Principia*. As the 18th century progressed, as a consequence of the renovation of the courses in natural philosophy, the courses of mathematics that since then had been the places were the experimental philosophers had found room to teach their physics, were restructured and were devoted to the teaching of pure mathematics only.

In France the situation was different. The traditional courses in natural philosophy were not replaced by courses of experimental physics. They however changed in courses where mathematics had a great role, in the wake of the traditional mixed mathematics. This kind of teaching was dominant in the second half of the 18th century. As a result, the experimental physics found little place in the courses of natural philosophy, at least in the colleges, and were often taught outside the main curriculum, in the vacation [37].[148]

In northern Italy a chair in experimental physics, endowed with a cabinet of physics, seems to have been established in Pavia in 1730. The next in Padua in 1738. Other neighboring universities gradually followed—Pisa (1746), Turin (1748), Modena (1760) and Parma (1770)—whereas in 1787 the facilities at Pavia, where the chair was held from 1778 by the young experimenter, Alessandro Volta (1745–1827),

[146]Tome II, pp. 128–131.

[147]pp. 65–66.

[148]p. 66.

were improved with the opening of a purpose-built physics theater [37].[149] In Padua the new course was entrusted to Giovanni Poleni (1683–1761), a professor who had long shown an interest in providing visual tuition in the natural sciences [45]. At that date the Paduan laboratory was purportedly the best equipped in Europe. In the *Institutionum philosophiae experimentalis specimen* of 1741 [91], a short booklet, written in the occasion of the opening of the cabinet of physics, Poleni sustained that experimental physics should be of support both to physics and mathematics. The reading of Poleni's booklet could be still useful for historians because of the huge amount of citations of works in experimental physics, also of today relatively unknown authors.

Everywhere the emergence of experimental physics on the one hand, and on the other hand the consolidation of an approach mathematically founded, especially of mechanics, shaped a dichotomy, both in the teaching and research, in physics. The experimental physics was characterized by a low level of theorization and mathematics used; the theoretical physics, in the form of mixed mathematics, to the contrary was characterized by a strong theoretical connotation and led to a massive use of mathematics. There were social barriers to overcome, because experimentalist and mathematicians belonged to different communities that had little to share with each other [64].[150] Gradually however experimentalists started to carry out quantitative studies and consequently called for the help of mathematics to interpreter the results they obtained. A fundamental role, both from a theoretical and social point of view to break the division between the two approaches to physics, was the foundation of the École polytechnique in the revolutionary France, at the end of the 18th century.

The other part of the traditional teaching of natural philosophy, that is natural sciences, biology and chemistry were generally delivered in the medical faculties. This in part could be explained by the fact that these sciences, were descriptive in nature, like botany which already had its place in the teaching of medicine. There were also similarity with anatomy, where reference to the parts of bodies resembled the reference to parts of animals, stone, etc. And chemistry was useful to prepare drugs. As the new courses in the different branches were established, they were occupied by specialist professors and working scientists. However the connection between teaching and research was still weak and many scientists were not professors.

As a final comment it can be said that the new science of physics was a creation of the university world. Left to the new scientific academies, physics might have remained associated with traditional natural philosophy and may be disappeared. When the Académie des sciences de Paris was organized into sections in 1699, the new sciences were divided into six categories: three mathematical (geometry, astronomy, and mechanics) and three medical (anatomy, chemistry, and botany). Physics was not among them. Nearly a century later in 1795, when the academy was reconstituted as the first section of the Institute, the new classification of scientific knowledge

[149] p. 64.

[150] p. 66.

mirrored the developments in the universities. Physics constituted a separate section and a clear distinction was made between the mathematical sciences to which physics was considered a part and the experimental or classificatory sciences [37].[151]

2.7 The Technology of Scientific Instruments

Spreading of experimental physics determined a virtuous cycle with the technology of physical models and measuring instruments. On the one hand the request of these objects by the professional researchers and the itinerant lecturers determined the stabilization of a crafted profession, especially in London, which could offer increasingly reliable instruments. On the other hand the possibility of using these instruments enlarged the front of physical research especially in the emerging fields of electricity, magnetism, chemistry, terminology and meteorology.

A clarification needs to be made, the term *scientific instruments* even though quite diffused today, made no sense in the 18th century. A part from that the same words *science* and *scientists* were not in use yet, it also obscures the production and use of instruments for a wide range of activities that would not be considered scientific today. Instruments were employed in many everyday, professional and leisure activities, rather than solely being scientific apparatuses. Early modern instruments were therefore not classed as scientific, but as optical, mathematical, or philosophical. Or even more narrowly for use in individual subjects such as astronomy and natural philosophy or in surveying and navigation. Most mathematical instruments such as drawing and geometric tools, sextants, and globes had a graduated scale for performing calculations or for measuring angles and distances; the use of verniers, already known since the 16th century, was providential. Optical instruments employed lenses or mirrors and included microscopes, telescopes, eyeglasses. Philosophical instruments were used in the demonstration or investigation of natural phenomena, including magnetism, electricity and the attributes of air [8].

Improving instruments meant to improve the easy of use, the endurance and especially the precision (the figures one could read) and the exactness (that is the proximity to the unknown 'exact value'). For what precision is concerned, because the values of any measurements required reading a graduate scale, its increase required the possibility to divide the scales in alway smaller portion. For what exactness is concerned, the situation was more complex. It was of course well known that many factors disturbed the measurement process: the weather, in particular the temperature, the physiology of human body, the eye in particular, the variation of uncontrolled boundary conditions, the accuracy of crafting. To improve an instrument both precision and exactness should be increased; the increase in precision only became embarrassing, because the more precise the measures the easier they where different from each other, and thus doubts on the exactness arose.

[151] pp. 83–84.

A field where the problem of the accuracy of measurements was present since the antiquity was astronomy. By the middle of the 18th century at least one statistical technique was in frequent use here: the arithmetic mean among a collection of measurements made under essentially the same conditions. Astronomers averaged measurements they considered to be equivalent, observations they felt were of equal intrinsic accuracy because the measurements had been made by the same observer, at the same time, in the same place, with the same instrument, and so forth. The problem appeared in a different way when previsional models started to be used, in the 18th century based on Newton's universal gravitation law. These models gave the position of a celestial body or a point in the space in different instant of times, with respect to a limited amount of parameters. The evaluation of these parameters by means of the direct measure of positions is called indirect measurement; it is usually reached by introducing a redundant system of equations (that is more equations than unknowns), that was not easy to solve.

The idea of experimental errors and the application of statistical procedures were developments of the 19th rather than the 18th century. In this latter century there was the spread idea that using a good enough instrument the experimental errors could be reduced to a minimum, close to zero, and that in case of errors the fault fell on the experimenter who was not very accurate and thus guilty. Attention was thus focused on the search for always more precise instruments. Moreover, in the 18th century the experimenters in reporting quantitative data often, in the indirect measurements, uncritically presented long strings of digits when in fact these were merely products of their numerical computations starting from direct measurements possibly of not very high precision (and exactness). Sometimes they announced general conclusions on the basis of astonishingly small bodies of empirical data. Coulomb, for example, in his determination of the law of force between electric charges, presented only three sets of experimental data, see Sect. 4.3.9.2, which did not even fit his proposed inverse-square law very well [64].

Among the first to suggest a systematic solution for indirect measurements were the 'mathematician' Leonhard Euler and the astronomer Tobias Mayer (1723–1762). Over a period of two years Mayer, made numerous observations useful to evaluate the characteristic parameters (3) of the orbit of the moon, whose number was much lower than the number of measurements (27) he made of moon positions. From a mathematical point of view his problem was to find the solution of a system of twenty seven linear equations with three unknown only. Using an *a hoc* method he made use of all the 27 measurements.

Leonhard Euler was concerning instead in the measurements of the three body system Jupiter, Saturn and sun. The parameters to be evaluated were six and the measurements available were seventy five sets composed each of seven values (latitudes, longitudes and so on). He had thus to solve a system of seventy five equations with six unknowns. Differently from Mayer, Euler worked with small sets of equations (usually as many as the unknowns) and only accepted solutions when different small sets of equations yielded essentially the same results [110].[152]

[152] pp. 16–28.

Only at the turn of the 19th century a consistent procedure was found: the least squares method. The first clear and concise exposition of this method was published by Adrien Marie Legendre (1752–1833) in 1805 [75].[153] In 1809 Carl Friedrich Gauss (1777–1855) published his own method of calculating the orbits of celestial bodies. In that work he claimed to have been in possession of the method of least squares since 1795. This naturally led to a priority dispute with Legendre [109].

The problem of an accurate time measurement, very important both from the point of view of pure science and practical applications, was solved in a satisfactory way in the 18th century. The greatest merit in this field is due to the English horologist John Harrison (1693–1776) who in 1735 succeeded in making a satisfactory marine chronometer (controlled by two rockers oscillating in opposite sense, capable of counterbalancing the movements of the ship) to be used for the determination of longitude.

Another area where remarkable progress was made is that of optical devices: just remember that in 1757 the English astronomer and optician John Dollond (1706–1761) managed to construct an achromatic objective, which represented a milestone in this field of technology. The achromatism had already been made object a few years before of extensive studies by Euler that defended against the opinion of Newton the possibility of constructing achromatic lenses, but he did not apply his ideas. Some important improvements were also introduced, in the construction of the microscopes, by the German Ulrich Theodor Aepinus (1724–1802) and by the Dutch officer François Gerardzoon Beeldsnyder (1755–1808). Also the manufacture of reflecting telescopes, already begun in the 17th century underwent important improvements especially toward the end of the 18th century by the great astronomer William Herschel (1758–1822) who made his famous astronomical discoveries precisely with a device of this type.

Turning from optics to thermology, it should be remembered that precisely at the 18th century dated the introduction of the thermometric scales still used today, due to the Dutch Gabriel Daniel Fahrenheit (1686–1736), the French René Antoine Ferchault de Réaumur (1683–1757) and the Swedish Anders Celsius (1701–1744). This introduction was made possible by the discovery that the water boiling temperature is constant if the atmospheric pressure remains constant and the freezing point is substantially invariant with atmospheric pressure. Very advantageous was the employment—operated for the first time in a systematic form by Fahrenheit—of mercury in place of alcohol as thermometric liquid; it allowed the construction of smaller and more manageable thermometers, with which it was possible to determine with greater accuracy than before the course of the heating. Among the many other devices designed for the experimental study of thermology, to remember the first calorimeter built around 1750 by Joseph Black (1728–1799), professor at the University of Glasgow, to measure the amount of heat absorbed in the changes in state, and dilatometers built to determine the dilatations of the metal rods to be used in the construction of watches.

---

[153] The least squares method is referred to in a very clear way in a short appendix with the title: *Sur la méthode de moindres quarrées*, pp. 72–25.

John Harrison had also the merit of having built the first precision scale. Other scales even more sensitive and accurate were built by the French mechanical engineer Pierre Bernard Megnie (1751–1807) to order by Lavoisier . They will constitute the fundamental instrument with which Lavoisier operated his revolutionary discoveries [56].[154]

The development of instruments of measure needed both a manual and a theoretical capacity. Many of the manufacturers that introduced new ideas were 'physicists' or in any case had good knowledge of physics. This is an example of a synthesis between technology and theoretical science. This is also true for the instruments to be used in electricity and magnetism, where the problem at the beginning was to decide which magnitudes should be measured. For instance it would have been impossible to translate the qualitative circulating-vortex theory of magnetism that was widely accepted during the first half of the eighteenth century into a quantitative theory and to decide which were its representative magnitudes, since the then available mathematical hydrodynamics would not have been up to the task. Franklin's theory of electricity, despite its success in rendering the Leyden experiment comprehensible, was insufficiently coherent to sustain quantification. Only when the basic principles of electricity had been cleaned up and rendered mutually consistent by Aepinus, sometimes in ways far removed from Franklin's own conceptions, did electricity become a candidate for quantitative treatment and measurements. Yet even then, because Aepinus could not prove that the law of force between charges was inverse-square in form, he failed to advance beyond a semi-mathematical formulation. Henry Cavendish (1731–1810) in 1771 and Charles Augustin de Coulomb (1736–1806) in the 1780 s took the process somewhat further, but the development of a fully quantitative theory of electricity had to await the work of Siméon Denis Poisson (1781–1840) in the early years of the 19th century [64].[155]

2.8 Quotations

B.1 Mr. Boyle est mort, comme vous seaurez déja sans doute. Il paroit assez étrange qu'il n'ait rien basti sur tant d'expériences dont ses livres sont pleins; mais la chose est difficile, et je ne l'ay jamais cru capable d'une aussi grande application qu'il faut pour establir des principes vraisemblables.

B.2 In questo mentre con la sagacità del suo ingegno invento quella semplicissima e regolata misura del tempo per mezzo del pendulo, non prima da alcun altro avvertita, pigliando occasione d'osservarla dal moto d'una lampada, mentre era un giorno nel Duomo di Pisa; e facendone esperienze esattissime, si accerto dell'egualitàdelle sue vibrazioni, e per allora sovvennegli di adattarla all' uso della medicina per la misura della frequenza de' polsi, con stupore e diletto de' medici di que' tempi e come pure oggi si pratica volgarmente: della quale

[154] vol. 3, pp. 193–195.

[155] p. 273.

invenzione si valse poi in varie esperienze e misure di tempi e moti, e fu il primo che l'applicasse alle osservazioni celesti, con incredibile acquisto nell'astronomia e geografia.

[...]

In questo tempo, parendogli d'apprendere ch' all'investigazione delli effetti naturali necessariamente si richiedesse una vera cognizione della natura del moto, stante quel filosofico e vulgato assioma ignorato *motu ignoratur natura*, tutto si diede alia contemplazione di quello: et allora, con gran sconcerto di tutti i filosofi, furono da esso convinte di falsità, per mezzo d' esperienze e con salde dimostrazioni e discorsi, moltissime conclusioni dell'istesso Aristotele intorno alia materia del moto, sin a quel tempo state tenute per chiarissime et indubitabili; come, tra l'altre, che le velocitàde' mobili dell'istessa materia, disegualmente gravi, movendosi per un istesso mezzo, non conservano altrimenti la proporzione delle gravita loro, assegnatagli da Aristotele, anzi che si muovon tutti con pari velocità, dimostrando ciòcon replicate esperienze, fatte dall'altezza del Campanile di Pisa con l'intervento delli altri lettori e filosofi e di tutta la scolaresca.

B.3 Hebbe pochissima quantitàdi libri, e lo studio suo dependea dalla continua osservazione, con dedurre da tutte le cose che vedea, udiva o toccava, argomento di filosofare; e diceva egli ch'il libro nel quale si dovea studiare era quello della natura, che sta aperto per tutti.

B.4 Or questo è appunto quello che l'anima va tentando nell'investigazione delle naturali cose; e a ciò bisogna confessare che non v'ha miglior mano di quella ella geometria, la quale dando alla bella prima nel vero, ne libera in un subito da ogni altro più incerto e faticoso rintracciamento. Il fatto è, ch'ella ci conduce un pezzo innanzi nel cammino delle filosofiche speculazioni, ma poi ella ci abbandona in sul bello: non perchèla geometria non cammini spazj infiniti, e tutta non trascorra l'università dell'opere della natura, secondo che tutte obbediscono alle matematiche leggi onde l'eterno Intendimento con liberissimo consiglio le governa e le tempera, ma perchè noi di questa si lunga e sì spaziosa via per anche non tenghiamo dietro che pochi passi. Or quivi ove non ci è più lecito metter piede innanzi, non vi ha cui meglio rivolgersi che alla fede dell'esperienza; la quale non altrimenti di chi varie gioie sciolte e sconnesse cercasse di rimettere ciascuna per ciascuna al suo incastro, così ella adattando gli effetti a cagioni e cagioni ad effetti, se non di primo lancio, come la geometria, tanto fa che PROVANDO E RIPROVANDO le riesce talora di dar nel segno.

B.5 Si quelqu'un de cette humeur vouloit entreprendre d'écrire l'histoire des apparences celestes, selon la methode de Verulamius, & que, sans y mettre aucunes raisons ny hypotheses, il nous décrivist exactement le Ciel, tel qu'il paroist maintenant, quelle situation a chaque Etoile fixe au respect de ses voisines, quelle difference, ou de grosseur, ou de couleur ou de clarté, ou d'estre plus ou moins étincelantes, &c.; item, si cela répond àce que les anciens astronomes en ont écrit, & quelle difference il s'y trouve (car ie ne doute point que les Estoiles ne changent tousiurs quelque peu entr'elles de situation, quoy qu'on les estime fixes); aprés cela qu'il y adjoustast les observations des Cometes, mettant une petite table du cours de chacune, ainsi que Tycho a fait de trois ou quatre qu'il a observées; & enfin les variations de l'ecliptique & des apogées des Planetes: ce seroit un ouvrage qui seroit plus utile au public qu'il ne semble peut estre d'abord, & qui me soulageroit de beaucoup de peine.

B.6 La Chymie, par des operations visibles, resout les corps en certains principes grossiers & palpables, sels, soufres, &c. Mais la Physique, par des speculations delicates, agit sur ces principes, comme la Chymie a fait sur les corps, elle les resout eux-memes en d'autres principes encore plus simples en petits corps mus et figures d'une infinite de façons: voila la principale difference de la Physique & de la Chymie.

B.7 NEWTONIANISME, s. m. ou Philosophie Newtonienne, (Physiq.) c'est la théorie du mechanisme de l'univers, & particulierement du mouvement des corps célestes, de leurs lois, de leurs propriétés, telle qu'elle a été enseignée par M. Newton. Voyez Philosophie.

Ce terme de philosophie newtonienne a été différemment appliqué, & delàsont venues plusieurs notions de ce mot. Quelques auteurs entendent par la là philosophie corpusculaire, telle qu'elle a été réformée & corrigée par les découvertes dont M. Newton l'a enrichie. Voyez Corpusculaire. C'est dans ce sens que M. Gravesande appelle ses élémens de Physique, Introductio ad philosophiam newtonianam. Dans ce sens, la philosophie newtonienne n'est autre chose que la nouvelle philosophie, différente des philosophies cartésienne & péripatéticienne, & des anciennes philosophies corpusculaires. Voyez Aristotélisme, Péripatétisme, Cartésianisme, &c. D'autres entendent par philosophie newtonienne la méthode que M. Newton observe dans sa philosophie, méthode qui consiste à déduire ses raisonnemens & ses conclusions directement des phénomenes, sans aucune hypothèse antécédente,àcommencer par des principes simples,à déduire les premieres lois de la nature d'un petit nombre de phénomenes choisis, & à se servir de ces lois pour expliquer les autres effets. Voyez Lois de la Nature au mot Nature. Dans ce sens la philosophie newtonienne n'est autre chose que la physique expérimentale, & est opposée à l'ancienne philosophie corpusculaire. Voyez Expérimentale. D'autres entendent par philosophie newtonienne, celle oùles corps physiques sont considérés mathématiquement, & où la géométrie & la méchanique sont appliquées à la solution des phénomenes. La philosophie newtonienne prise dans ce sens, n'est autre chose que la philosophie méchanique & mathématique. Voyez Méchanique & Physicomathématique.

D'autres entendent par philosophie newtonienne, cette partie de la Physique que M. Newton a traitée, étendue, & expliquée dans son livre des Principes.

D'autres enfin entendent par philosophie newtonienne, les nouveaux principes que M. Newton a apportés dans la Philosophie, le nouveau système qu'il a fondé sur ces principes, & les nouvelles explications des phénomenes qu'il en a déduites.

B.8 Mais cet abus n'est rien en comparaison des inconvéniens où l'on tombe lorsqu'on veut appliquer la Géométrie & le calculà des sujets de Physique trop compliquez, à des objets dont nous ne connoissons pas assez les propriétés pour pouvoir les mesurer; on est obligé dans tous ces cas de faire des suppositions toûjours contraires à la Nature, de dépouiller le sujet de la plépart de ses qualités, d'en faire un être abstrait qui ne ressemble plus à l'être réel, & lorsqu'on a beaucoup raisonné & calculé sur les rapports & les propriétés de cet être abstrait, & qu'on est arrivé à une conclusion toute aussi abstraite, on croit avoir trouvé quelque chose de réel, & on transporte ce résultat idéal dans le sujet réel, ce qui produit une infinité de fausses conséquences & d'erreurs.

B.9 Toute le matière s'attire en raison inverse du carré de la distance, & cette loi générale ne paroît varier, dans les attractions particulières, que par l'effet de la figure des parties constituantes de chaque substances; parce que cette figure entre comme élément dans la distance.

B.10 PHYSIQUE. Cette science que l'on appelle aussi quelquefois Philosophie naturelle, est la science des propriétés des corps naturels, de leurs phénomenes & de leurs effets, comme de leurs différentes affections, mouvevemens, &c. Voyez &c. Voyez Philosophie & Nature. Ce mot vient du grec φύσιζ, nature.

References

1. Anstey PR (2000) The philosophy of Robert Boyle. Routledge, London
2. Anstey PR (2002) Locke, Bacon and natural history. Early Sci Med 7(1):65–92
3. Anstey PR (2005) Experimental versus speculative natural philosophy. In: Anstey PR, Schuster JA (eds) The science of nature in the seventeenth century. Springer, Dordrecht, pp 87–102
4. Anstey PR (2012) Francis Bacon and the classification of natural history. Early Sci Med 7(1–2):11–31
5. Anstey PR, Vanzo A (2012) The origins of early modern experimental philosophy. Intell Hist Rev 22(4):499–518
6. Anstey PR, Vanzo A (2016) Early modern experimental philosophy. In: Justin S, Buckwalter W (eds) A companion to experimental philosophy. Wiley-Blackwell, Malden Massachusetts, pp 87–102
7. Bacon F (1828) Of the proficience and advancement of learning, divine and human. Dove, London
8. Baker A (2012) Precision, perfection, and the reality of British scientific instruments on the move during the 18th century. Revue de la Culture Matérielle, pp 74–75
9. Bazerman C (1988) Shaping written knowledge. The University of Wisconsin Press, Madison
10. Beretta M (2000) At the source of Western science: the organization of experimentalism at the Accademia del cimento (1637–1667). Notes Rec R Soc Lond 54(2):131–151
11. Bertoloni Meli D (2019) Mechanism. A visual, lexical, and conceptual history. University of Pittsburgh, Pittsburg
12. Birch T (1756–1757) History of the royal society of London (4 vols). Millar, London
13. Boas M (1952) The establishment of the mechanical philosophy. Osiris 10:412–541
14. Boas M (1958) Robert Boyle and seventeenth-century chemistry. Cambridge University Press, Cambridge
15. Boas Hall M (1991) Promoting experimental learning. Experiment and the Royal society 1660–1727. Cambridge University Press, Cambridge

16. Boas Hall M (1992) The library and archives of the Royal society 1660–1990. Arrowsmith, Bristol
17. Boerhaave H (1733) Elementa chemiae (2 vols), 2nd edn. Cavelier, Paris
18. Boerhave H (1715) Sermo academicus de comparando certo in physicis. Vander, Leyden
19. Boerhave H (1741) A new method of chemistry. Including the history, theory, and practice of the art (2 vols), 2 edn. Translated into English and edited by Shaw P and Chambers E. Longman, London
20. Boschiero L (2007) Experiment and natural philosophy in seventeenth-century Tuscany. Springer, Dordrecht
21. Boyle R (1661) The sceptical chymist or chymico-physical doubts & paradoxes. Cadwell, London
22. Boyle R (1666) Hydrostatical paradoxes made out of new experiments. Hall, London
23. Boyle R (1666) The origine of formes and qualities, according to the corpuscular philosophy, Illustrated by considerations and experiments. Hall, Oxford
24. Boyle R (1682) A defence of Mr. R. Boyle's explications of his physico-mechanical experiments, against Franciscus Linus. In: Boyle R (ed) New experiments physico-mechanical touching the air: Whereunto is added a defence of the author's explication of the experiments against the objections of Franciscus Linus and Thomas Hobbs (third edition), Flesher, London, pp 1–89
25. Boyle R (1682) New experiments physico-mechanical touching the air: Whereunto is added a defence of the author's explication of the experiments against the objections of Franciscus Linus and Thomas Hobbs (third edition). Flesher, London
26. Boyle R (1690) The Christian virtuoso; shewing, that by being addicted to experimental philosophy, a man is rather assisted, than indisposed, to be a good Christian. Jones, London
27. Boyle R (1725) The philosophical works of the honourable Robert Boyle (3 vols). Innys, London
28. Boyle R (1772) The experimental history of colours begun. In: Birch T (ed) The works of honourable Robert Boyle (6 vols). A new edition, vol 1, Rivington et als, London, pp 668–788
29. Boyle R (1772) Free inquiry into the vulgarly received notion of nature. In: Birch T (ed) The works of honourable Robert Boyle (6 vols). A new edition, vol 2, Rivington et als, London, pp 106–142
30. Boyle R (1772) The general history of the air designed and begun. In: Birch T (ed) The works of honourable Robert Boyle (6 vols). A new edition, vol 5, Rivington et als, London, pp 609–750
31. Boyle R (1772) Medicina hydrostatica: or, hydrostatics applied to the materia medica. In: Birch T (ed) The works of honourable Robert Boyle (6 vols). A new edition, vol 5, Rivington et als, London, pp 453–505
32. Boyle R (1772) Some considerations touching the usefulness of natural philosophy. In: Birch T (ed) The works of honourable Robert Boyle (6 vols). A new edition, vol 2, Rivington et als, London, pp 1–201
33. Boyle R (1772) The works of honourable Robert Boyle (6 vols) A new edition. Birch T (ed). Rivington et als, London
34. Boyle R (1999–2000) In: Hunter M, Davis EB (eds) The works of Robert Boyle (14 vols). Pickering & Chatto, London
35. Boyle R (2019) Robert Boyle Project, the website also includes a section, Boyle's manuscripts online, which presents digital images of the most important volumes of the Boyle papers at the Royal society. http://www.bbk.ac.uk/boyle/
36. Boyle R (2020) Requisites of a good hypothesis, the Robert Boyle Project. http://www.bbk.ac.uk/boyle/papers/
37. Brockliss L (2003) Science, the universities, and other public space: teaching science in Europe and Americas. In: Porter R (ed) The Cambridge history of science. Eighteenth-century science, vol 4. Cambridge University Press, Cambridge, pp 44–66
38. de Buffon GLL (1749–1789) Histoire naturelle, générale et particulière, avec la description du Cabinet du roi (36 vols). Imprimerie Royale, Paris

39. de Buffon GLL (1797) Buffon's natural history (10 vols). Sysmond, London
40. Capecchi D (2014) The problem of motion of bodies. Springer, Cham-Dordrecht
41. Capecchi D (2018) The path to post-Galilean epistemology. Springer, Cham-Dordrecht
42. Cohen IB (1956) Franklin and Newton. Harvard University Press, Cambridge
43. Cusanus N (1996) Nicholas of Cusa on wisdom and knowledge. Translated into English by Hopkins J. Benning, Minneapolis
44. Dallowe T (1735) Elements of chemistry: being the annual lectures of Herman Boerhaave (2 vols). Translated into English and edited by Dallowe T. Pemberton et als, London
45. De Fouchy G (1763) Eloge de M. le Marquis Poleni. Histoire de l'Académie Royale des Sciences de Paris, pp 151–163
46. Dear P (1995) Discipline & experience. Cornell University Press, Ithaca
47. Desaguliers J (2019) Mactutor history of mathematics archive. http://www-history.mcs.st-and.ac.uk/index.html
48. Desaguliers JT (1727) An account of a book entitul'd vegetable staticks: or an account of some statistical experiments on the sap in vegetables. Philos Trans R Soc Lond 34(398):264–291
49. Desaguliers JT (1734–1744) A course of experimental philosophy (2 vols). Longman et als, London
50. Descartes R (1964) Oeuvres de Descartes; nouvelle édition complétée (1896–1913) (11 vols). In: Adam C (ed) Tannery P. Vrin, Paris
51. Ducheyne S (2017) Different shades of Newton: Herman Boerhaave on Newton mathematicus, philosophus, and opticochemicus. Annals of Science 74(2):108, 125
52. Encyclopédie (1751–1772) Encyclopédie ou dictionnaire raisonné des sciences, des arts et des métiers, par une societè de gens de lettres (17 vols). Briasson-David-Le Breton-Durand, Paris
53. de Fontenelle B (1669) Physique/chimie. Mémoires de l'Académie Royale des Sciences depuis 1666 jusqu'en 1699, pp 53–54
54. Galilei G (1890) Opere di Galileo Galilei linceo (4 vols). Eredi del Dozza, Bologna
55. Galilei G (1890–1909) In: Barbera FA (ed) Le opere di Galileo Galilei (National edition) (20 vols)
56. Geymonat L (1983) L'esigenza di una più ampia sperimentazione nelle scienze della natura. In: Geymonat L (ed) Storia del pensiero scientifico e filosofico. 9 vols, vol 3, Garzanti, Milan, pp 192–215
57. Guerlac H (1973) Hales, Stephen. In: Gillispie CC (ed) Complete dictionary of scientific biography. Scribner, New York
58. Hales S (1727) Vegetable staticks; or, an account of some statistical experiments on the sap in vegetables. Innys, London
59. Hales S (1738) Statistical essays (2 vols), 3rd edn. Innys, London
60. Hall R (1970) Desaguliers, John Theophilus. In: Gillispie CC (ed) Complete dictionary of scientific biography. Scribner, New York
61. Hattab H (2011) The mechanical philosophy. In: Clarke DM, Wilson C (eds) The Oxford handbook of philosophy in early modern Europe. Oxford University Press, New York, pp 71–95
62. Hauksbee F (1708) Touching the difficulty of separating two hemispheres upon the injecting of an atmosphere of air in their outward surface, without withdrawing the included air. Philos Trans R Soc Lond 25(305):2415–2417
63. Heilbron JL (1979) Electricity in the 17th and 18th centuries. University of California Press, Berkley
64. Home R (2003) Mechanics and experimental physics. Eighteenth-century science. In: Porter R (ed) The Cambridge history of science. Cambridge University Press, Cambridge, pp 354–374
65. Hooke R (1665) Micrographia. Martyn & Allestry, London
66. Hunter M (ed) (1994) Robert Boyle reconsidered. Cambridge University Press, Cambridge
67. Hunter M (1995) Science and the shape of orthodoxy: intellectual change in late seventeenth-century Britain. Boydell, Woodbridge

68. Hunter M (2007) Robert Boyle and the early Royal society: a reciprocal exchange in the making of Baconian science. Br J Hist Sci 40(1):1–23
69. Huygens C (1888–1950) Oeuvres complètes de Christiaan Huygens (22 vols). Nijhoff, The Hague
70. Kepler J (1937) Johannes Kepler Gesammelte Werke (23 vols). Beck, Munich
71. Kerker DH (1955) Herman Boerhaave and the development of pneumatic chemistry. Isis 46(1):36, 49
72. Knoeff R (2002) Herman Boerhaave (1668–1738). Calvinist chemist and physician. Royal Netherlands Academy of Arts and Sciences. Amsterdam
73. Kubbinga H (1988) Newton's theory of matter. In: PB S, G D (eds) Newton's scientific and philosophical legacy, Kluwer, Dordrecht, pp 321–342
74. Kuhn T (1952) Robert Boyle and structural chemistry in the seventeenth century. Isis 43(1):12–36
75. Legendre A (1805) Nouvelles méthodes pour la détermination des orbites des comètes. Didot, Paris
76. Linch W (2001) Solomon's child. Method in the early Royal Society of London. Stanford University Press, Stanford
77. Lindeboom G (1970) Boerhaave, Hermann. In: Gillispie CC (ed) Complete dictionary of scientific biography. Scribner, New York
78. Lyon J (1976) The 'Initial discourse' to Buffon's "Histoire naturelle": the first complete English translation. J Hist Biol 9(1):133–181
79. Magalotti L (1691) Saggi di naturali esperienze fatte nell'accademia del Cimento. Second edition. Cecchi, Florence
80. Magalotti L (1841) Saggi di naturali esperienze. Tipografia Galileiana, Florence
81. Mariotte E (1740) Essais de physique, ou mémoires pour serviràla science des eboses naturelles. Second essai. De la nature de l'air. In: Mariotte E (ed) Oeuvres de Mariotte. Nouvelle edition, vol I, Neaulme, La Hat, pp 148–182
82. McClauglin T (1996) Was there an empirical movement in mid-seventeenth century France? Experiments in Jacques Rohault's Traité de physique. Revue d'Histoire des Sciences 49(4):459–481
83. Metzger H (1974) Newton, Stahal. Boerhaave et la doctrine chimique, Blanchard, London Paris
84. More H (1659) The immortality of the soul, so farre forth as it is demonstrable from the knowledge of nature and the light of reason. Flesher, London
85. van Musschenbroek P (1731) Tentamina experimentorum naturalium captorum in Academia del Cimento sub auspiciis serenissimi principis Leopoldi Magni Etruriae Ducis et ab ejus academiae secretario conscriptorum: ex Italico in Latinum sermonem conversa. Quibus commentarios, nova experimenta, et orationem De methodo instituendi experimenta physica addidit Petrus van Musschenbroek. Verbeek, Leyden
86. Newman W (1996) The alchemical sources of Robert Boyle's corpuscular philosophy. Ann Sci 53(6):567–585
87. Newman W (2014) Robert Boyle, transmutation, and the history of chemistry before Lavoisier: a response to Kuhn. Osiris 29(1):63–77
88. Newton I (1730) Opticks: Or, a treatise of the reflections, refractions, inflections and colours. Innys, London
89. Nollet J (1754–1765) Leçons de physique expérimentale (6 vols). Arksté and Merkus, Amsterdam & Leipzig
90. Paciucchi M (2010) Il lessico della meccanica dei solidi fra settecento e ottocento. Aracne, Rome
91. Poleni G (1741) Institutionum philosophiae experimentalis specimen. Stamperia del Seminario, Padua
92. Roger J (1977) Buffon and mathematics. In: Miles R, Serra J (eds) Proceedings of the Buffon bicentenary symposium on geometrical probability, image analysis, mathematical stereology, and their relevance to the determination of biological structures, Paris, pp 29–36

93. Rohault J (1671) Traité de physique. Thierry, Paris
94. Rohault J (1723) System of natural philosophy. Illustrated with Dr. Samuel Clarke's notes (2 vols). Knapton, London
95. Roller DH (1960) Stephen Hales and quantitative mechanism. Bios 31(4):195, 204
96. Roux S (2013) Was there a Cartesian experimentalism in 1660s France? In: Dobre M, Nyden T (eds) Cartesian empiricism. Springer, Dordrecht, pp 47–88
97. Rowlinson J (2002) Cohesion. A scientific history of intermolecular forces. Cambridge University Press, Cambridge
98. Rowning J (1737-1743) A compendious system of natural philosophy (4 parts). Harding, London
99. Royal Society of London (1940) The record of the Royal Society of London for the promotion of natural knowledge. Morrison and Gibb, London
100. s' Gravesande J (1747) Mathematical elements of natural philosophy confirmed by experiments, or an introduction to Sir Isaac Newton's philosophy (2 vols), 6 edn. Innys et als, London
101. Schofield R (1970) Rowning, John. In: Gillispie CC (ed) Complete dictionary of scientific biography. Scribner, New York
102. Segre M (1989) Viviani's life of Galileo. Isis 80(2):206–231
103. Shapin S (1988) Robert Boyle and mathematics: reality, representation, and experimental practice. Sci Context 2(1):23–58
104. Shapin S, Shaffer S (1985) Leviathan and the air-pump. Hobbes, Boyle, and the experimental life. Princeton University Press, Princeton
105. Sorrenson R (1996) Towards a history of the Royal society in the eighteenth century. Notes Rec R Soc Lond 50(1):29–46
106. Spink A (2018) The experimental physics of Jacques Rohault. Br J Hist Philos 26(5):850–870
107. Sprat T (1667) The history of the Royal society of London for the improving of natural knowledge. Martyn & Allestry, London
108. Spuit L (2019) The transit of science and philosophy between the dutch republic and italy: the case of newtonism, inTRAlinea. Transit and translation in early modern Europe, Special issue
109. Stigler S (1981) Gauss and the invention of least squares. Ann Stat 9(2):465–474
110. Stigler S (1986) The history of statistics. The Belknap Press of Harvard University Press, Cambridge Massachusetts
111. Thomson T (1930) History of chemistry (2 vols). Colburn and Bentley, London
112. van Besouw J (2017) Out of Newton's shadow: An examination of Willem Jacob 's Gravesande's scientific methodology. PhD thesis, Vrije univerity of Brussel
113. Waller R (1684) Essayes of natural experiments made in the Academie del cimento. Alsep, London
114. Webster C (1968) The origins of the Royal society. Hist Sci 6:106–127
115. Wojcik JW (1997) Robert Boyle and the limits of reason. Cambridge University Press, Cambridge

Chapter 3
Classical Mechanics

Abstract The evolution of *mechanics* towards what is now called *classical mechanics* is explored in depth. The exposition of fundaments and applications in various treatises and papers are examined, especially those of continental scientists and in particular the syntheses of Euler and d'Alembert, who described the way mechanics could be made a rational discipline, like mathematics, based on Calculus. The complete mathematization of mechanics only occurred at the end of the 18th century however, with Lagrangian synthesis which is briefly summarized, concentrating more on the relations between physics and mathematics than to the technical aspects, which, however, are very important. The justification by the scholars of the 18th century of the foundations of mechanics, required substantial involvement in metaphysics and epistemology to introduce fundamental notions: the nature of space, time, force, constitution and properties of bodies, nature of motion. However, this effort was pursued not with the classical and organic approach of canonical philosophy but with the pragmatism of mathematical philosophers. Reading the chapter, for the nature of the subject, requires a basic knowledge of mathematics; to to make the text smoother the more technical aspects are left to notes.

3.1 Mechanics and Natural Philosophy at the Turn of Century

The term *mechanics* at the turn of the 18th century had already a meaning close to the modern one: the science of equilibrium and motion. Still the ancient meaning of science of simple machines, or statics, persisted and the science of motion was considered separately, as *dynamics*, a term Leibniz contributed to spread. The logical status of mechanics had always been particular among the traditional mixed mathematics. Aristotle, for instance, added it to the list of subordinate sciences (the term he used for mixed mathematics) at last, as a limiting case. There were sociological reasons. Mechanics, although in the purified version that mathematicians gave it, referred to practical or technological activities, not highly esteemed by the 'intellectuals' of the time. Which was not true for astronomy and music that were reserved for the elite, and not even for optics, which had light as reference, the noblest of

D. Capecchi, *Epistemology and Natural Philosophy in the 18th Century*, History of Mechanism and Machine Science 39,
https://doi.org/10.1007/978-3-030-52852-2_3

the substances. There were epistemological reasons. Astronomy and optics studied, all considered, geometric objects: lines and curves. Music was based on arithmetic ratios between the pitch of sounds. The insights that gave raise to these disciplines had their origin in the senses; sight for optics and astronomy, hearing for music.

Mechanics referred to a quality that was certainly revealed to us, but in a more abstract way: weight. Of it we could have intuition for the interaction that aroused on our body; with touch when a heavy object weighs on us and with muscular effort when we try to lift this object. Weight however could hardly be reduced to geometry or arithmetic. Even the science of motion, the modern part of mechanics, presented similar problems; it required the use of a somewhat even more mysterious and difficultly quantifiable magnitude: time. And until the 16th century motion remained substantially in the sphere of natural philosophy and was not studied by mathematicians. In the 16th century they began to take an interest in the subject for various reasons related in large part to technological development. Among the most interesting contributions were those of Niccolò Tartaglia—with his ballistics studies—and Giovanni Battista Benedetti—with his studies on the natural motion of heavy bodies. In the 17th century Galileo succeeded in bringing back an important part of the science of motion, the study of the fall of bodies, in the field of mathematics, with an approach that however was long object of criticisms both of merit and method. With Newton finally, mechanics (old mechanics and dynamics) left the restricted area of the traditional mixed mathematics that dealt with particular problems only, even if complex, and became a discipline with a cosmological character with the ambition to face all the problems of motion (and equilibrium) of both sublunar and celestial worlds. In this way the traditional natural philosophy was completely replaced by the new science in the study of motion.

Paradoxically, mechanics, the least mathematical among classical mixed mathematics, changed its status in the 18th century and became the closest to mathematics. This was possible not only for a better understanding of mechanical processes, but also for a change in the epistemology of science. This saw a gradual shift from the typical approach of mixed mathematics (with a partial exception of astronomy and music) that starting from 'certain' principles—either because derived from more or less direct experience or from the conclusions of natural philosophy and metaphysic reasonings—passed to an approach in which certainty was replaced by plausibility. An approach that will developed in the modern hypothetical deductive method. The justification of the principles—and in this astronomy had been a model—was based on the 'truth' of the theorems derived from these principles through mathematics. Sometimes the interest of mathematicians for the adherence to reality of their theorems passed into the background compared to elegance and rigor they obtained. Moreover, a theoretical core was enucleated, based on general relations, always valid. These general relations were made special only when one wanted to study concrete problems, such for instance the fall of heavy bodies and the motion of the planets around the sun.

Most mathematicians of the 18th century had a good knowledge of the philosophy of nature; they came from schools where there was a teaching largely based on philosophy of nature, if not properly the Aristotelian, at least the mechanist one. They could move so as natural philosophers. Less careful, compared to the old canonical natural philosophers, to the coherence and elegance of their constructions, that were sometimes a miscellany of the various schools, from which they picked up the elements considered more suitable to the development of their mechanical and physical theories. Thus, not always philosophy drew mechanics, but sometimes was mechanics which drew philosophy. Main interaction of mechanics with philosophy was not however with natural philosophy but rather with metaphysics and epistemology: the role of mathematics, concepts of space, time and force, the nature of constraints, and so on.

At the turn of the 18th century mechanics had not yet a well defined orientation, as well as not even its ambits were very clear. Three main paths can be enucleated. The first, that can be traced back to Galileo (and Newton), according to which the object of mechanics was the study of the variation of velocity of a body after an impressed force. When the motion develops on the three dimensional space this approach gave raise to a vector mechanics which saws as a main character the vector (modern term) force. The mathematical theory more suitable for completing the study of an isolated or a set of mass points was Calculus. The second path, not completely distinct from the first, focused on the phenomenon of impact among either hard or elastic bodies. It was still a vector mechanics, which had as its main object the vector velocity; its interest was motivated by the dominant mechanical philosophy of the time. The suitable kind of mathematics was (linear) algebra. The third path, saw at its center work and energy (modern terms). The kind of mathematics suitable for its study was algebra, but also Calculus was called for non completely elementary situations.

3.2 The Spreading of Calculus

Mathematicians of the 18th century who gave a substantial contribution to the development of mechanics were a restricted handful; a main reason for this is that in the particular historical moment the development of mechanics was strictly connected with the new born Calculus and the mathematicians who could handle it were very few.

If in physics the 18th was the century of electricity, in mathematics it was the century of differential and integral calculus, simply *Calculus*. Even though its fundaments were laid in the previous century it was the 18th century to improve and use it in a massive way. Great was its influence on the development of physics; this is quite natural because the kind of mathematics influenced always the representation of the external world, at least into two ways. On the one hand mathematics, as an instrument, describes the reality in the limits possible for its language. On the other hand, when the instrument is powerful, as it happens for Calculus, the temptation to apply it to all the aspects of reality is great. And where there are difficulties to describe the reality, this is changed and adapted, by introducing models suitable to mathematics. In the case of Calculus the main imprinting was for sure the character of continuity that should have the model to be treated analytically, both in time and space. Besides this aspect connected to profound properties of Calculus, there was another aspect, apparently trivial but that instead revealed to be fundamental: its algorithmic nature. Calculus in its essence has an algebraic nature and is based on relations of equality which leads to equations. The solution of these equations, that are derived from the principles of a science reduced to mathematical expressions, is carried out with well codified algorithms and all considered simply to apply, though the calculations involved may be tremendous. They sometimes led to unexpected solutions that the use of geometry could not predict because here, for any problem, the solutions were partially prefigured by visual intuition.

By comparing the treatises and papers of mechanics of the 18th century is immediately evident the denial of the rhetorical exposition typical of the 17th century, where apparently there was not so great difference in form of reasoning between natural philosophy and mathematics (geometry). The exposition of the principles has now as its main purpose the setting up of one or more differential equations. Most of the text is made of mathematics.

The received view ascribes the invention of the infinitesimal calculus (both differential and integral) to the most famous mathematicians of the 17th century, Newton and Leibniz; well known is the quarrel on the priority in which the two scholars where involved. This is of course an oversimplification and it should be more correct to say that Newton and Leibniz gave a very important contribution to Calculus. Indeed the infinitesimal calculus found its roots in ancient times and not only in the West but also in the East, India and Arabia. For the western civilization it saw its origins in the difficulties to deal numerically with the continuous magnitudes of geometry and in the method of exhaustion by Eudoxus and Archimedes. More recently with the method of indivisibles of Cavalieri, then in the works of Torricelli and later of Pascal, Descartes, Roberval, Barrow and especially Pierre de Fermat (1601–1665). For a general view of the history of infinitesimal calculus see the old but still convincing [23].

Newton and Leibniz had the chance to arrive last and the great merit to understand the potentiality of Calculus and to give it an algorithmic aspect. They developed approaches based on similar concepts but with a different language. Newton came first, but the route traced by Leibniz prevailed in the long run. Both for intrinsic and substantial reasons. Leibniz's version was simpler to handle with, especially for the introduction of the concept of partial derivatives, necessary to deal with continuous systems. Moreover it found in Jakob and Johann Bernoulli careful readers that clarified and expanded the quite cryptic Leibniz's writings. Other mathematicians completed the work, among which Varignon, Hermann, Euler, d'Alembert and others. Newton did not find analogous supporters and the Calculus at the beginning of the century was a continental affair. Its spreading in Europe was largely due to Johann Bernoulli and Malebranche. The former for his technical ability, the latter for his ability as cultural manager.

Nicolas Malebranche (1638–1715) has been considered by modern historians essentially as a philosopher with a strong metaphysical vocation. Actually, as for many other scholars of the 17th and 18th centuries, such as Descartes and Leibniz, he was a more articulated character. He had a good mathematical culture and greatly appreciated the role of mathematics in philosophy [138]. For this attitude he was the reference for many French mathematicians, the so called Malebranchians, or the members of Malebranche's circle; among them: de l'Hôpital, Reyneau, Jaquemet, Byzance, Bernard Lamy, Varignon, Carré, Rémond de Montmort, Sauveur, Saurin, Guisnée, Renau d'Élisagaray, Fontenelle and Polignac, Nicole, Privât de Molières, etc. They united their efforts against the traditional Cartesians, taking Leibniz for reference and Johann Bernoulli "à leur service" [151].[1] Indeed if the true inventors of Calculus were the Leibnizians, the Malebranchians were the popularizers and helped to spread Calculus in the whole Europe.

In particular two brilliant members of Malebranche's circle, Pierre Varignon (1654–1722) and Guillaume François Antoine de Sainte Mesme, marquis de l'Hôpital (1661–1704) were directly taught by Johann Bernoulli. Who also taught Pierre Maupertuis, Alexis Clairaut, Christian Goldbach, Samuel-Henri König and Euler. Jakob Hermann (1678–1733) was taught instead by Jakob Bernoulli. In particular de L'Hôpital, under Johann Bernoulli's payed tutoring,[2] wrote a fundamental treatise *Analyse des infinitement petits, pour l'intelligence des lignes curbes*, published anonimously in 1696. The treatise was successful and followed by a second edition in 1716. Varignon made clarification and addition publishing a his version.

L'Hôpital treatise is limited to differential calculus. For what concerned integral calculus de Hôpital in his preface, explained that he had also intended to present this, however, Leibniz, having written him that he was working on the subject, he "took care not to deprive the public of such a beautiful work" [2].[3] Unfortunately, Leibniz never completed his job.

In the following I will present a short account of l'Hôpital's book so that the reader can have an idea about the knowledge on the matter at the turn of the 18th century. To start with, the content:

1. The rules of this calculus.
2. Use of the differential calculus[4] for finding the tangents of all kinds of curved lines.
3. Use of the differential calculus for finding the greatest and the least ordinates, to which are reduced questions de maximis & minimis.
4. Use of the differential calculus for finding inflection points and cusps.
5. Use of the differential calculus for finding evolutes.
6. Use of the differential calculus for finding caustics by reflection.
7. Use of the differential calculus for finding caustics by refraction.
8. Use of the differential calculus for finding the points of curved lines that touch an infinity of lines given in position, whether straight or curved.
9. The solution of several problems that depend upon the previous methods.

---

[1] p. 287.

[2] On the relationship between l'Hôpital and Johann Bernoulli see for instance [128], pp. VII–XVI.

[3] Preface.

[4] Note that l'Hôpital said "calcul des differences" instead of differential calculus.

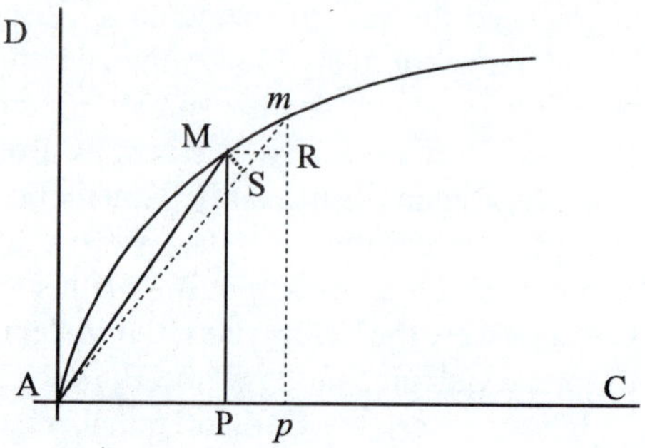

Fig. 3.1 Definition of differential. Redrawn from [2], Plance 1, Fig. 1

10. A new method for using the differential calculus with geometric curves, from which we deduce the method of Messrs. Descartes and Hudde [2].[5]

The fundamental and primitive concept of the differential calculus is not that of derivative as it is today, but that of differential. It is defined in a intuitive way, without recourse to the concept of limit:

> The infinitely small portion by which a variable quantity continually increases or decreases is called the Difference. For example, let AMB be an arbitrary curved line (Fig. 3.1) which has the line AC as its axis or diameter, and has PM as one of its ordinates. Let *pm* be another ordinate, infinitely close to the first one. Given this, if we also draw MR parallel to AC, and the chords AM, A*m*, and describe the little circular arc MS of the circle with center A and radius AM, then P*p* is the differential of AP, R*m* the differential of PM, S*m* the differential of [the straight line] AM, and M*m* the differential of the arc AM. Furthermore, the little triangle MA*m*, which has the arc M*m* as its base is the differential of the segment AM [the region contained by the straight line AM and the arc AM], and the little region MP*pm* is the differential of the region contained by the straight lines AP and PM, and by the arc AM [2].[6] (C.1)

Note that the definition of the differential is very general; it refers to variation of any quantities, angles, arcs, areas, segments. There is not the distinction between dependent and independent variables, but only *variables quantities*, indicated with the letters x, y, z &c, and constants. There is not a clearly defined coordinate system and not an explicit concept of function. In the frequent geometrical representations, used mainly for explanatory purpose, there are two axes. The coordinates on the horizontal axis are usually referred to as abscissas, but they are not alway the independent variable (a concept made not explicit). The coordinates on the other axis, usually perpendicular to the horizontal axis, are named ordinates; although oblique ordinates were occasionally arranged at a different angle to the axis.

L'Hôpital then gave the rules for the differential calculus:

1. The differential of a constant is 0.
2. The differential of the sum $x + y$ is $dx + dy$.
3. The differential of the product xy is $ydx + xdy$.

---

[5]Table of content. English translation in [128].

[6]p. 2. English translation in [128].

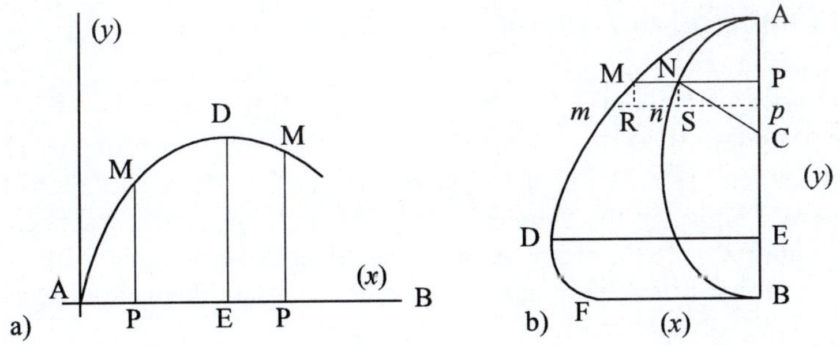

Fig. 3.2 Maxima and minima. Redrawn from [2], Plance 4, Figs. 35 and 36

4. The differential of the quotient x/y is $(ydx - xdy)/yy$.
5. The differential of a power x^n is nx^{n-1} [2].[7]

Only polynomial expressions are present in the treatise; sinusoidal functions and logarithms, well known at the time, are missing.

One of the most important problems dealt with in the l'Hôpital's treatise is the one concerning maxima and minima (the greatest and the least ordinates); attention is focused on geometrical curves. With reference to Fig. 3.2a, the problem is formulated as follows: "Given the nature of the curved line MDM, we wish to find a value AE of AP such that the ordinate ED is the greatest or the least of its similar ordinates PM" [2].[8]

The solution is associated[9] to points where the differential of the ordinate, dy is zero.[I] In the case of point D of Fig. 3.2b, where a modern sees a maximum assuming the vertical coordinate as independent variable, l'Hôpital imposed $dy = \infty$ instead of $dx = 0$, as a modern could have done.

Of a certain interest is the introduction and use of higher order differentials, which presents some difference with respect to the modern use, at least for notation. This is the definition

> Definition I. The infinitely small portion by which the difference of a variable quantity continually increases or decreases is called the difference of the difference of this quantity, or else its second difference [2].[10] (C.2)

The second order differentials are indicated either as ddx or ddy (modern use d^2x or d^2y), to distinguish them from the power of a differential, dx^2 or dy^2.[II]

---

[7] pp. 3–7.

[8] p. 41.

[9] Labels (x) and (y) are mine.

[10] p. 55.

3.2.1 First Uses in Dynamics

First applications of Calculus to dynamics are commonly attributed to Newton. Probably not at the time of the first edition of the *Principia*, but shortly after in the attempt to bring back into the algebraic language of fluxions the differential geometry of his masterpiece. Leibniz in the *Tentamen de motu coelestium causis* appeared in the Acta Eruditorum in 1689, two years after the release of Newton's *Principia* [122], presented the first differential equation of motion in the language of Calculus; he was not able to solve it however.[III]

It is not far from truth saying that the first serious attempts to write down and solve the equations of motions using the language of Calculus are due to Varignon, who moved in the route that will affirm in the 18th century. He knew very well Calculus, even though probably did not contribute much to improve it. This is Johann Bernoulli's biting comment: "I am not sure who, after L'Hôpital's death, is now versed in mathematics in France. Certainly I do not know anyone there today who excels in the most profound questions, except perhaps Varignon, from which, however, one should not expect much progress. He understands other people's results and knows how to perfect them, but is not able to invent. You would say that he is a good commentator, but not a true author" [126].[11]

Varignon's reference in dynamics, at least officially, was more Galileo than Newton. Of Galileo two are the propositions used by Varignon as well as by many scholars at the turn of the 18th century: the proportionality of speed with respect to time, and the law of the squares of times for the height of fall. These two relations were often generalized to the case of forces other than weight, by introducing the ratio between force and weight. This generalization was made simple by the introduction of Calculus. In particular the proportionality between speed and time, $v \propto t$, becomes $dv = adt$ with a a constant, by assuming the proportionality holds good also for infinitesimal intervals of time; it can easily be integrated to give $x = 1/at^2$. This result is obtained in a straightforward way, inside a well established algorithm, without the use of complex reasonings involving the nature of space and time. If the cause of motion is different from gravity but due to a force of intensity f, it is immediate and intuitive to replace a with f and write $dv = f/mdt$, with m the weight. It should be said that the forces other than weights were essentially of two kinds in the 18th century, the forces due to elasticity (and the spring model was very present) and the centripetal force among the celestial bodies, which is proportional to the mass but depends on the mutual distance.

In three memories probably written around 1693, Varignon published results somehow connected to the applications of Calculus to dynamics; the way of exposition recalls Wallis in his *Mechanica sive de motu* of 1669–1671 [27],[12] who is indeed praised. In the first memoir, *Regles du mouvement en general* [163], he stated the following principle or *fundamental rule*:

---

[11] Johann Bernoulli to Leibniz, 21st June 1704, Band III/2, p. 755.

[12] pp. 235–240.

> In all motions [...], either they are uniform, accelerated or retarded [...], the summation of forces which give motion is always proportional to the summation of the spaces passed by all points [163].[13] (C.3)

The principle is given without any justification, but considering that here for force of a body Varignon intended mv, that is the product of mass and speed,[14] the fundamental rule can easily proved by using the integral calculus. The sum of force is $\int mvdt$, which of course is equal to ms and thus proportional to s, the space passed.

The second paper *Regles du mouvement accelerez suivant toutes les proportions imaginables d'acceleration ordonnes* [162] is a simple generalization of the fundamental rule. More interesting is the third paper *Application de la regle generale des mouvements à toute les hypotheses possible d'accelerations ordonnes dans la chute des corps* [161]. Even here the recourse to Calculus is hidden, but there is an important result that will be the basis of subsequent work. In the paper Varignon introduced a different interpretation of the force/weigth, or at least an ambiguous one, arriving to the equality [161]:[15]

$$eblf^2d^2 = gakh^2c^2$$

which holds between two different bodies falling from two different inclined planes, to read with the following table:

body	mass	weight	length	height	time
1	e	a	f	k	c
2	g	b	h	l	d

where *length* and *height* refer to the inclined planes and *time* is the time needed for the body to come down.

The previous relation can be easily justified using the Galilean law of fall. Mass and weight are apparently redundant, because they are not required in the Galilean law; but they reveal fundamental when Varignon attempted to apply his expression to forces more general than weight.

In the years 1700–1701, Varignon presented four memoirs on dynamics to the academy of Paris more directly connected with Calculus [19];[16] of them I will discuss *Maniere generale de determiner les forces, les vitesses, les espaces, et les temps, une seule de ces quatre choses étant donnée dans toutes sortes de mouvemens rectilignes variés à discretion* [166] and *Des forces centrales ou des pesanteurs necessaires aux planetes pour leur faire decrire les orbes qu'on leur a supposez jusq'ici* [165]. In the first memoir Varignon came back to Proposition 39 of the first book of the *Principia*, using the language of Calculus. Here Newton had given diagrams relating space,

[13]p. 226.

[14]In Sects. 3.2 and 3.3 the variation of space for unit of time is called speed and not velocity, becasue we are concerned with a scalr magnitude.

[15]p. 359.

[16]p. 119.

force, velocity, time, for different expressions of force as function of distance using a geometric approach, in which the solution was first given and then checked, with an inverse procedure (modern meaning) [140].[17]

Varignon started his enterprise by expressing the instantaneous speed v as the ratio between the infinitesimal space dx and infinitesimal time dt: $v = dx/dt$, an apparently innocuous move but with a great heuristic power. The expression is differentiated by Varignon, assuming dt as constant, arriving to the expression $dv = ddx/dt$. Then he passed to a mechanical relation, the generalized Galilean law: "For the spaces passed by a body pressed by a constant and continuously applied force [y], as we usually conceive gravity, being in the composite ratio of this force and the square of times, one will have $ddx = ydt^2$" [166].[18]

Two things should be noted here; first the assumption that spaces passed are proportional to forces impressed. This requires a concept of mass, which however is kept hidden because not necessary to deal with, in the case of the motion of a single mass point. Second the Galilean law is applied to an infinitesimal interval of time dt—so that y can be assumed constant—and, with a certain nonchalance, the space passed is qualified as a second order differential and the factor $1/2$ is ignored. In the end Varignon could write two general rules of motion along a straight line:

$$v = \frac{dx}{dt}, \qquad y = \frac{dv}{dt} = \frac{ddx}{dt^2}$$

These rules were used in different situations. For instance if the relation $v - x$ is known, given by the Galilean law $v = \sqrt{x}$ for instance, one has the equation $v = \sqrt{x} = dx/dt$, which can be rewritten as $dx/\sqrt{x} = dt$, that integrated gives $x \propto t^2$. In the case of variable forces, Varignon combined his two rules, to obtain the relation $ydx = vdv$, which he integrated to give $vv = 2\int ydx$, or $v = \sqrt{2\int ydx}$, "as Newton proved in the first part of this 39th proposition" [166],[19] which furnishes a relation between x and v. To obtain a relation between x and t, Varignon made recourse, as in the previous case, to the relation: $dt = dx/v = \sqrt{2\int ydx}^{\,-1} dx$, which integrated furnishes the searched expression.

In the second memoir, Varignon introduced explicitly the mass in his equations. The memoir deals with the motion of bodies, planets, under central forces (both centripetal and centrifugal), on curvilinear orbits. In the first part of the memoir he reproduced with the language of Calculus the geometrical reasoning carried out by Newton to obtain the expression of the central force acting on planets while rotating about the sun.[IV]

In the final part of the paper Varignon assumed the motion of two different bodies-planets, whose mechanical and geometrical quantities can be read in the following table:

---

[17]pp. 120–122.

[18]p. 23.

[19]p. 27.

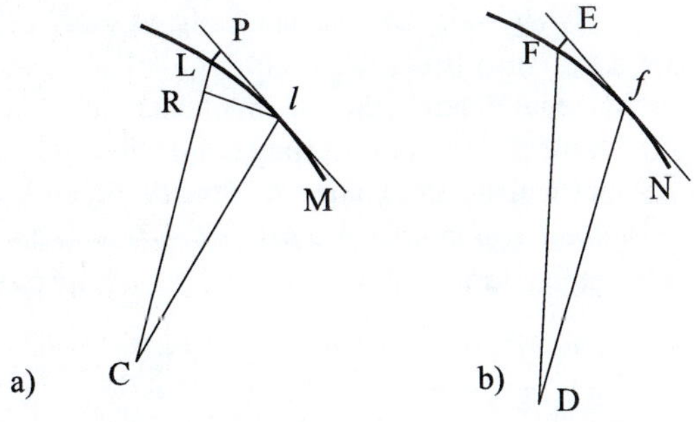

Fig. 3.3 Bodies under central forces. Redrawn from [165], Figs. 6 and 7

body	mass	weight	length of fall	time of fall	central forces
L	m	p	l	t	f
F	μ	π	λ	θ	φ

Varignon said that in one of his memoirs of 1693 [165],[20] he had proved the relation:[V]

$$ml\varphi\theta^2 = \mu\lambda f t^2$$

This relation is differentiated, and after some passages that I avoid to comment, the expressions is obtained: $m\ dl\ \varphi\ d\theta^2 = \mu\ d\lambda\ f\ dt^2$, where d means differential.

At this point with reference to Fig. 3.3, where the thickest lines represent the orbits of a mass point and C and D the centers of the forces, Varignon identified dl with PL and $d\lambda$ with EF (parallel to lC and fD respectively), and arrived to the expression:

$$\frac{f dt^2}{PL \times m} = \frac{\varphi d\theta^2}{EF \times \mu}$$

Which assuming as constant one of the term of the equality (in particular equal to 1), gives:

$$f = \frac{PL \times m}{dt^2}$$

Geometry shows that PL is an infinitesimal of the second order, so that the previous relation could be interpreted as $f = mddx/dt^2$, which is probably the first complete expression in the language of Calculus of the second law of motion (modern term). Varignon seemed however not to be conscious of the fact. Firstly he did not say explicitly that PL is a second order differential of the radius r, and secondly he ended his paper by stating that because the mass m is constant for a given body it can be neglected and instead of the previous equation one can use the simpler one

[20] p. 241. Actually for the sake of coherence in this equation and in the following I changed p with f and π with ϕ.

$f = \mathrm{PL}/dt^2$. Varignon is justified however because he was not searching the relation we see in his results. Varignon did not see, but others probably did.

More interesting, though much later, are the applications of Calculus by Hermann; not so much in his masterpiece the *Phoronomia* of 1716, which is largely based on geometrical argumentation *à la* Newton, but in a paper of 1727, *Theoria motuum qui nascuntur a potentiis in corpora indesinenter agentibus*, published in 1729 [106]. The paper starts with the *general lemma*, proved in an unconvincing way:

> A force [Hermann said *power*] (P) acting continuously in the element of time (dt) is equal to the mass of the body (M), to which the force is applied, multiplied by the element of speed (dC) produced in the small interval of time (dt) [106].[21] (C.4)

That is $Pdt = MdC$. This is a nearly verbatim expression of Newton's second law, according to the standard interpretation, expressed in the language of Calculus.

With his general lemma, Hermann could integrate the equations of motion associated to different expressions of force as function of the position of the body to which is applied. The purpose, limited to one dimension, is to find first the expression of the speed C as function of the space S. This is made by changing the equation $Pdt = MdC$, in the other, more used at the time, $PdS = MCdC$ and introducing the *force of a molecule* p, such that $P = pM$, that is the force of unity of mass (he so did not accept here the Newtonian locution *accelerative force*). In such a way Hermann's relations are simplified as $pdt = dC$; $pdS = CdC$. This last equation is easily integrated by giving $CC = 2\int pdS$. Once one has found C as function of S, he can find the relation $t = \int dS/C$.[VI]

If Varignon and Hermann were first to introduce the Calculus into the equations of motions, Johann Bernoulli's contributions were the most interesting before Euler. Bernoulli faced more complex problems than those treated by Varignon: the skew impact of extended bodies, the motion of cycloidal and conical pendulums, the oscillation of a floating body, the vibration of a chain, the motion of a bi-pendulum, the motion of a heavy wedge on a smooth surface due to the descent of a body, the motion of a heavy wedge under impact, the motion of a body contained in a rotating tube, the motion of fluids and so on. Dating Bernoulli's writings on dynamics is not a simple task, but most are after Varignon's, and probably after 1710, the year Bernoulli was involved in the solution of the problem of the motion of planets about the sun [99]. In the following I will give some hints of the works preceding the publication of the *Mechanica sive de motu* by Euler in 1736.

Bernoulli solved his mechanical problems partly with the principle of conservation of living forces, partly using the "ordinary principle of mechanics". In the following quotation he made clear the difference between principles like that of conservation of living forces, classified as indirect principles and the ordinary principles of mechanics. The former are not principles in the strict sense; they are true propositions but not yet proved in a rigorous way; the latter instead are evident in themselves and accepted by all mathematicians.

---

[21]pp. 139–140.

In the book Hydrodynamics which my Son published not long ago, he undertook that subject under luckier auspices, but he relied upon an indirect foundation, namely the conservation of live forces, which is most certainly true and was proved by me as well, but is still not accepted by all Philosophers. It was I who first presented this hypothesis in the Dynamics of solids (after Huygens used a similar principle to determine the center of oscillation), and from that hypothesis I firmly exhibited the same solution for a water-course which is given by the ordinary principles of dynamics accepted by all Geometers [15].[22] (C.5)

The work in which Bernoulli made an use of the differential calculus moving in the same path of Varignon and Hermann is the *Hydraulica* [15], composed not before 1738 but artfully antedated by him 1732 for reason of priority over his son Daniel who published an *Hydrodynamica* in 1738.

Hydraulica starts with some definitions that are summarized below:

The space traveled by a body	$= x$
The mass of the propelled body	$= m$
The motive force within the limit of the region traveled	$= p$
The speed acquired	$= v$
The time through x	$= t$ [15][23].

which are followed by the well known relations: $dt = \dfrac{dx}{v}$, $\dfrac{pdx}{mv} = dv$ and therefore [15]:[24]

$$\int pdx = \frac{1}{2}vv$$

To notice that among the accelerating force there is gravity; it is indicated by Bernoulli, as presently, with g. The motive force (the weight) of a body of mass m is thus mg [15].[25]

I am not interested here to discuss the various and interesting suggestions and some incoherences in the *Hydraulica*, but only intend to show as Bernoulli applied the infinitesimal calculus (Fig. 3.4).[26] Let consider the conduit ABCFDE of Fig. 3.5, composed of two cylindrical pipes of different size, ACDE and CBCF, of which the former has a base GD open at the orifice GF through which it connects to the narrower pipe BF. A homogeneous and weightless liquid flows with a stationary motion; the fluid is pressed by a motive force p at EA. Because the speed of water increases passing from the left side to the right one, in a limited zone IG the fluid is object to an acceleration. A part of the fluid in the zone IDF remains at rest instead.

Bernoulli studied the motion of a layer of water having a surface defined by the varying height $y = ML$ and the infinitesimal thickness dt=Ll; the width is not named. The fluid in IG is endowed with an accelerating force γ, while the speed with which it moves is given by $u = vm/y$, being v the speed of the fluid in the

[22] vol. 4, p. 392. English translation in [17].

[23] vol. 4, p. 395. English translation in [17].

[24] p. 395.

[25] vol. 4, p. 394.

[26] For comments on Bernoulli's *Hydraulica* see [158].

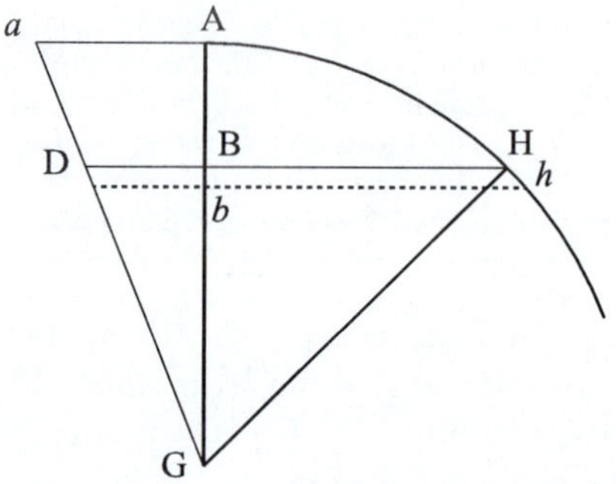

Fig. 3.4 Central forces varying linearly. Redrawn from [106], Table 13, Fig. 2

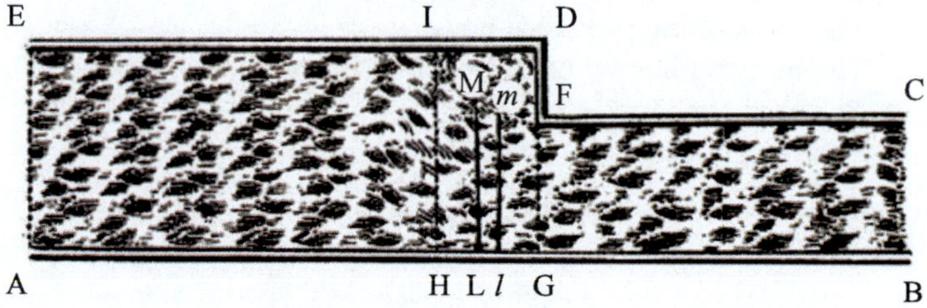

Fig. 3.5 Flow of a fluid through pipes of different sections. Redrawn from [15], vol. 4, Table 89, Fig. 1

narrower pipe and m = FG, its height. For the law of dynamics he has introduced at the beginning, Bernoulli can write $\gamma dt = udu$. The motive force is obtained by multiplying the accelerating force (Bernoulli term) γ by the mass on which it acts, proportional to y, resulting in the expression $y\gamma dt = yudu$. From now on I could not follow Bernoulli reasoning. It seems to me, but I am not sure, he wanted to find the whole motive force by summing up the motive force for all the layers, that is by integrating the accelerating force along the portion IG. The result of this integration if furnished by Bernoulli as:

$$\frac{hh - mm}{2h} vv$$

with h = IH the height of the larger pipe, to which Bernoulli referred as the motive force or pressure p of the fluid [15].[27]

---

[27] vol. 4, p. 400.

3.3 Scalar Approaches to Mechanics

Approaches to mechanics that had at its center living forces, work and action, classified as scalar magnitudes from a mathematical point of view, were quite central in the 18th century. Today these approaches are part of what is called *scalar mechanics* because it did not use geometric vectors. This is a modern category and cannot be applied to the 18th century, when the use of a scalar approach was restricted to solve particular problems. Only with Lagrange, at the turn of the 19th century, a comprehensive scalar mechanics became an alternative to mechanics based on geometric vectors.

If a scalar approach dated back to time immemorial, one can however look at Descartes and his followers as recent reference. Without claiming that Descartes had exactly these notions, in his writings many scholars read two principles of conservation, referred to two different entities but both identified with the term *force*. In the first sense, and using a modern term to simplify, force is measured by the work made by a force, usually gravity. In the second sense the word force is associate to a body in motion and is measured by the product $m|v|$, where $|v|$ is its speed (being v the vector of velocity) and m its mass. Both forces conserve in their ambit of competence; statics and dynamics respectively. Leibniz in a his well known work ***Brevis demonstratio erroris memorabilis Cartesii***, published in 1686 [120], noted that this homonymy was misleading. Starting from the metaphysical principle that force should be conserved, he claimed that $m|v|$ was a wrong measure for the dynamic force [120].[28] The writing of Leibniz, in itself not particularly new, had the great merit to stimulate a discussion on the true measure of force, that lasted for nearly a century; the so called *living forces quarrel*, well known and documented by historians; see [22, 40, 101, 104, 118]. And if in some cases the discussion was sterile, it gave raise to a debate about foundation of mechanics.

Leibniz someway was successful in changing his natural philosophy into a mixed mathematics. His most interesting theoretical works in this respect are ***Dynamica de potentia at legibus naturae corporum***, ***Tentamen de motuum coelestium causis*** of 1689, ***Essay de dynamique*** of 1692, ***Specimen dynamicum*** of 1695 and ***Essay de dynamique sur les loix du mouvement*** of 1699 [121–125].

In the *Essay de dynamique* Leibniz, proposed two kinds of force, or better two ways to measure force: static and dynamic, though in not very clear way. The static force is taken to be the product of the weight of a body and the height from which it falls (mgh), the dynamic force is given by the product of the mass/weight and the square of speed the body possesses (mv^2), named *living force*, a term not yet introduced in the ***Brevis demonstratio erroris memorabilis Cartesii***, which had great success for the whole 18th century [108]. The *Essay de dynamique* seems to suggest that, all considered, there is only one kind of force and that work can be transformed into living force and *vice versa* [27, 62].[29] This reading, that could be not historically

---

[28] p. 163.

[29] pp. 290–293.

correct, was made by Johann and Daniel Bernoulli, arriving to the formulation and use of the principle of conservation of living forces.

In the following for the sake of space I will focus only on two scholars, Johann Bernoulli and Maupertuis with the hope to furnish a view on the conception about mechanics of the first half of the 18th century, where the link with the philosophy of nature of the 17th century was still important.

3.3.1 Johann Bernoulli's Forces and Energies

The leading actors in the history of mechanics of the first half of the 18th century can be divided into three groups. The mathematicians of the Académie des sciences de Paris; the Basel mathematicians and the English mathematicians. This last group, apart from an initial influential role with Maclaurin, Keill, Taylor, choose to move on the path of geometrical approach traced by Newton, that turned out to be unsuccessful in the long run. The mathematicians of the Continent adopted instead the algebraic approach introduced by Leibniz. They disregarded, for the most part, the logical and conceptual problems associated with the manipulation of infinitesimals and exploited the power of the algebraic notation to solve new and old canonical problems. Scientific journals published these solutions alongside the empirical results of chemists, astronomers, and anatomists.

A fundamental character of this enterprise was Johann Bernoulli, the greatest mathematician of the period and the owner of the keys of Calculus. As already commented in a previous section he and his brother Jakob spent a lot of time to decipher the synthetic Leibniz's writings published on the Acta Eruditorum in the 1680s. Differently from Leibniz they were interested more in the concrete development of the algorithm than on philosophical worries.

Johann Bernoulli (1667–1748) was born in Basel. Johann's father Nicolaus had made a fortune as a *Spezierer*, a merchant of imported drugs, spices, paint and the like. At the age of fourteen, Johann was sent to Neuchatel to learn French and commerce and was enrolled fifteen in 1682 at Basel University. He got his *prima laurea* in 1684 with the *Dissertatio de effervescientia et fermentatione* which was reviewed nothing less than by Leibniz in the Acta Eruditorum of 1691. He wrote his second dissertation for the academic degree as doctor of medicine: *De motu muscolorum meditationes mathematicae* in 1694.

Bernoulli went on to Paris at the turn of 1690 where he won a good place in the mathematical circle of Nicolas Malebranche (1638–1715), as a representative of the new Leibnizian calculus. Here he gained immediate success with his mathematical solution to the curve of a chain, the catenaria, reached with Calculus. He met l'Hôpital (1661–1704) and Varignon. Both became his disciples though they were older than him. In 1693 he began his exchange of letters with Leibniz, which resulted into the most extensive correspondence ever conducted by the latter.

On the recommendation of Christiaan Huygens, Bernoulli had been offered the chair of mathematics at Groningen. After Jacob death in 1705, Johann succeeded him at the mathematics chair in Basel. And upon Newton's death in 1727 he was considered the unchallenged leading mathematical preceptor to all Europe. Among his many important achievements is the discovery, in 1694, of a general development in series by means of repeated integrations by parts; the series subsequently named after him. He made fundamental contributions to differential geometry (inverse problem of tangents), to the calculus of variations and to the systematization of the differential calculus (integration of algebraic functions), and coined the term *integral* in mathematics. Bernoulli was actively involved in the priority dispute between Leibniz and Newton about Calculus, supporting the former. He became a member of the scientific academies of Paris, Berlin, London, St. Petersburg and Bologna [91].

Bernoulli writings, published in the Acta Eruditorum and in the memoirs of the Académie des science de Paris, are collected in his *Opera Omnia*, published in 1742 in four volumes [16] and in the *Virorum celeber. Got. Gul. Leibnitii et Johann Bernoulli commercium philosophicum et mathematicum*, of 1745 into two volumes [127]. Other writings are the object of a modern editorial project, of which only a few volumes are published, such as *Der Briefwechsel von Johann I Bernoulli* of 1955, 1988 and 1992, and a volume on mechanics published in 2002, Band 6 of *Die Werke von Johann I und Nicolaus II*. Most of his writings concern mathematics; but there are some on mechanics and a few on philosophy of nature. Bernoulli's passion for communicating his idea is clear from his numerous scientific correspondence, about 2 500 letters, exchanged with some 110 scholars.

Often Johann Bernoulli, thanks to his leading role in mathematics and his Cartesian and Leibnizian philosophy, is charged to have slowed down the diffusion of Newtonianism with great damage to the development of science in the Continent. Personally I think that this judgment, a part from being ungenerous, is misleading. If it is intended that he contrasted some aspects of Newton's natural philosophy, particularly the idea of action at distance and defended Leibniz against Newton on the priority of Calculus, then the charge is correct. But if it is intended that he slowed the diffusion of Newtonian mechanics, this is not absolutely true. Certainly, Bernoulli did not embrace completely Newtown's *Principia*; but he was probably the first to read and understand it in full in the Continent and got from Newton what was essential in mechanics; from the axioms of motion—he used a concept of force very close to the Newtonian one, even though assumed for it a Leibnitzian name, dead force—to the law of universal gravitation, its mathematical expression only of course. He discussed and integrated the equation of motion of a planet by showing with a different approach than Newton, that it is an ellipse. He shared with Newton the corpuscular conception of light.

Bernoulli lived a few generations after Newton and thus he moved from a higher point of view. Besides the dynamics of the material points, to which he contributed with the introduction of differential calculus, Sect. 3.2.1, he also addressed other problems that became the benchmarks for new theories of mechanics at the start of the 18th century. For the purpose he used also theoretical tools not related to Newton's. In particular he developed and used the principle of living forces and the principle of virtual velocities (modern—Lagrangian—term).

3.3.1.1 Philosophy of Nature

It may be strange but the man who after the dead of Leibniz and Newton was the greatest mathematician of the whole Europe, graduated as a doctor in medicine at the university of Basel. This is less strange if one reflects that studies at the universities at the time were conceived differently than today and that his dissertations *Dissertatio de effervescientia et fermentatione* and *De motu muscolorum meditationes mathematicae*, though connected to medical issues, should be framed into the tradition of physico-mathematica of the end of the 17th century. Here Bernoulli developed on the one hand themes typical of natural philosophy, on the other hand he applied, in the second dissertation, his recent acquisitions of Calculus to solve the mechanical problems associated with the working of muscles. In the following the evaluation of the two dissertations that Bernoulli himself expressed in a letter of 1695 to Pierre Chirac (1650–1732), professor of medicine in Montpellier:

> I regret to have published it [Dissertatio de effervescientia] in my youth, seeing now that I should have explained my thoughts in a more geometrical way. There is in this piece of work no analysis, neither new nor old, and, therefore, the demonstrations cannot detain anybody except perhaps by the obscurity of the words. As far as the dissertation on the movement of muscles is concerned, I must recognize that I had not the leisure of dealing thoroughly with such a curious matter, having been distraught by the troubles which other matters had given me then. It is true that I use a new analysis but it would be impossible to turn the demonstrations which I draw from it in the usual way of common geometry, for, if that could be done, Mr Borelli who, although one of the greatest geometers of this time, did not know anything of this new calculus, no doubt would have told the truth better on these hypotheses which he established for the theory of the movement of muscles.[30]

Though Bernoulli considered his dissertations as scarcely interesting, they reveal his epistemology. In the *Dissertatio de effervescientia et fermentatione*, to explain the process of fermentation he assumed the interaction between two different kinds of corpuscles. Though I attributed a definite shape to the particles of acid and alkali, said Bernoulli, that this could not be demonstrated by any experiment because of the smallness of the particles which could be seen neither by the naked eye nor with the help of a microscope. It is enough—continued Bernoulli —for me if the attribution of such a shape to these particles so as to explain at best the nature of effervescence does not oppose reason or experience. The astronomers adopt that hypothesis of the system of the world which explains at best the celestial phenomena and the movement of the heavenly bodies although they cannot demonstrate its truth by unquestionable and unshakable arguments. They retain it so far until another one more likely and convenient appears [18].[31] This kind of epistemology is influenced by the Cartesian conception of mechanical explanations: what matters is to find a possible explanation; its actual existence has less interest. But Bernoulli often flanked to the Cartesian approach that of the mixed mathematics; the assumed hypothesis should be verified against the experimental data and changed if contradicted.

---

[30] Quoted from [18], pp. 9–10.

[31] p. 87.

In 1730, with the paper *Nouvelle pensée sur le Système de M. Descartes*, Bernoulli won the prize proposed by the Académies des sciences de Paris on the cause of the elliptic shape of the orbits of planets about the sun and other aspects. Here he tried to defend the theory of vortices by Descartes against the charge of incongruence raised by Newton and Huygens [12]. In 1734, he won another prize of the same academy with the paper *Essai d'une nouvelle physique celeste* [14], where he presented his own conception on the composition of matter and the nature of gravity, that was a significative variant with respect to Descartes's.

For Bernoulli at the moment of creation the supreme being (l'Etre souverain) produced two kinds of matter starting from a primeval preexisting undifferentiated matter. A part of the undifferentiated matter was left unchanged, it was going to constitute a completely fluid medium divisible at infinity and infinitely subtle—the element of first kind; the remain part of matter was used to make corpuscles that piked up because of their motion form small masses (*massules*) whose part are coherent without the need they are perfectly hard—the element of second kind [14].[32]

Bernoulli is not very clear about the behavior of the undifferentiated matter; on the one hand he asserted that it is so subtle to not impede in any way the motion of the *massules*. On the other hand he suggests that it can transport the whirlpools of massules he supposed fill the universe. Bernoulli universe is made up of a huge quantity of the element of the first kind; the elements of the second kind though may be close to each other, leave intervals that are very large compared to their dimension. As a matter of fact the volume occupied by the element of the second kind is infinitely small with respect to the volume of the first element.

The sun (and the stars) is made up of the matter of the first kind very compressed. It, together with the element of second kind (the massules), at the moment of creation acquired a motion of rotation about its center. Because of this motion the *massules* may leave the sun and reach the earth. This is the origin of light. To explain gravity Bernoulli assumed that each star emanates a huge number of massules. Made reference to sun for instance. The stars surrounding sun emanate a lot of small masses, that are very tiny and can pass without any difficulty all the material bodies, the planet for instance. They however, because their great number, inevitably clash with the massules emanated by the sun. In the impact, the particles merges to form greater masses, the *pelotons*. These are pressed by the massules coming from the stars and directed toward the sun. When they meet a planet, the pelotons being grater than the massules, cannot pass through the matter of the planet and push them toward the sun. This is the origin of gravity. That gravity follows the law of the inverse square of the distance is explained by Bernoulli by considering that the density of pelotons varies with the inverse of the square of the distance, because the surface of a sphere varies with the square of radius as well. The pelotons that fall in the sun reintegrate the matter that left it as massules. In the explanation Bernoulli is obliged to introduce an impact between hard bodies, while at the beginning of his expositions had supposed the massules not necessarily hard; this is of course an incoherent position.

---

[32] p. 276.

In an appendix to the *Du discours sur le loix de la communication du mouvement* of 1724, devoted to the explanation of the cause of elasticity, Bernoulli assumed that the presence of the subtle matter is the main responsible for the elasticity of bodies. To explain the fact, consider for example a vessel of any shape, filled with subtle matter; we can admit that this matter, which passes easily through the pores most narrow of all bodies, will cross with the same facility the pores of the vessel. We suppose that in addition to this subtle matter there is a lot of corpuscles too coarse to escape through the pores of the vessel, but which swim freely in the subtle matter; leaving between them intervals so spacious that all these corpuscles pick up in a heap, might not occupy a hundred thousandth of the volume of the vessel. Each of the corpuscles, because of the different shape, will move differently from each other. In this way the corpuscles form circles of various diameters which tend to expand due to the centrifugal force; the expansion is opposed by the walls of the vessel. A body, which plays the role of the vessel described above, has a huge number of pores; when it is compressed, they decrease and the the diameters of the circles tend to shrink, so the centrifugal force tends to grow and to oppose the narrowing; the elastic effect is thus generated [10].[33]

3.3.1.2 Metaphysical Principles. Living Force and Virtual Velocities

Both the principles of conservation of living force and virtual velocities were proposed by Bernoulli in a very clear form, even though the proofs were based on scarcely evident metaphysical principles. Only a mathematician could follow this approach. Neither Leibniz, that a great mathematician, was able to do this because bridled by his philosophy. The introduction of the principle of conservation of living forces is carried out by Bernoulli in an important paper of 1724 (published 1727), *Du discours sur le loix de la communication du mouvement*, presented for a contest to the Académie des sciences et belles lettres de Berlin won by Maclaurin. The approach followed by Bernoulli was not usual in the traditional mixed mathematics and would not have been approved by Newton and only slightly tolerated by Euler, because based on metaphysical questionable assumptions.

Principle of living forces.
Bernoulli proved, in more ore less rigorous way, as a theorem, that the vis viva should be measured after Leibniz. Here the statement of the theorem: "The living force of a body is proportional to the square of the speed and not to the simple speed [and to the mass]" [10].[34] The term *living force* can be traced back to Leibniz and indicates the force of a body in motion, or to Galileo and his force of blow. It represents the capacity of action, or the force, of a moving body, for the simple fact that it is in motion. This force, which is not defined clearly and distinctly, is treated essentially as a substance. It transforms and does not destroy itself. In order to argue this thesis, Bernoulli made use of the metaphysical principle, typical of the pre-Cartesian philosophy of nature:

[33] pp. 91–97.

[34] p. 53.

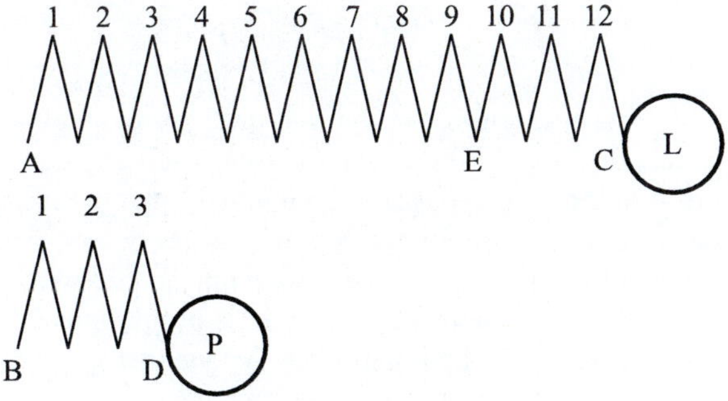

Fig. 3.6 Two assemblies of equally compressed springs. Redrawn from [10], Plate 1, Fig. 6

"In effect anybody consider it as an incontestable axiom, that no part of an efficient cause can be lost, unless it produces an effect equal to its lost" [10],[35] generally not accepted by the community of mathematicians.

The proof of the theorem on the true measure of living forces was quite articulated and made use of springs and mass points. The first phase of the argumentation considered two sets of elementary springs, as shown in Fig. 3.6.

The first set (system A) is composed by twelve elements, the second (system B) by three elements, all the elementary springs are equal so that the first set is four times long the second, if they support the same *pressure* (Bernoulli's term). Two equal masses, L and P are appended at the free end of the springs. When the springs are left free to extend, they apply a force on the masses putting them in motion. The pressure/dead force is thus transformed into living force. Bernoulli concluded this first phase of argumentations by asserting that the living force acquired by the two masses, at a certain time, or better at the end of the expansion of the springs, is proportional to the length of the two set of springs, because it should be proportional to the number of elementary springs contained in any set.

In the second phase of his argumentation Bernoulli studied the dynamics of the two masses L and P of Fig. 3.6, passing from rest to motion. This is made by recurring to the "known law of acceleration", $dv = pdt$, which is rewritten as $vdv = pdx$ [10].[36] Notice that Bernoulli has replaced a relation of proportionality, $dv \propto pdt$, with a relation of equality $dv = pdt$, by neglecting the constant of proportionality, in such a case the mass.

Integrating the previous relation from the instant (or position) of release to the instant (or position) when the spring stops to act, with Bernoulli's notation, gives $1/2vv = \int pdx$. This general expression is particularized for the two masses L and P. For P the speed is still indicated by v, while for L it is indicated by z. For the latter the following relation holds true, $1/2zz = n \int pdx$—being n equal to the ratio between the number of springs of the systems A and B, and also the ratio between their length; in this particular case $n = 4$—because $\int pdx$ clearly adds on the springs. In the end Bernoulli could write the proportion:

[35] p. 56.

[36] p. 44.

$$vv : zz = \int pdx : n \int pdx = 1 : n$$

For what discussed in the first phase of the argumentation, the ratio of lengths n equals the ratio of living forces, so also $vv : zz$ equals the ratio of living force; the proof of the theorem is thus concluded. That the living force is proportional also to the mass m is given for granted, what was not considered as problematic and no need was felt to specify that instead of p one should have considered p/m.

Bernoulli considered, just after his proof, and as a corollary, the case of force of gravity, which can be imagined to be due to a spring of infinite length which as such acts with a constant pressure. In such a case the integral $\int pdx$ gives ph where h is the height of fall of a heavy body, and p is its weight. In such a way Bernoulli could connect his reasoning to that of Leibniz, who to establish the true measure of living forces made recourse to the model of a heavy body in its fall.

The issue of the conservation of living force, using for it the 'true' measure, is brought into the field of the impact of bodies. In particular the impact of two mass points that clashed frontally. On this issue it must be remembered that Bernoulli's essay concurred for a contest based on the question: "What are the laws according to which a perfectly hard body, put into motion, moves another body of the same nature either at rest or in motion, and which it encounters either in a vacuum or in a plenum?" [10].[37] According to Bernoulli there are two possible views about hard bodies; either as a limit case of plasticity with infinite strength or as limit case of elasticity with infinite stiffness. The first possibility is to disregard, for him, because it contradicts the principle of continuity postulated by Leibniz which he accepted, according to which in the impact between two bodies there cannot be an abrupt change in speed, "natura non operatur per saltum" [10],[38] as it would occur for plastic bodies. Instead hardness due to elasticity and infinite stiffness is welcome in physics. And "even if the existence of hard bodies from this point of view would be physically impossible, it is not less certain that one can always consider these bodies as one considers prefect lines and surfaces in geometry and inflexible levers without weight in mechanics" [53].[39]

For elastic bodies the three properties hold true in the impact:

1. The relative speed is conserved.
2. The quantity of direction [that is quantity of motion] is conserved.
3. The living force is conserved.

The first two are proved by Bernoulli by using the ordinary laws of mechanics. The third property, known as the principle (referred to by Bernoulli as theorem) of conservation of living forces, or more simply the principle of living force, could be proved, as a theorem by the first two with simple algebraic passages, and Bernoulli

[37] p. 8.

[38] p. 5.

[39] vol. 3, Eloge de Jean Bernoulli, p. 355. English translation from [153]. Of course Bernoulli's thesis was not accepted by the academy of science; he was disqualified and Colin Maclaurin won [153].

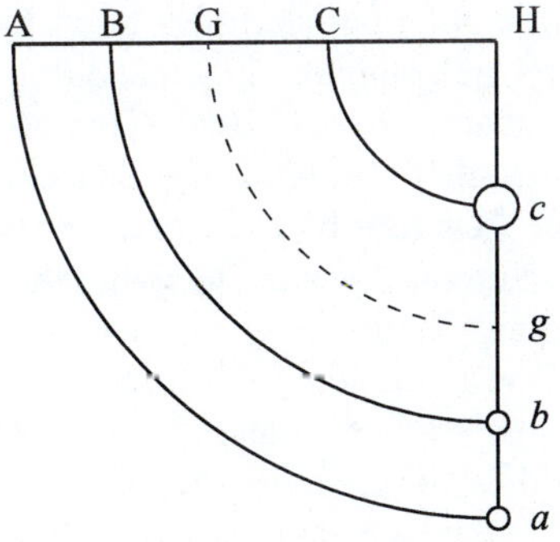

Fig. 3.7 Equivalent length of a compound pendulum. Redrawn from [10], Plate 4, Fig. 15

did this. To notice that he quoted Huygens while he did not quote Leibniz that already noticed the fact that of the three relations referred above, only two were independent [27].[40]

But for Bernoulli his algebraic proof was not necessary, because the conservation of living force is logically necessary and derives from the indubitable metaphysical principle of the reciprocity between cause and effect. This is what Bernoulli said.

> But now that this truth is put in its light and beyond all reach, one has reason to admire the perfect conformity which reigns between the laws of nature and geometry, which it observes so constantly. In all circumstances, it seems that nature has consulted geometry by establishing the laws of the notion. For if it had been possible for the forces of the bodies, which are in motion, not to have been due to the products of the masses by the squares of speed, and that nature had made them in another ratio; it would have contradicted itself; the order of geometry would have been violated. The quantity of living forces, the only source of the continuance of motion in the universe, would not have been preserved; consequently, there would be no equality between the efficient causes and their effects; in a word, all nature would have fallen into disorder [10]. [41] (C.6)

Bernoulli concluded his essay by applying the principle of conservation of living forces to a theme that had long stimulated the curiosity of mathematicians, the search for the length of a simple pendulum equivalent to a compound pendulum, that he himself had studied using the ordinary principles of mechanics [9]. He based on this assumption (or hypothesis): the living force of a compound pendulum made of n masses connected to a rigid thread, dropped from a given position, when it reaches the lowest position, equals the summation of the living forces of the n masses assumed as free from the rigidity constraint, that is belonging to many simple pendulums all dropped from the same position of the compound pendulum. With reference to Fig. 3.7, the whole compound pendulum HABC and the single masses A, B, C forming the simple pendulums HA, HB, HC are dropped from the horizontal position.

Bernoulli knew that his hypothesis is scarcely convincing and justified it by asserting that however it is more convincing than the hypothesis assumed by Huygens "who supposed that the center of gravity of a composite pendulum, felt from an assigned

[40] p. 298.

[41] vol. 3, p. 56.

height, does not raise to an hight greater from which it felt if the single weights which composed the pendulum tore loose when the pendulum has reached the vertical position" [10].[42] This because both for the compound pendulum and for the single pendulums there is a common cause of their descent: gravity.

Under this hypothesis the search for the length of the equivalent pendulum is straightforward. The living force of the compound pendulum, less a constant of proportionality, is given by $aaA + bbB + ccC$, with the upper case letters indicating the masses and the lower case letters indicating their distances from the point of suspension H, that are proportional to speeds. In the case of isolated simple pendulums, the living forces of each mass A, B, C are proportional to their heights of fall, which are in turn proportional to the distances a, b, c. Without any justification, I do not see it, Bernoulli, most probably referring to results found in his *Nouvelle theorie du centre d'oscillation* of 1714 [9],[VII] assumed that the sum of the living forces of the simple pendulums is $xaA + xbB + xcC$, being x the sought length of the equivalent simple pendulum. By equation the two expression of the living forces $aaA + bbB + ccC = xaA + xbB + xcC$, allows to evaluate x, which assumes the well known value.

If in the *Du discours sur le loix de la communication du mouvement* Bernoulli spoke about the conservation of living force clearly only in the case of the impact among elastic bodies, in subsequent writings he extended the idea of conservation in a clear way also when there is a variation of the apparent living force due to the action of external forces, due either to springs or gravity.

In the *De vera notione virium vivarum, earunque usu in dynamicis*, printed in 1735 but probably written before [13], Bernoulli assumed that for the conservation of living force one does not consider only the free living forces, expressed by the mass multiplied by the square of speed, but also the (living) force, which is stored for instance in deformed springs. The living force is "something real and substantial", which has its own value, "it flows but always preserves. This is what we call conservation of living forces" [13].[43] Let us consider for instance a compressed spring with a body placed at one end. In this case the force of the spring is completely inside the spring, whereas the body is at rest. But, if we suddenly release the compressed spring, the body then gradually acquires a speed and the spring is deprived of any force. Thus, without any exterior transfer of force, we have converted the force of the spring into the living force in the body. The procedure is clearly reversed when a body traveling with uniform speed is brought to rest by an initially uncompressed spring. If the spring could push two or more bodies, when released, then the sum of the living forces gained by the separate bodies, each one endowed with its own speed, is equal to the force initially accumulated within the spring. This is true because, as a principle, there is equality between the cause and the effect [13].[44] Notice that Bernoulli seems to speak of living force of a spring; but never he said exactly this; he spook rather of the force of the spring which can be measured by the living force it

[42]p. 78.

[43]p. 240.

[44]pp. 241–242. Adapted from the paraphrase in [19].

Fig. 3.8 Living forces from a compressed spring. Redrawn from [13], Table 49, Fig. 2

can impart to a body attached to it. His son Daniel instead spook clearly of the living force of the spring referring to the living force which is actual but at a microscopic level in the perpetual motion of aethereal particles which are responsible for the elasticity [27].[45]

Bernoulli took again the problem of the true measure of the living forces, using a new point of view. Let consider the two unequal masses A and B of Fig. 3.8, at the end of a set of equal and equally compressed elementary springs. Let then the springs allowed to extend; the masses A and B are pressed in the opposite direction, until the springs exhausted their force, and start moving with speeds a and b respectively. Bernoulli then stated:

1. The two masses are pushed by the same forces.
2. The increments of speeds a and b are inversely proportional to A and B.
3. The final speeds a and b stay in the ratio $A : B = b : a$.
4. Consequently the quantity of motion of the two masses are equal, that is $Aa = Bb$.
5. The time the two masses reaches the final speed is the same.
6. The position of the center of the common center of gravity C remains at rest.

For the last point instead of a single set of spring one can considered two independent sets AC and BD, fixed at C. At this point Bernoulli made an assertion that at first appears unjustified. Name f the living force acquired at the end of the expansion of the springs, by the mass A and φ the corresponding one of the mass B, said Bernoulli, then the equality holds: $f : \varphi = a : b$. This assumption appears clear however by referring to the *Du discours sur le loix de la communication du mouvement*, where it is proved that the living force of a set of equal and equally compressed springs equals the number of springs and so the length of the set. In the case of Fig. 3.8, at any instant the speed (as well as the displacement) of the extremity of the two sets of springs AC and BC is proportional to the number of springs. Thus as living force is proportional to the number of springs and the number of springs is proportional to speed, it follows that living force is proportional to speed.

By considering together the two relations, $Aa = Bb$, and $f : \varphi = a : b$, by multiplying the numerator of the second term of the proportion by Aa and the denominator by Bb (which equals Aa), one obtains:

$$f : \varphi = aaA : bbB$$

that is the living forces of a body in motion is proportional to the product of its mass and the square of its speed.

---

[45] p. 326.

Bernoulli applied the principle of living forces to solve a certain number of problems in the *Theoremata selecta pro conservatione virium vivarum demonstranda et esperimenta confirmanda* (of 1727) [11], published in 1729, that logically should follow the text published in 1735 [27].[46]

Principle of virtual velocities

The formulation of the principle of virtual velocities by Bernoulli, who referred to it as the *principle of energies*, is more difficult to explain. Largely because he did not make it the subject of any specific treatise or article, but only of some letters in the years 1714–1715, to abandon completely the matter, until fifty years later, Lagrange revived and made the principle the fulcrum of his analytical mechanics. The letters in object are from and to Varignon and Bernard Renau d'Eliçagaray (1652–1719) and had as an occasion the comment on a treatise of d'Eliçagaray about the motion of ships, *Théorie de la manoeuvre des vaisseaux* of 1689 [56].

The first time Bernoulli suggested a solution based on the principle of virtual velocities is in a letter to d'Eliçagaray of 12th August 1714, which I was not able to read, but whose content is partially quoted in [148] and in a letter of Varignon to Bernoulli of 22th January 1715. Here Varignon said that Bernoulli considered as the greatest and first principle of mechanics the following statement, known after Lagrange as the *principle of virtual velocities*:

> In each equilibrium there is an equality of the energies [what the Latins (and Galileo) called momentum] of the absolute forces, that is among the sum of the products of the forces by the virtual velocities [20].[47] (C.7)

The meaning of virtual velocities is explained in the following quotation:

> I mean with virtual velocity the only tendency to move the forces have in a perfect equilibrium, where they do not move actually [...]. Wherefore to avoid ambiguity, instead of saying that their powers or forces are as the products of the masses by their velocities you might have done better to express yourself well, the energies of powers or forces are as the products of these powers or forces by the virtual velocities [148].[48] (C.8)

The greatest and first principle of mechanics[49] was stated in a more precise form in the letter of Bernoulli to Varignon of February 26th, 1715 (written shortly after the letter of Varignon to Bernoulli, above referred to), which is well known because reproduced in the *Nouvelle mécanique ou statique* by Varignon, published posthumous in 1725, with the wrong date February 26th, 1717 (a typo?):

> Conceive several different forces acting along different trends or directions to balance a point, line, surface, or body; conceive also to impress to the whole system of these forces a small motion either parallel to itself in any direction, or around a fixed point whatsoever: you will be glad to understand that with this motion each of these forces will advance or retire in

---

[46]pp. 321–324.

[47]Letter of Varignon to Bernoulli of 22th January 1725.

[48]p. 18.

[49]Bernoulli gave no name to his principle; it was named by Lagrange *principle of virtual velocities* and by Varignon principle of energies.

> its direction, unless someone or more forces had their trends perpendicular to the direction of the small movement, in which case this force or these forces, neither advance nor retire anything. These advancements or retirements, which are what I call virtual velocities, are nothing but than what each direction increases or decreases by the small movement [...]. All this being understood, I form this general proposition: *In any equilibrium of any forces in any way they are applied and following any directions, either they interact with each other indirectly or directly, the sum of the positive energies will be equal to the sum of the negative energies taken positively* [167].[50] (C.9)

In this letter Bernoulli furnished the applications of the principle to various problems.[51] Bernoulli gave neither justification of his principle of energies nor a hint from what it could be derived. Two ways appear possible; a static derivation or a dynamic derivation. Varignon suggested the static alternative and made reference to Descartes and his letters to Constantijn Huygens of 1637 [26].[52] Varignon said that Bernoulli virtual velocities are nothing but the *chemins instantanées* of Descartes and that the Cartesians had already deduced from his principle the same equality of moments, or energies, or quantities of motion:

> Cartesians, according to the letter I cited of their Master,[53] had already deduced from his principle the same equality of Moments or energies, or the quantity of motion, that you use, for two powers in equilibrium on simple machines, and in fluids, from the incipient motion that Mr. Descartes prescribes in this letter. But you are the only, for what I know, who extended the equality of energies to as many powers you like, acting in any direction and in equilibrium with themselves. This point is very nice, but (as I have already said) it supposes the equilibrium among them and does non prove it [20]. (C.10)

But it is possible a dynamic derivation based on the notions of living and dead force. For Bernoulli the living force came from the accumulation or sum of pulses of dead force; it is not difficult to assume the accumulation is made by summing the energies pdx, where p is the dead force and dx is the small displacement in the direction allowed by constraints. If there are competing dead forces-pulses of contrary tendencies, they destroy and in some circumstance there will be equilibrium. It is natural to assume that the destruction of all pulses is satisfied when the sum of all energies vanishes; which gives Bernoulli's rule. The possibility of this derivation was clearly seen and justified, quite convincingly, by Vincenzo Riccati in his *De' principi della meccanica* of 1772 [26].[54]

---

[50] vol 2, pp. 175–176.

[51] For a thorough discussion about the genesis and meaning of the virtual velocities principle see [26].

[52] p. 167.

[53] [57], letter 73, vol. 1, pp. 327–346.

[54] pp. 230–233.

3.3.2 Maupertuis and the Role of God in Mechanics

Pierre-Louis Moreau de Maupertuis (1698–1759), after private schooling went to Paris at the age of sixteen. In 1717 he began to study music, but soon developed a strong interest in mathematics. Since 1722 he was living in Paris becoming friendly with the dramatist, novelist and journalist Pierre Carlet de Chamblain de Marivaux, the playwright Antoine Houdard de La Motte, and the mathematicians Joseph Saurin, François Nicole, and Jean Baptiste Terrasson. His early interest in mathematics now blossomed and, with instruction in the higher reaches of the subject from these men, he soon acquired a deep understanding. Maupertuis became an adjoint of the Académie des sciences in 1723 (associé in 1725) and in the following year he produced his first paper *Sur la forme des instruments de musique* which studied the effect of the shape of an instrument on the characteristics of the sound it produced.

In 1728 Maupertuis made a trip to London that was to exert a major influence upon his subsequent career. From this time on, Maupertuis was the foremost proponent of the Newtonian movement in France and a convinced defender of Newton's ideas about the shape of the earth. In order to extend the range of his mathematical and scientific knowledge Maupertuis went to Basle to study under Johann Bernoulli. He matriculated in Basle on 30 September 1729 and spent the session living in Bernoulli's home. At the University of Basle he received an outstanding education and training. He learnt of Descartes's vortex theory model of the solar system and of Leibniz's views on mechanics from his teacher Johann Bernoulli who was perhaps the strongest supporter of these theories.

Back in Paris in 1730, Maupertuis began writing papers on mechanics in which he used the expertise he had already developed in mathematics. In May 1735 the Paris academy sent an expedition to Peru to make measurements of the Earth. It was headed by Charles Marie de La Condamine (1701–1774) and had Pierre Bouguer (1698–1758) and Louis Godin (1704–1760) as members. A second expedition was sent to Lapland headed by Maupertuis. In a meeting of the academy on 20 August 1737 he referred to that his results confirmed that the earth was oblate. He was elected to the Académie Française in 1743.

Maupertuis was invited to Germany by Frederick II of Prussia in 1740 as part of his aim of bringing top philosophers and scientists to Berlin. In 1746 Maupertuis was officially appointed as the president of the Berlin academy, a post which he was to hold for eight years. Although he tried very hard to make a success of his role as president of the Berlin academy, things were rather against him. Partially because he did not speak German and, although the official business of the academy was conducted in French or Latin, Maupertuis was rather cut off from the day to day administration which was conducted in German. Anyway he published on many topics including mathematics, geography, moral philosophy, biology, astronomy and cosmology [98, 107]. The career of Maupertuis shows as a working mathematician could manage about philosophy. Maupertuis was successful in both the two fields, to the point that starting from the age of forty eight, in precarious health he no longer carried out original works in mathematics but shifted his attention to metaphysics.

In 1751 Maupertuis was accused by the mathematician Samuel König (1712–1757) to have derived from Leibniz his ideas about the most important result he thought to have given to mechanics: the principle of least action [24].[55] The evidence which was put forward to support the claim was a letter of 1707 from Leibniz to Jakob Hermann, even though this letter could not be exhibited in the original and the accusation by König was quite difficult to sustain [147].

The contrast between König and Maupertuis became the pretext for keep on a cultural battle, already going on for years, which saw on the one side the supporters of the new science, Euler in the lead, but also d'Alembert, lined up in favor of Maupertuis; on the other side the supporters of the claim that it was up to canonical philosophers to dictate theoretical principles and the method of their disciplines to mathematicians and physicists. They were headed by Wolff and his school (but also the progressive Voltaire supported König). That of the Wolfians was a retro-battle; physics, mechanics, astronomy, theory of matter, theory of light were now autonomous disciplines, based on axioms often of an empirical nature, far from the old school formulas concerning the substance (form and essence), which in some way Leibniz had resumed and renewed [35].[56] The battle was won; even Frederick II supported the president of his academy, but Maupertuis's failing health collapsed under the strain and he left Berlin for Paris in 1753

Of Maupertuis I will discuss here only his papers concerning the least action: *Loi du repos des corps* of 1740 (published in 1742), *Accord de différentes loix de la nature qui avoient jusqu'ici paru incompatibles* of 1744 (published 1748), *Les loix du mouvement et du repos, déduites d'un principe de métaphysique* of 1746 (published 1748). The way Maupertuis dealt with the subject is very interesting because it often a view of the epistemological position about mechanics at be beginning of the 18th century.

In his paper of 1740, the *Loi du repos des corps*, he did not speak about action. He referred only to the minimum of a certain function of forces. Quite interesting are his considerations about the different nature of the principles to adopt in physics.

> If the sciences are based on certain, simple and clear principles, from which all the truths which are the object of them depend, there are still other principles, less simple, to be honest, and often difficult to discover, but which, once discovered, are very useful. These in some way are the laws that nature follows in certain combinations of circumstances, and teaches us what it is on such occasions. The first kind of principles hardly need demonstration, by the evidence which they make when the mind examines them; the latter cannot be considered as physical demonstrations in the strict sense, because it is impossible to go through all the cases in which they take place. Such, for example, is the principle so well known and useful in the ordinary statics; that *in all body assemblies, their common center of gravity drops as low as possible*. Such is that of the preservation of the living forces. Never have we given a general demonstration of the rigor of these principles; but no one, accustomed to judge in the sciences, and who will know the power of the induction, will doubt their truth. When we have seen that on a thousand occasions nature acts in a certain way, there is no man of common sense who believes that in the millennium it will follow other laws.

[55] pp. 337–343.

[56] p. 131.

> As for the *a priori* demonstrations of these kinds of principles, it does not appear that physics can give them; they seem to belong to some higher science. However, their certainty is so great, that many mathematicians do not hesitate to make the foundations of their theories and use them every day to solve problems, whose solution would cost much more trouble without them. Our mind being as small as it is, it is often too far for it [the distance] from the first principles to the point where it wants to arrive, and it gets tired or departs from his path. These laws, of which we speak, exempt it [our mind] from a part of the way: one leaves with all his strength, and often has only a few steps to go in order to reach the place where he wishes [135].[57] (C.11)

The idea expressed in this quotation echoes that expressed at the beginning of the 20th century by Albert Einstein in an article published on The London Times in 1919, about the difference between theories based on principles and constructive theories:

> We can distinguish various kinds of theories in physics. Most of them are constructive. They attempt to build up a picture of the more complex phenomena out of the materials of a relatively simple formal scheme from which they start out. Thus the kinetic theory of gases seeks to reduce mechanical, thermal, and diffusional processes to movements of molecules, i.e., to build them up out of the hypothesis of molecular motion. When we say that we have succeeded in understanding a group of natural processes, we invariably mean that a constructive theory has been found which covers the processes in question. Along with this most important class of theories there exists a second, which I will call "principle theories". These employ the analytic, not the synthetic, method. The elements which form their basis and starting point are not hypothetically constructed but empirically discovered ones, general characteristics of natural processes, principles that give rise to mathematically formulated criteria which the separate processes or the theoretical representations of them have to satisfy. Thus the science of thermodynamics seeks by analytical means to deduce necessary conditions, which separate events have to satisfy, from the universally experienced fact that perpetual motion is impossible. The advantages of the constructive theory are completeness, adaptability, and clearness, those of the principle theory are logical perfection and security of the foundations. The theory of relativity belongs to the latter class [64].[58]

After some premises Maupertuis tempted to formulate a principle of minimum, general enough to pair the principle of living forces:

> Let a system of bodies that weigh, or that are drawn to some centers by forces that act on each body, as the power n of their distances from the centers. In order that all these bodies may remain at rest, the sum of the products of each mass by the intensity of the force and by the power $n + 1$ of the distance from the center of the force (that we can call the sum of the *forces of rest* [emphasis added]) make a minimum or a maximum [135].[59] (C.12)

A modern reader can see here a particular case of the minimum of potential energy theorem, limited to central forces depending on a power of the distance and a system of mass points (remember that the potential energy of a force proportional to r^n is proportional to r^{n+1}); constraints are not named but are implicitly assumed. The principle is proved for same particular cases using the equation of equilibrium of statics, with a procedure similar to that used to prove the principle of virtual velocities; and

[57] pp. 170–171.

[58] p. 228.

[59] p. 171.

probably Bernoulli's and Varignon's writings that Maupertuis knew were of inspiration. Maupertuis's principle though interesting was however scarcely appealing for who wanted to give a general law of nature.

In the paper of 1644, *Accord de différentes loix de la nature qui avoient jusqu'ici paru incompatibles*, devoted to the refraction of lights Maupertuis formulated his principle of minimum introducing the word action—a word that for the same admission by Maupertuis, goes back to an analogous definition of Leibniz [27, 110][60]—and basing it on the metaphysical principle according to which "nature for the production of his effects always operates with the simplest means" [136].[61] That is assuming the validity of final causes in physics.

Before applying the principle of minimum to refraction of light, Maupertuis referred to Fermat, praising him for the brilliant idea but also criticizing him because of his recourse to a wrong principle. Fermat assumed that a ray of light to pass from a point A of a given medium, where light moves with speed a, to a point B of another medium where light moves with speed b, with the two media separated by a plane, makes its way according to a path that needs the minimum time. If the first medium is more rarefied than the second, then $a > b$ and the angle on refraction is less than the angle of incidence, as experience shows [1, 92].

Even though the result is correct, the approach was wrong for Maupertuis, because for him the speed of light is the greater the greater the density of medium. This position, that we know as wrong, was the position of both Descartes and Newton and was natural to assume it for Maupertuis. If applied to the 'right' nature of light propagation Fermat's principle would give a wrong result. Maupertuis thus proposed to make minimum not time, but the effort that nature makes, which he called action. It, according to Maupertuis, "depends on the speed of the body and the space it passes, but it is neither speed nor space taken separately [...], it is rather proportional to the sum of the spaces multiplied by the speeds with which they are passed" [136].[62]

This definition of action is in no way justified. The suspect for a cunning reader is that Maupertuis chose an *a hoc* expression on the basis of result to obtain, known in advance. With reference to Fig. 3.9, where V is the speed of light in the more rarefied medium and W that in the denser medium, and the positions of points A and B are given, the action is defined by $V \times \mathrm{AR} + W \times \mathrm{RB}$. It should be made minimum by varying the position of R. Of course the result obtained is the correct one.

Maupertuis concluded his article by recalling the hostility of most mathematicians to the recourse to final cause, asserting that he himself partially agree with these criticism, even considering the errors on which one can fall by using it, as Fermat and Leibniz did. But for him "it is not the principle in itself that led them to error, but rather the harry [with which they applied it]" [136].[63]

In the paper of 1746 *Les loix du mouvement et du repos, déduites d'un principe de métaphysique*, Maupertuis extended his principle to mechanics; statics and dynamics

[60] p. 297; p. 425.

[61] p. 412.

[62] p. 423.

[63] p. 425.

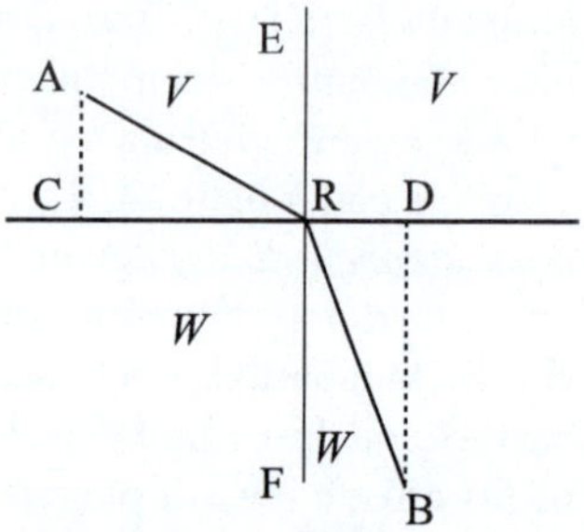

Fig. 3.9 Action and law of refraction. Redrawn from [136], p. 424

(limited to the impact of bodies). An important part of the paper is the perspective with which the principle of least action is presented. Instead to refer simply to nature which operates with the minimum effort, Maupertuis brought into play nonetheless than God himself and presented the principle of least action as a proof of the existence of God, to the extent that the original title of the paper was *The laws of motion and rest deduced from the attributes of God* [157].[64] Actually God had a twofold role. From the one hand the existence of God, which is certain, with the attribute of infinite wisdom, makes reliable the principle of least action; on the other hand the trueness of the principle of least action deduced by experimental and theoretical results is the proof of the existence of God.

The introduction of the principle of least action to mechanics is however much more complex than the application to refraction. Maupertuis has full consciousness of the fact and consulted Euler, to whom he recognized greater skill in mathematics, exchanging some letters with him on the matter. Euler appreciated Maupertuis' work and found in it some suggestions for its applications to his mechanics and to develop the calculus of variations. Below Maupertuis' principle:

> GENERAL PRINCIPLE.
> *In any change in nature the Quantity of Action necessary for this change, is the less as possible.*
> The *Quantity of Action* is the product of the mass of bodies, by their speed and by the space they travel. When a body is transported from one place to another, the action is the greater, the larger the mass; the faster the speed and the longer the space by which it is transported [137].[65] (C.13)

What leaves the greatest perplexities in the paper by Maupertuis is the choice of the mathematical expression of the action to be minimized. Apart from the arbitrariness in the choice, this definition is not clear mainly because it leaves as undefined the evaluation of the space to be considered; a problem that did not exist in the case of refraction. From the applications it appears that Maupertuis considered the space passed in the unity of time; which is the same as speed; thus the action is proportional to the square of speed.

---

[64]p. 270.

[65]p. 290.

The first situation to be examined is the impact between bodies of mass A and B moving with an initial speed a and b. If the velocities after the impact of the two bodies are in de order α and β, the change in speed is respectively $a - \alpha$ and $\beta - b$. Maupertuis, with a reasoning not fully perspicuous—at least for me and criticized by Euler also [157][66]—defined the action assuming the change of speed instead of the speed itself, as it should be according to his definition; thus the action is given by as $A(a-\alpha)^2 + B(\beta - b)^2$.

For the solution of the problem one has to introduce a constitutive relationship; in particular to specify if the two impacting bodies are hard or elastic. Using the mathematical language it can be said that the problem of least action is a problem of conditioned minimum. Thus even though admitting to establish the expression of the action without any ambiguity, it remains the fact that it is not sufficient alone to solve a given mechanical problem, sharing the fate with the principe of conservation of living forces.

In the case of hard bodies it was know from experience and natural philosophy, that after the impact the bodies move with the same speed that is $\alpha = \beta$. If x is such a speed, the action assumes the expression $A(a-x)^2 + B(x-b)^2$, which is a function of a single variable x and can minimized with the ordinary means of Calculus. In the case of elastic bodies, it is known from mechanics—and Maupertuis added no comment on this fact, by substantially minimizing the use of a principle independent of his—that the relative speeds of the two bodies are the same before and after the impact, that is it is $a - b = \alpha - \beta$, which makes determinate the problem of minimum [137].[67]

In the second situation Maupertuis passed to examine static problems, which were the object of his paper of 1740. For the sake of simplicity he limited to the case of the lever. Let A and B the masses/weights appended at the extremities of a lever whose length is assigned, it is wanted the position of the fulcrum C for the equilibrium. By indicating with z the distance CA and $c - z$ the distance CB, Maupertuis defined the action as $Azz + B(c-z)^2$. The choice of this expression is justified by Maupertuis by asserting the z and $c - z$ are proportional to the speed of A and B respectively [137].[68] By minimizing the action with respect to x, the law of lever is obtained.

3.4 Euler's Natural Philosophy and Vector Analysis

Leonhard Euler (1707–1783),[69] born in Basel, the first child of Paul Euler and Margaretha, spent his beginning youth not in Basel but in the nearby Swiss countryside Riehen. His parents were his first teachers. Margaretha, is thought to have instructed him in beginning reading. His father gave an elementary education in mathematics.

---

[66] pp. 171–172.

[67] p. 291–293.

[68] p. 294.

[69] This section report an updating of a previous paper [29].

Like his teacher Jakob Bernoulli, Paul Euler taught his young son mathematics not as an isolated discipline but as underlying all natural knowledge, interrelated with other fields.

Euler, who required instructional preparation completed, was sent by his parents to Basel, perhaps as early as the age of eight. By the second decade of the 1700s Basel was no longer in its golden period; nonetheless the presence of Jakob and Johann Bernoulli made it a center for scientific and mathematical research. In 1715 Paul Euler hired as a private tutor Johannes Burckhardt (1691–1743), a young theologian with a tolerable background in mathematics. At the time Burckhardt was supporting Johann Bernoulli in arguments with Brook Taylor and other members of the Royal society of London over which was superior, Leibniz's differential calculus or Newton's.

After completing his gymnasium education in 1720, Euler registered at the university of Basel for courses in the philosophical faculty, which covered fields of learning outside recognized professions. It was the equivalent of the modern secondary school. The philosophical faculty imparted a general education before a student chose a specialty for a higher degree. Through hard work and an astonishing memory, Euler mastered all of his subjects. During his first two years he was enrolled in Johann Bernoulli's class for beginners in geometry as well as practical and theoretical arithmetic. He spent two years earning his *prima laurea*, roughly equivalent to a bachelor's degree, receiving it in 1722 at the age of fifteen. In 1723 he passed the examination of the philosophical faculty for the master of arts degree, and officially received it in 1724 at the age of seventeen.

Euler's increasing attention toward mathematics and natural philosophy did not please his father, who obliged him to register in the theology faculty in 1723 in preparation for taking holy orders. He felt fortunate to continue Saturday meetings with the stern Johann Bernoulli. While likely concerned about employment for his genial son, Paul Euler accepted a Johann Bernoulli's request for Euler's shift out of theology.

All suggests that Euler examined such classics as the second edition of Copernicus's *De revolutionibus orbium coelestium* as well as Kepler's *Astronomia nova* and Galileo's *Dialogo sopra i due massimi sistemi del mondo*. Euler's master's lecture shows that he was studying Descartes's *Principia philosophiae* and *La géométrie*. He possibly also read Rohault's Cartesian masterful *Traité de physique*, a major physics text of the late 17th century, which appeared in 1671.

The interests of Euler's teacher make it probable that he read a range of works relating to the new Calculus and its applications, beginning with the *Analyse des infiniment petits pour l'intelligence des lignes courbes* published in 1696 in Paris, attributed to de L'Hôpital. Euler possibly examined two texts of Varignon, a correspondent, disciple and friend of Johann Bernoulli: the *Projet d'une nouvelle méchanique*, published in 1687, and the *Nouvelles conjectures sur la pesanteur* of 1690. Varignon's *Nouvelle méchanique ou statique* and *Éclaircissemens sur l'analyse des infiniment petits*, both from 1725, may also have been available. Euler must have studied Jakob Bernoulli's articles on the theory of infinite series, published from 1682 to 1704 and reprinted in 1713; his *Ars conjectandi* on probability, with a preface by Nikolaus I, published posthumously in 1713; and Jakob Hermann's *Phoronomia, sive de viribus et motibus corporum solidorum et fluidorum* of 1716. Other books which Euler referred to are John Wallis's *Arithmetica infinitorum* (1656) and Brook Taylor's *Methodus incrementorum directa et inversa* (1715). No confirming evidence exists on whether Euler saw Johann Bernoulli's pioneering articles for the Académie des sciences de Paris in 1718 on what would become the calculus of variations [24].

Euler was called to the Academia scientiarum imperialis petropolitanae of St. Petersburg by his friend Daniel Bernoulli in 1727. He remained there until 1741 to reach the Académie royales des sciences et belles lettres de Berlin. Here he had some problems with Frederick II that finally pushed him back to St. Petersburg in 1766 where he remained until his death in 1783, at the age of 78 and almost completely blind. One of the most prolific writers, among the greatest mathematicians of all time, Euler treated all the themes of physics, with an approach, however, more as a 'mathematician' than an experimental physicist; from astronomy to optics, from electricity to magnetism, from hydraulics to mechanics, to music, leaving an indelible mark in all sectors. He also wrote many pages of philosophy of nature, still of interest today because written with the sober language of the mathematician. His writings, more than twenty books and pamphlets and about 800 papers, are collected in his *Opera omnia* [90]. It remains to complete the correspondence, written in several languages. Most are in Latin and French, some in German and Russian (Euler also had competence in Italian, Spanish, Chinese and Japanese). Despite his strong religiousness Euler was an illuminist philosopher-scientist and his faith in reason transpires in all his writings.

3.4.1 Philosophy of Nature

Since entered the university of Basel, Euler, as discussed just above, had been deeply interested in the studies of theology and natural philosophy and only seventeen he gave a lecture for his master degree comparing the natural philosophies of Descartes and Newton.

Certainly he was not a canonical philosopher but he reflected at length on the classical themes of philosophy of nature. He was not interested in discussions of an abstract character and even for that he was generally hostile to the approach to the natural philosophy and mechanics of Leibniz's school, represented in his time by Christian Wolff (1679–1754), a colleague of his at the academy of Berlin. Studying Euler's philosophy is challenging not so much to understand his works on mechanics or physics (modern sense), but rather to see how natural philosophy was being transformed into his hands. It was indeed his mathematics that influenced his philosophy and not *vice versa*; in particular his research on Calculus influenced his conception of matter and space. Euler had no particular reason to stand on one side or the other in the philosophical debates of the period though he had a profound estimate of Newton referring to him as the *Summus Neutonus*; not being a canonical philosopher he could limit himself to accepting those ideas that seemed closer to his sensitivity as a mathematician without looking deeply into their consistency.

Euler's ideas about philosophy of nature are scattered everywhere. The natural references are however the *Anleitung zur Naturlehre* [89] of the 1750s but published only in 1862, referred in the following as the *Anleitung* the and *Lettres à une princesse d'Allemagne* [86] of 1760s, referred in the following as *Lettres*. The *Anleitung* did not spread greatly, even because of his late publication. The *Letters* had instead a different fate. Michael Faraday (1791–1867) for instance read the letters [24][70] but not the *Anleitung* which was probably read instead by Bernhard Riemann (1826–1866) [155].[71]

Euler's philosophy of nature was generally well received, praised by Voltaire for instance. According to Alexandre Koyré the *Lettres* may be included among prominent Newtonian popularizations [111].[72] But also by Leibniz, even though here and there Euler criticized his philosophy of nature. It should be cited however a strong criticism by three scholars well connected with Euler, d'Alembert and Lagrange and Daniel Bernoulli. They judged him severely. Lagrange wrote to d'Alembert, "Our friend is a great analyst but quite poor a philosopher" [116].[73] D'Alembert replied indignant: "It is incredible that such a great genius as him on geometry and analysis is in metaphysics so inferior to the smallest schoolboy, not to say so flat and so absurd, and it is the case to say: *Non omnia eidem Dii dedere*" [116].[74] Daniel Bernoulli in turn complained Euler's lake of competence in philosophical matter with the Swiss astronomer Johann Jakob Huber (1733–1798): "Mr. Euler is an admirable man, when the principles are well established; but I do not ordinarily like him in the examination of principles. The physical and mainly metaphysics are out of his reach and it is a great misfortune for this excellent man to confuse his strength and his weakness"[75]

3.4.1.1 Anleitung Zur Naturlehre

The *Anleitung*, has been received incomplete. The end of Chap. 5 and the beginning of Chap. 6 are missing (which probably occupied about 7 pages). There are reasons to think that, even with these pages, the work could not be not complete and we can therefore ask why Euler did not finished it [155]. In its current form the book consists of 21 chapters that can be grouped as follows:

1. Chapters 1–5, the role of natural philosophy and the main properties of matter;
2. Chapters 7–9, general introduction to the principles of mechanics;
3. Chapters 10–11, apparent motion, general rules of motions;
4. Chapters 12–18, the property of matter;
5. Chapters 15–18, possible forms of matter;
6. Chapters 19, gravity;

---

[70] p. 565, note 35.

[71] p. 46.

[72] p. 18.

[73] vol. 13, p. 135. Letter of Lagrange to d'Alembert, 16th June 1769.

[74] vol. 13, p. 148. Letter of d'Alembert to Lagrange , 7th August 1769.

[75] Letter of Daniel Bernoulli to Johann Jakob Huber, 27th July 1756. Quoted from [147], p. 455.

7. Chapters 20–21, principles of hydrostatics and hydrodynamics.

The opening of the *Anleitung* suggests a traditional treatise on natural philosophy or natural science, even of Aristotelian mould. According to Euler, *natural science* (the name he gave to natural pilosophy) aims to explain the causes of changes that occur on material bodies. Wherever there is such a science, he continued, it is very incomplete, since it is able to state with certainty the causes of only very few changes. Moreover, changes are restricted to those occurring on inorganic or material bodies, thus distinguishing natural science from the science of the mind, which aims to explain mental changes. Nothing is said about living beings [89].[76]

All changes involving material bodies must arise from the essence and from the properties of bodies themselves. What is common to all material bodies without exception is called a property of the bodies and therefore all things not sharing this property are excluded from the domain of material bodies. The general properties of material bodies, are those shared without exception. The essence of material bodies is a property that is not only shared by all of them but such that all things having this property must of necessity be considered material bodies [89].[77]

Following the reading of the *Anleitung* it is understood however that the causes and the properties of which Euler spook are those useful for a mathematical treatment of the philosophy of nature, adopting a mechanicistic approach; they are extension (that is the position or configuration), mobility (velocity and acceleration), persistence or inertia (mass) and impenetrability (force, because forces are due to impenetrability). Euler, wanted to found a coherent system, able to describe all the inorganic nature, starting only from some principles of mechanics and a few other hypotheses. Even Newton, another 'mathematician' who wondered about the relationship between philosophy of nature and mathematics, in some of his *Queries* (28–31) had sketched some ideas in this direction, but they remained only queries.

A general property of material bodies is extension, and anything that has no extension cannot be regarded as a material body. We are not only convinced by our experience that all material bodies that we know possess extension, but our concept incorporates extension in such a way, that we can exclude all things without extension from the category of material bodies. It follows then that whatever can be said of extension *per se*, can without exception also be said of material bodies. Everything with extension is divisible, and divisibility can be continued ad infinitum; therefore all material bodies are infinitely divisible [89].[78] However Euler did not enter the question of the actual corpuscular (atoms) or continuum constitution of matter. What he said is that one knows from experience that the actual subdivision of many material bodies can be carried out to an astonishing degree, and that our tools and senses are too blunt to permit this subdivision to be carried even further. He was not discussing what can actually be done, but rather the merely the possibility of taking the subdivision even further.

For liquid matter Euler however denied the existence of elementary indivisible particles or atoms.

---

[76] vol. 2, p. 449.

[77] vol. 2, pp. 450–453.

[78] vol. 2, pp. 453–455.

> A liquid matter cannot be formed from a number of small particles that are solid and hard, for whatever the shape and arrangement of the particles might be, for it is not possible that a pressure that acts at one location will propagate in all directions with equal force [89].[79] (C.14)

An important concept that Euler considered worthy to be exposed in his foundational text is what today is called potential energy or possibly work, to which he referred to as the *effectiveness*, and to which he also assigned the name *effort des forces sollicitantes* [82].[80] The effectiveness of a force is the integral quantity that is found if one multiplies the force with the differential of the distance to which it pushes the body, and then integrates. This concept is of the utmost importance, because (1) the sum of the effectiveness of all forces $\int Pdx + \int Qdy + \int Rdz$ always has the same magnitude—given by the increase of living force—for any three different coordinate planes, that is, in modern term, it is invariant with the change of coordinates. (2) The whole theory of equilibrium is based on it. For it can be shown that equilibrium cannot occur, if the sum of the effectivenesses is not a minimum, or occasionally a maximum: "This marvelous theorem was first derived by the world renowned President de Maupertuis and is closely connected with the other general principle of frugality. From this we see at the very least that the effectiveness has a major influence on all motions that can be produced by forces, and that *it deserves to be given a special name* [emphasis added]" [89].[81] It also must be remarked however that Euler seems not to be conscious that he had to limit the nature of forces in his definition; only for what are now called conservative forces—for which the integrals appearing in the definition of effectiveness are independent of the path—Euler considerations has any meaning.

It may seem strange that in a treatise dealing with general aspects, Euler had also found ample space to exquisitely technical aspects concerning the laws of hydrostatics and hydrodynamics (Chaps. 20–21). It must be kept in mind however that for Euler fluids were not only the ordinary ones, for example water and air; also the aether (see next sections), the medium through which light propagates, was a fluid and as such subjected to the laws of hydraulics.

David Speiser, with a bit of whiggism, suggests that Euler was the first to imagine a unified field theory, an idea and a hope that are at the center of the ambitions of physicists [155].[82] In fact, according to Euler all physical phenomena, with the possible exclusion of gravity (see below), can be reduced to the interactions of four scalar fields (the densities and pressures of gross matter and aether) and two vector fields (the velocities of matter and aether), interconnected and governed by partial derivative equations. Seen as an attempt, Euler's way of proceeding with hydrodynamics is not very different from that used in modern physics to unify field theories,

---

[79] vol. 2, p. 526. Translation into English by Hirsch E.

[80] p. 287. It must be said that Daniel Bernoulli in his work of 1750, *Remarques sur le principe de la conservation des forces vives pris dans un sens général*, introduced something like the effectiveness [7], p. 359.

[81] vol. 2, p. 493. Translation into English by Hirsch E.

[82] p. 45.

despite the present more in-depth knowledge of the structure of matter and the huge amount of accumulated empirical data. Even today, the basic differential equations of a field of physics, whose theory is known, are also considered as fundamental for other sectors, in which a theory is still lacking. In the case of Euler, the basic differential equations are those of hydrodynamics and the radical hypothesis is that of an unique aether [155].[83] Despite the considerations of Speiser, which seem convincing, in the *Anititung* there is however no mention of the aethereal fluid, neither for what concerns electricity nor for what concerns magnetism. Topics that will be treated with plenty in the *Letters*.

3.4.1.2 *Lettres à* une Princesse d'Allemagne

At the time of the publication of the *Letters*, Euler stood at the peak of his career. He was the director of the mathematics class (division) of the St. Petersburg academy and a well known mathematician. Although he sent the individual letters between 1760 and 1762, he had derived them from articles dating back to 1720s. Thus, notwithstanding the didactic character, they offer an outline of his scientific ideas and the modifications occurring in them along a long period of time.

The letters, 234 in number (a thousand pages), originated from lessons delivered to the princess Charlotte Ludovica Luisa, a relative of Frederick II of Prussia, but were conceived more generally for young students. They met with prodigious success; by the turn of the 19th century they had been translated from the original French into eight other languages: Russian, German, Dutch, Swedish, Italian, English, Spanish and Danish. The numbering of the letters is different from edition to edition, here that of 1770–1774 is considered [86].

Euler's letters essentially concern philosophy of nature seen from different points of view. That of the canonical philosopher, with considerations of *physica generalis* and that of the philosophers emerging at the time. It is not easy to group them by themes; it can be said that the first letters (about 140) have a more general character; the others are more technical and deal with current topics in physical research, among them at least seventeen letters are devoted to electricity and nineteenth to magnetism.

Among the first letters there are five dedicated to music (4–8). We are in a period in which the musical theory was undergoing somehow a revolution with Jean-Philippe Rameau (1683–1764) among the protagonists. Euler, only twenty four, had written an interesting paper on music, the *Tentamen novae theoriae musicae* of 1730–1731 [69], which was criticized by Rameau. The two 'musicians' exchanged some letters about consonances in the 1750s [24].[84]

Letters (17–44) follow, that deal with optics. The first four of the theory of propagation of light with particular reference to Newton; only a nod is made at Descartes; the others face problems of geometrical optics, of brightness and color of bodies, and also of the structure of the human eye. Euler compared the 'hemanatistic' (pro-

[83] p. 46.

[84] pp. 362–363.

jectile) theory attributed to Newton with his vibrational theory, reported in particular in the *Nova theoria lucis* of 1744 [72]. He did so without mentioning Huygens who had proposed a theory in which light spread in a medium as pulses.[85] According to Euler, to explain the nature of light and color, it must be admitted celestial spaces filled with a subtle matter, he called aether. A fluid matter like air but much thinner; nevertheless, with much greater elasticity (using a modern term, stiffer than air). And as in the air sound is propagated, so light propagates in the aether [86].[86]

In the letters from 45 to 79 there are general considerations on mechanics. They concern gravity in a general sense, the law of universal gravitation, to conclude with the explanation of gravity, associated with the elasticity of the aether, without however going into details. It is here that one sees that a mature Euler is not fully satisfied of his mechanical explanation of gravity. Indeed, in the letters 46 and 54, for instance, he wrote that "philosophers have warmly disputed, whether there actually exists a power which acts in an invisible manner upon bodies; or whether it be an internal quality inherent in the very nature of the bodies, and, like a natural instinct, constraining them to descend" [86],[87] "On this question philosophers are divided. Same are of opinion that this phenomenon is analogous to impulsion; others maintain, with Newton, and the English in general, that is it consists in attraction [86].[88] Thus declaring his incapacity of a definitive choice.

Letters from 80 to 87 concern aspects of natural philosophy, bodies and spirits. In the letter 76 Euler criticized Wolff's conceptions of dynamics. He did this by referring substantially to what he had already written in his *Gedancken von den Elementen der Cörper* of 1746 [72] (see below).

Letters 88 to 114 address issues of ethic, psychology and logic. Particularly in the letters 103–105 the diagrams of Euler–Venn are used to explain some logical relations. The simplest case is that reproduced in Fig. 3.10.

It is true that the idea of representing graphically logical relationships was not completely new; this had happened in some treatises of the 18th century. But it was to Euler that the mathematician John Venn (1834–1923) referred to in his studies on logic more or less a century later.

Letters from 115 to 132 concern essentially epistemology, from which the influence of Descartes and Locke and perhaps even of Condillac transpire—without explicit reference to them—and letters of metaphysics in which Euler repeated the criticism toward Leibniz's monad system.

To illustrate in a clear way Euler's reasoning in philosophical themes, in the following it is fully reported the letter 115, which discusses the different types of knowledge: rational, empirical and moral. It should be noted that when Euler named rational knowledge, he referred explicitly only to geometry. Nothing is said about mechanics, that in his treatises he always regarded as a purely rational discipline.

---

[85] About of the difference between the conceptions of Euler and Newton and for a justification of the absence of the name of Huygens in Euler's writings of optics see [100].

[86] [vol. 1, Letter 19.

[87] vol. 1, Letter 46, p. 195.

[88] vol. 1, Letter 54, p. 230.

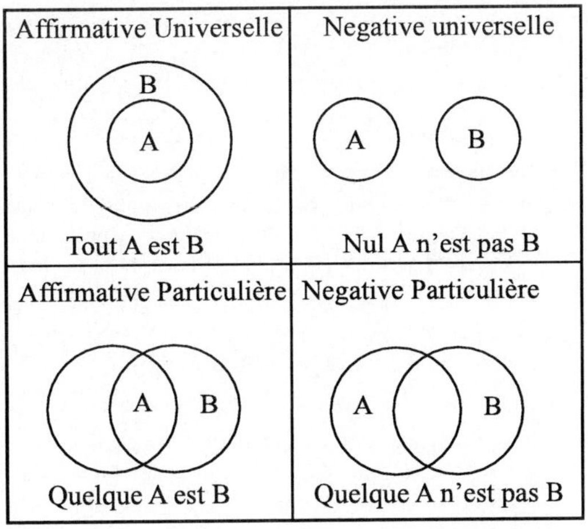

Fig. 3.10 Simple diagrams of Euler–Venn. Redrawn from [87], p. 101

Letter 115. *The true Foundation of human Knowledge. Sources of Truth, and Classes of Information derived from it.*

Having taken the liberty to lay before you my opinion respecting the most important article of human knowledge, I flatter myself it will be sufficient to dissipate the doubts which naturally arise out of the subject, from want of exact ideas of the liberty of spirits.

I shall now have the honour of submitting to your consideration the true foundation of all our knowledge, and the means we have of being assured of the truth and certainty of what we know. We are very far from being always certain of the truth of all our sentiments; for we are but too frequently dazzled by appearances, sometimes exceedingly slight, and whose falsehood we afterwards discover. As we are, therefore, continually in danger of deceiving ourselves, a reasonable man is bound to use every effort to avoid error, though he may not always be so happy as to succeed.

The thing to be here chiefly considered is the solidity of the proofs on which we found our persuasion of any truth whatever, and it is absolutely necessary that we should be in a condition to judge if they are sufficient to convince us or not. For this effect I remark, first, that all truths within our reach are referable to three classes, essentially distinguished from each other.

The first contains the truth of the senses; the second those of the understanding; and the third those of belief. Each of these classes requires peculiar proofs of the truths included in it, and in these three classes all human knowledge is comprehended.

Proof of the first class are reducible to the senses, and are thus expressed:

This is true for I saw it, or am convinced of it by the evidence of my senses.

It is thus I know that the magnet attracts iron, because I see it, and experience furnishes me with incontestable proofs of the fact. Truths of this class are called sensible, because they are founded on the senses, or on experience.

Proofs of the second class are founded in ratiocination; thus:

This is true, for I am able to demonstrate it on principles of just reasoning, or by fair syllogisms.

To this class, principally, logic is to be referred, which prescribes rules for reasoning consequentially. It is thus, we know, that the three angles of a rectilinear triangle are together equal

> to two right angles. In this case I do not say I see it, or that my senses convince me of it; but I am assured of it's truth by a process of reasoning. Truths of this class are called intellectual, and here we must rank *all the truths of geometry, and of the other sciences* [emphasis added], in as much as they are supported by demonstration. You must be sensible, that such truths are wholly different from those of the first class, in support of which we adduce no other proofs but the senses, or experience, which assure us that the fact is so, though we may not know the cause of it. In the example of the magnet, we do not know how the attraction of iron is a necessary effect of the nature of the magnet, and of iron; but we are not the less convinced of the truth of the fact. Truths of the first class are as certain as those of the second, though the proofs which we have of them are entirely different.
>
> I proceed to the third class of truths, that of faith, which we believe, because persons worthy of credit relate them; or when we say:
>
> *This is true, for several creditable persons have assured us if it.*
>
> This class, accordingly, includes all historical truths. You believe, no doubt, that there was formerly a king of Macedon, called Alexander the Great, who made himself master of the kingdom of Persia, though you never saw him, and are unable to demonstrate, geometrically, that such a person ever existed. But we believe it on the authority of the authors, who have written his history, and we entertain no doubt of their fidelity. But may it not be possible that these authors have concerted to deceive us? We have every reason to reject such an insinuation, and we are as much convinced of the truth of these facts, at least of a great part of them, as of truths of the first and second classes.
>
> The proofs of these three classes of truths are extremely different; but if they are solid, each in it's kind, they must equally produce conviction. You cannot possibly doubt that Russians and Austrians have been at Berlin, though you did not see them: this, then, is to you a truth of the third class, as you believe it on the report of others; but to me it is one of the first class, because I saw them, and conversed with them, and as many others were assured of their presence by means of other senses. You have, nevertheless, as complete conviction of the fact as we have. 31st March, 1761 [88].[89]

From letter 133 onwards, physical problems are addressed. In particular letters from 138 to 154 deal with electricity and letters from 169 to 186 with magnetism.

Apart from the *Letters*, the only Euler's left written work on the theory of electricity is a comment of a paper by Aepinus [83]. This notwithstanding he considered electricity a very interesting field to be explored:

> The subject which I am now going to recommend to your attention almost terrifies me. The variety it presents is immense, and the enumeration of facts serves rather to confound than to inform. The subject I mean is electricity, which for some time past has become an object of such importance in physics that everyone is supposed to be acquainted with its effects [86].[90] (C.15)

Euler was primarily interested in the electric field, not in the charge (he was among those who rejected the idea that there were two types of electricity). For him electricity is due to the presence of one electric fluid, the aether which in normal circumstances fill the pores which travers all bodies in all directions. The electric qualities of a body depend on how easy or how difficult the pores are to open. In conductors (modern term)—metals and such—the aether can easily flow in and out; in this case, the pores

[89] vol. 1, pp. 448–451.

[90] Letter 138, vol. 3, p. 277. English translation in [154].

are always open, so to speak. With other materials a certain strength is required to open the pores through friction: these are the insulators (modern term).

To magnetism instead Euler dedicated some papers and letters, see for instance [74, 75, 149]. Magnetism also is associated to a fluid. Originally—in the *Anleitung* for instance—in Euler's opinion the same aether was responsible for gravity, magnetism, electricity and light. In the *Lettres*, in particular in Letter 176, instead he was of the opinion that the aether associated to magnetic forces was a special kind of subtle matter and not simply the subtlest part of a single all pervading aether [86].[91] He seems never to have had any doubt that the light aether is responsible for electric phenomena as well.

3.4.1.3 Matter and Its Properties

Euler dealt at length with physics and metaphysics of bodies, matter and related properties here and there in his scientific works. However, there are some papers in which he focused specifically on the subject. Among them: *Dissertatio de igne, in qua ejus natura et proprietates explicantur* written in 1737, *Gedancken von den Elementen der Körper* of 1746,[92] *Recherches physiques sur la nature des moindres parties de la matiere* of 1746, *Réflexions sur l'espace et le tems* of 1748, *Recherches sur l'origines des force* of 1750, *Lettres à une princesse d'Allemagne sur divers sujets de physique & de philosophie* of 1760–1761, *Anleitung zur Naturlehre* of 1750s.

According to Euler, bodies are characterized by the following general properties: extension, mobility, inertia and impenetrability. Among them the fundamental is impenetrability which as such should be considered the very essential property of bodies.

> Whatever is impenetrable belongs to the category of bodies, and therefore the essence of bodies is their impenetrability, on which therefore all the other properties must be founded [89].[93] (C.16)

A body has extension in common with space, mobility with moving images projected on a wall, both of which however do not possess the property of impenetrability.

In [97], it is suggested that the idea that impenetrability constitutes the essence of body depends upon two Euler's claims: it is unique and necessary to a body and it is irreducible, in the sense that conceiving a body as impenetrable is a primitive clear notion. Both Boscovich in the *Theoria philosophiae naturalis* of 1758 and Kant in the *Metaphysische Anfangsgründe der Naturwissenschaft* of 1786, thought impenetrability of bodies as an obscure idea requiring elucidation in terms of a more basic notion, that of repulsive force. For them, repulsive force was the physically primitive notion not requiring further elucidation, just as attractive force was physically

---

[91] Letter 176, vol. 3, pp. 118–119.

[92] Euler published anonymously *Gedancken von den Elementen der Körper*, an anti-Wolffian work. It was immediately recognized as bearing his signature however.

[93] vol. 2, p. 472.

primitive [97].[94] Considering that Euler was deriving force from impenetrability (see below), his is exactly the opposite path than that of Boscovich and Kant.

According to Euler, from impenetrability, extension and mobility follow. His argumentations on these points have more a rhetorical than demonstrative value, but are equally interesting. From the very convincing thesis (a) no-extension → no-impenetrability, from the rule of classical logic, it follows (b) impenetrability → extension. More difficult is to argue (c) impenetrability → mobility; indeed a thing can be at rest though impenetrable, thus impenetrability is not sufficient for actual motion; Euler maintained it was sufficient instead for possible motion, thus implication (c).

Inertia, called persistence (Standhaftigkeit) in the *Anleitung*, is also associated by Euler with impenetrability. But the derivation (d) impenetrability → persistence, has more problems than derivations discussed previously. Euler's strategy is to prove the implication (e) mobility → persistence, that with the implication (c) would give (d). The derivation of the implication (e) implies the principle of inertia—"because when a thing is mobile, it should also have persistence, since otherwise any change would occur without sufficient reason" [89][95]—that Euler, like d'Alembert, based on the principle of sufficient reason, which has been the object of criticism by many scholars. Because implication (d) would derive from two weak implications (c) and (e) and it would consequently be two times weak. For a different view on this point see [97].[96]

Bodies are portions of matter, thus Euler devoted a lot of attention to the properties of matter. The *Recherches physiques sur la nature des moindres parties de la matiere*, a relatively early work, began with the following question: "It is an important question in physics and metaphysics to establish wether or not the smallest part of matter are similar to each other" [73].[97] Euler tried to answer this question coming to the conclusion that all the smallest parts of bodies, referred to as the *smallest molecules* of matter, are made of the same stuff. In order to speak about the smallest molecules it is necessary however to distinguish between the properly said matter (groben Materie), *gross matter*, characterizing the inertia of bodies to which only the term molecules is referred to, from another type of matter, which is usually not recognized as such, called *subtle matter*, interposed among molecules (Notice that notwithstanding Euler used the term molecules, he said nothing about a corpuscular conception of matter; that is the molecules may be not corpuscles).

Following these premises, repeated some years later in the *Anleitung*, Euler argued that also the subtle matter probably is of one kind only; so in the world there are probably only two kinds of matter; a subtle matter and a gross matter. The different apparent nature and density of bodies depends on a different combination of subtle matter and gross matter, with the subtle matter that fills the pores of the gross matter, that may be more or less numerous and great. The molecules of gross matter have no

[94] p. 140.

[95] vol. 2, p. 472.

[96] pp. 141–142.

[97] p. 289.

pores, and even if they cannot be divided in act, being extended they can be divided with the imagination.

> There will be thus at least two main kinds of matter; one [the gross matter] which gives the fabric of sensible bodies, whose the particles have all an unchangeable degree of density,[98] which is even greater than the apparent density of gold; the other [the subtle matter] kind of matter will be that of which the subtle fluid which cause the gravity is composed, and we call the aether [73].[99] (C17)

The thesis that there are only a few kinds of matter is carried out by Euler with both physical and metaphysical arguments. From a metaphysical point of view he invoked a principle of economics. While for the gross matter Euler arrived at a more or less convincing argument that it is only of one kind, for the subtle matter he had less certainty.

To prove the uniqueness of gross matter, after making considerations based on daily experience that lead to believe that there is a single type of (gross) matter, Euler passed to examine two laws of empirical nature. The first law concerns the fact that the force of gravity varies with the inverse of the square of the distance, as Newton has shown (Euler's words) and thus the weight of bodies decreases with the square of their distance from the center of the earth. The second empirical law concerns the fact that bodies of different weight fall with the same temporal law (in vacuum).

From the law of the inverse of square, Euler proved that the density and thus the pressure of the aether towards the centre of the earth which causes gravity (see below) must decrease with inverse proportion of the distance. Thus, if x is the distance from the center of the earth, the pressure of the aether should vary as $h - A/x$, being A a constant of proportionality, and h the pressure of the aether when at rest, as it is for $x = \infty$. From this expression of the pressure, it is easy for Euler to prove that the weight of a body is proportional to the volume of the gross matter (what Euler called the *true size*) it contains. Then, from the independence of the temporal law of fall for bodies of different weight, and from the law of mechanics (force and mass are proportional), it derives that weight and mass are proportional. But for Euler mass is the measure of persistence; this means that if one takes two whichever piece of gross matter with the same true volume, they have the same values for all the four essential properties of matter—extension by assumption, mobility and impenetrability (not measurable magnitudes) because they are shared by all bodies, and persistence as just proved—thus they are perfectly equal, and thus all gross matter is of the same stuff [89].[100]

A first draft of the hypothesis of the existence of a subtle matter or aether can be found in the *Dissertatio de igne, in qua ejus natura et proprietates explicantur* that Euler wrote in 1737 for the award of 1738 proposed by the Académie des sciences de Paris, which was assigned to him [78]. To explain the phenomenon of combustion, Euler claimed that the flame arises from the sudden explosion of molecules of "materiae subtilis igneae compressae" and to illustrate his conjecture

[98] The density does not vary in time, that is gross matter is incompressible [89], vol. 2, p. 510.

[99] p. 300.

[100] pp. 542–545.

he used a corpuscular model: "Let's imagine a great quantity of glass spheres, full of air strongly compressed, and this mass of spheres is the material we want. Suppose that a force intervenes just sufficient to break a single sphere; it is clear that both the impetus of the air and the projection of the fragments of glass will produce a similar effect in the nearby spheres and then from these in all the others, until all are broken, emitting with immense roar the air that they enclosed" [79].[101]

Euler went on to argue that the heat given off by the fire, the rapid spread of the flame in all directions seems to violate the ordinary laws of mechanics and must be explained in another way; to maintain the laws of mechanics it is necessary to resort to a further postulate: the existence of an aether, a substance distinct from the igneous particles. The flame does not instantly disperse in the air due to the resistance of this aether, much more elastic and rarefied than the igneous matter, which surrounds and holds the flame in an unstable equilibrium. The continuous shocks that are created at the limit between the aether and flame give place to vibrations that generate light and transmit it in a straight line [79].[102] To postulate an aether was certainly not new in the 18th century and many variants were formulated. For a discussion of the various hypotheses discussed in the period 1740–1750 by physicists see [25].

The properties of subtle matter had been discussed in depth in the *Recherches physiques sur la nature des moindres parties de la matiere* and in the *Gedancken von den Elementen der Körper* (to be taken again in the *Anleitung*). In the second part of the former of the two papers, Euler first denied the possibility of the existence of vacuum with metaphysical considerations; scarcely convincing indeed. While he did not mention a physical more convincing (for us) motivation, namely that the presence of vacuum would not justify the oscillatory theory of light proposed in his *Nova theoria lucis*, where propagation is conceived, in analogy to that of sound, as elastic waves propagating in the medium.

The subtle matter is very different from the gross matter; in particular it is not gross matter made of very tiny corpuscles separated by vacuum, but rather a continuous substance or, with a technical modern word, a continuum. In the *Anleitung*, Euler spoke at length about his continuous *aether* and did not feel embarrassed to introduce a concept, continuity, which was alien to the dominating mechanical philosophy where the aether had usually a corpuscular nature. Euler's aether fills the whole space, leaving no emptiness. Air consists essentially of aether with small corpuscles of gross matter. While the gross matter is incompressible, the aether is compressible:

> Gross matter is therefore not capable of any change other than in the appearance of its shape, which, if appropriate forces are available, can be changed in arbitrary ways. [...] It does not appear to be true that subtle matter also has always and everywhere the same density, such that it could through no force be driven into a smaller space. Instead an important difference between gross and subtle matter seems to be that the latter can be compressed [89].[103] (C18)

The aether although has the property to expand or shrink and behaves like an elastic medium, has a certain density that is proper to it. However, in no case it is possible to

[101] p. 10.

[102] pp. 18–19.

[103] vol. 2, pp. 511–512. Translation adapted from Hirsch E.

reduce a portion of aether to a point, that is to effectively annihilate it [89].[104] Euler clearly stated that the compressibility of subtle matter does not imply the violation of its property of impenetrability.

Euler did not propose a convincing, at least for a modern, explanation of the force/pressure caused by the elasticity of aether which is not deducible from any of the typical properties of matter, extension, inertia, impenetrability, mobility, and therefore should be treated as a primitive (essential?) concept. Forces on gross matter are explained in Euler's mechanics by means of impenetrability and inertia. For instance, suppose an elastic body in motion that impacts with a rigid surface; it receives a force and rebounds because of the impenetrability of the surface and the resistance to change its motion from the inertia of the ball. The pressure of the aether against a rigid surface cannot be explained in the same way because it acts on static situation also and the inertia of the aether is assumed to be negligible: "because experience shows that the celestial bodies do not suffer effects with their impact with the aether that fills the universe" [89].[105] Certainly one can see an analogy with gases; but in this case he (Euler himself) can resort to a mechanicistic explanation; this was what Daniel Bernoulli did in his *Hydrodynamica* of 1738: the pressure of a gas is determined by the impact of the particles on the walls of the vessel that contains it [6].[106] But for Euler the aether is not formed by particles.

Mature Euler also doubted that there is on kind only of subtle matter: "Whether there are several kinds of this subtle matter, some of which are denser than others, we shall not be discussing here, but if indeed there are several kinds, we shall refer to them collectively as subtle matter. As long as the explanation of what occurs in nature does not require several such kinds, it would be bold and against the rules of a sound science of nature if, merely following our imagination we were to increase the number of kinds of subtle matter" [89].[107]

According to him, the aether that is found in the universe is compressed far beyond its natural state and consequently exerts a great elastic force on the bodies formed by the gross matter [89],[108] able to explain the resistance of the materials, their elasticity and the force of gravity.

The resistance of solid bodies is associated with the hydrostatic pressure that the subtle compressed matter exerts on the various particles of ordinary gross matter. An explanation similar to that suggested by Galileo with the use of horror vacui. When two smooth grains of gross matter come into contact with each other, the elastic force of the aether makes them adhere strongly and the breaking strength is maximum. If, on the other hand, the contact between the grains is not complete, then the breaking strength will be the minor the minor the contact surface.

To show this more clearly, let the two bodies, or better two portions of a body, of Fig. 3.11, ABCD and ABEF, be joined at the surface AB such that between them the

[104] vol. 2, p. 513.

[105] vol. 2, p. 542.

[106] pp. 200–203.

[107] p. 509.

[108] vol. 2, p. 517.

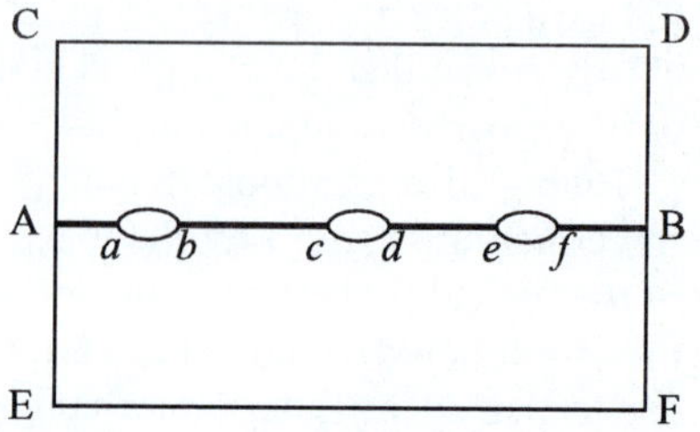

Fig. 3.11 Strength of bodies [89], vol. 2, Table XX, Fig. 235

cavities *ab*, *cd*, *ef* are filled with the aether. The body ABCD on the one hand, will then be forced against the other body by the aether that presses on the planes CD and EF; on the other hand it will be pushed back by the aether in the cavities *ab*, *cd*, *ef*. Therefore the force with which the body ABCD is pressed against the other body ABEF is the result of the two contrasting forces.

The lack of resistance in fluids is explained by the fact that here gross matter is so suffused by the aether that the particles have nowhere immediate contact. To see how this could happen, imagine that every particle of gross matter is surrounded by subtle matter, and the particles consequently never approach each other so closely that there would not remain some subtle matter between them. If every particle were at rest, such a mixing would be hard to understand; but if the subtle matter is in motion so that it continuously flows between the gross particles, then in this way immediate contact can be impeded [89].[109]

The elasticity is explained by admitting that inside bodies there are cavities (pores) filled with subtle matter that do not communicate with the outside. When a body is deformed, either compressed into a smaller space or expanded into a larger one, then there will be change in the size of its pores, some being expanded, but others being compressed, with the elastic aether contained in the pores that follows the same fate contrasting the change of size [89].[110] The elasticity of commonly experienced bodies is thus explained by means of the elasticity of the aether, which is clearly a vicious circle. Thus what Euler actually explained was not the elasticity itself but rather why bodies made by an incompressible matter exhibit elasticity.

The account above referred about elasticity explains the behavior of the various bodies. The difference in elasticity lies in the diversity of the dimensions and arrangement of the pores. The loss of elasticity as a result of heating is explained by assuming that the subtle matter contained in the pores is set into motion by heat, opening up access to previously closed pores; at the end the matter has less and larger pores than before. Therefore even though before heating there were large forces, these can vanish after heating.

The weight of the bodies, that is gravity, is explained by admitting that the density, and therefore also the pressure of the aether surrounding the earth, increases by increasing the distance from its center. Euler is very precise in his explanation.

[109] vol. 2, p. 535.

[110] vol. 2, pp. 537–538.

Gravity should be associated to the gradient of pressure in the aether around the earth that determines the movement of the aethereal fluid downward with the consequent downward drag of the gross matter, determining the heaviness of the bodies [89].[111]

The section devoted to the heaviness of bodies ends with the consideration that even if one does not know why the pressure decreases in proximity to the planets, to stay with this doubt is always better than not knowing anything as when one says that the heaviness is due to attraction.

> Although we have to stop here and hardly can hope ever to find the cause of the diminution of the elastic force of the aether, it is easier to resign to this than to merely maintain that all bodies are by their nature endowed with a force to attract each other. For since one cannot even form an understandable concept of this attraction, one can by way of contrast at least understand how it is possible that the elastic force of a liquid matter is reduced, and one also understands that this can occur in a way that is in accordance with the laws of nature [89].[112] (C19)

3.4.1.4 The Origin of Forces

Force in the 18th century was the fundamental magnitude of mechanics and was introduced in various ways, each with its own problems. In his youthful treatise of 1736, the *Mechanica sive de motu analytice exposita*, Euler introduced the concept of force in a classical and rather generic way. The force [potentia] is "an action on a free body that either leads to the motion of the body at rest, or changes the motion of that body" [66].[113] It is worth noting that Euler had two Latin terms to indicate force: *potentia* and *vis*. To potentia he gave the technical meaning of cause of the variation of the motion of a body, to vis a generic meaning, either technical or common, for example he can speak of *vis gravitatis* as an example of technical use, as well as of vis inertiae and the vis of impenetrability, as an example of common use.

Many years later, when he wrote the *Theoria motus corporum solidorum seu rigidorum*, with a much clearer understanding of mechanics, Euler proposed the definition:

> Definition 12. What induces to change the absolute state of bodies is named a *force* [vis]; it should be due to external causes, since the body will remain in its own state due to internal causes [84].[114] (C20)

Is there any change in the status of force (in the ontology); from a being has it became a name, and instead of a potentia a vis? The question mark is justified also because in the *Anleitung*, preceding of some years the *Theoria*, Euler in a succinct proof of the laws of motion had written that force is measured by the effect it produces. Here he wrote: "A force twice as big must in the same time produce twice as big an

---

[111] vol. 2, p. 542.

[112] vol. 2, p. 547. Translation into English by Hirsch E.

[113] p. 78.

[114] p. 44.

effect, because just in view of that do we consider it twice as big" [89].[115] A similar position is also reiterated a few pages after Definition 12 in the *Theoria*, as will be commented in a subsequent section. In any case the above definition hides some ambiguities, always present in the scholars of the 18th century when they dealt with force, Newton included.

In the following in the attempt to decipher the complex argumentations on the concept of force I will assume that *force* is for Euler a primitive concept, which in particular is reduced to the anthropomorphic idea of pressure. What Euler tried to explain is not what is force or pressure but rather what is its origin. The fact then that Definition 12 defines force as the name given to the external cause of the change of motion does not regard ontology but simply serves to delimit the broader concept that force has in the natural philosophy of the period, somehow a hypostatization of any cause, also internal.

In the *Recherches sur l'origine des forces* of 1750, Euler to explain forces had paused on the role of impenetrability. He observed that despite the inertia, or the general property of bodies by virtue of which each body tends to preserve its state, we see that bodies that fall under our observation continually change their state. So these bodies are subject to forces that must necessarily be external [80].[116]

To illustrate how forces are generated Euler considered two bodies A and B. A is at rest while B is moving toward A. After the impact, in order to avoid its own penetration A acts to change the state of B and in turn B acts on A. In other words A applies a force to B (due the impenetrability of A, which is an external cause) and B to A. Euler concluded that these forces see their source in the impenetrability only [80].[117] That is the forces due to impenetrability are passive forces (modern term); they are not determined neither for their quantity nor for their direction; they only are 'obliged' to act when the impenetrability of a body is threatened, otherwise they do not produce any effect.

It is true that a greater action would prevent penetration as well, but it is more natural to assume a minimal value; that is forces have at all times value and direction strictly necessary to prevent penetration, so that if a minimum force is sufficient for the purpose only it acts [80].[118] In this criterion of "least action" Euler took inspiration from Maupertuis and in it one can find a justification of the principle of least action. To Maupertuis he made explicit reference in the letter 78 of the *Lettres* [87].[119]

Euler argued that the forces that two bodies exchange are equal and opposite to each other. For Euler this equality of forces, which is commonly known as the principle of action and reaction, is a necessary consequence of the impossibility of penetration. So, for him, it is a rational and not an empirical law.

---

[115] vol. 2, p. 476.

[116] p. 423.

[117] p. 427.

[118] p. 430–431.

[119] vol. 1, p. 304.

> This equality of forces, hence the great principle of equality between action and reaction, is a necessary consequence of the nature of penetration. For, if it would be possible for the body A to penetrate the body B, body A would be equally as much penetrated by the body B; therefore, since the damage that these bodies that penetrate, is equal, these two bodies must also employ equal forces to resist penetration. So, as much as the body B is solicited by the body A, that will be solicited equally by this, both deploying exactly as much force as necessary to prevent penetration. But these two bodies acting one on the other by any force, will be in the same state as if they were compressed together by the same force [80].[120] (C21)

Euler was also able to analyze the evolution of the force due to impenetrability in the case of impact. Differently from many scholars of the period, and in agreement with Johann Bernoulli (and Leibniz), he assumed that impact is not an instantaneous process: "The instantaneous impact would not be in accordance with the always respected law of nature, for which nothing can occur instantaneously, and as for a leap. According to this law a so great change, as sometimes is that occurring in the impact of bodies cannot occur without corresponding to a some interval of time" [71].[121]

In the hypothesis that the impacting bodies are linear elastic Euler can furnish an explicit expression of the force of impact. If P is such a force, it is given by the expression $P = Fz/k$, where F is the force necessary to get a penetration k, while z is the current value of the penetration during a static experiment; F/k, today known as stiffness, was named by Euler degree of hardness (degree de dureté) [80].[122] For reasons of space, and also because it is not relevant for the present text, I do not repeat the mathematical steps involved in the full solution of the problem of impact, for instance the evaluation of the maximum value reached by P. I limit myself to say that the approach and the results are essentially the same Euler had obtained in his youthful work *De communicatione motus in collisione corporum* of 1731 [67]. The difference is that now Euler had a clear concept of mass, whereas in 1731 he had not. But the equations are the same, demonstrating how mathematics may be more powerful than physics.

In subsequent works Euler associated force not only with impenetrability but with inertia also. Imagine, for example, stopping a moving ball by opposing it a hand; if inertia would have no role in the explication of force one could be free to assume the ball without inertia. In this case the ball would be rejected or stopped but the hand would not notice anything; which is contrary to what happen in the real world. To obtain an effect it is necessary that the ball is equipped with the ability to resist changes of motion, that is that it has not zero mass (inertia). In the *Theoria* Euler limited to say that forces originate from impact of bodies because of their impenetrability. He wondered if all the forces have this origin. Without excluding the existence of other forces, he was content to say that a very important class of forces has this origin.

---

[120] p. 434.

[121] p. 36.

[122] p. 441.

Here's how Euler describes the role of impenetrability: "The cause of those forces by which the state of a body is changed may be agreed to lie not in inertia alone but in inertia coupled with impenetrability. Indeed, seeing that only bodies can be said to be impenetrable and since bodies are necessarily endowed with inertia, impenetrability as such involves inertia, so that impenetrability alone is rightly considered the source of all forces by which the state of bodies is changed. It will therefore be proper to consider this property more exactly as being the origin of all forces" [84].[123] He thus can still pretend that only impenetrability counts, because inertia is given when impenetrability is given.

The account of the force of impact between the two bodies A and B considered in Sect. 3.4.1.4, may be completed as follows: when B comes into contact with A, it experiences a force which we would normally term A's force of resistance to change of state. Note that the force is not in any sense in A: what is in A is its inertia, that is not a force because it only maintains A's state. But this internal principle is experienced by B as a force. There is, therefore, an external force acting on B which is not internal to A. Nor does it act at a distance because it is a prior condition of there actually being a force that A and B be impenetrable and that they be in contact [97].[124]

Euler underlined in several occasions that the forces of impact are external to the bodies and that inertia is not a force in the sense that it is not responsible for the change in the motion of a body, it has only the role of preventing it. In his work of 1746 *Gedancken von den Elementen der Körper* [72], Euler criticized Leibniz's, or rather Wolff's, philosophy of forces and monads. Euler's main criticism concerned Wolff's thesis that the monads, which are the ultimate components of reality, are endowed with an internal force that intends to continually change their state. Euler believed that there was indeed an internal force, which however tended to maintain the state and not to change it, and concluded: "One must therefore stipulate that two particular well differentiated classes of things exist in the world, to one of which belong the corporeal things, whose essence consists in the force [ability] steadily to maintain their state. The other however comprises the souls and ghosts, which are endowed with a force to change their state" [72].[125]

3.4.1.5 Concepts of Time and Space

Euler discussed the concepts of space in his technical works such as the *Mechanica sive de motu analytice exposita* and *Theoria motus corporum solidorum seu rigidorum*, and in some more philosophical ones such as the *Lettres* and *Anleitung*. But the text in which he did it in a systematic way is *Réflexions sur l'espace et le tems* of 1748 [77]. Euler's discussion is very interesting; he had perhaps less philosophical culture than a professor of philosophy but he knew much more directly the topic

[123] p. 46.
[124] p. 147.
[125] p. 15.

he is dealing with. Euler's paper did not escape to Immanuel Kant (1724–1804) who read it a fifteen years after its publication. Kant observed that the mathematical and empirical considerations of motion and space furnished many data to guide the metaphysical speculation in the track of the truth and avoid the void speculations of the philosophers of his time [36].[126]

Euler defended his thesis of absolute space against the 'metaphysicians' who believe that only the concept of relative space made sense. The metaphysicians in question were naturally the followers of Leibniz and Wolff, not explicitly named however. That he referred to them is evident when he stated "I strongly doubt [...] that the equality of the spaces should be judged by the number of monads that fill it" [77].[127] He addressed his thesis of absolute space with metaphysical and physical reasons, although he recognized that in fact it is not possible to individuate any absolute system. The metaphysical reason consists in affirming that a fertile concept in physics cannot be empty: "One should instead assert that both the absolute space and the absolute time, such as mathematicians look at, are real things which submit outside our imagination, because it would be absurd to sustain that pure imaginary objects could be assumed as foundation of the real principles of mechanics" [77].[128] The physical reason is that only by conceiving absolute space can one explain the inertia of bodies, empirically detected.

Euler argued with a simple example that inertia cannot depend on the presence of nearby bodies to which refer motion by considering a body that floats on still water, remaining at rest in turn. If the water begins to flow, an observer, rigidly linked with it, sees the body move without any force being applied to it. Or if he sees the body remaining at rest with respect to the water, a careful examination shows him that this is due to the effect of water dragging. So Euler can conclude: "I strongly doubt that metaphysicians dare to sustain that bodies maintain their position with respect to other bodies thanks to their inertia, because it could be easy to show the falsehood of this explanation because of the consideration I discussed on bodies close to their neighbor" [77].[129]

Euler then dealt with the case in which the position of a body is referred to fixed stars. Being them at rest—a possibility that is not certain for Euler [84][130]—the absolute space of the mathematicians would coincide with that of the metaphysicians. But, he wondered, how the fixed stars so far can determine inertia? This is impossible.

This thesis, rejected by Euler, is endorsed by some modern scientists, who also refer to the reflections on the subject by the epistemologist Ernst Mach.

> As Ernst Mach has pointed out, it cannot be a coincidence that the fixed stars appear indeed fixed relative to inertial frames, and hence that it is reasonable to consider inertia as a force exerted on local objects by the totality of the objects in the entire universe. Thus, Newton's law may best be interpreted as a consequence of the basic axiom that the sum of the forces,

[126] p. 351.

[127] p. 331.

[128] p. 326.

[129] p. 328.

[130] p. 38.

> including the inertial force, acting on a particle should be zero and of the constitutive law of inertia, which states that this inertial force should be given by $-ma$, where a is the acceleration relative to an inertial frame [142].[131]

In essence Euler concluded that the preservation of the state (motion or rest) of bodies is explained only by conceiving place according to the criterion of mathematicians in an absolute way and not in relation to other bodies. And nobody can say that the principle of inertia is based only on something that exists in our head.

The reality of space is confirmed by the principle of preservation the direction of motion. If space were a relative concept, what sense would it have to speak of direction? It must be a direction of an absolute space that can be conceived in a natural way by abstraction.

Euler ended his considerations on space by reiterating the metaphysical argument that one cannot say that a principle accepted as true in physics by nearly all scholars (the principle of inertia), is founded on a thing that exists in our imagination only (the absolute space), and from this it must be concluded that the mathematical idea of absolute space cannot be in any way imaginary, but that there is something of real in the world which corresponds to this idea [77].[132]

According to Ernst Cassirer the *Réflexions sur l'espace et le tems* set up in fact not only a program for the construction of mechanics but a general program for the epistemology of natural sciences. It sought to define the concept of truth of mathematical physics independently of the concept of truth of the metaphysicians. The considerations of Euler rested entirely on the foundations on which Newton had erected the classical system of mechanics. His concepts of absolute space (and absolute time, see below) were revealed not only as the necessary fundamental concepts of mathematical-physical knowledge of nature, but as true physical realities. To deny these realities on philosophical grounds, means to deprive the fundamental laws of dynamics—above all the law of inertia—of any real physical significance. In such an alternative, the outcome cannot be questioned: the philosopher must withdraw his suspicions concerning the "possibility" of an absolute space and an absolute time as soon as the reality of both can be shown to be an immediate consequence of the validity of the fundamental laws of motion [36].[133]

Once established the reality of space, Euler went on to examine time. Beginning with the observation that the ideas of space and time always go together [77].[134] Arguing on the reality of time is however more complicated. Euler distinguished between the idea of time and time itself. He had no objection to the idea that time is conceived as a succession of changes. But this idea is not sufficient to assert that two time intervals are the same, as one has to say in mechanics. Thus, according to Euler, time cannot be reduced to the idea of succession of changes; there must be something else: absolute time. Euler concluded by saying that he realized that his considerations could only be taken on by philosophers who are willing to give some

[131] p. 9.

[132] p. 329.

[133] p. 351.

[134] p. 331.

sense of reality to time and motion, while they will not make the slightest impression to those who consider everything as relative [77].[135]

3.4.2 Mechanics and Mathematics

Euler's role in mechanics has been overshadowed by the historians of mechanics such as Mach, Dugas, Montucla, Dühring, very careful to the fundaments, who see him more as a mathematician than a mechanician. After all, more than 60% of his work deals with pure mathematics, and even those whose object is mechanics and astronomy contain many sections that can be classified as mathematics. Today Euler's role, even for what concerns the fundaments of mechanics, is re-evaluated. Some see him as the founder of modern classical mechanics as well as the one who threw the germs of the theory of relativity and quantum mechanics [156].

Euler was (lucky enough to be) born in a period when all the mechanics and all the Calculus was to build starting from the foundations of the 17th century. He has often been seen as the successor of Newton and Leibniz. Regarding the legacy left by Euler we must distinguish between the actual one, that is, the influence that he had on his contemporaries and the influence he could have had if all of his writings had been published and understood. And also the influence it could have on modern mathematicians and physicists if they were still studied. On this last point I limit myself to note that many of Euler's scientific writings could be well understood by a modern reader and that it is still possible to identify promising lines of research now abandoned.

In any case the relevance of Euler's thought, although interesting, is not a priority of this book. The reference to predecessors and contemporaries in mechanics is considered only on specific points, the main objective being to expose and explain his ideas. In this section I will focus almost exclusively on the fundaments of Euler's mechanics, largely neglecting his elaborations in the areas of rigid body dynamics, theoretical astronomy, theory of elasticity and fluid dynamics, even though they represent the greatest contribution of Euler to mechanics.

In my analysis, I will undergo a critical examination of the current view, due in large part to Clifford Ambrose Truesdell's studies who edited (vol. 12, 13) or co-edited six volumes of the collected works of Euler, and wrote appreciated articles and essays on him. He was Euler's greatest advertising agent, making him a 18th-century hero; a genius like Newton and perhaps superior to him. The current view can be summarized as follows:

1. Euler translated Newton's mechanics into Leibnizian language.
2. He introduced the use of vector calculus in mechanics.
3. He defined precisely the mechanics of the mass point, starting from the uncertain Newtonian mechanics.

[135] p. 333.

4. He defined precisely the mechanics of the rigid body.
5. He made fundamental contributions to the systems of deformable bodies.
6. He founded modern hydrodynamics.
7. He introduced the concept of observer.

I will not stop to question these theses, limiting myself to say that it is one of the many myths of historiography that sees the evolution of science due to the work of isolated geniuses. It is hard to imagine that one man, not even one with exceptional memory and great workmanship and intelligence like Euler, could have done so much on his own. However, I believe that it is true that Euler was able to gather and summarize the ideas of his time and for this reason, when one does not care about the genesis of the various ideas, but wants to take stock of a certain era, the study of his researches must be considered as essential.

The reconstruction of the basis of Euler's mechanics is carried out here by studying his published works—articles and books—and some letters. The main text which has ben considered is *Theoria motus corporum solidorum seu rigidorum* (hereinafter *Theoria*), written in 1760 and published in 1765 [84]. That is, I will start from a text in which the foundations of mechanics have already been laid. Of course, I will also refer to the *Mechanica sive de motu analytice exposita* (herein after *Mechanica*), published in 1736 and to articles and books that show both his researches and conceptions of natural philosophy, among which the most important is the *Anleitung*.

Euler had his own clear idea of how to develop a mechanical theory able to solve the problems that were then still waiting for a solution. Already in his early treatise, the *Mechanica*, he presented the work program that engaged him throughout his life: developing all the mechanics starting from the laws of motion of the mass point, a concept that he was among the first to specify. Here is Euler's program as reported in the general scholium of Chap. 1 of the *Mechanica*:

1. In the first place, very small bodies which can be considered as points are referred to.
2. Then the approach follows for these bodies of finite magnitudes which are rigid and are not allowed to change their shape.
3. In the third case flexible bodies will be considered.
4. Fourthly, those which allow extension and contraction.
5. Fifthly the motion of bodies, constrained by others.
6. In the sixth case the motion of fluids are in the agenda [66].[136]

Never was Euler explicit about his epistemologic conceptions about mechanics. In particular, although in practice he treated mechanics as a purely rational discipline, widely described with the language of Calculus, nowhere made his position explicit and never equated the status of mechanics with that of mathematics. Probably this silence was not due to an oversight but to some form of uncertainty. The only explicit statement he made about the role of mathematics in mechanics was his declaration of the preference given to the analytic treatment over the geometric one. For Euler only the former allows a systematic approach to all problems, while the latter involves

[136] p. 37.

the search for new routes for every problem. But this is of course true not only for mechanics, but for mathematics in general, and it was for this reason that in the 18th century in every field of mathematics and physics, geometry was replaced by algebra and Calculus, bringing the discipline of geometry to a state of decadence, from which it will recover only in the 19th century. Here is what Euler wrote about this point in the preface of the *Mechanica*:

> Thus, I always have the same trouble, when I might chance to glance through Newton's *Principia* or Hermann's *Phoronomia*, that comes about in using these [synthetic methods], that whenever the solutions of problems seem to be sufficiently well understood by me, that yet by making only a small change, I might not be able to solve the new problem using this method. Thus I have endeavored a long time now, to use the old synthetic method to elicit the same propositions that are more readily handled by my own analytical method, and so by working with this latter method I have gained a perceptible increase in my understanding. Then in like manner also, everything regarding the writings about this science that I have pursued, is scattered everywhere, whereas I have set out my own method in a plain and well-ordered manner, and with everything arranged in a suitable order. Being engaged in this business, not only have I fallen upon many questions not to be found in previous tracts, to which I have been happy to provide solutions, but also I have increased our knowledge of the science by providing it with many unusual methods, by which it must be admitted that *both mechanics and analysis are evidently augmented more than just a little* [66].[137] (C22)

The last sentence of the above quotation, in which Euler declared that the developments of mechanics and mathematics are closely linked, is worthy of note; mathematics allows to solve problems of mechanics, mechanics suggests cues to mathematics to treat unresolved problems.

3.4.2.1 *Theoria Motus Corporum Solidorum Seu Rigidorum*. The Motion of Mass Points

The first part of the *Theoria*, the *Introductio* (103 pages over a total of 527, preface excluded), reports a summary of the main results on the dynamics of the mass point, largely taking up from *Mechanica*, with many clarifications and updatings.

What a modern reader notices first, in reading the *Theoria*, though to a less extent to what happens for the *Mechanica*, is the verbosity of the exposition, common at the times, that in many cases instead of helping confuses and makes it difficult a reading because it does not allow to distinguish what is strictly necessary from what is superfluous. The sense of frustration is increased by the fact that in what is presented as a formal and generally axiomatic exposition there is little or no distinction between theorems and principles. Principles are almost never declared; the main exception concerns the principle of inertia.

Here is what Euler wrote about this point in the *Lettres*, a text published shortly after the *Theoria*:

> This principle [of inertia] is commonly expressed in the two following proposition: A body once at rest will remain eternally at rest, unless it be put in motion by some external or foreign

[137] Preface. Translation into English by Bruce I.

cause: Secondly, A body once in motion will preserve it eternally, in the same direction, and with the same velocity; or will proceed with an uniform motion, in a straight line, unless it is disturbed by some external, or foreign cause. In these two propositions consists the foundation of the whole science of motion, called mechanics [87].[138] (C23)

While in the *Mechanica* the principle of inertia is not explicitly referred to as such, in the *Theoria* it is widely introduced with two propositions, qualified as axioms (see below). At no point Euler referred to as a principle what a modern would treat as such; that is the metaphysical principle of sufficient reason that plays a fundamental role in Euler's proofs.

The way of introducing and using principles becomes less mysterious if one reflects that the 18th century mathematicians understood the term principle differently from us. They referred to a principle as a proposition placed at the foundation of a theory, requiring nothing else. Then the principle can be either first, that is, primitive, evident in itself or true empirically, or it can be a second principle if it can be demonstrated by first principles. Among the principles of mechanics there are, for example, the principle of minimum action and that of the conservation of living forces. In Euler's time the logical status of these principles was not exactly defined; it was thought that they could be proved by first principles (and Euler will in part do so), but they still were not. Here is what Euler wrote about principles:

Although the principles in question are new, as they are not yet known or spread by the authors who have treated of Mechanics, it is understood, however, that the foundation of these principles cannot be new, but that is absolutely necessary that these principles should be deduced from first principles, or rather axioms, over which the doctrine of motion is established [78].[139] (C24)

The term *axiom* is used by Euler in the *Introductio* only three times to extend the validity of the properties from relative to absolute motion and to introduce the concept of inertia, with the classical meaning of a self-evident proposition. The term *theorem* is used eight times; there are then fifteen definitions and nineteen propositions qualified as problems. These propositions could generally be formulated as theorems because they provide the solution of the problems also. Euler's operation, perhaps similar to that of Euclid, serves to make the discussion less abstract.

The *Theoria* opens with the definitions of rest and motion:

Definition I. Just as a body at rest remains perpetually in the same place, so a body in motion continues to change its position. Clearly a body that is observed to adhere always to the same place is said to be at rest: but that body which advances by gliding from place to place in time is said to be moving [84].[140] (C25)

The primitive concepts necessary to understand this definition, those of space and time, are introduced soon after, at a discursive level into two Explications and one Scholium. Space is introduced as an absolute entity: "Now we can only conceive a notion about this space itself by abstraction, by considering all the bodies to be

[138] vol. 1, p. 309. English translation in [88].

[139] p. 194.

[140] p. 1. Translation into English by Bruce I.

removed, and what is left we decide to call space" [84],[141] "a concept which was argued at length by philosophers (and uselessly)".

But the idea of absolute space, admitted Euler, cannot be used in practice neither to define the position of bodies nor their motion. They must be referred to nearby bodies which must be able to maintain invariable in time positions relative to one another in order to fulfill the purpose. To determine the location of a point P it is necessary to know its distance from at least four non-coplanar points A, B, C, D [84].[142]

After introducing the reference system, or using a modern term the observer, Euler went on to define the notions of relative motion and rest. A body is at rest with respect to A, B, C, D if its distance from them does not vary, otherwise it is in motion (Definitons 2, 3) [84].[143] Note that Euler did not use the term relative, but "with respect to".

In a few definitions Euler introduced the concepts of path (Definiton 4), uniform and difform motions (Definiton 5) and velocity (Definiton 5), limited to uniform motion. Some comments are dedicated to the definition of velocity as the relationship between two heterogeneous quantities. The difficulty, inherent in the geometric concept of proportion and magnitude that requires comparison between uniform magnitudes, is solved by introducing units of measurement for space and time; in this way it is possible to define (the numerical value of) the velocity as the ratio between the numerical values that represent the measurements of space and time. Definiton 7 concerns the direction of motion in the rectilinear and curvilinear cases, assumed as that straight line in which the motion is occurring or, if it is along a curve, as the tangent to the curve.

After these definitions, Problem 1 follows, which deals with the use of the differential calculus to evaluate the space traveled by a point moving on a straight line with a velocity v assigned as function of time. Euler assumed that for a small dt (the element of time), the passed space ds, an elementary space indeed, is traversed with constant v velocity defined by $v = ds/dt$, in which v and s are explicitly thought of as generic functions of time, without particular restrictions on their regularity. This assumption is referred to as Proposition 3 in the *Mechanica*. "As indeed in geometry the elements of curved lines are considered to be the elements, non-uniform motion is resolved into an infinite number of uniform motions. This is not the case, but it can be ignored without error. In either case the truth of the proposition is therefore apparent" [66].[144]

The idea of considering constant any magnitude (in the present case the velocity) on condition of referring to a very small spatial or temporal intervals is what gives Calculus a great heuristic power. If the velocity of the body v is given at individual instants, then the distances s passed in time t can be obtained with the help of the relation $ds = vdt$, the integral of which gives $s = \int vdt$. Similarly if the speed v

[141] p. 1.

[142] p. 2.

[143] pp. 3–5.

[144] p. 12. Translation into English by Bruce I.

corresponding to a distance s is known, then the time t, in which the distance s is passed, is given by the differential equation $dt = ds/v$, which gives $t = \int ds/v$.

More than one historian concedes that this approach to motion contains elements of novelty. Meanwhile, velocity is defined as a derivative (modern term, actually for Euler it was a ratio of differentials) of a function $s(t)$ which is considered as a generic mathematical function and therefore an abstract concept not necessarily connected to a geometric curve. Then time is treated in the same way as any other physical quantity, and it does not have a particular role as it happens for example in Newton, with his idea of fluxions [156].[145]

I do not pronounce on this judgment about Euler's originality. Certainly a similar treatment was not possible before the introduction of Calculus; furthermore it was established that Euler contributed a lot to the generalization of the concept of function. If one compares the way velocity is introduced in the *Theoria* and in the *Mechanica*, written a twenty years earlier, he notices greater ease due to greater confidence both with the Calculus and the concept of function. It should be kept in mind that Euler's text also had a didactic function and the maturation of the concepts referred to was not only his but mainly of the contemporary readers who did not need to be reassured about the validity of the procedure.

Problem 2 considers the motion of a mass point on a plane curve. It is reduced to the study of two mono-dimentional motions projecting the velocity into two directions not necessarily orthogonal, even if, added Euler, it would be better if they were such, for reasons of simplicity of calculations. For the motion in space (Problem 3) the projection is on to three axes. It should be noted that this is only a kinematic problem where the problem of the legitimacy of decomposition does not arise.

Studying motion with respect to fixed orthogonal coordinate axes instead of mobile intrinsic or natural coordinates is considered a fundamental turning point of Euler's mechanics, a choice that appears first in his work *Recherches sur le mouvement des corps célestes en général* of 1747 [76]. The use of intrinsic coordinates, the standard approach at the turn of the 18th century, had two drawbacks. On the one hand it required skill to find the right frame of reference. On the other hand the approach was too difficult for bodies moving into a three dimensional space. To be carried out consistently it needed concepts of differential geometry such as curvature, torsion and so on, fully developed only in the 19th century. The use of fixed coordinates does not present these problems. Although the equations may be complicated, they can be written in a standardized way. Even in this case, however, it must be said that Euler was not the first to project the equations of motion on two or three axes. According to Lagrange the first was Colin Maclaurin (1698–1746) in his *A treatise on fluxions* of 1742 [115].[146]

> If the curve FM be described by powers directed in any manner whatsoever, and the force at any point M, resulting from the composition of these powers, act in the direction MK, and be measured by MK; let MK be resolved into the force MO in the direction of the ordinates MP (= y), and the force OK parallel to the base AP (= x); then, the time being supposed to

[145] p. 132.

[146] p. 243.

flow uniformly, or the velocity at M being represented by the fluxion of the curve FM, the force MO will be measured by $\ddot{y}$ and the force OK by $\ddot{x}$ [132].[147]

Johann Bernoulli also had done it according to [134, 159, 160].[148] But only with Euler, the use became systematic; indeed it became the only one.

Decomposing the motion into two or three directions can be seen as a step toward modern vector algebra. A same magnitude, a geometric vector, can give rise to different pairs (or triads if in space) of components as the coordinate system varies. The choice of the coordinate system can be arbitrary. To the decomposition of the motion in several directions Euler gave the name *resolution*. The motion is said to be resolved, provided that the small interval traversed in the element of time is considered as the diagonal of a parallelogram or parallelepiped [84].[149] Euler also considered other types of coordinates and therefore of decomposition. In problems 5 and 6 polar coordinates are introduced in the plane and in the space respectively.

In Chap. 1 of the *Introductio* Euler carried out only kinematic considerations. In the following chapters he went on to discuss the causes of motion. They are classified internal and external:

1. Internal causes. Responsibile for the reason either for rest or motion of a body, with the exclusion of all external causes. They are able to contribute anything to change the state of motion or rest.
2. External causes. Responsible of the change of the state of motion and rest of a body.

In Chap. 2 Euler explained the nature and characteristics of internal causes, that is inertia. To achieve his purpose he introduced the concept of absolute motion. In essence he said that, if there is an absolute space, of any body one can say if it is at rest or in motion with respect to this space; motion and rest which are classified as absolute. This obvious fact can be expressed through an axiom:

Axiom 1. Every body, even without being relative to other bodies, either remains at rest or moves, that is, it is either at absolute rest or in absolute motion [84].[150] (C26)

Two more axioms express the principle of inertia, a term which Euler has not yet introduced.

Axiom 2. A body, which is absolutely at rest, if subjected to no external actions, will persist indefinitely in the state of rest. Axiom 3. A body, which is absolutely in motion, if subjected to no external actions, continues to move uniformly along the same direction [84].[151] (C27)

These axioms, which Euler called the principles of internal motion, are justified in a simple way, basing on the principle of sufficient reason. For the state of rest Euler argued for instance, that because all the external causes of motion have been

---

[147] vol. 1, p. 298.

[148] p. 210; pp. 184–236; p. 112.

[149] Definiton 8, p. 21.

[150] p. 30.

[151] pp. 32, 33.

withdrawn, no reason is present, that a body should begin to move in one direction rather than in any other: "This truth depends on the principle of sufficient reason" [84].[152] A somewhat more sophisticated reasoning concerns the permanence in the state of uniform motion. No change in the direction can indeed occur, since there is no reason it should be deflected from that, in one rather than in all the other directions; "clearly it surely maintains the same direction, for the principle of sufficient reason". About speed [the modulus of velocity] it can be said that "unless it always remains the same, either it increases or decreases, of which neither absurdity can be said; for if it is either being increased or decreased, it must happen to follow a certain law; but what this law may be cannot be conceived in any way, since nothing surely will be agreed upon [...] Therefore nothing is relinquished, unless as we have stated, the speed too always stays the same, and the direction likewise" [84].[153] The reader is asked to reflect on the validity of these demonstrations.

With the introduction of the three axioms, the first theorem (Theorem 1) appears very simple in reality. It asserts that the axioms valid for absolute motion also apply to relative motion, provided it is referred to a body (to an observer) that is at (absolute) rest or in uniform rectilinear motion [84].[154]

In Definition 11 the reason of the validity of axioms 2 and 3, that is the internal cause of motion, is given a name, *inertia* (the persistence) of the *Anleitung*):

> That quality of bodies, the reason for persisting in the same state present within a body itself, is called inertia, and also sometimes the force of inertia [84].[155] (C28)

with the concept of state (absolute) introduced as follows: "While a body is either absolutely at rest or moving uniformly in a direction, it is said to persist in the same state" [84].[156]

Inertia indicates that property of bodies, whereby being at rest means that they will continue to be at rest, therefore as if they oppose to motion; but since bodies set up in motion themselves equally oppose all to be changed, either on account of the speed or direction; the name *inertia* seems a good choice. Sometime, said Euler, it is called *force of inertia*, because the body is resistant to change the state; but because often force is defined as the (external) cause which is changing the state of the body, force of inertia is not acceptable with this meaning—though Euler occasionally did. Whereby, as confusion should arise, it is better to omit the name force and refer to it by the simpler name of inertia [84].[157]

Chapter 3 is devoted to the external cause of motion; to it also is given a name, *force*, with Definition 12. Though it has already been introduced in Sect. 3.4.1.4, I rewrite it for the sake of clarity:

---

[152] p. 32.

[153] p. 34.

[154] p. 37.

[155] p. 35.

[156] p. 35.

[157] p. 36.

What induces to change the absolute state of bodies is named a *force* [vis]; it could be due to external causes, since the body will remain in its own state due to internal causes

After the introduction of external and internal causes of motion, Euler could go on to demonstrate what is now known as the Newtonian equation of motion, for the one-dimensional case. In essence Euler considered as a necessary truth, that is rational, the law that Newton, at least officially, considered contingent, that is empirical. The demonstration takes place into two steps.

Theorem 2a. The small space [$d\omega$], through which a given body at rest is advanced in the assigned small interval of time dt by different forces, is proportional to the forces [84].[158] (C29)

Theorem 3. If equal forces act on unequal bodies at rest, the effect [$d\omega$] produced in the same small time intervals [dt] is inversely proportional to the inertias of the bodies [84].[159] (C30)

The proof of Theorem 2a is very simple, based on the assumption of the additivity of the effect of different forces. Basically, said Euler, if a small body, a corpuscle, is pulled forwards by a force equal to p in the short time interval dt through the small space equal to $d\omega$, and if another force equal to p is acting on the same body along the same direction, the body progresses through another equal small interval equal to $d\omega$. Thus this corpuscle acted on by a force equal to $2p$ in the same interval of time dt is pulled through the small distance equal to $2d\omega$. Similarly if n forces—or equivalently a force np—act on the corpuscle at rest for the same time interval dt, they move the body through the interval equal to $nd\omega$.

The demonstration is not convincing however. Euler assumed indeed that the effect of the sum of two forces (two causes) is the sum of their effects, that is he is assuming a principle of superposition and this is not granted.[160] And even if this is conceded the proof would be only a trivial theorem of arithmetics (at Euler's time) for which if two quantities increase of the same amount they are proportional. Moreover the demonstration follows a reasoning apparently different from that carried out in the *Anleitung*. Here (*Theoria*), in the proof of Theorem 2a, force is considered as measured *a priori*, in the sense that it is independent of the formulation of dynamics, there (*Anleitung*), force was measured from effect and the proof of the theorem would become a simple truism. On this point see Sect. 3.4.2.2.

Euler's writing is substantially contemporary to one of Daniel Bernoulli, the *Examen principiorum mechanicae, et demonstrationes geometricae de compositione et resolutione virium* of 1726 [5], in which the latter posed the problem of the logic status of the law that links acceleration to force, assuming force as defined *a priori* concluding that the law is empirical and not rational. Bernoulli said that the part of mechanics that deals with the balance of forces can be deduced from the principle of composition of forces, as Varignon has shown. If another principle is added to this

[158] p. 55.

[159] p. 56.

[160] The principle of superposition presupposes for its validity that the two cause do not interact to give a cause different from their sum.

principle, namely that according to which the increases in velocity are proportional to the increments of time multiplied by the force, mechanics is completed, relative to the motion. Galileo used this principle. Bernoulli believed that the principle of composition of forces is a necessary truth, while that of Galileo is a contingent truth: "Nature could have made the increases in velocity in the bodies proportional to the increments of time multiplied by any function of the pressure, so that said t the time, p the pressure and v the velocity, it was not $dv = pdt$, but for example $dv = p^2dt$ or $dv = p^3dt$" [5].[161]

The proof of Theorem 3 is analogous. Basically it is said that if one joins two bodies of equal mass under a given force the effect is halved. With some details, following Euler's reasoning: Let consider a corpuscle having an inertia A, which at rest is moved by a force equal to p for the short interval of time dt through the small space $d\omega$; if another corpuscle B equal to A is acted on by a force also equal to p along the same direction, it moves in the same dt by the same small space $d\omega$. If the two corpuscles are joined together into one resulting in a body with an inertia $2A$ (it is an implicit assumption by Euler), it acted by the force equal to $2p$, in the time equal dt moves still of the small space $d\omega$. Thus a force $2p$ on the body of inertia $2A$ produces the displacement $d\omega$. Similarly a force np applied to a body nA produce the same displacement $d\omega$. Thus for the Theorem 2a, the force p on a body nA produces a displacement $d\omega/n$.

Note that inertia is not equivalent to the Newtonian quantity of matter, at least as it is conceived in the first edition of the *Principia* (see Sect. 1.2.2.1), it is rather a property of a body whose measurement is defined in an operational way as the constant of proportionality between force and displacement $d\omega$. Euler can thus introduce the concept of mass in the following way, with Definition 15:

> The *mass* or the *quantity of matter* of a body is the name given to the amount of the *inertia* which is present in that body, by which just as it tries to continue in its own state so it tries to resist all changes [84].[162] (C31)

where also the term quantity of matter is redefined by means of the concept of mass, inverting the usual approach where the mass is defined by the quantity of matter. The property of additivity is implicitly attributed to the mass as clear from the proof of Theorem 3. That is, if one joins two bodies with mass m_1 and m_2, he gets a body with mass $m_1 + m_2$.

The operational (or dynamical) definition of mass used by Euler was not common in the 18th century; almost all mathematicians treated mass (and inertia) as proportional to the quantity of matter, taking for granted the meaning of this term (for instance the number of atoms or a volume), a geometric definition all considered). The operational definition of mass will be given a foundation role in the 19th century by Mach in his *Die Mechanik in ihrer Entwicklung historisch-kritisch dargestellt* written at the end of the 19th century [131].

In the *Mechanica* Euler had given a different formulation of mass and it is after a theorem—and not a definition—(Proposition 17) that the amount of inertia (vis

[161] pp. 126–127.

[162] p. 57.

inertia) is proportional to the amount of *matter or mass* [66].[163] The amount of matter, however, was defined in a vague way, like the set of points (atoms?) that make up a body. Only that not all the points have the same mass. More precisely the points can be taken having the same mass when the same force exerts an equal effect on them (that is they have the same inertia). Thus the demonstration of the theorem on the equality between inertia and mass of the *Mechanica* is the result of a vicious circle of the type: the amount of matter (geometry) is proportional to the amount of inertia (dynamics) so the amount of inertia is proportional to the amount of matter.

At this point, by putting together Theorems 2a and 3 and using the definition of mass, Euler could enunciate Theorem 4, according to which:

> Theorem 4. If corpuscles at rest with masses in an unequal ratio, are acted on by some singular forces, the small intervals through which they are thrust forwards in the same short time intervals will be composed in the direct ratio of the forces and the inverse of the masses [84].[164] (C32)

That is, for a force p and a mass A, it can be written $d\omega \propto p/A$.

In Problem 9, Euler started to replace the elementary displacement $d\omega$ measured from the rest with the variation dv of velocity. Coming to the relation (with his symbols):

$$dv = \frac{\lambda p dt}{A}$$

where λ is a proportionality constant necessary to move from a proportion to a formula, which is the natural way to work with algebra and Calculus.[165] Euler basically said that the increment $d\omega$, for a given interval of time dt is proportional to the element of velocity dv and therefore for Theorem 4 the above relation follows.

A modern reader sees in the previous formula the second law of motion of Newtonian mechanics in the monodimensional case. To discuss whether Euler's was the first formulation of the equation of motion—it was not indeed—is not relevant here; certainly it was one of the first times the second law of motion was written in a very clear and unequivocal form. The only thing missing is the meaning of the constant λ; but this concerns only the choice of units of measurement and Euler made a choice later.

What leaves a little 'surprised', especially if reading the *Theoria* immediately after the *Mechanica*, is the fact that dv is placed directly proportional to the time interval dt, Theorem 4 authorizes only to say that dv is proportional to p and inversely proportional to A. The proportionality between dv and dt had been treated as a theorem in the *Mechanica* (Proposition 15), attributing its discovery to Galileo, who actually saw it as a plausible hypothesis to be verified experimentally, or if not it, its consequence, that is the law according to which spaces vary with the square of times. "Galileo was the first to use this theorem in the investigation of falling heavy

[163] p. 57.

[164] p. 59. English translation adapted from Bruce I.

[165] Incidentally if A is the weight, as Euler always assumes, then λ coincides with the acceleration of gravity (modern meaning).

bodies. Indeed he did not give a demonstration of it, however because of the strong agreement with phenomena, nobody doubted it anymore" [66].[166]

Euler's demonstration is very simple but circular. This is how it works: consider n infinitesimal time intervals $d\tau$. The increase in velocity du is the same in each interval of time $d\tau$, assumed equal to each other, because by hypothesis the force, that is the cause of motion is the same. So after n equal intervals $d\tau$ the velocity becomes $dv = ndu$, from which it follows that dv is proportional to $dt = nd\tau$ [66].[167] The circularity is to admit the constancy of the increase in velocity over time, which is the same as saying that the increase in velocity is proportional to time.

The reasons for omitting the proof of proportionality of dv with dt in the *Theoria* are not clear, at least to me. It is possible that everything derives from Euler's, and others, studies on finite differences, in fashion in the early 18th century. These studies had shown that the variation of a function y of x, in a certain small interval dx, is necessarily proportional to dx itself, that is $\Delta y \sim \psi dx$, width ψ a proportionality constant, generally dependent on x. Euler assumed it natural that the variation of velocity v in a given interval of time, once force and mass are fixed, cannot but vary linearly with the infinitesimal interval of time dt, and therefore $dv = \psi dt$. The physical nature of the problem tells us that the variations of velocity in a certain interval dt are always the same ($\psi =$ const.), or it varies as $\psi = \psi(p, A)$, but the case $\psi = \psi(t)$ is not contemplated.

The motion of a body moving along a plane curve (Problem 13) following a force acting on the plane is treated by resolving the force into two (orthogonal) components and then studying two separate one-dimensional motions. It should be noted that unlike the case presented in Problem 2, which was purely kinematic in nature, the idea of resolution or projection hides the underlying physical principle, the parallelogram rule, for which the projections of forces are forces themselves that act independently of each other (without interaction). This is one of the cases in which the power of the mathematical instrument hides the physical nature of the problem. Somehow Euler was aware of the problem and said that a force is traced back to three forces P, Q and R acting in the (orthogonal) directions x, y and z, with the "static resolution criterion" [84].[168]

The equations of motion eventually take the form [84]:[169]

$$ddx = \frac{2gPdt^2}{A}; \qquad ddy = \frac{2gQdt^2}{A}; \qquad ddz = \frac{2gRdt^2}{A}$$

in which A is the mass/weight while the symbol g denotes the height through which a heavy body drops in the time of a second, in a specified region of the globe, in case mass and weight are assumed to have the same numerical value: "Let the mass of the

166 p. 53.
167 p. 51.
168 p. 77.
169 p. 85.

corpuscle be equal to A, which clearly indicates as well the weight, if the corpuscle is situated in a region of the earth chosen to evaluate an absolute measure [84].[170,171]

In these equations the constant λ introduced in Problem 9 has been replaced by $2g$. That is a coefficient that had only the role to transform a proportion in an algebraic equation is given a mechanical meaning, the space traveled in one second. Still in his work of 1749 Euler had assumed $\lambda = 1/2$, which in modern terms, corresponds to assume an acceleration of gravity $G = 1/2$ and thus $g = 1/4$. The choice had derived from the desire of giving a simple expression to the speed of fall from the height h, which is simply given by $v^2 = h$, instead of $v^2 = 2Gh$. No inconsistency is at play; it is just a different choice of units of measurement.

To note that the equations of motion Euler wrote are in accord with the rule of Calculus of the 18th century, where the concept of differential was fundamental. Using modern notation, Euler normally wrote the equation of motion as $mdv = fdt$, and not $ma = f$, where the symbols are the usual. However, he also used the acceleration a in a technical way, as the ratios dv/dt or dds/dt^2. This is what he wrote when integrating the equation of motion: "[In dynamics] the effect should be measured by the acceleration or the change in the speed, that is impressed on the body in a given time: this is proportional to that force divided by the mass of the body. [...] dv is equal to the product of the acceleration and the element of distance travelled" [66].[172]

3.4.2.2 Measurement of Forces

The discussion of the origin of forces given in Euler's writings on natural philosophy, in the *Researches sur l'origine des forces* in particular, does not provide any criterion for their measurement. The force due to impenetrability is not measurable in itself as it exists only at the moment of impact. What, at least in principle, is declared to be measurable, is its effect, which for Euler according to the Newtonian approach, is expressed by the variation in velocity in an assigned small interval of time, which in the case of the impact should be much less than the duration of the contact between the bodies that collide.

In the writings specific on mechanics, such as the *Mechanica* and the *Theoria*, the problem is discussed a little more. In the first treatise, Euler showed no particular difficulty to introduce the measurement of forces. It could be obtained by the rules of statics, a discipline at the time considered well founded. These rules indeed allows to measure a force as function of a sample force, for instance a weight. It is enough to evaluate how many unities of the sample force should be summed (the additivity of forces is granted, for instance by the rule of parallelogram) to reach the equilibrium when they act in the opposite direction with respect to the force to be measured. This

[170] p. 77.

[171] Note that g is one half the value of the acceleration of gravity G, this follows from the relation $s = 1/2Gt^2$: By assuming $t = 1$ it follows $g = 1/2G$.

[172] p. 84.

means that the measure of a force is *a priori* and independent of dynamics, that is independent of the dynamical effect it produces.

Euler is more careful in the *Theoria*, here he distinguished between a static and a dynamic measure; of course when two forces measured with the two approaches give the same numerical value for Euler the forces are equal under all the aspects, that is there are not dynamic and static forces but only different ways to measure them; the dynamic way is used only when static measurements are not possible, as for instance it occurs for gravitational forces of astronomy.

Euler had established his equations of motion, Theorems 2a and 3, referring to the increment of the displacement of a body due a force starting from the rest, because only in this case he had an independent criterion to measure force, the static one. In the case of motion Euler said that nothing is known concerning the measurement of forces and the way to measure them is left to us: "Since in statics, from which we draw the measurement of forces, the bodies to which the measurements are applied, may be considered in a state of rest, and thus nothing is defined concerning the measurement of these forces when they act on the body in motion" [84].[173]

In dynamics the only way left for the measure of forces should be searched in the measure of the effects, that is they can be measured a posteriori only: "Therefore the magnitude of these forces is determined not by the impenetrability, which clearly cannot be quantified, but from the change of the state which must be effected lest the body penetrate each other" [84].[174]

Euler, however, did not dwell at length on the problem of measuring forces and did not relate the dynamic and a posteriori measure of force, based on motions, with the static and a priori measure, based on the equilibrium. In such a way he avoided to make clear if the foundation of his mechanics is dynamic or static. Euler left comments on the dynamic measure of force only in some Explicationes of the *Theoria*, as he would not to compromise himself with strong declarations. Here he proposed the convention that if a force acting on a body in motion causes a displacement shift σ in the direction in which it acts, that force is assumed to be equal to the force that would cause the same displacement σ from the rest, evaluated, for instance, according to Theorem 2a (that is if $\sigma = kp$ is the law derived from Theorem 2a, and the shift of displacement measured for a body in motion is $\overline{\sigma}$, the force associated to this variation of motion is given by $\overline{p} = \overline{\sigma}/k$).

> For the forces, then, by which bodies already in motion are acted upon, we set up this ground of measuring, so that we shall judge these equal to those which would have executed the same effect on the same bodies at rest in the same time. This ground, however, does not require proving, because it rests upon a definition and thus it was open to us to establish it. For if for any motion the small space $s\sigma$ should be equal to small space $S\sigma$, through which the same small body at rest is brought forward in that same little time by force p, we also call those forces equal [84].[175] (C33)

---

[173] p. 53.

[174] p. 50.

[175] p. 53. English translation in [146].

Admitting of measuring a force through its dynamical effect, it seems that the Newtonian law of motion could be considered as a simple definition and the concept of force could become superfluous. Only a weak ontological substratum would remain which allows to attribute some reality to force; but it is a link that can be broken without substantial changes in the formal development of mechanics.

Actually, things are a little different; to better understand it, consider the relation $f = ma$. If the mass m is assumed as a primitive magnitude, defined for example in a geometric way as a volume, then $f = ma$ can be seen either as an operational definition of f or as an empirical law—in this last case f should be defined a priori. If, on the other hand, the mass m is defined operationally (or dynamically), then the force f must necessarily be a primitive magnitude. Moving from epistemological and ontological aspects to purely mathematical ones in which only the numerical values of the physical quantities, that is their measures, are concerned, then the different definitions of force and mass flatten out. For example, considering force as a primitive physical magnitude, the natural way to measure it is the use of the laws of statics. But the force can also be measured by the relation $f = ma$, without it becomes a defined quantity because a measure does not affect the ontology of f. The mass either defined operationally (dynamically) or directly (geometrically) can always be measured with weight. In the first case because there is an empirical law—the acceleration of heavy bodies is constant—which allows to affirm the proportionality between mass and weight; in the second case because there is sufficient empirical evidence that the weight is proportional to the amount of matter (volume). As for the historical aspects are concerned, it can be said that Newton did not pronounce clearly on the nature of mass and force, hesitating between a priori definition (apparently preferred) and an operational one. Euler and d'Alembert resolved this ambiguity differently. The former considered mass as defined operationally (and force as primitive), the latter mass as primitive (and force as defined operationally).

3.4.3 The Apparent Motion and the Observer

Although Euler believed in the existence of an absolute space, he nevertheless believed that motion, as a matter of fact, could only be studied in a relative space, defined by a certain conventional reference frame. This is, for instance, what Euler wrote in his *Mechanica*, and in other occasions:

> Because of the immense nature of space and of its unbounded nature [...] we are unable to form a fixed idea of this. Thus, in place of this immense space and of the boundaries of this, we are accustomed to defining a finite space and the limits within which bodies can move, from which we can indicate the states of motion and of rest of bodies. Thus, we are accustomed to say that a body that keeps the same situation with respect to its boundaries is at rest, and truly that which changes with respect to the same, to be in a state of motion [66].[176] (C34)

[176] vol. 1, p. 2. Translation into English by Bruce I.

In the *Anleitung* Euler posed himself the problem to see what happens by changing the reference frame, in particular by considering two frames in motion with respect to each other, a problem which had ancient origins. Galileo founded his mechanics referring to a thought experiment; imagining to be in the cabin of a moving ship. Huygens also set out on a uniform motion to study the problem of collision. Euler was part of this tradition, also followed by some mathematicians at the beginning of the 18th century, but as usual, he specified and enlarged it. A fairly detailed analysis of his modus operandi can be found in [21, 133].

Already in one of his works of 1739 in which he addressed the problem of the influence of the finite speed of light in astronomical observations, Euler faced the purely kinematic problem to correct the observational data to take into account the fact that the observers were based on the earth mobile with respect to the fixed stars [68]. In Chap. 10 of the *Anleitung* Euler studied instead the influence of a change of reference in dynamics and introduced the concept of observer, to which he referred to with the German term *Zuschauer*, which is properly translated as spectator. Basically a fixed or mobile coordinate frame (without clock). There are some important results that clarify what today is known as the Galilean principle of relativity. Thanks to the use of Cartesian coordinates and Calculus, the mathematical passages that Euler had to face appears very simple to a modern.

The first proposition states:

> If the observer moves at constant speed along a straight line and he estimates directions correctly, then all bodies that are either at rest or are moving at constant speed in a straight line, will appear to him to remain in the same state [89].[177] (C35)

Notice that it is not specified very clearly if the observer is considered to be moving with respect to the absolute space. The second proposition basically states that the equations of motion are the same for an absolute system and for a system that moves uniformly with respect to it:

> If the observer moves uniformly in a straight line and if he judges directions correctly, that is according to parallel running lines, then the maintenance of the apparent movement requires the same forces as the true movement, however much the apparent movement may differ from the true movement [89].[178] (C36)

In the two propositions the statement asserting that the direction are judged correctly means in modern term that the observer is moving without rotation.[VIII]

In a third proposition, perhaps the newest one, Euler considered an observer which transaltes (without turning) unevenly. In this case, he concluded that fictitious forces must be considered alongside real forces.

> But if the observer does not move uniformly in a straight line, *but does estimate directions correctly* [emphasis added], then to maintain the apparent motion of all bodies, in addition to the forces that are actually acting on the bodies, further forces are required that will at every instant and in every body produce the change that takes place at the location of the observer, but acting in the opposite direction [89].[179] (C37)

---

[177] vol. 2, p. 497. Translation into English by Hirsch E.

[178] vol. 2, p. 497. Translation into English by Hirsch E.

[179] vol. 2, p. 498. Translation into English by Hirsch E.

Therefore, apart from the forces *P*, *Q*, *R* that actually act on each body, three additional forces are needed that in each body produce the same change that occurs at the location of the observer, but in the opposite direction.[XI]

The above proposition of Euler found its applications on several occasions. In the *Theoria*, for example, he applied it to astronomy, in order to eliminate the non-inertial motion of the observer's frame based on the earth [84].[180] In a his important work on water wheels Euler studied the action of a jet of water coming out of a rotating pipe [81]. To determine this action, the observer is located in a frame rigidly linked with the pipe [21, 133].[181] An approach that will be followed by the 19th century engineers involved the study of turbines. In a more general study of 1755, Euler arrived to formulate relations that have analogies, a part from some mistakes in calculation, with the famous relations that Gaspard-Gustave de Coriolis (1792–1843) found almost a century later [21, 133],[182] with reference to a problem similar to that faced by him, that is a rotating wheel.

3.5 D'Alembert Science and Philosophy

Jean Le-Rond d'Alembert (1717–1783), the son of an officer and an aristocrat, was left in the steps of the Saint-Jean-le-Rond de Paris church, from which he took his name. Raised by a commoner, he later had a pension from his father who also recognized him. He however continued living for many years in the home of his foster mother. D'Alembert communicated his first memoirs to the Académie des Sciences de Paris, of which he was elected as an *adjoint astronomer* in 1741 and *associé géomètre* in 1746. In 1746, he also earned his admission to the Berlin academy of sciences. The friendship with Diderot and the beginning of the collaboration with the Encyclopédie dates back to 1749. The editing of the *Discours préliminaire* and numerous items to 1750s. He took a lively part in the battles for the Encyclopédie and the recovery after the closure of the 1752 was largely due to his prestige. In 1754 Frederick II of Prussia invited him to preside over the Berlin academy, but d'Alembert refused (and so again in 1759 and 1762). The same year he was elected member of the Académie Française. The personal attacks of the opponents gradually removed him from the Encyclopédie, until the complete breakdown of 1758. Even relations with Diderot then became colder; with Rousseau the break was total. D'Alembert went ever closer to Frederick II of Prussia and Voltaire, with both entertaining a dense correspondence; but he refused any offer to leave France, just as he refused to become tutor to the son of Catherine II of Russia. He preferred to live modestly in Paris, attending the lounges of Mme Geoffrin, Mme Du Deffand and Mlle de Lespinasse, for whom he had a deep affection. In 1772 he was elected perpetual secretary of the Académie Française. Sick and tired, he gathered talented young people around

[180]pp. 120–121.

[181]p. 342; pp. 322–326.

[182]p. 326; pp. 342–342.

him, including Condorcet and came publishing his *Opuscules de mathématiques*. D'Alembert was also a Latin scholar of some note and worked in the latter part of his life on a superb translation of Tacitus, for which he received wide praise including that of Diderot [34].

On the assessment of d'Alembert, surely one of the most intelligent and creative mathematician of the 18th century, weighed the lack of interest on the part of scholars, either trained in science or in letters. Considered by historians of literary and philosophical education a mere scientist, a slightly original follower of Locke, or at most a 'precursor of positivism', he has been confused among others in the sensationalist current of the second 18th century. Historians with scientific education have considered instead d'Alembert a brilliant scientist but a superficial one; distracted by his literary interests.

His real originality as mathematician, epistemologist, ideologist of the Enlightenment is best placed to light by recent studies and his figure of activist *philosophe* and promoter of the alliance between philosophers and enlightened rulers emerges in the foreground, besides those of Diderot and Voltaire. Among modern historians educated in science interested in the work of d'Alembert should be mentioned Clifford Ambrose Truesdell and Thomas Hankins. Truesdell, known for his preference for Leonhard Euler and for his habit to judge scientists of the past with modern standards, gives sometimes a fierce judgment of d'Alembert, though all in all arriving to a favorable overall assessment. Here examples of his sarcasm: "one searches for the little solid matter [in d'Alembert's work] as a sparrow pecks some nutritious seeds from a dung heap a task not altogaether savory" [90].[183] "D'Alembert is a notorious schizograph: the elegant directness of his belles-lettres, often seen also in the preface to his scientific works, never enlightens the thick penumbra of his mathematical exposition" [159].[184] Not even Hankins is very tender. He stresses the superficiality of d'Alembert referring testimonies of Clairaut, Daniel Bernoulli and Condorcet on this aspect. Hankins also expressed a negative opinion on the late d'Alembert's scientific production. D'Alembert would have contrasted these criticisms, especially those of Truesdell, by replaying: "The important thing is to discover; there will be never the lack of textbook-makers" [102].[185]

An important present-day editorial project concerns the complete works of d'Alembert [55, 145]. The project contemplates the publication of a forty volumes and should be completed in the next few years to heal the current gap in d'Alembert's literature. In fact until today there were only two editions, 19th-century, of d'Alembert's complete works, which, however, exclude all works of a scientific nature. For the latter the relatively recent publications are fragmentary. An examination of the manuscripts for the preparation of the complete works contributed among other things to debunk a mite, fueled by d'Alembert himself, according to which since 1760, his scientific literature would have suffered a sharp decline in favor of

[183] Sect. 2, vol. 12, p. CXVII.

[184] pp. 186–187.

[185] p. 63.

publicist and philosophic activities, due to depressive phases and illnesses. The mood of d'Alembert would be well represented in the following:

> What bothers me a lot is the fact that geometry, which is the only occupation that really interests me, is something I cannot do. Everything I do in literature, although very well considered (it seems to me) in our public sessions of the French Academy, is for me only a way to pass the time, for the lack of nothing better to do [116].[186] (C.38)

In fact, during the last years of his life, d'Alembert wrote at least 4000 pages of scientific subject, some very important: on vibrating strings, on hydrodynamics, on probability. Some of these works were published in the *Opuscules mathématiques*, many others remained unpublished.

D'Alembert received his education at the Jansenist Collège des Quatre Nations in Paris where he entered as *gentilhomme*. The choice of the Quatre Nations was lucky, as it was the only school in Paris that devoted an entire year to the study of mathematics. The instruction offered by the college was elementary, but the library contained 2500 volumes only on mathematics: a rich mine for the curious d'Alembert. In 1735 he received the baccalaureate. After a year without success studying medicine he directed his attentions to mathematics. In fact he must have kept this interest alive, because on 1739 he presented his first mathematical document to the Académie des sciences de Paris.

His mathematical sources were all products of the circle around Malebranche and the philosophy course he attended at Quatre Nations was a Cartesian one that besieged him with premonition, innate ideas and vortices that grew up to despise. This did not, of course, make Alembert a Cartesian or Malebranchian, particularly since he read their papers mainly for the mathematics they contained. In any case, ideas of Malebranche and Descartes emerged in many points in all his philosophical works.

D'Alembert must have been familiar from the earliest times with the popular side of Newtonianism as it appeared in the *Eloge* of Fontenelle and in the *Lettres philosophiques* and *Elemens de la philosophie de Newton* of Voltaire. However d' Alembert made a serious study of Newton's *Principia* only after he had worked his way through simpler texts. Some time after 1739 he read the annotated edition published by Thomas Le Seur and François Jacquier [141] and wrote a short commentary on the first book. D'Alembert published his first important work, *Traité de dynamique*, in 1743 (he was only 26). It was concerning mechanics with new methods and foundations. He had just started to read it at the regular sessions of the academy when Clairaut started reading a his own treatise, with methods similar to his. D'Alembert thus rushed his treatise to the publisher and it appeared substantially unfinished with some misprintings; indeed all his books went to the publisher before they were really ready. It was a move that d'Alembert repeated frequently through his career to forestall any mathematicians who was working on the same subject.

D'Alembert did not take up the pen for philosophy until 1751, after the acquaintance with Diderot, Condillac and Rousseau, three of the most brilliant minds of the

[186] vol. 13, p. 331. Letter of d'Alembert to Lagrange, September 22th, 1777.

French Enlightenment. The only information we have concerning these d'Alembert's early friendships comes from Rousseau's *Confessions*. It appears that Rousseau introduced Condillac to Diderot and then Diderot brought d'Alembert into the circle. Diderot was a closest friend, Condillac the one whose philosophy agreed most closely with his own. These were the men who turned d'Alembert's head to philosophy.[187]

D'Alembert proved to be good philosopher, strict meaning, not a philosopher of nature however but rather an epistemologist; a natural role for a scholar trained in mathematics who charged himself for founding the new science. As an epistemologist—he was familiar with the empiricist philosophy of Locke and Condillac— he influenced the view on modern mathematical physics to an extent that should be still clarified by historians, very great however. It must be said, in any case, that the reflections on the epistemology of science made explicit by d'Alembert after the 1750s had its roots in his mathematical works of the 1740s, also and especially in his youth work, the *Traité de dynamique*. After all, d'Alembert was not a schizophrenic with two personalities, that of the philosopher and that of the mathematician.

He lived in a period where the Aristotelian-Euclidean epistemology was challenged from many sides. He knew the logic of Port Royal for having studied it at the Collège des Quatre Nations; he was also an admirer of Newton's empirical philosophy, to which in the meantime Locke had attempted a solid philosophical basis. The Lockian and Newtonian conceptions are evident in his works on general physics. Here the principles have a clear empirical value; they are true because they are as they are in fact and not because they have to be as they are. Nevertheless, they are undoubted.

The Port-realist conception surfaces from mathematics and mathematical physics, where the distinction between first and second order principles, echoes the conventionality in the choice of the principles. However, d'Alembert's position is not well defined; on the one hand, he maintained that notions obtained by abstraction from sensations become clear and distinct and provide the basis for apodictical reasoning; on the other hand (this is explicitly stated for geometry and indirectly for mechanics) he argued that many basic notions could not be clear and distinct, or at least one could not provide a precise definition of them. This is, for example, the case of the straight line. Indeed we do not have any specific definition for it. Perhaps for this reason d'Alembert introduced a concept that referred to that of the nominal essence of Lockian mold, and spook of definition as intermediate between nominal and real definitions as commonly accepted, in whose understanding the genetic method plays an important role.

D'Alembert questioned the position of Descartes, Pascal and the Port-realists, for which man is able to produce correct reasoning in a natural way. According to him, we express our notions through language but language is imperfect so that our reasoning can hardly have that degree of apodicticity that we would like to attribute it, at least in the sciences. Only for mathematics, especially for algebra, precise languages have been developed that can make reasoning conclusive. In any case,

[187] Some bibliographic information is drawn from [102].

even if the language is imperfect, the philosopher and the mathematician learn to extricate themselves with it. Science so conceived loses its traditional structure of a series of deductively linked reasonings. Rather it is a chaining based on a rationality that is not reduced to deduction only but also recalls, for example, analogy.

The problematic of the apodeictic validity of the conclusion of a reasoning is evident also in the considerations d'Alembert developed on the reduction to the absurd and for his interest in the art of conjecture. Below what he wrote on the reduction to the absurd, by doubting that *not-not* equals *yes*, appearing *ante litteram* a supporter of non-classical logic, that so much space is conquering in modern epistemology.

> But if the number of our certain knowledge is very small, that of our direct knowledge is even more so. We ignore, for a large number of objects, what they are and what they are not. And we have only negative ideas about others; that is, we know better what is not of what it is. Happy, however, in our indigence, to possess this imperfect and truncated knowledge, which is nothing but a more reasoned and sweeter way of being ignorant. Now in all these cases we will be forced to resort to indirect demonstrations, the main demonstrations of this type are known under the name of *reduction to the absurd*. They consist in proving a truth for the absurdity that would follow if it were not admitted [to be true]. In this category all the demonstrations concerning the incommensurable must be put, that is, the *magnitudes that have no common measure*. In fact the idea of infinite necessarily enters this kind of quantities; now we do not have of the infinite but a negative idea, since we do not conceive it but by the negation of the finite [51].[188] (C.39)

The art of conjecture prefigures what today goes under the name of inductive logic; it has the task of drawing conclusions where the premises are uncertain or deriving the causes given the effects. Here is what D'Alembert wrote on the matter:

> In the art of conjecture we can distinguish three branches. The first [is] what mathematicians call probability analysis in gambling [...]. The second branch deals with [...] different issues related to common life, such as those concerning the duration of the life of men, maritime insurance, inoculations [...]. The third branch deals with the sciences in which it is rare or impossible to reach the proof and in which the art of conjecture is nevertheless necessary [52].[189] (C.40)

The main method d'Alembert saw in the art of conjecture in the third branch, that of the empirical sciences, was the analogy among facts which have something common.

At length it has been debated whether d'Alembert should be considered as a Newtonian or a Cartesian, with conflicting opinions. An interesting synthesis is suggested by Michel Paty who sees in d'Alembert the one who reorganized the Newtonian dynamics on the basis of Cartesian rationality: "Cartesian intelligibility for a Newtonian program" [144].[190] Granted it makes sense to consider the dichotomy Newton-Descartes—indeed it did not exist, or at least was not so clear at the turn of the 18th century—this approach could not capture the essence of the problem. D'Alembert entered strongly into the vein of mixed mathematics as Newton did and as Descartes did not instead, to propose himself as a new philosopher of nature.

[188] pp. 166–168.

[189] vol. 2, pp. 83–84.

[190] p. 4.

From this point of view I believe that with Euler and Clairaut, d'Alembert should be considered as one of the first great mathematicians of the Continent to enter the new course of mixed mathematics, and in such I agree with the common view of historians with scientific background. Even a superficial reading of d'Alembert's scientific works makes it clear that they are enormously closer to those of Newton than to those of Descartes. Both Newton and d'Alembert, after interesting preliminaries of philosophy of nature, epistemology and metaphysics, that differ perhaps profoundly from each other, moved in the same direction. D'Alembert, like Newton, placed the law of universal gravitation at the basis of his astronomy and as a technical tool he assumed what in the previous sections has been called a vectorial approach based on the so-called Newton's second law.

Even outside mechanics, in what Alembert called physical-mathematics, the methodology is the classical one of mixed mathematics that refers to an empirical approach to knowledge. This is the judgment d'Alembert wrote about Newton in the *Discourse preliminaire*: "That great genius saw that it was time to banish conjectures and vague hypotheses from physics, or at least to present them only for what they were worth, and that this science was uniquely susceptible to the experiments of geometry" [54].[191]

3.5.1 The Way to Knowledge

D'Alembert assumed the disposition of all our knowledge according to a single great chain. This has been seen as the generalization of the Newtonian empirical approach which from the physical phenomena manages to go back to a few principles from which then to descend to obtain all the particular truths. And also the generalization of Locke's approach, according to which, starting from simple notions, one can follow the pathways of every science. This is true, but d'Alembert modified the conceptions of his illustrious predecessors, which remained all in all within the main stream of the Aristotelian-Euclidean epistemology. His modifications concerned both the very conception of science and the organization of theories.

For d'Alembert the principles of our knowledge, in physics for instance, are the simplest properties that the observation shows us. These properties depend on the essence, the intimate nature of bodies, which we do not know and never will know. Their certainty depends on the fact that they are human products resulting from an activity whose phases can be reconstructed; these are abstractions that explain the nature of the object as it appears to us and not as it is.

The results of our observations depend also on the intimate nature of our soul that is even more unknown to us. "The human spirit that has long been searching for these primitive truths, making a thousand attempts to reach and never finding them [...] looks like a criminal imprisoned in a dark place, trying in every way to find a

[191] p. 100.

way out [...] most often glimpsing a faint light for some narrow and tortuous cleft that strives in vain to enlarge [52].[192]

D'Alembert was in a privileged position; he was a great mathematician, a professional physicist and had an excellent philosophical culture; characteristics that was combined in very few people. His mathematical preparation allowed him to develop a synthesis between the rationalist approach, still strong in France also for the persistence of the Cartesian heritage, and the empiricist approach. A synthesis that today is unanimously recognized as indispensable in dealing with those sciences he called *physico-mathematique*—in the following I will use the modern term *mathematical physics* which is close to d'Alembert meaning—adopting and specifying a word introduced in the 17th century.

Aristotle believed that there were more sciences, each one characterized by its own principles. This point of view was accepted, or at least not challenged by Newton. D'Alembert instead supported the uniqueness of the sciences and the possibility, even if only ideal, to identify a single principle for it. This position is documented in this very significant passage:

> If we knew why things exist we would probably be a long way to solve the problem of how different things exist. Because, most likely, everything is bound in the universe more intimately than we think; and if we knew this first reason, this reason so embarrassing for us, we would hold the end of the chain that forms the general system of beings and we would have nothing but to develop [this principle] [52].[193] (C.41)

According to the majority of historians, the radicalization of d'Alembert in the application of a single principle and of a single methodology is to be found in his Cartesianism. According to Hankins [102], for instance, d'Alembert's vision, and also Condillac's, on the structure of the sciences and consequently on the chain of knowledge/being is essentially the vision of a mathematician: "There can be few doubts about the origin of this vision; it is undoubtedly the rational philosophy of René Descartes" [102].[194] As proof of his statements, Hankins cites the *Discours de la méthode* in which Descartes refers to a sequence of reasoning as a chain, on the model of mathematical reasoning. According to Descartes all the sciences are connected like a chain and none can be grasped completely without the others follow. Moreover, "Those long chains of simple and easy reasons, of which the geometers routinely use to bring their demonstrations to the end, had made me imagine that all the things likely to fall under human knowledge, follow one another in the same way" [58].[195]

Hankins also claims that d'Alembert had no doubt that a chain exists. As far as created beings are concerned, this is documented by many statements by d'Alembert, see for example the entry Cosmologie of the *Ecyclopédie*. As for the chain of deductions, its sensationalist philosophy implies that the structure of our knowledge reflects

[192] vol. 2, pp. 64–65.

[193] vol.2, p. 64.

[194] p. 117.

[195] p. 21.

the structure of the natural world; if this is continuous, even that is continuous. Here is what D'Alembert wrote:

> All beings and therefore all the objects of our knowledge have an interconnection that escapes us [51].[196] (C.42)

The concept and the term chain of being has been re-enacted by Arthur Lovejoy (1873–1962), in his *The great chain of being* of 1936 [130]. Lovejoy gave an overview of the evolution of this concept, which he considered contemplates the principles of fullness, continuity and hierarchy. In the 18th century this idea was particularly popular near the naturalists and Buffon among them. However, the idea of the chain of being lent itself to being applied also in the sphere of ideas and reasoning to describe their consequentiality.

Paolo Casini [32, 33] expresses greater prudence, stating that if there is a Cartesianism in d'Alembert, it has been filtered by the cultural debate of the epoch. For him the characteristic accent placed in logical consequentiality introduces a common intellectual necessity in France, even among those who share the anti-Cartesian critique carried out in Newton's *Principia*. Is it a precise Cartesian heritage or a generic *forma mentis*? [32].[197]Moreover insisting on the Cartesianism of d'Alembert and contrasting it with his Newtonianism does not make great sense, if it is not first cleared how the two *isms* differ, since they are historically much more connected than generally accepted.

Also in the work where one sees more d'Alembert's Cartesianism, that is the *Traité de dynamique*, together with the undeniable influences of Descartes, both on the structure of the work and on the individual concepts, one could see a continuation and purification of mechanics, in the spirit of Newton rather than a flattening on Descartes's rationalism. One should not forget that d'Alembert was mainly a mathematician and a mathematician does not need the permission of a philosopher to attempt the mathematization of the world. It is true that Newton's *Principia* were simplified and purified of many residue of old metaphysics and natural philosophy, but there was still room for phenomenological laws, of a contingent nature and thus not necessary to reason. Among them, the law of universal gravitation and the phenomena of friction—in the study of the motion of fluids.

Casini maintains that d'Alembert changed over time his point of view. From an optimistic position in the *Discourse preliminaire* he would have moved to a more skeptical position in the *Essai sur les elemens de la philosophie*. In this mature work d'Alembert expressed the idea that it was impossible to know the whole chain of knowledge; we could only see pieces. The chain for d'Alembert has a methodological function. It is not a concrete reality like Buffon's chain of creatures, but it is an operational background that serves to justify the economic character that characterizes his conception of sciences [33].[198]

---

[196] p. 24.

[197] p. 85.

[198] p. 42.

This thesis of two phases of thought, however, appears to be a little forced, in my opinion. In fact, that the idea of the chain of being has a methodological function is clear not only from the late works of d'Alembert, but also from the *Discourse preliminaire*; thus may be there are not two phases. In the *Discourse preliminaire*, after developing an ideal genetics of all knowledge, d'Alembert recognized that ideas have developed in a different way in historical reality. Mainly not in a linear way, but following failed attempts more than once before arriving at the right solution. This fact certainly diminishes the reality of the chain of being in the development of thought, a reality that instead obviously continues to exist in the natural world, where the errors of nature always leave in some way a permanent trace.

3.5.1.1 D'Alembert's Sensationalism

To understand d'Alembert's scientific epistemology, one must examine his conceptions of knowledge in general; how it originates and what it is based on. Following Locke, d'Alembert believed that the origin of our knowledge should not be sought in innate ideas but rather in sensations:

> All our immediate knowledges are reduced to those we receive through the senses; it follows that we owe all our ideas to sensations [54].[199] (C.43)

It is undeniable the influence of Condillac, who already in his *Essai sur l'origine des connaisances humaines* of 1746 had simplified the conception of Locke according to which simple ideas come from sensation and reflection, admitting the primacy of sensation.

This conception, continued d'Alembert, is now generally accepted and the innate system of ideas is rejected by everyone. To prove that sensations are the origin of all our knowledge, it is sufficient to show that they have the possibility of being so. This possibility can be considered a proof, because it explains the origin of our knowledge in the simplest way, without assuming anything that comes from certain a priori knowledge.

> The first thing that our sensations make us to know, and that is not even distinguished from them, is our existence [...]. The second knowledge we owe to our sensations is the existence of external objects [...]. The effect that these innumerable objects produce on ourselves is so strong, so continuous and unites us so much to them, that, after the first moment in which our ideas press us within ourselves, the sensations that besiege us from every side, force us to escape from [...]. The involuntary affections that make us feel, compared to the voluntary act that presides over our reflected ideas [...] determine in us an invincible inclination to affirm the existence of the objects to which we refer these affections, seeming to be the cause of them [54].[200] (C.44)

[199] p. 14.

[200] p. 15–16.

In this passage two statements must be underlined:

1. Our existence is not distinguished from our sensations.
2. We come to know the external bodies because they determine in us involuntary affections.

The first statement seems to reflect the conceptions of Condillac, which at the beginning of the *Essai sur l'origine des connaisances humaines*, had written: "Whether we raise ourselves up to the heavens, or descend into the abysses, we never go out of ourselves and do not perceive anything other than our thinking" [39].[201]

The second statement presents a personal solution by d'Alembert to the thorny problem of the existence or absence of external bodies. A solution that is not definitive however; in a letter to Voltaire he indeed expressed his skepticism on the matter:

> *I swear that the only reasonable way in all these metaphysical shadows is skepticism* [emphasis added]. I do not have a distinct idea and even less a complete idea, of matter and everything else [...]. I am led to believe that what we see is only a phenomenon that does not contain anything outside of ourselves as we imagine it; and I always return to the question of the Indian king: *Why do things exist?* Because this is really the most surprising thing of all [53].[202] (C.45)

Condillac had tried an answer in the *Traité des sensations*. But his proposal was not accepted by d'Alembert who was essentially convinced, as also Condillac was, that we cannot prove in a convince way the existence of bodies external to us. It seemed to him only a most reasonable hypothesis; in fact, if the existence of bodies is supposed, the sensations we would feel of them could not be more alive than they actually are. Therefore, for a reason of economy, it is reasonable to accept the existence of external bodies. In any case, a serious philosopher should not raise these doubts but rather turn to the understanding of how we form the idea of external bodies, including ours.

The object of the philosopher must indeed be the study of phenomena trying to understand them as much as possible. The method to follow to order our knowledge must be genetic; of every notion we must analyze the way it was formed, the simple notions from which it derives, but not the mechanism for which ideas are formed in individuals. Thus in d'Alembert's genetic analysis of knowledge there are no concessions to psychologism.

3.5.1.2 The Organization of Theories

A fairly detailed exposition of the genetic method—the method of establishing ideas by following their formation—is reported in the Éclaircissement I of the *Essai su les elemens de la philosophie*. Here d'Alembert introduced first the Lockian concept of simple ideas, defined as those ideas or notions that we do not know how to decompose into other notions. There are two types of them, the notions that derive immediately from the senses and the abstract notions, which are acquired by abstraction, in which

---

[201] pp. 1–2.

[202] vol.5, first part, p. 186. Letter to Voltaire, 29 August 1769.

Simple notions Are undefinable	Primitive notions acquired through senses Abstract notions acquired by genralization or decomposition of objects
Composite notions Can be defined. Simple notions are derived from their analysis	
Definitions	They explain the composition of notions. That is they develop the simple notions they contain They must be clear and concise; composite notions can be used to shorten, provided they are defined beforehand They are more nominal than real definitions; they explain the nature of the objects as we conceive them and not as they actually are

Fig. 3.12 Simple, composite notions and definitions in d'Alembert. Adapted from [119], p. 70

we consider in a single object one (or a few) properties, without paying attention to the others. The process of abstraction in d'Alembert has the function of extracting from the sensations the most important characteristics. For this reason his abstract notions, including the notions of extension and of body, resemble the primary qualities of Locke, while notions derived directly from the senses, such as smells, tastes, etc., refer to secondary qualities, that would have no meaning without the perceiving subject [32] (Fig. 3.12).[203]

The abstract notions of d'Alembert are therefore by no means concrete; indeed they are the 'facts' that the science of nature and metaphysics must deal with; simple and recognized facts, which do not presuppose others and which can neither be explained nor disputed. In physics they are the phenomena that observation discovers in the eyes of all; in geometry the sensitive properties of the extension; in mechanics the impenetrability of bodies; in metaphysics the results of our reflections.

Alongside simple abstract notions there are abstract composite notions, characterized by the fact that simple notions can be derived from their analysis. All compound abstract notions need to be defined, while attempts to define simple abstract notions can only cause confusion. According to d'Alembert, from simple notions we can derive not only composite notions but also other simple notions, for generalization of them. They cannot be further generalized and are not capable of definition [52].[204]

[203]p.41.

[204]vol. 2, p. 49.

As an example of generalization, d'Alembert reported the case of simple abstract notions of extension and duration. They contain the most general notion of part. The parts subsist together in the extension and follow each other over the duration. But the notion of part is no longer definable than those of extension and duration [51].[205] The mechanism of generalization is not well explained by d'Alembert. It resembles definition, but here the generalized notion is only suggested by the juxtaposition of simple notions, with a process in which the intervention of intuition is decisively required.

For d'Alembert, composed (and even simple very general) notions, can be grasped very well with the genetic method, which must often be preferred in philosophy to give a proper definition. The genetic method allows to analyze the composite notions and to identify the simple components, making them clear, better than the definition can. One of its main advantages is to guarantee us from the error in which we will fall on abstract notions by referring them to concrete objects. From the following passage, which draws inspiration from geometry, transpires those that according to d'Alembert are the advantages of the genetic method; instead of saying that the straight line is an extension without width or thickness; the surface an extension without thickness, the bodies an extension with width, length and thickness, one would prefer to proceed in the following way: let suppose to have a solid body; immediately distinguish three things, the extent, the limitation in all directions and the impenetrability. Making abstraction from this last one it remains the abstract notions of extension and limitation. These ideas constitute the geometric body, which differs from the physical body by the idea of impenetrability. Then make an abstraction from the extension or the space that this body occupies, only considering its limitation in all directions. And this limitation gives the idea of surface, which is reduced as it is clear to the extension in two dimensions. Finally, in the idea of surface still make abstraction from one of the two dimensions that compose it and still you have the idea of line [52].[206]

Simple and complex notions, with definitions, are the bricks with which to build the elements of knowledge. *Philosophy*, that for d'Alembert is the general term for rational knowledge, has as its *elements* the primitive parts which we can suppose the whole is formed of. The elements are also the principles from which the explanations of the various aspects of philosophy originate in the explanation of phenomena. D'Alembert specified his concepts on the elements of philosophy in various texts. In the following I will refer mainly to the *Discourse preliminaire* to the *Encyclopédie* (1751), to the *Essai sur les elemens de la philosophie*—collected in the *Melanges de littérature, d'histoire et de philosophie* [47], a first edition in 1759 and a second in 1767 with the addition of Eclaircissements [51]—and to some entries to the *Encyclopédie*, among which in particular: Élemens, Cause, Système, Expérimental, Cosmologie. Here is what D'Alembert wrote at the beginning of the *Essai sur les elemens de la philosophie*:

---

[205] pp. 30–31.

[206] vol. 2, p. 53.

> In the multitude of truths that the Encyclopédie embraces, and which in vain we would try to know all together, there are some that raise and dominate above the others, like some rock top of an immense sea. If these truths that are most interesting to know, are gathered in the elements of the philosophy that will be used in the Encyclopédie as an introduction, the usefulness of this great work will undoubtedly become more general and more secure [51].[207] (C.46)

Since philosophy concerns all knowledge, the elements of philosophy must contain the fundamental principles of all human knowledge. According to d'Alembert, this knowledge is of three kinds, either facts, or feelings, or discussions. Only this last type for him would belong entirely to philosophy, even if the other two are close to it for some of the aspects under which they can be seen.

> But even though these sciences are different from each other, both for their extension and for their nature, there are nonetheless general points of view that must be followed to treat the elements; moreover there are different nuances in the way of applying these general points of view to the elements of each particular science; this is what needs to be developed [51].[208] (C.47)

In order to understand better the idea of an element, imagine, d'Alembert said, a science that is completely defined and transcribed into a single text. Moreover, in this text all the propositions are concatenated, that is, "they form an absolutely continuous series" so that each proposition depends solely on the preceding. In this case all propositions are nothing but the "translation of the first, presented under a different aspect; all these were reduced to this first proposition" [51].[209] Proposition which is for d'Alembert the element of the science in question. If all the sciences, d'Alembert continued, follow this pattern, it would be possible to unite all the existing elements in a few pages. In this way one could also have a unique point of view on them and the elements of all the sciences would be reduced to a single principle, the main consequences of which would be the elements of each particular science.

It is useful to assume two kinds of elements, those who are at the top of a chain of deductions and those that belong to a node formed by two or more branches of the chain. D'Alembert referred to the former as principles of first order and to the latter as principles of second order, that are only improperly principles, because derived from others. They could be assumed as principles because of their key role in the derivation of many propositions of the theory [51].[210] However according to d'Alembert even the so called first order principles, may be, are not such and could depend on others not yet known (Fig. 3.13).

The principles of the second order are in general the theorems of a science, obtained by considering more than one principle of it. These theorems, however, are of a very general nature and a large part of the propositions of a science can be derived from them without resorting to first order principles. For many scientists of the 18th century, a (first) principle was a principle in the Aristotelian sense, that is

[207] pp. 8–9.

[208] p. 24.

[209] p. 24.

[210] pp. 35–37.

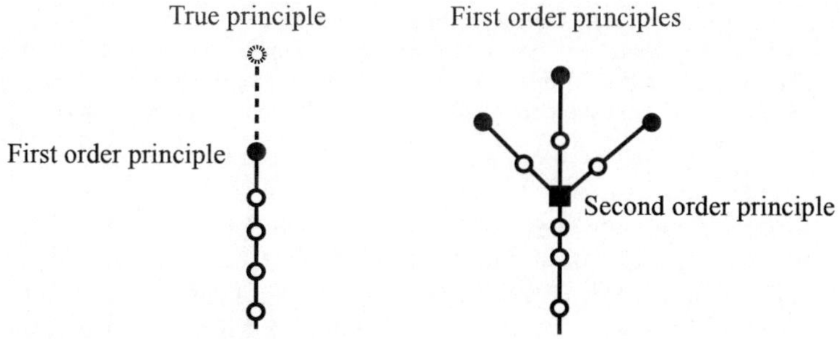

Fig. 3.13 First and second order principles

evident in itself, indubitable. A second order principle was a principle only because it was placed at the beginning of a theory but was not evident. It was not necessarily a theorem in the modern sense, because it is not said that general demonstrations were available, but there was a shared conjecture that it could be demonstrated or it was simply true.

If it is desirable that the elements of philosophy be reduced to one only, when it comes to exposing and explain them, as in the *Encyclopédie*, it is advisable to follow a different point of view; otherwise the elements would be reduced to almost nothing and their use and application would be too difficult. All kinds of knowledge concerning revealed religion must be excluded from the elements of philosophy. They are absolutely alien to the human sciences for their object, for their character, for the very type of convictions that they produce in us, suitable more, as noted Pascal, for the heart than for the spirit [51].[211]

3.5.2 *The Parts of Science*

Science, that is the knowledge clear and distinct based of principles evident in themselves [65],[212] is the term d'Alembert used to name a specific part of knowledge. According to d'Alembert, the three great objects of knowledge are: God, man, nature, which give raise to three main sciences. In the *Essai sur les elemens de la philosophie* he followed this order to present the elements of the various sciences, presenting first the sciences of the spirit, which deal with God and man, then those of nature. To all, however, preceded logic, or the "art of reasoning". It is the first science to be dealt with in the elements and "forms its title and entrance". Then metaphysics follows that deals with the problem of existence and divinity and moral that deals with man. Grammar, which plays a transversal role like logic, closes the review of the sciences of the spirit.

[211]p. 20.

[212]Article *Science*.

In the following a summary of the sciences related to the third great object of knowledge, nature, are reported with some details. Here d'Alembert adopted the order: algebra, geometry, mechanics, astronomy, hydraulics, optics, acoustics, general physics. This order has changed below for editorial and logical reasons; moreover for the sake of space no room is left to optics and acoustics.

3.5.2.1 Mathematical Sciences: Algebra and Geometry

For d'Alembert, the study of nature is that of the properties of bodies which depend on motion and shape. So mechanics and geometry that deal with these two things are the two necessary keys. Geometry, which comes first is simpler but must be preceded by algebra, a more general science that deals with the properties of magnitudes. Two reasons give this science a prominent place among the ingredients of philosophy. The first is that it helps geometry and physics and is absolutely necessary for the 'transcendent' part of the two sciences. The second reason is that it is among the most certain knowledge we have. And even if there are still obscure things in algebra, this, according to d'Alembert, depends more on those who study it than on the discipline itself. Furthermore, algebra must be distinguished from mathematical analysis. The first is the science of the calculation of quantities in general, the second concerns the way of solving problems:

> The principles of algebra are addressed only to purely intellectual notions, to the ideas that we form in ourselves by abstraction, simplifying and generalizing primitive ideas. Thus these principles properly contain only what we have put in them and what is most simple in our perceptions. They are somehow our work; so how can they leave something to be desired as far as the evidence is concerned? [51].[213] (C.48)

The 'inventors' of algebra have formulated a series of rules that are used by most without knowing the origin, how the language is spoken without knowing its grammar. The philosopher, however, must investigate the elements of algebra and the examination of the simplest operations is sufficient for the purpose (the most complicated operations will indeed give only technical information). In this way algebra will not take much room in the elements of philosophy, but by restricting it in such a small space, it can be presented in an almost entirely new form.

The use that mathematics makes of algebra to determine unknown by known magnitudes is what distinguishes it from logical analysis, which is nothing but than discovering what is not known by means of what is known. All algebraists use logical analysis to begin with and to conduct the calculations; but at the same time the help of algebra greatly facilitates the application of this analysis to the solution of problems [51].[214]

Geometry "is the science of the properties of extension, when one considers it as simply extended and endowed with form" [51].[215] The truths of geometry are purely

[213] pp. 154–155.

[214] p. 157.

[215] vol. 4. p. 158.

hypothetical when applied to the world. Nevertheless, these truths are not useless, having regard to the practical consequences that result from them. For instance, in geometry one knows about curved lines that continually approximate a straight line without ever meeting it, and that nonetheless, being traced on a paper, are sensibly confused with this straight line leaving less than a very small space. It is the same of other propositions of geometry; they are the intellectual limit of physical truths, the term to which they can be approximated as much as desired, without ever arriving exactly. But if the geometrical theorems do not take place strictly in nature, at least they serve to solve with sufficient precision for the practice, the different questions that can be placed on the extension [51].[216] In order to demonstrate in a rigorous way the 'truths' of the figures of the bodies, one is obliged to suppose in them an arbitrary perfection. Once the 'truths' are found, they can be transferred in an approximate way to the outside world. Geometry is used every day in this way: from speculative geometry to practical geometry and *vice versa*.

According to d'Alembert, the 'method of inventors', that is the analytic method, is the most suitable way to present algebra and its elements. Because algebra is a purely intellectual science and can be treated in a simple and rigorous way, it is sufficient to follow the natural order of the spirit, avoiding only the useless or false attempts in which some inventors have incurred. For geometry the matter is different; here the method of inventors can be used to illustrate it to ordinary people; the philosopher should instead follow the synthetic method.

In the *Essai* d'Alembert expressed his regret that the scholars who deal with elements are even the least prepared ones. The masters of art, who after a long and diligent study, have seen difficulty and subtleties known, disdain to retrace their steps, either because they are committed always to open new frontiers, or because they consider it less productive to deal with elements. But yet only those who know a science in detail can get to formulate the true elements. Of course, these considerations by d'Alembert are very apt for an era like ours where the gap between education and research is very high.

According to d'Alembert the fundaments of geometry are the definitions, the propositions and the demonstrations; a modern reader is quite surprise not to find in the fundaments the primitive notions, a concept to which d'Alembert devoted large space, and to which a great relevance is given in modern axiomatic theories. They are given for granted. Among the fundaments, however, the definitions are the most delicate ones. For example, abstract notions of straight line and surfaces are accepted as primitives. These 'primitive' notions can nevertheless be defined. For example, the straight line can be defined as the shortest line that can be carried from one point to another. But definitions of this type do not completely contain the primitive idea that we form of the straight line; and similarly for the plan. A definition cannot make these notions clearer; also because of the imperfection of language. For this reason, in general, the definitions are those that deserve the greatest attention.

According to d'Alembert, the elements of geometry immediately lead to the geometry of curves, a fundamental discipline in mechanics. Although in some cases it can

[216]vol. 4, p. 159.

also be studied with the method of the ancients, it can more easily studied with algebraic methods and sometimes only with these. Those who consider the method of the ancients more rigorous than the algebraic one, find a kind of consolation to consider useless what they do not know [51].[217]

3.5.2.2 General Physics

Most of the *philosophes*, even those of litterary inclination, confronted with empirical sciences. Voltaire somehow metabolized Newtonian physics, Diderot dealt with physiology and chemistry, Rousseau with chemistry and so on. Everyone wanted to break with the optimistic Cartesian physics, according to which the world is a great machine whose mechanism can be thought out and unveiled, considering instead a most complex issue. Condillac, in the *Traité des systèmes* of 1749, criticized the great systems of the 17th century. According to Condillac, the physicist must renounce to explain the mechanism of the whole universe and limit himself to searching for constant relations.

For a correct understanding of d'Alembert's conceptions of general physics, it is better to see them alongside those of Diderot, which are clearer. Diderot essentially believed that mathematics had reached its peak and that further progress in science could only be achieved through experiments. In the *De l'interprétation de la nature* of 1754, he expressed the conception that only experiments can open access to the knowledge of nature; mathematics, for its part, tends to remain enclosed in itself and not to provide new knowledge.

> Abstract sciences have occupied the best minds for too long and with too little fruit; neither one did study what was important to know, nor he put choice, nor ideas nor method in his studies. Words have multiplied endlessly, and the knowledge of things has remained behind. [...] Facts, of whatever nature, are the true wealth of the philosopher. But one of the prejudices of rational philosophy is that he who does not know how to number his crowns, will be scarcely richer than one who has only one crown [60].[218] (C.49)

According to Diderot, in order to become active and fruitful, the scientific method must be brought to full autonomy and freed both by the constraints imposed by the old metaphysics and by the constraints imposed by mathematics. This position is also somehow accepted by the 'mathematician' d'Alembert, who wrote:

> The more useful you can get from applying Geometry to Physics, the more you have to be careful in this application. It is to the simplicity of its object that Geometry owes its certainty; as the object becomes more complex, certainty darkens and moves away. We must therefore settle on what is unknown, nor believe that the words of Theorem and Corollary, make the essence of proof, by secret virtue [42].[219] (C.50)

---

[217]pp. 176–177. D'Alembert, in the *Essai sur les elemens de la philosophie* and in other writings, expressed a harsh judgment on the decadence of English mathematics, which influenced by the ambiguous example of Newton considered only the way of the ancients.

[218]vol. II, pp. 18–19.

[219]pp. XLII–XLIII.

For Diderot the ideal of science is the descriptive one. And for him the most interesting disciplines are biology, physiology, botany, chemistry, etc. Here experience teaches to rebuild the chain of being, without the help of mathematics. Naturally the study must have its own systematic nature, which must not, however, be derived from a priori reasoning, but deduced from its particular objects. Diderot also criticized the taxonomic approach of Linnaeus, because of the rigid division of the plant world into genus and species, contrasting it with the approach of Buffon, according to which in nature only individuals exist and only what is described exactly is known.

D'Alembert expressed similar ideas in what he called *general physics*. According to him, for those phenomena for which mathematics is of little use, one must follow mainly a 'method'. Facts are what a physicist must try to know well; the more he collects them, the closer he will be to see their concatenation. His purpose must be to put order until it is possible. To explain as far as possible from one side to the other, to grasp both the main trunk and the units.

But d'Alembert moved in a broader perspective than Diderot's. Next to general physics he put mathematical physics (sciences physico-mathematiques) [51],[220] that is experimental sciences, in which however mathematics plays a fundamental role, moving in the mainstream traced by his colleagues mathematicians since the ancient Greece. For him, there are large parts in physics where only one experience, or even one observation, an empirical principle, serves as a basis for a complete theory. These parts belong to mathematical physics and consist in the application of geometry and Calculus to phenomena. Often, to be successful, the finer geometry is required. And the geometry of the ancients could not go much further.

At the beginning of the paragraph of the *Essai sur les elemens de la philosophie*. dedicated to general physics, d'Alembert took up the distinction already introduced in the *Discourse preliminaire*, between common physics and experimental physics. According to him, the study of physics should be divided into two parts that should not be confused; observation and experience. The observation is less refined and precise, it is limited to the facts that are before our eyes. Instead, experience seeks to penetrate nature more deeply. "The observation can be called the physics of the facts, or rather the vulgar and palpable physics, while one should reserve for the experience the name of occult physics" [51].[221]

Observation, due to inevitable voids it leaves and the curiosity it provokes, leads to experience. Similarly, experience leads to observation; thus one can consider experience and observation as the continuation and the complement of each other. The experiments or observations of many phenomena are indispensable also for the mathematical physical sciences and also for the same mechanics, which is a completely a priori discipline. When the conclusions of a mathematical physical theory refer to an ideal situation (for example, when friction is ignored in the motion of bodies), the only way to take into account the phenomena not attributable to the idealized theory is to make use of experience:

[220] p. 282.

[221] p. 270. The distinction between observation and experience is not new. It can be also found in the texts of experimental philosophy of the 17th century [28], pp. 400–406.

> The only experimental utility that the physicist can draw from observations, on the laws of equilibrium, on those of motion and on the primitive affections of bodies, is to carefully examine the difference between the result that the theory provides and the one that provides the experience; and to use this difference with skill, to determine, for example, in the effects of the impulse, the alteration caused by the air resistance; in the effects of simple machines, the alteration caused by friction and other causes [...] so then experience will no longer simply serve to confirm the theory, but by differing from it without causing it to fall, it will lead to new truths that theory alone could not have reached [51].[222] (C.51)

The first object of general physics is the examination of the properties of bodies that observation makes us to know, so to speak, in general, but of which only experience can measure and determine the effects. According to d'Alembert, among these effects there are the "phenomena of heaviness". No theory has been able to explain the law of falling bodies; but once this law has been found through experience, all that belongs to the motion of heavy bodies is reduced to calculation. The role of general physicist has ended and that of mathematical physicist starts. Moreover, d'Alembert added, when we release physics from wanting to explain everything, we are far from condemning the spirit of conjecture and the spirit of analogy. These two precious and rare gifts are rarely mistaken when soberly used. In any case, a wall must never be raised between nature and the human spirit. And waring, pay attention not to be wary of excess. How many modern discoveries of which the ancients had no idea? And how many others that we will judge impossible, are reserved for our posterity?

Here is a very famous image, which will be repeated with some variations in many of the writings of d'Alembert on physical sciences:

> Nature is an immense machine whose main springs are hidden from us and we see this machine only through a veil that hides the play of the most delicate parts. Among the most visible parts, and perhaps we can say the most gross, this veil allows us to glimpse or to discover, that many are set in motion by a single spring, and it is mainly this that we must try to unveil [42].[223] (C.52)

In the following I briefly mention the historical evolution of physics as reported in the *Essai sur les elemens de la philosophie*, which echoes that of the *Discourse preliminaire*; a synthesis that is useful for understanding both the conceptions of history and physics of d'Alembert. According to him, the ancients had not neglected the study of nature. Their physics was not as limited or unreasonable as many believe. Examples include the medicine of Hippocrates and the atomistic theories of Democritus. "Nevertheless, the ancients seem to have studied the physics we call vulgar, rather than what we have called occult physics and which is properly experimental physics", "The wisest among them have made the picture of what they saw; they did well and stopped there" [51].[224] D'Alembert believed that the true physical taste of the ancients is to be found in Aristotle's *De historia animalium*.

After the science of classical Greece, there was the Middle Ages "those dark times", in which, according to d'Alembert, only Roger Bacon and Gerbert [of Auril-

---

[222] p. 284.

[223] p. XXIX.

[224] p. 270–272.

lac] gave some light. Moreover the small number of the great geniuses that studied nature, until the Renaissance, did not take much care of experimental physics. Chemists, rather than physicists; rich in an infinity of useful or curious but isolated knowledge, they ignored the laws of motion, those of hydraulics, the heaviness of the air and many other things.

A very important role is given to Francis Bacon, who first embraced an extremely vast field: "He saw the general principles that were to serve as a foundation for the study of nature, he proposed to recognize them through language, he announced a large number of discoveries [51].[225] Then d'Alembert quoted Descartes, giving him great merits in philosophy. Less however in experimental physics where he "opened some way, but more by recommending it than by practicing it. He had the courage to give the laws of motion first; but the experience, or rather as we will say later, the reflection on the most common observations would have shown him that his laws were unsustainable" [51].[226]

Despite its limitations, the spirit of experimental physics that Bacon and Descartes had introduced expanded considerably. The school of Galileo, Boyle, Mariotte and many others carried out a large number of successful experiences. Little by little, the physics of Descartes replaced that of Aristotle, or rather that of his commentators, in schools. "Finally, Newton first showed the art of introducing geometry into physics and combining experience with calculation." Thus forming an "exact, profound, luminous and new science" [51].[227]

3.5.2.3 Mathematical Physical Sciences: Astronomy and Hydraulics

D'Alembert introduced the term *mathematical physical sciences* giving it a meaning very close to the modern one: "Physico-mathematiques. We thus call the parts of physics, in which we combine observation and experience with mathematical calculus, and where we apply this calculation to the phenomena of nature [...]. Mathematical physical sciences are as numerous as there are branches in mixed mathematics" [65].[228]

From the previous definition one could imagine that the mathematical physical sciences are nothing but the mixed mathematics of the 16th and 17th centuries, apart from the use of the Calculus as main mathematical approach. Probably this is not the case however. Indeed in the entry *Physico-mathématique* of the *Encyclopédie*, d'Alembert furnished the following list for his mathematical physical sciences: astronomy, mechanics, statics, hydrostatics, hydrodynamics or hydraulics, optics, catoptrics, dioptrics, airometry, music, acoustics, etc. While the list he gave in the entry *Mathématique* for the mixed mathematics is: mechanics, optics, astronomy, geography, chronology, military architecture, hydrostatics, hydrodynamics, naviga-

[225] pp. 277–278.

[226] p. 278.

[227] p. 280.

[228] Entry: Physico-Mathématique.

tion, etc. That is geography, military architecture, chronology, navigation seem to be considered as mixed mathematics—because they use mathematics—but not physical mathematical disciplines. This, most probably, because d'Alembert did not count them as physical matters, or because mathematics is used only on some aspects of the discipline. Notice that the physico-mathématique of d'Alembert is different from the physico-mathematica of the 17th century which indicated both mixed mathematics and experimental physics when treated with the same strictness as that used in mathematics [28].[229]

It is puzzling to see that in the list of the mathematical physical sciences there is also mechanics (as well as statics), which d'Alembert considered a strictly mathematical discipline. This is most probably due to the fact that while writing for the *Encyclopédie* he had to take account non only of his own conceptions but also the most common ones.

In most parts of natural philosophy the physical mathematical approach can be carried out only for some aspects. In some cases the phenomena are too complex to be described by mathematical relationships and cannot therefore serve as a starting point for a mathematical theory. For d'Alembert, for example, this is the case of the application of the laws of hydrodynamics to the study of the circulation of blood in the veins or arteries of men. In other cases it may happen that, even if "starting from reliable empirical data, the predictions of some phenomena obtained with mathematical elaborations are found to be contradicted by experience. Such a contradiction, according to me, can only come from certain purely analytical assumptions that the application to physics makes necessary. In this case, I believe, we must renounce every theory [...] and treat it as one of the issues in which the calculation cannot have any hold" [42].[230] This statement by d'Alembert is not easy to decipher. Most probably he meant that to apply the mathematical analysis, it is necessary to resort to simplifications. For example d'Alembert believed that not all functions (modern meaning) are likely to be treated with mathematics, but only those analytical expressions constructed with the procedures of algebra and differential calculus. If the empirical data are represented by non analytical expressions, analysis cannot give reliable results.

Astronomy, apart from mechanics, is the most certain of all the parts dealing with physical objects. It is divided into two parts: proper astronomy, that is, geometrical astronomy or observational astronomy and physical astronomy. Geometrical astronomy is not a pure mathematical discipline because it needs the empirical laws about the motion of stars derived from observations. D'Alembert believed that it could be better explained by the method of 'inventors', that is by means of analysis. For this purpose he imagined a man fallen from heaven to earth, with only geometrical notions, who begins his empirical observations before following the enunciation of the first laws. Among the first observations are those concerning the motion of the planets. For these it is natural to suppose first circular orbits around the earth, to

[229]pp. 240–243.

[230]pp. 37–38.

finally arrive after many observations and refinements of the theory to true laws, those that prescribe an elliptical orbit with the sun in a focus.

Physical astronomy is also very important, "I mean here by physical astronomy, not the chimera of vortices, but the explanation of astronomical phenomena by the admirable theory of gravitation. See gravitation, attraction, Newtonianism. If Astronomy is one of the sciences that do the most honor to the human spirit, Newtonian physical astronomy is one of those that makes it the most" [65].[231] In fact, the knowledge of the true laws of celestial phenomena also allows us to improve the knowledge of geometrical astronomy: one can in fact foresee motions that he did not imagine existed and then actually find them.

D'Alembert stated that the ancients had already formulated more or less all the hypotheses that are known today. However, he added, at least from a scientific point of view, their hypotheses do not have a fundamental value, because they are are "vague and ill-tested". In science, vague hypotheses do not count, but only precise statements based on true facts. According to d'Alembert, the first carefully developed system of the world was that of Descartes, based on vortices. But this system was based on inaccurate assumptions and tended to explain everything and nothing. For this reason Newton had to introduce the *universal gravitation*, which perhaps less seductive, however, ceased to be a hypothesis for its admirable agreement with the celestial phenomena.

Below a passage of relevant epistemological interest, in which in a maybe antihistorical way, one could glimpse a falsificationist approach:

> *Among the different hypotheses we can imagine to express an effect, the only ones worthy of our examination are those that by their nature provide us with infallible means to verify if they are true* [emphasis added]. The system of gravitation falls into this category and deserves only for this the interest of philosophers. There is no fear here of that abuse of calculation and geometry, in which physicists have fallen too often to defend or to fight hypotheses. The planets being supposed to move, either in the void, or at least within a non-resistant space, and being known the forces with which they interact, determining the phenomena that must occur is a purely mathematical problem. There is therefore the rare advantage of being able to judge irrevocably the Newtonian system [...] it would be desirable that all matters of physics could be decided so incontestably. Thus the system of gravitation can be considered true only after having assured itself, by means of precise calculations, that it responds perfectly to the phenomena [51].[232] (C.53)

And the agreement that was found between the celestial phenomena and the calculations based on universal gravitation, an agreement that occurs more and more every day, "seems to have made definitively decide the philosophers [mathematicians] in favor of this system, even if it is not explained in any way" [51].[233]

It should be said that around the middle of the 18th century there had been a heated debate about the validity of the law of universal gravitation, which ended with the triumph of the Newtonian theory. In 1747, Clairaut read a memory at the Académie des sciences de Paris, where he asserted that the motion of the moon could

[231] Entry: Physico-Mathématique.

[232] pp. 229–230.

[233] p. 231.

not be explained by the Newtonian theory. Clairaut suggested a corrective function to the quadratic law of Newton, active only at a small distance. Among other things, according to Clairaut, in this way the phenomena of capillarity were better explained. Euler was on Clairaut's thesis; d'Alembert suggested caution and a better check of the calculations, while Buffon sharply contrasted Clairaut's thesis. Finally with a refinement of his calculations Clairaut succeeded in explaining the anomalies he had observed in the motion of the moon, maintaining the quadratic dependence.

D'Alembert, at the end of his considerations on astronomy posed the problem if the theory of universal gravitation, undoubtedly valid in astronomy, also applies to the terrestrial bodies. In this regard, he maintained a prudent position: "We only recognize that the effects of this force have not yet been traced back to any of the known laws of mechanics. We do not imprison [therefore] nature within the narrow limits of our intelligence" [51].[234] After some consideration on the magnetic and electric forces, and more generally on the forces that are exerted among the elements of matter, d'Alembert added: "Let us look carefully at precipitating our judgment about nature and even on the existence of an attractive force among terrestrial bodies [...] we cannot at all conclude that attraction is a universal principle, until we are forced by phenomena" [51].[235]

D'Alembert devoted a great deal of attention to hydraulics throughout his life; particularly to hydrodynamics. Among his studies on the subject, are fundamental: the *Traité de l'équilibre et du mouvement des fluides* of 1744 (followed by a second edition in 1770) and the *Essai d'une nouvelle théorie de la résistance des fluides* of 1752 [42, 45]. In addition there are many entries to the *Encyclopédie* and a large session of the *Essai sur les elemens de philosophie*. The two treatises of 1744 and 1752 are mainly considered here. It should be noted that in the second treatise there are considerations of an epistemological character which, later, will be taken entirely in the *Essai sur les elemens de philosophie*, however, moving them from the chapter dedicated to hydraulics to the one dedicated to general physics.

In his discussion on hydraulics d'Alembert concentrated on the role of geometry and algebra in physics and warned of their abuse. According to d'Alembert, hydraulics could also be studied with the laws of mechanics, developed in the *Traité de dynamique*. If we knew the shape and the mutual position of all the particles composing the fluid, claimed d'Alembert, the principle of mechanics could be sufficient to study their rest or motion. But because the shape and position of the particles of the fluid are unknown one has to deal with an indeterminate problem, at least from a mathematical point of view [45].[236] Only the experience can instruct us in detail on the laws of hydraulics, that the most sophisticated theory could never have made us to suppose. For the need to the recurse to empirical, and thus contingent, principles hydraulics is considered by d'Alembert as a mathematical-physical science and not a pure mathematical one, as mechanics.

[234] p. 236.

[235] p. 243.

[236] p. VII.

D'Alembert devoted considerable attention to the nature of the principles of hydraulics, with considerations also valid for the empirical sciences in general. For him one must be careful not to be influenced by his mathematical training in choosing the principles: "I have looked for the principles of fluid resistance as if analysis had nothing to do with it; and once these principles have been found, I have tried to apply analysis" [42].[237] It is natural that this methodological choice could be not easy, added d'Alembert, because after having sacrificed the simplicity of calculation to the certainty of the principles, naturally one must expect that the application of calculations to these principles could be very difficult. But there are no other ways [42].[238]

There are few empirical principles of very general nature and, according to d'Alembert, indubitable, in hydrostatics; they are:

1. Fluids are made up of particles.
2. The pressure inside fluids is the same in all directions.

> I suppose only, that no one can contradict me, that a fluid is a body composed of very small particles, separated and able to move freely. [...] Since philosophers cannot immediately and directly deduce the laws of their equilibrium from the nature of fluids, they have at least reduced them to a single principle of experience; the equality of pressure in all directions. [...]. In fact, condemned as we are to ignore the first properties, the inner structure of bodies, the only resource that remains to our sagacity is at least to try to make in each subject the analogy between the phenomena and bring them all to a small common denominator of primitive and fundamental facts [42].[239] (C.54)

Hydrodynamics is more complex; it can be studied by flanking the principle of dynamics of the *Traité de dynamique*, in particular d'Alembert principle (see below), with other indubitable empirical principles:

1. When a fluid flows from a vessel its superior layer remains always sensibly horizontal.
2. The velocity of the particles of each layer are equal and parallel to each others [45].[240]

If these conditions are not satisfied mathematics cannot be applied to hydrodynamics.

One of the advantages, d'Alembert claimed, of his approach is the possibility to prove that the principle known as *conservation of living forces* is valid both for solids and fluids and to show and avoid its drawback. D'Alembert praised Daniel Bernoulli to have used the principle of conservation of living forces in his *Hydrodynamica* of 1738, in a very elegant way, but blamed him for not having proved it in a convincing way [45].[241]

---

[237] p. XI.

[238] pp. XXI–XXII.

[239] pp. XXV–XXIX.

[240] p. XIII.

[241] pp. XVI–XVII.

3.5.2.4 Mathematical Physical Sciences: Music

Even though musical theory had been since Pythagoras subject of inquire of mathematicians, by the 18th century what was called *musica theorica* was absorbed in treatises of *musica practica*, increasingly written by musicians for their colleagues and more concerned with empirical aspects than abstract theories; moreover some parts of music relating to theoretical aspects of acoustics where detached from their musical context. D'Alembert represented a partial exception; for him indeed music played an important role in his conception of science, as it was the case for Ptolemy's *Harmonics* [28].[242] He wrote about thirty musical articles for the Éncyclopédie; a little number when compared with about the 1600 entries that portent his march, but however important, in the period 1750–1757. Moreover he made meaningful interventions in his role of editor with the articles presented by the 'musician' Jean Jaques Rousseau (1712–1778), [37].

The involvement of d'Alembert in musical issues has its roots in 1749, when Jean Philippe Rameau (1683–1764) submitted to the Académie des sciences de Paris, for approval, a manuscript containing the elements of his musical theory. It was published in a revised form as the *Démonstration du principe de l'harmonie* in 1750 [150], a title that was subsequently subject to d'Alembert's criticisms. Indeed the term *démonstration* did not appear in the memoir presented and approved by the academy in 1749. The title was, as reported in the register of the academy, *Mémoire où on expose les fondemens d'un systéme de musique théorique & pratique*. M. Rameau, noticed d'Alembert, after the approval of the academy, believed he could assigns to his system the quality of demonstration even though the academy declared repeatedly that he could not claim to assume his principle as proved [50].[243]

In his writings Rameau argued that music is largely a mathematical subject, that can be comprehended under a unique principle, an experimental result indeed, clear and evident to him, with an explicit reference to Cartesian epistemology. By making some musical experiments he was struck as by a *flash of light* by the acoustical phenomenon for which a vibrating string will normally generate not only its fundamental frequency but also higher frequencies, in particular the perfect twelfth and the major seventeenth (the meaning of these terms will be discussed later). Rameau called such a vibrating system a *corps sonore* [150].[244]

When d'Alembert read Rameau he had no theoretical knowledge of musics. In any case he was shortly able to understand and greatly appreciated him, to the point he assigned Rameau a relevant position among the geniuses of science and arts in his *Discourse préliminaire de l'Éncyclopédie* [54].[245] However, even though d'Alembert praised Rameau's theory he did not appreciate his prose, and for this reason he felt himself impelled to write a more readable treatise than that proposed by Rameau himself, which originated the *Élémens de musique* of 1752, written anonymously as

---

[242] pp. 34–42.

[243] p. XVI. Footnote.

[244] p. 8.

[245] p. 122.

a form of respect [3], which was followed in 1762 by a second edition, this time with the name d'Alembert in the cover [50].

The *Élémens de musique* was not simply a didactical work; it was rather a propagandistic pamphlet for spreading d'Alembert's rational epistemology. "By reformulating Rameau's brilliant but ineptly articulated theory into a rigorous writing d'Alembert was able to provide both a vindication as well as an advertisement for his own peculiar scientific epistemology" [38].[246] A treatise on music was a perfect means to illustrate the merits of his professed empirical-rational metodology of physical sciences because more simple of his *Traité de dynamique*, for instance

D'Alembert exposed the general ideas of his epistemology of music in the Discourse préliminaire to the second edition of the *Élémens de musique*, in a period in which his conceptions of philosophy of science were mature. To start with he declared music a physical mathematical science, like hydraulics and astronomy and unlike mechanics which is purely rational like mathematics. Thus for music too, as for the other physical mathematical sciences one should not seek a striking evidence, which is the characteristic of the works of mathematics alone, and which is found so rarely in those of physics.

According to d'Alembert, there will always be into the theory of musical phenomena "a kind of metaphysics, which these phenomena implicitly suppose and which carries its natural obscurity there" [50].[247] It is for this reason that one should not expect to find in this matter what is called demonstration; it is enough to reduce the main facts to a well-interlocked and well-consistent system deduced from a 'single experience', and to have established on this simple foundation the most well-known rules of musics. "At the same time we doubt that it is possible to bring a greater light to these matters" [50].[248]

Thus, continued d'Alembert, although most of the phenomena of musical art seem to be deduced in a simple and easy way from the resonance of sound bodies, one should perhaps not yet hasten to affirm that this resonance is demonstratively the unique principle of harmony. At the same time, it would not be less injurious to reject this principle, because certain phenomena do not seem to be deduced from it as easily than others. From this it is can only be concluded, either we will perhaps be able, by new research, to reduce these phenomena to a single known principle; or that there should be some other unknown principle, more general than that of the resonance of the sound body, of which the latter is only a branch. Or finally, that we should perhaps not try to reduce all musical science to one and the same principle.

There are indeed physical mathematical sciences which demand only one experience, or one principle; there are others which necessarily suppose several, the combination of which is necessary to form an exact and complete system. At this point d'Alembert stated that music belongs to this last category of sciences [50].[249]

[246] p. 419.

[247] p.XIII.

[248] p. XIV.

[249] p. XVII.

This is in contradiction with what claimed by Rameau, who asked for one principle only.

D'Alembert in *Élémens de musique* made indeed recourse to more that one experience. Besides that referred to by Rameau, which remained the fundamental one, he considered two more experiences in the edition of 1752 and only one in the edition of 1762. In the following reference will be to the 1762 edition only; for the sake of completeness the two experiences refereed here, that due to Rameau and that added by d'Alembert, are resumed below.

> Experience I. If one excites a sonorous body, beyond the fundamental sound and the octave, two others sounds are heard, one of which is the perfect twelfth, that is the octave of the [interval of] fifth, the other the major seventeenth, that is the double octave of the [interval of] major third [50].[250]

D'Alembert added that this experience is particularly evident by making the larger strings of the cello vibrate; the fundamental sound is called *generator*, the others, the octave included, its *harmonics*.

> Experience II. Anybody notices the similarity between a sound and its octave. These two sounds are perceived by the ear as almost equal [50].[251]

In order to make the meaning of these experiences comprehensible to a non-musician, some definitions need to be recalled. The interval between two notes of frequencies f_0 and f_1 is given by the ratio $n = f_1/f_0$; the sum of two intervals is the interval defined by the product of the two intervals. An interval of third major is characterized by $n = 5/4$, an interval of fifth is characterized by $n = 3/2$, an interval of octave by $n = 2$. An interval of two octaves above a third major is thus characterized by the product of the interval of a third major and two intervals of octave, that is, by $n = 5/4 \times 2 \times 2 = 5$; an interval of one octave above a fifth is defined by $n = 3/2 \times 2 = 3$.

Basically what Experience I says is that together with the generating or fundamental frequency f_0, one hears also notes corresponding to frequencies $3f_0$ and $5f_0$. This result does not surprise a modern reader who knows that though a vibrating string is capable of all the frequencies $f = nf_0$, with $n = 1, 2, 3, \ldots$, even frequencies as well as higher frequencies are difficult to excite and therefore cannot be heard. The result could not surprise even d'Alembert, who knew acoustics very well, only that for him the experience in question was not of acoustic; that is, it did not concern the knowledge of the frequencies that a body can potentially emit, but only those that the ear perceives in the ordinary conditions in which a musician operates.

Indeed d'Alembert warned philosophers not to waste their time looking for physical explanations of musical phenomena, explanations always vague and insufficient; still less must they consume themselves in efforts to rise in a region more distant from their gaze, and to lose themselves in a labyrinth of metaphysical speculations

[250] p. 14.

[251] pp. 16–17.

on the causes of the pleasure that harmony makes us experience. It is not a question here, of the physical principle of the resonance of sound bodies, which at the moment are unknown, and which perhaps we will seek for a long time in vain; it is even less a question of the metaphysical principle of the feeling of harmony, a principle even less known, and which according to all appearances will always remain covered with clouds. It is only a question of showing how it is possible to deduce from a single experience the main laws of harmony, which musicians have found only by trial and error [50].[252]

The two experiences in themselves cannot be considered as principles of harmony; they become such by adding a metaphysical principle, for which what is conform to nature should be treated as a guide by a composer; and since nature says that by exciting a sound body we obtain a chord formed with the base frequencies f_0 and the two harmonics $3f_0$ and $5f_0$, this chord must be considered as perfect to the maximum degree. Furthermore, the perfection is maintained, albeit at a lower level, if a note of this chord is substituted with that corresponding to an upper or lower octave.

As an example of the application of his musical principles, d'Alembert gave the example in which the fundamental note was *do* (or *ut*, according the 18th century nomenclature). The harmony suggested by nature is the chord formed by *do* followed by a note one octave above the fifth and a note two octaves above the major third. The two notes that follow *do* can be replaced by two others, scaling these below by one or two octaves, therefore considering the major third and the fifth which in the note scale adopted by Rameau correspond respectively to *mi* and *sol*. The chord *do mi sol do* is thus obtained which can be played more easily than that suggested by Experience I and that if it is not quite perfect, it is the one that comes closest to the perfection of nature. With similar reasonings, more or less perfect chords can be built.

D'Alembert ended his Discourse préliminaire, by warning against an abuse of mathematics in music, even though it is largely a mathematical science. "As a professional geometer, I believe I have some right to protest here (if I may be allowed to express myself strong) against this ridiculous abuse of Geometry in Music" [50].[253] And noticed that the perfect number used in the musical theory should be interpreted with prudence. The ratio of the octave as 1 to 2, that of the fifth as 2 to 3 that of the major third such as 4 to 5, etc. may not be the true relationships of nature; but only approximate ratios such as experience has made them known. Because experience gives them only approximatively. But luckily these ratios are sufficient, even though they are not not exactly true, to give reason of phenomena that depend on the ratio of sounds [50].[254]

[252] p. XXVII.

[253] p. XXX.

[254] pp. XXX–XXXI.

3.5.3 *Mechanics as a Mathematical Science*

In the 18th century mechanics was the science par excellence. Just as it had been in the previous two centuries and as it will be in the next century. This is true even for d'Alembert. According to him, mechanics is an entirely deductive science, such as algebra and geometry, accepting what asserted by Euler in the *Mechanica sive motus scientia analytice exposita* of 1736, facing other scientists of the time, of whom perhaps the most representative was Daniel Bernoulli, who considered mechanics an empirical science (for these arguments see [59]).

As for Euler, in the mechanical writings of d'Alembert one can identify two parts, physically separated. A first part in which the principles are discussed and justified. Here d'Alembert worked as a philosopher; he chose the main notions and justified the principles that regulate them. In the second part, he used the principles as a starting point to develop their consequences by means of mathematical analysis. This second part is often the one that takes up more space and at first glance makes the treatise of dynamics to seem a mathematical one. However, it is not a simple application of an already finished mathematical theory; this is not given in its entirely and the solution of various problems requires the development of new mathematical tools.

3.5.3.1 Clear and Distinct Notions

In the following I present an exposition of the main notions about mechanics of d'Alembert derived from both the scientific treatises of the 1740s, the *Treatise de dynamique* at first place, and the philosophical writings of the 1750s, such as the *Essai sur les elemens de la philosophie*. As already noted above, although the epistemological texts are posterior, up to ten years, to its main scientific treatises, these do nothing but explain what is implicitly contained in those.

As discussed in the previous sections, for d'Alembert in a science only notions that have adequate clarity and distinction can be accepted. And d'Alembert identified only two basic abstract notions that possess these characteristics: the notions of space and time and consequently those of impenetrability and motion; that are so the unique 'objects' of mechanics.

Space, time and motion

Space and time according to d'Alembert's terminology are not principles, because principles for him should be propositions (possibly 'true'). They are instead simple (abstract) notions, and as such undefinable. Motion, the fundamental object of mechanics, is a composite notion, to be defined from the simple notions of space and time. Motion is nothing but the transport *in time* of a body from one place to another, places that can be confronted with parts of the space assumed as "fixed and immobile" [48].[255] Or otherwise, motion is: "The reiterate applications of a mobile to the different parts of the undefined space, that we imagine as the place of bodies"

[255] p. VI.

[48].[256] For d'Alembert a motion can be studied in geometry too, but here it does not occur in time and the consideration of (physical) time is what distinguishes physical from geometrical motion [48].[257]

> Thus in mechanics or science of the motion of bodies, neither space nor time must be defined, because these words contain only a simple notion; but we can and must define motion, although the notion is quite familiar to everyone, because the idea of motion is a composite notion which contains two simple ones, that of the space traversed, and that of time employed to browse it [65].[258] (C.55)

Indeed the definition of motion, besides the two simple notions of space and time also requires those of place and body, that however can be obtained by the former, and the principle of superposition, which assures the possibility to perform reiterate applications. A *body* is a portion of extension which enjoys the property of impenetrability. Impenetrability is the main property of bodies for which they are distinct from indefinite space. The *place* of a body is the part of the space it occupies; that is, the part of the space that coincides with the extension of the body [48].[259]

D'Alembert indirectly criticized Descartes's concepts of space and motion, by declaring that metaphysical considerations are scarcely relevant for the study of mechanics, and distancing from his conception of extension. For d'Alembert there are two kinds of extension, "one that should be regarded as impenetrable which constitute what we properly call bodies; the other which is considered simply as extended, without examining if it is penetrable or not" [48].[260] D'Alembert had read Newton and knew his ideas of absolute time and space, but he did not make them his own; in this regard he took a pragmatic approach. While space is assumed in principle as indefinite—but not infinite—and, like Newton, in some absolute way, it is in practice treated as relative, that is referred to parts assumed as fixed. A similar discourse holds true for time; after the introduction of time as an absolute notion, which "by his nature flows smoothly" [48],[261] attention is then concentrated on its measure, by treating it as relative. For some more comments on the notions of time and space see [119].[262]

The measure of motion depends on two different magnitudes, associated respectively to space and time, that are heterogeneous; for d'Alembert they can however be compared if a unit of measure is assumed for them, so that both of them give raise to real numbers, which actually can be compared [48].[263] This specification reveals that d'Alembert, as well as Euler and other mathematicians of the time, had still some problems to deal with heterogeneous magnitudes when moving from geometry to algebra.

[256] p. V.

[257] p. VII.

[258] Entry: Elemens.

[259] pp. 1–2.

[260] p. VI.

[261] p. V.

[262] pp. 37–49.

[263] p. 16.

Matter and Mass
Mass is a term and a notion that appears but without any comment in d'Alembert's mechanics. It is unclear whether he used a rhetorical artifice, introducing a fundamental notion of mechanics by minimizing the thing so that not to deprive mechanics of its predetermined geometric character, or if he fitted it into the tradition that goes from Descartes to Newton, which identified mass with the amount of matter. This second way could allow d'Alembert to include mass in the geometric order, because a clear and distinct notion. In fact, for him, matter is nothing but extension with impenetrability and therefore volume. D'Alembert expressed his views on mass in the entries Masse, Densité, Divisibilité of the *Encyclopédie*. For example he wrote: "The density of a body is the ratio of its mass (ie, *the space which it would occupy if there were absolutely no pores* [emphasis added]) to the space that it actually occupies" [65].[264]

D'Alembert in his treatises on mechanics, never referred to mass as inertia. He defined the latter term in some entries of the *Encyclopédie*. For example, he wrote:

> The force of inertia, is the property common to all bodies to remain in their state, either of rest or motion, unless some foreign cause makes them to change. [...] It is called *resistance* when one wants to speak of the effort a body makes against what tends to change its state; and we call it *action*, when we want to express the effort that the same body makes to change the state of the obstacle that resists it [65].[265] (C.56)

But the entries of the *Encyclopédie* did not necessarily reflect d'Alembert's own idea, but rather the most widespread ones. The difficulties inherent in the notion of mass will become clear only since 19th century when giving a rigorous formulation of the laws of mechanics [109].

Force
The various notions of force as moving cause or power, are to be rejected for d'Alembert because not clear and distinct; force is a "doubtful dark and metaphysical being, which is only capable of spreading darkness on a science otherwise clear to itself" [48].[266] D'Alembert's position was accepted by some, including Lazare Carnot and rejected by others. In general the majority of scholars referred to the notion of force, as perceived by common sense. Ignoring d'Alembert's criticisms.

Skepticism toward force did not originate directly from d'Alembert; it was a direct consequence of the acceptance of the early mechanical philosophy. Even Maupertuis, had this same attitude. The skepticism was already present in Descartes and especially in Malebranche; two scholars well present to d'Alembert.

The notion of force as power was eliminated by Malebranche not only because its intake limits the power of God, but also because it is not well defined. According to Malebranche force cannot be observed or measured directly; it appears a simple word, invented by philosophers to conceal their ignorance. Berkeley, in *De Motu* wrote sentences in which d'Alembert would recognize, though it seems unlikely that

---

[264] Entry: Densité.

[265] Entry: Force.

[266] p. XVII.

he knew his work, published only in England, in the 1720s. Even Hume was contrary to the notion of force as well as to the notion of cause in general; however in this case it is possible that there was some influence, because Hume enjoyed a good popularity among the *philosophes*.

According to d'Alembert there are only two kinds of causes, and thus 'forces' in mechanics:

1. The causes that result from the mutual action of the bodies because of the impenetrability, which is one of the "major causes of the effects" that we observe in nature.
2. Causes not reducible to impulse. These causes are considered as separate, although one should be convinced of their reducibility to impulse and just do not know how to prove this reduction.

The causes of the first type have well known laws. What we cannot say for those for those of the second type. We know them only through the effects; we speak of a cause only because we see an effect. Among the causes of the second type there is gravity, which because cannot be reduced to impact, and then to geometry, could not obey the necessary laws of mechanics and thus gives raise to contingent laws. That is what d'Alembert wrote: "[...] because it is not yet possible to explain the celestial phenomena with impact and also because the impossibility to explain them with this principle [the impact] is based on convincing evidence, if not on demonstrations" [65].[267] But, if one should choose between the hypothesis of gravity generated by the impact of aether particles (as frequently used in the time) or generated by an innate force he would choose the latter. Said with Sherlock Holmes: "Eliminated the impossible, the improbable remains".

D'Alembert went on to say that, in fact, also the causes of the first type, which seem obvious, are so only improperly. In the entry Cause of the *Encyclopédie* he wrote:

> What we call causes, even those of the first type, are such improperly; they are indeed effects from which other effects result. A body pushes another or a body is in motion and meets another, there should necessarily occur changes in the status of bodies *on this occasion* [emphasis added], because of their impenetrability. The laws of these change are determined through certain principles, and accordingly the impacting bodies are considered as causes of the motion of impacted bodies. But this way of speaking is improper. The metaphysical cause, the real cause is unknown [65].[268] (C.57)

In the same entry, d'Alembert argued against the metaphysical principle: effects are proportional to their causes. Basically he said that if we are dealing with 'real' causes, the principle does not make sense because metaphysical causes cannot be measured (we cannot for example, say that one sensation is double of another or that white is twice the color of red). If instead a cause is considered an effect, then the principle makes no sense. D'Alembert also contrasted another commonly accepted metaphysical principle, the final causation. As examples of final causation: "Water

[267]Entry: Attraction.

[268]Entry: Cause.

goes back into the pumps because nature has a horror of vacuum", the principle of least action for the reflection and refraction of light. He argued their inconsistency, but leaves a glimpse of the final causes as a pragmatic method, when the results obtained with them agree with the conclusions obtained from a mechanicistic model. Thus he saved in some way the principle of least action of Maupertuis, perhaps out of friendship.

In essence d'Alembert, despite the rationalistic system of his mechanics, reaffirmed his empiricist belief. We have only to deal with the effects. In mechanics we call cause of effect another effect; the real causes are unknown. A chain of explanations of cause-effect is nothing but than a relationship between effects, which, however, can be connected by necessary laws.

Eliminated the force (cause) as possible notion of dynamics, the motion can be described by geometry alone, through the recruitment of impenetrability added to extension. Thus d'Alembert did apparently not leave room in his mechanics to what now goes by the name of Newton's second law:

> So why should we resort to the principle that everyone uses today, that for which the accelerating or retarding force is proportional to element of velocity? [...] We will not examine at all if this principle is a necessary truth [...] not even, as some Geometer [think], a purely contingent truth [...] we will be content to observe that, true or doubtful, clear or obscure, it is useless in mechanics and consequently it must be banned [48].[269] (C.58)

Though the usual notion of force is not clear and distinct, d'Alembert cannot avoid its use, because it was an integral part of the approaches to mechanics of his time; it and a relation expressing the proportionality of force and acceleration-variation of velocity has already been used by Hermann, Euler, Varignon and the Bernoullis. The only way remained to d'Alembert was to make clear the notion; and he did so by introducing it as a definition. The word used by d'Alembert for 'force' was *motive force* (force motrice) φ, to indicate the product of mass by acceleration or, sometimes, the element of velocity du, given by the product of acceleration a—named *accelerating force* (force acceleratrice) and very often indicated by d'Alembert with the same symbol φ used for the motive force—by the element of time dt [48].[270] In the end d'Alembert wrote the relation:

$$\varphi dt = mdu$$

This relation can be used—and was used by d'Alembert – as Newton/Euler did. The motive force is known in an empirical way, by collecting data concerning motion and its derivative in a given situation. For different situations (that is different initial conditions) the actual motion can be found by integrating a second order differential equation. This is the way d'Alembert explain the meaning of the previous relation:

> We have just seen that in whatever way the motion is accelerated or retarded, the differential equation of the curve will always be of this form $\pm dde = \varphi dt^2$. Now if one wants to make use of this equation, as well of the equations $\varphi dt = \pm du$ and $\varphi de = \pm udu$ to determine in any motion the relation between u, t, e he could think that for the purpose the knowledge of

[269] p. XII.

[270] Sect. 22.

> the cause which accelerates or delays the movement would be necessary. The object of the remark is to show that this is not the case, but that φ is always given by the very definition of the motion in question; thus, according to this remark, if one wanted to make use of the equations $\varphi dt^2 = \pm dde$, $\varphi dt = \pm du$, & $\varphi de = \pm udu$ to determine the relation of spaces, velocities, and times in a motion the law of which will be given, it will suffice to substitute in these equations instead of φ a quantity suitable for expressing the law according to which it will be assumed that velocity increases or decreases: when we assume, for example, that instantaneous velocity decreases are like squares of velocity, we write $gu^2 dt = -du$, $gu^2 de = -udu$ (g being a constant coefficient), & so on [48].[271] (C.59)

It should be noted that the relationship $\varphi dt = mdu$ is more than a tautology; even if d'Alembert glances on the fact. The two members of the equality not always have the same meaning. Indeed in one case φ may be unknown and u known, in another case φ may be known and u unknown. For example, proceeding like Newton, one can find that in the motion of a planet around the sun represented by u, φ can be evaluated as $\varphi = mdu/dt$. Once the empirical relation of φ has been found, for instance it is found that it depends on the square of distances, one assumes the principle of determinism according to which φ is invariable over time, in a certain portion of the universe and therefore also applies to new motions different from those with which it was obtained. These new motions can be obtained by solving the differential equation $mdu = \varphi$, with the appropriate initial conditions.

Indeed in many important applications, as in his astronomical or hydraulic studies, d'Alembert used a mechanic of forces and hardly can it be distinguished by that of Euler for instance. Only in the part devoted to fundaments d'Alembert is the use of the term force precise.

3.5.3.2 *Traité* de Dynamique

D'Alembert delineated his design to reduce mechanics to mathematics in the *Traité de dynamique*, which saw two editions in 1743 and 1758 [44, 48]; the second edition is longer (a 30% more) and contains many footnotes, due to Étienne Bézout (1730–1783) but there were no substantial changes, a part from a long addition to the preface; for a comparison of the two editions see [41]. To note the change in the use of personal pronouns, "we" (nous) of the first edition is replaced by "they" (ils) in the second edition to mark difference on some positions [48, 119].[272]

When d'Alembert wrote the *Traité de dynamque* in 1743, his first printed work, he was only 26. Then he was not yet involved in the activity of 'professional' philosopher which will characterize his mature works, and his epistemology of science was still unripe. However he had already clear the role of mechanics. Moreover in the second edition of 1758, when he had already finished his *Essai sur les elemens de la philosophie* (published 1759) he maintained unchanged the approach.

In the following I do not want to stress the defects of his text, object of severe criticisms more ore less convincing, by many historians, even though it was appreciated

[271] p. 25.

[272] p. II, p. 28.

by d'Alembert's contemporaries, such as Euler and Daniel Bernoulli [102].[273] I only want to illustrate the way d'Alembert imagined to develop his design of rationalization of mechanics. For instance, if d'Alembert said he had proved a theorem I give it for granted, without any criticism for his eventual (and indeed frequent) vicious circles.

Reading the *Traité de dynamique* is not a very simple task. The fact cannot be explained and dismissed as done by Truesdell, by saying that d'Alembert did not have clear ideas. As noticed, the treatise saw two editions more than ten years one from the other that did not substantially changed the structure and therefore it can be considered the fruit of in-depth reflections. Thus, d'Alembert most probably had clear ideas even though they are not such for us. The treatise consists of a preface and two parts. In the preface d'Alembert described his conception of mechanics and the plan of his work. In the first part he presented his principles and argued to justify their status as fundament of mechanics. This part is written, according to d'Alembert, to be accessible to beginners. The second part opens with the statement of the so called *d'Alembert principle* and is followed by three chapters of applications. Here d'Alembert treated questions already raised by Jakob Hermann, Leonhard Euler, Jakob, Johann and Daniel Bernoulli and others.

The preface states that there are three principles of mechanics, the principle of inertia, the principle of the composition of motions and the principle of equilibrium. Then d'Alembert spook of a general simple and direct method of solution [48].[274]

In the first part of the treatise there are four definitions concerning bodies and place (Definiton 1) and motion (Definitons 2, 3, 4), two laws (Lex I and Lex II) concerning the persistence in the state of rest and of motion. In the second part, what was classified as a method in the premise, is referred to as "general principle to find the motion of many bodies" [48].[275] All the other propositions of the treatise are named theorems or corollaries. Thus theorems are also the *principle* of composition of motion and the *principle* of equilibrium. Below, laws, principles, theorems assumed at the basis of d'Alembert mechanics.

1. *Lex I.* A body at rest will persist, unless that one external cause move it.
2. *Lex II.* A boy once put in motion by any cause, must always to move along a straight line, until a new cause, different from that had caused the motion, will act on it.[276]
3. *Definition* [masked principle]. The motive force [that is force] is equal to the product of mass by accelerating force [that is acceleration].
4. *Theorem.* If two any powers acts together on a body at point A to move it, one from A to B uniformly during a given time, the other from A to C, uniformly during the same time, and if one considers the parallelogram ABCD; I say that

[273]p. 45.

[274]p. XXXII.

[275]p. 72.

[276]Note that also Descartes in the *Principia philosophiae* presented the principle of inertia into two steps.

the body A will move along the diagonal AD uniformly, in the same time it passed AB and AC.[277]

5. *Theorem*. If two bodies whose velocities are in the inverse ratio of their masses and have opposite direction, so that one cannot move without facing the other, there will be equilibrium among the two bodies.
6. *Principle*. Let A, B, C, &c. be the bodies of the system and suppose they are impressed the motions a, b, c, &c. and that they are forced by their interactions to change in the motion a, b, c, &c. It is clear that the motion a impressed to body A can be regarded as composed of the motion a and of another motion α. In the same way it is possible to consider the motions b, c, &c. composed of the motions b, β, c, κ, &c., from which it follows that the relative motions of the bodies A, B, C, &c. would be the same if instead to give them the impulse a, b, c it would be given the couples of impulses a, α; b, β; c, κ, &c. Now, because of supposition, bodies A, B, C, &c. took the motions a, b, c,&c, then the motions α, β, κ, &c. must not to disturb in any way the motions a, b, c, &c. That is if the bodies had received only the motions α, β, κ, &c. they should have destroyed themselves mutually and the system remain at rest. From this it results the *following principle* [emphasis added] to find the motion of any bodies that interact each to other. Decomposed each of the motions a, b, c, &c. impressed to each body in other two motions a, α; b, β; c, κ, &c., such that if only the motions α, β, γ, &c. were impressed to the bodies, the system should have remained at rest, it is clear that a, b, c, &c will be the motions these bodies will assume because of their actions [48].[278]

For sure there is a lack of precision in the treatise, some misprintings, but there are no serious contradictions. Some modern historians accuse d'Alembert of having made confusion and abuse of language by calling for instance principles what actually are not such, see for instance [61]. A careful reading of d'Alembert, assuming that already in 1743 he had similar ideas to those expressed in 1759 in his *Essai sur les elemens de la philosophie*, makes clear that the meaning of the term principle is not that of Aristotelian-Euclidean mould, as proposition evident per se. Rather it is a 'truth' that lies at the top of a chain of deductions of a science, a first proposition. It is often more properly a theorem as it can be demonstrated from simple notions; this is a common use in the 18th century, as already commented in previous sections. For example the laws of inertia are 'demonstrated' using the compound notion of motion, the implicit assumption of the isotropy of space and the principle of sufficient reason. Law I and II are justified by d'Alembert in two very different way. The former is considered as nearly evident in itself; this is d'Alembert argumentation: "Because a body cannot impress a motion by itself, *being no reason to move from one side instead from another* [for the isotropy of space]" [48],[279] The proof of Lex II requires

[277]D'Alembert presented a proof of this principle in a new different and more convincing way, in a memoire of 1761 [49].

[278]pp. 71–75.

[279]p. 4. To note that Lex I is assumed as an axiom (VI) in *La logique de Port Royale*, and is derived directly from the more general axiom III: "The nothingness cannot be the cause of something", [4], p. 336. To note also that in the edition of 1743 the sentence evidenced in italic was missing.

a more involuted treatment and is substantially circular. Indeed it uses a presumed corollary of Lex I, for which a body cannot modify the motion by itself, which as a matter of fact prove Lex II. But the assumed corollary is not such and in no way can be inferred from Lex I.[280]

Thus d'Alembert, notwithstanding the weakness of his proofs—for modern standard at least—is justified in the use of principle and theorem as synonyms; although we would have preferred, for reasons of clarity, to reserve the term theorem to the propositions demonstrated starting from first propositions; but this is a problem of ours.

D'Alembert did not use the term axiom instead, that refers to a proposition evident for itself (see the entry Axiome of the *Encyclopédie*), as in Euclid, or in fact true, as in Newton, and unlike Newton, he did not consider interchangeable the two terms law and axiom.

D'Alembert in the *Essai sur les elemens de la philosophie*, written some year later than the *Traité de dynamique*, introduced the distinction between principles of first order and principles of second order, the latter being derived from two or more first order principles. It is no clear d'Alembert would have considered the three principles of dynamics—inertia, composition of motion and equilibrium—as principles of first order, because they are at the top of the chain of deductions in dynamics or if they are principles of second order, because they are proved simply by reflecting on the notions of motion and impenetrability. D'Alembert's principle would have instead considered by him for sure as a principle of the second order, because proved by principles of first order. Indeed the principle of inertia and the decomposition of motions are explicitly referred to in the proof, even though for the last there is no explicit reference to the way the decomposition is made, that is, it is not explicitly stated that one should use his second principle of dynamics which provides the parallelogram rule (items 4 of the previous list). The principle of equilibrium is only hinted at by asserting that there are motions to be destroyed. From the applications we see however that the destruction of the motions is regulated by his third principle of dynamics and its consequences; using as a matter of fact the criteria for the balancing of simple machines.

Another aspect that surely hits the modern reader is that the three principles of d'Alembert seem different from Newton's. There are various reason for this, indeed. First, what are now called Newton's principles (or laws) were not known at d'Alembert time, and hardly could Newton himself recognize them as his. They were presented for instance by Euler in a clear way more or less in the same period of the publication of the *Traité de dynamique*. There was not yet a well-defined method for dealing with the various problems of mechanics. Not even in astronomy, where much could be done and much had be done by Newton, limiting to the motion of the free mass point. There were thus several methods carried out by a small group of mathematicians who felt competing with each other and held to their originality to the utmost, trying to devise methods different from others only to distinguish themselves.

280For a discussion on the problematic in any statement of the law of inertia see [139].

There were thus no compellent reasons for a free mind to adopt the ideas of another person, not even Newton who d'Alembert considered "for sure the greatest physicist of the century" [45].[281] Moreover, a careful reading of the *Traité de dynamique* reveals that from a technical point of view there is not a great difference between Newton's (even in the form suggested by Euler) and d'Alembert's mechanics. There is the law of inertia; the principle of equilibrium has strong similarity with the law of action and reaction. Lastly, there is also a notion of force and the second law of motion, even though it is not presented as a principle but as a definition.

A modern reader, accustomed to different formulations of mechanics, based either on the notion of force, or work, or energy, would be tempted to think that d'Alembert is simply doing a make-up operation. In fact, once one has a working theory available, he does not take much to restructure it—as today it takes little to move from a force-based mechanic to an energy-based mechanics. It is just a matter of doing some mathematical or rhetorical steps. After all, d'Alembert almost always used the relationship $\varphi dt = mdu$, as one does in a force-based mechanics. But if there is something true in this consideration and there is any make-up in d'Alembert's mechanics it is not dictated by a desire for originality but by a different sensitivity that is authentic and profound.

In the second edition of the *Traité de dynamique* of 1758, d'Alembert added a comment long enough, and all considered attractive, on whether or not the laws of mechanics are necessary. For d'Alembert the laws of mechanics deducted according to his notions of space and motion would 'necessarily' be those that matter left to itself would follow. But nothing prohibits, according to him, that God could implement direct intervention in order to change them (it would be what today we might call a continuous miracle).

According to d'Alembert, the proposed question is therefore reduced to knowing whether the laws of equilibrium and motion that are observed in nature are different from those that the matter abandoned to itself would have followed. If they are different, the laws of statics and mechanics, as given by experience, are contingent truths, since they are the result of a particular will expressed by the supreme being; if, on the contrary, the laws provided by experience agree with those that reasoning alone finds, the observed laws are necessary truths [48].[282]

The addition of the comment was motivated by the prize launched in 1756 by the Académie royale des sciences et belles lettres de Berlin on the subject: "If the laws of statics and mechanics are necessary or contingent truths". D'Alembert participated to the competition with a lost text, but the competition was postponed and eventually canceled. It is reasonable that the lost text contained the additions to the second edition of the *Traité de dynamique* of the aforesaid comment [119].[283]

The reasoning of d'Alembert appears not entirely coherent to a modern reader and even to some contemporary ones. A scholar of mechanics substantially contemporary to him, Daviet de Foncenex (1734–1798), a student of Lagrange, will object to him

---

[281] p. V.

[282] pp. XXV–XXVII.

[283] p. 25.

in a convincing way. In his work *Sur les principes fondamentaux de la mécanique* of 1760 [93], Foncenex noted that if the laws are necessary in no way they could be contingent:

> It seems, moreover, that the action of the Creator on the bodies do not make in any way hypothetical the application of the laws of mechanics to the universe, as M. d'Alembert seems to suggest in his introduction [to the *Traité de dynamique*]. Because there will always be allowed to consider this action of God as a new force acting on the bodies; then whatever the motions which result, they will never be contrary to the laws of mechanics that need to be immutable in themselves [93][284] (C.60)

3.5.3.3 The Principle of D'Alembert and Its Applications

In the previous section I underlined the similitude between d'Alembert's and Newton's principles. Now I speak about the differences. D'Alembert's mechanics, much more than Newton's, was a mechanics of impact, like that of Descartes, at least for what foundations are concerned. D'Alembert was not fully clear, but he said that the effect of forces (d'Alembert meaning), acting continuously has been studied quite in depth (may be he was referring to Euler); the effect of the force of impact were instead poorly known and studied; thus he intended to fill the gap [48].[285]

The bodies of d'Alembert are usually mass points, even though on this aspect he was not consistent, and *hard*, in the sense generally adopted in the 18th century, that is rigid and deprived of elasticity. In the direct impact of a hard mass point with an obstacle, assumed as hard, the body does not bounce and loses all its speed. Mass points, besides hard are also assumed smooth. In a oblique impact only the component of velocity orthogonal to the obstacle is lost; the parallel component is maintained because of the absence of friction. This simplification that d'Alembert adopted is necessary to carry forward his project of mathematization of mechanics. D'Alembert is conscious of the fact; his mechanics was a pure science and as such not suitable to deal with many practical problems, as for instance those for which frictions is important. But he had no choice. Like Torricelli who in his ballistic studies declared to write for mathematicians and not for the gunners, he was not bothered by the fact [102],[286] his interest too was to write for mathematicians.

The first principle (nomencalture of the Preface) of d'Alembert's dynamics, (law of inertia) concerns the motions of bodies before and after the impact. The second principle (decomposition of motion) is functional to the decomposition of the velocity of a mass point into two components, one orthogonal and the other parallel to an impacted body. The third principle (the law of equilibrium), is only apparently a statical law. It concerns more than one body and introduced tacitly for the first time the concept of mass. Actually its role is to find the law of impact between two bodies (mass points), by evaluating both the motion cancelled and that remained in the

[284] pp. 299–300.

[285] p. 73.

[286] p.173.

impact. The principle also deals with 'static' forces (common use), that is equilibrium strict sense. The extension was quite straightforward because for d'Alembert static forces were nothing but mass times virtual motion[287]: "Mr. Leibniz assumed that the dead force [static force] is equal to the product of mass by the virtual velocity" [65].[288]

The principle of equilibrium considers not only free bodies, mass points, but also constrained bodies. In such a case the laws of equilibrium are those valid for the simple machines found in classical statics: the lever, the inclined plane, the pulley and so on. Nowhere d'Alembert referred to the principle of virtual velocities as a criterion of equilibrium, though he should have known it from Johann Bernoulli and Varignon.

The so called d'Alembert principle studies the motion of many bodies that impact directly or through rigid threads. A problem that was not faced by Newton and that instead became an important theme of research for the mathematicians of the 18th century. This principle has been the object of much criticism, even fierce, many interpretations about its meaning, many hypotheses about its origins [61, 63, 94–96, 131, 152, 159].[289] I believe that for it is worth the saying "many enemies much merit". The criticisms are only partly motivated and depend on a not careful reading of d'Alembert's works and the lack of contextualization. The same applies to origins and interpretations.

With regard the criticisms, I refer to my exposition of the principle, that I hope should make possible for a careful reader to get his own idea. I just want to say that with his principle d'Alembert was able to solve many complex problems, Lagrange placed it at the base of his dynamics. Certainly the principle does not present itself as a well-defined algorithm; in particular, the methods of 'destruction' of the lost motions, that is the modalities with which to impose the equilibrium are not specified. But the use of the principle becomes more clear if one reads the whole work of d'Alembert and he studies its applications. The same perplexities may find a young student who begins to read modern treatises on classical mechanics. Without a training on practical applications and problem solving, he would not be able to fully grasp and use the theory he studied in the theoretical treatises. This process was made explicit for example by Thomas Kuhn in his *The structure of scientific revolutions* of 1962 [112] with his notion of paradigm that questions the algorithmization of any theory of modern physics. On the other hand, it would be anti-historical to accuse d'Alembert not to have operated, for example, as Euler, who starting from his first principles formulated, in the more or less convincing way, propositions of simpler application such as, for instance, the law of moment of momentum.

With regard to the genesis of d'Alembert's principle there are various opinions; they are commented on by Christophe Schmit [152] which also adds his own ideas.

[287] Note that for d'Alembert virtual velocity has a meaning different from the usual one, thai is possible velocity; it is better defined as the velocity of tendency or incipient velocity; see below.

[288] Entry:: Force vive. The force of impact due to the virtual motion was a concept known since Torricelli in the 17th century [27], pp. 175, 284–286.

[289] pp.238–292, 331–343, 172–183, 239–271.

The commonly accepted view, strongly supported by Truesdell [160],[290] sees the first formulation of it, though in a less general form, in a paper by Jakob Bernoulli of 1703, *Démonstration générale du centre du balancement ou d'oscillation tirée de la nature de la levier* [8]; the formulation had been reported already in 1686 in the Actes de Leipsick, before Newton published his *Principia* [160].[291] The context is similar; the reasoning and the wording the same; there are destroyed motions (Bernoulli used also terms like velocity, force, impulse) which should be equilibrated by the law of statics. Moreover for sure d'Alembert knew Jakob Bernoulli's work because summarized in a paper by Johann Bernoulli which he quoted [9]. Michel Paty instead sees the origin of d'Alembert principle in unpublished papers by d'Alembert, on hydraulics [152].[292] In Hermann's *Phoronomia* of 1716 there is a variant of Bernoulli's approach [105][293] and even Euler used some form of it in a paper of 1740 [30, 70].[294] In [152] it is suggested that the roots of the principle are spread in the various conception of equilibrium at the end of the 17th century, and in the rule of parallelogram due to Varignon published in 1687 [164].

Be that as it may, d'Alembert was a man of his time and knew many of the 17–18-century mechanical works; his principle should therefore be seen as a synthesis and a generalization of previous works. Nothing original at all but all very original, therefore. Lazare Carnot gave an 'algorithmic' form of it in his *Essai sur le machines in géneral* of 1803, in the case of impacting bodies, and Lagrange gave it the definitive form for continuously varying forces assuming the principle of virtual work as a criterion of equilibrium for the destroyed motions [26].[295]

Problems solved in the Traité de dynamique

Instead to insist in the difficult of understanding theory and applications for a modern reader of d'Alembert principle, I go to show how d'Alembert used it to solve some puzzling problems of the time and as he could extend it to study the motion of extended bodies and fluids. Of the fourteen problems studied by d'Alembert in the *Treatise de dynamique* I will present only a few.

> Problem I. Find the motion of a thread CR, fixed in C and charged by many bodies A, B, R as you like. Suppose that such bodies, if the thread would not impede their motions, will move in the same time, along the lines AO, BQ, RT, perpendicularly to the thread [48].[296] (C.61)

The motions (velocities) that the bodies (of mass) A, B, R of Fig. 3.14a would assume without constraints are AO, BQ, RT; or better these are the motions the bodies would assume if they were appended to the three independent pendulums CA, CB, CR.

---

[290] pp. 248–252.

[291] p. 252.

[292] p. 499.

[293] pp. 100–110.

[294] pp. 104–105.

[295] pp. 291, 242, 268.

[296] p. 96.

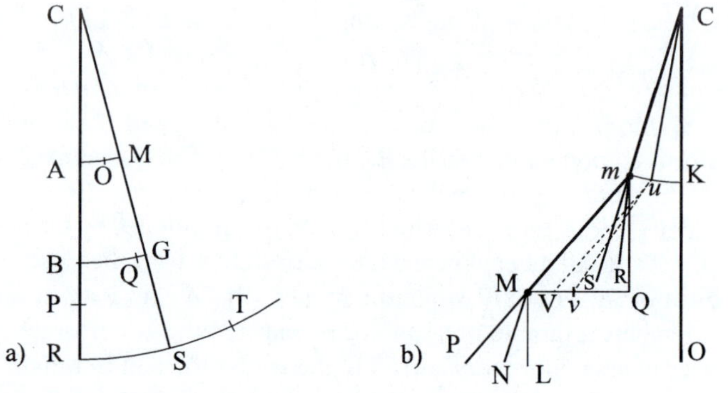

Fig. 3.14 Composite pendulums. Redrawn from [48], Plate II, (**a**) Fig. 22, (**b**) Fig. 31

Indeed there is some confusion between displacements and velocity in the exposition of the problem, because the drawing shows a path along an arc of circle, which implies one is assuming displacements. This however should be considered as one of the many d'Alembert's imprecisions. The correct interpretation should be velocity.

The rigidity constraints imposed by the thread CR make that the actual motions are AM, BG, RS. So that −OM, −QG, ST are destroyed motions. For d'Alembert principle they equilibrate somehow. The equilibrium is imposed by assuming the rule of the lever, that is in modern language by equating to zero the (static) moments of the destroyed motions of the masses A, B, C multiplied by their masses with respect to the fulcrum C. Which means that the following equality should be verified:

$$-\mathrm{A}\times\mathrm{MO}\times\ \mathrm{CA}-\mathrm{B}\times\mathrm{GQ}\times\mathrm{CB}+\mathrm{R}\times\mathrm{ST}\times\mathrm{CR}=0$$

From it is easy to find the motion of a point of the thread AR, for instance point R, as a function of the impressed velocities.

For a modern reader it is difficult to see as the solution of the problem could be afforded by means of d'Alembert principle. There are not indeed impacts, and thus one cannot understand how there could be destroyed velocities. The problem becomes clearer if instead of thinking to actual velocity one thinks about 'virtual' velocity, or better velocities of tendency as defined by d'Alembert. These velocities should be assumed infinitesimal. If u is a velocity, the virtual velocity is the element of velocity du; sometimes assumed coincident with acceleration du/dt. So AO, BQ, RT should be assumed as the virtual velocities that the bodies would acquire if they were not constrained by the thread, and similarly AM, BG, RS the 'actual' virtual velocities.

D'Alembert struggled enough about the concept of virtual velocity. If from a mathematical point of view the concept could be accepted (this was the same problem one could find in any treatment of Calculus); from a physical point of view the situation was different. Indeed d'Alembert, as Descartes, could not think of a velocity small as he pleased because for him a possible cause for the variation of velocity was the impact with other bodies (eventually by the aether), and may occur only for finite increment.[297]

That d'Alembert motion should be interpreted as virtual velocities appears clear if one considers the Corollary I to problem I. This is a simple extension of the solution to the case when instead of the impressed velocities one has motive forces F, f, ϕ applied respectively to the points A, B, R. They are defined as the product of the mass by the element of velocity. Previous relations remain still valid if one pose $\mathrm{A} \times \mathrm{AO} = F, \mathrm{B} \times \mathrm{BQ} = f, \mathrm{R} \times \mathrm{RT} = \varphi$, or equivalently $\mathrm{AO} = F/\mathrm{A}, \mathrm{BQ} = f/\mathrm{B}$, $\mathrm{RT} = \varphi/\mathrm{R}$. In the Corollary II d'Alembert credited Bernoulli to have already found a similar results when the impressed motions are those due to gravity. The Bernoulli in question is Johann and not Jakob, as one could think at first glance, the reference is *Nouvelle theorie du centre d'oscillation* of 1717 [9].

> Problem V. Let consider a wire CmM fixed in C and loaded by two weights m, M, being only slightly moved from the vertical line CO. Find the duration of the oscillations of the wire [48].[298] (C.62)

The application of d'Alembert principle to Problem V is puzzling; not so much for the approach itself but for the wording. Indeed here neither reference to impact nor to motion is present. The treatment is the same as that carried out by scholars that used explicitly the concept of force with the classical meaning. The external motive force of gravity (d'Alembert terms) acting on the body m of Fig. 3.14b is decomposed along the direction mu, the direction of the actual motion, and along another one, at the moment unknown direction mR, to be destroyed. Similarly the force of gravity on M is decomposed into the direction Mv and in the direction MP to be destroyed. For the equilibrium (destruction of forces) it is necessary that the efforts acting along mR and MP have a resultant in the direction of Cm (or mR); or equivalently they have equal and contrary components in the direction orthogonal to Cm (or mS). Thus there should hold the proportion: [the effort direct along MP] is to the effort along mR, as the infinitely small [sine of the] angle RmS is to the infinitely small [sine of the] angle MmS [48].[299] With this equilibrium d'Alembert was able to find the expressions of the forces not destroyed and so the motion of the bodies. He arrived to two ordinary differential equations.[300] In [103][301] it is discussed the method used by d'Alembert to integrate these equations and the introduction he made of many concepts of linear algebra as that of eigenvalue.

> Problem IX. A body whose mass is m and velocity is u, moves on the same line as another body with mass M and velocity U. Find the velocity of those bodies after the impact [48].[302] (C.63)

---

[297] For some comment on d'Alembert's idea on the matter see [102], pp. 195–213.

[298] p. 139.

[299] p. 140.

[300] For details see [96], pp. 148–149.

[301] pp. 4–7.

[302] p. 211.

This problem is very simple, and probably could have been exposed directly in the illustration of the law of equilibrium. It is also one of the few cases of impact explicitly dealt by d'Alembert in his treatise. The solution of the problem is straightforward. Let be v the velocity of the former body after the impact, V that of the latter. It will be, $u = v + (u - v)$; $U = V + (U - V)$ and $V = v$, with the terms in parenthesis represent lost motions. By d'Alembert principle the destroyed motions are equilibrated and so, $m(u - v) + M(U - V) = 0$; from which it is $v = V = (mu + MU)/(M + m)$.

Then d'Alembert as a corollary considered the case, $m \ll M$, $U \ll u$. This to simulate the impact of the small and fast particles of aether with common bodies. Previous result gives mu as the quantity of motion gained by the greater body M. Which allows d'Alembert to conclude: "The quantity of motion that the body loses or gains should be regarded as proportional to the impacting power [...]. In this case the effect of such a power is always the same, either the [impacted] body is at rest or in motion" [48].[303]

Problems solved in other treatises

D'Alembert applied his principle, or better some adaption of it, also in his treatises: *Recherches sur la precession des equinoxes* of 1749, *Traité de l'équilibre et du mouvement des fluides* of 1744 and *Reflexions sur la cause générale des vents* of 1746. In the treatise on the precession of the equinoxes d'Alembert applied his principle to an extended rigid body, the whole earth indeed. Here there are not distinct mass points; however the acquaintance with Calculus allows d'Alembert to replace the mass point with a small volume, an infinitesimal particle of the whole extended body. Each particle is endowed with a motive force which follows the Newtonian law of gravitation. D'Alembert principle assumes the form:

> Let there be a body that is moved with any motion, and of which each of the parts has a different velocity represented by the indeterminate u in any instant; and let there be as many accelerative forces as one pleases, Ψ, Ψ', etc., that act on the body, and in virtue of which the velocity u of each part in any instant is changed in the following instant into another velocity u', different for each part. I say that if one regards the velocity u as composed of the velocity u' and another velocity u'', which is infinitely small, the system of all the parts of the body, each with its velocity u'', must be in equilibrium with the forces Ψ, Ψ', etc. [43].[304] (C.64)

An exposition quite simple for a modern reader of the version of the principle, could be: In the time dt the velocity u of each particles of the body is changed in the velocity u', there is the destruction of motion by constraints and the creation of new motion by the 'forces' Ψ, Ψ', &c. This last would give the velocities Ψdt, $\Psi' dt$, &c. So a particle would tend to move with the speeds $u + \Psi dt + \Psi' dt$ + &c, but because of the constraint it takes only the velocity u', so that $u + \Psi dt + \Psi' dt + \&c - u' = u'' + \Psi dt + \Psi' dt$ + &c, are the lost motions to be equilibrated. But u'' can be interpreted as the element of velocity du (actually du is by definition the difference between new and old velocity, that is $du = u' - u = -u''$), which sometimes d'Alembert confounded with acceleration. So the new formulation of the princi-

[303] p. 212.

[304] p. 35. English translation in [168].

ple can be stated by asserting the quantities mu'' ($= -mdu$, or in modern notation $-madt$) are equilibrated with the motive forces Ψdt, $\Psi' dt$, &c.

It should be noted that in this exposition there is an abuse (by myself) of notation, for the sake of simplicity. In fact, writing the lost motion in the form $u - u' = u''$, presupposes a knowledge by d'Alembert of the modern vector calculus notation (symbol $\pm$) for the composition/decomposition according to the parallelogram rule, which was not the case. However, for the components of the motion, indicated as α, β, γ, d'Alembert can use the ordinary symbols of sum/subtraction because it is an operation between scalar magnitudes, and thus he can write $\alpha = \alpha' + \alpha''$, $\beta = \beta' + \beta''$, $\gamma = \gamma' + \gamma''$.

D'Alembert wrote the equation of equilibrium between the motion u'' and the motive forces Ψ, Ψ', &c, by assuming a system of coordinate with the origin in the center of the rigid body-earth, one of the coordinate plane containing the ecliptic, another coordinate plane orthogonal to the ecliptic passing for the terrestrial axis and the third plane orthogonal to the first two. The equilibrium equations consist in imposing the equality of the static moment of motion mu'' and motive forces Ψ, Ψ', &c. with respect to the coordinate axes.[X]

In the *Traité de l'équilibre et du mouvement des fluides* of 1744 [45], d'Alembert still adapted his principle to the new situation, giving the formulation:

> Let the velocities of the different slices of the Fluid at the same time are represented, in general, by the indeterminate v. Imagine that dv is the increment of v in the next instant, this quantity dv is different for different slices, positive for some, & negative for others; in a word, that is $v \mp dv$ expresses the speed of each slice when it takes the place of the one immediately below; I say that if each slice was supposed to tend to move with the only infinitely small speed $\pm dv$, the Fluid would remain balanced [45].[305] (C.65)

In the *Reflexions sur la cause générale des vents* of 1747 [46], D'Alembert presented his principle in the following form:

> Suppose, then, that point A moves according to AG; on any curve PAD, being animated with a real accelerating [motive] force = π and at the same time it is solicited to to move according to AG by a force = F, which for some reason is changed to π; I say that if this body A, would be solicited according to AP by a force = $F - \pi$ would remain at rest [46].[306] (C.66)

This last statement, and even though to a less extent the previous two, gives reason to the interpretation proposed by Lagrange, which has now became a standard, see below, Sect. 3.6.

3.6 Epilogue. Lagrangian Synthesis

Mechanics with Euler and d'Alembert had became an algebraic theory; it still remained geometric in some parts however, and based on the concept of vectors. A concept that today could be founded without any reference to classical geometry

[305] p. 70.

[306] pp. 19–20.

but that in the 18th century could not. An important and large step toward a complete algebrization of mechanics was taken by Lagrange with his *Mechanique analitique* of 1788 [114]. This treatise added little, at least in its first editions, to the development of mechanics and collected most results obtained by Lagrange himself since 1760s. But it was completely new for its way of exposition and its logical-epistemological conception. Mechanics from rational became analytical.

The term analytical mechanics has different acceptations in the history of mechanics. Today its more diffuse meaning is that of a mathematical theory based on algebra and Calculus, whose principles are formulated by means of scalar relations and this definition substantially covers Lagrange's use. For Truesdell analytic mechanics is limited to discrete systems (excluding thus continua, solid and fluid), and this limitation of the term is quite diffuse also [143]. Vectors were still present in Lagrange, but they were referred only through components and mainly there could be many different local systems of reference in the same mechanical assembly. According to the meaning assumed beforehand Euler's mechanics is not analytic, even though he himself referred to it as analytical mechanics.

Lagrange in the preface of his masterpiece said that his mechanical theory had became a branch of analysis. The principles of this branch were represented by general formulas. They were enough to solve any problem of mechanics, both in statics and dynamics. They however were considered to be given, and no serious interest was addressed to their derivation. From this point of view the *Mechanique analitique* should be considered one of the first modern text in modern mathematical physics, that is a mathematical theory based on principles that could not be necessarily true.

In the following, fo the sake of completeness, the whole preface of Lagrange's treatise is quoted, notwithstanding it is well known and commented in the literature, because of it shortness.

> I propose to condense the theory of this science and the method of solving the related problems to general formulas whose simple application produces all the necessary equations for the solution of each problem. I hope that my presentation achieves this purpose and leaves nothing lacking. In addition, this work will have another use. The various principles presently available will be assembled and presented from a single point of view in order to facilitate the solution of the problems of mechanics. Moreover, it will also show their interdependence and mutual dependence and will permit the evaluation of their validity and scope. I have divided this work into two parts: statics or the theory of equilibrium, and dynamics or the theory of motion. In each part, I treat solid bodies and fluids separately. No figures will be found in this work. The methods I present require neither constructions nor geometrical or mechanical arguments, but solely algebraic operations subject to a regular and uniform procedure. *Those who appreciate mathematical analysis will see with pleasure mechanics becoming a new branch of it* [emphasis added] and hence, will recognize that I have enlarged its domain [114].[307]

Lagrange's approach is greatly different from that of Euler in his *Mechanica motus scientia analitice exposita* or *Theoria motus corporum solidorum seu rigidorum*, not only because based on different principles. They differ for two other main points of

[307]Preface. English translation in [117].

view. First. Euler's mechanics has a substantially axiomatic structure, where concepts are introduced and explained, before being used. Lagrange's mechanics has not such an axiomatic structure. It uses concepts and even some principles that are assumed to be already known and accepted in mechanics, by its scholars. Moreover the general formulas Lagrange talked about are not a true formulas but rather rules that need interpretation. Second. Lagrange's mechanics differently from Euler's makes no reference to natural philosophy to justify its assumptions. 'Principles', where necessary, are justified with mathematical reasoning. Where it would be necessary to exit from this ambit, Lagrange was elusory. Instead of considerations drawn from natural philosophy or metaphysics, Lagrange referred to an historical account in his premise to the parts in which his work is divided (statics, dynamics, hydrostatics, hydrodynamics). And a well made account indeed, which has been of inspiration for many historians of mechanics [30, 31].

History gives a justification of the theory and replaces that based on considerations of philosophical nature. Lagrange thus intended to show that his point of view is nothing but the results gained in the history of mechanics, starting from the ancients, Archimedes *in primis*. In his history there is no room for the philosophers of nature, Descartes is a partial exception. He presented the contribution of his peers, the mathematicians.

The historian of mathematics Gino Loria (1862–1954) proposes a different explanation. He maintains that Lagrange did so because congenial for him. He was indeed a profound connoisseur of the history of mechanics and mathematics and considered useful to frame his contribution in a more general context. Below Loria's considerations on this point:

> He, besides to presenting himself as one of the most original and profound thinkers we know, also appears as one of the most learned mathematicians. The scope of his studies can well be said to embrace all mathematics, pure and applied; theoretical and practical arithmetic; algebra and infinitesimal calculus; mechanics and mathematical physics, with the main applications of such disciplines (for instance ballistics); finally astronomy, with all the disciplines connected to it, such as watchmaking and navigation. Well, to fight and win in all these wide and very varied fields he armed himself with all the science of his own century, making familiar to everyone, without exception, handling the logical and algorithmic procedures in use at the time. To prove it it is enough to note that, although a true analyst by temperament and habits, he was nevertheless able to handle with enviable ease the kinematic-geometric methods constantly used by Newton [129].[308] (C.67)

Thus the history of physics written by Lagrange could be seen as a further enlargement of his studies, carried out by his universal mind.

One of the *general formulas* of analytical mechanics of Lagrange is given by Johann Bernoulli's law of energies, named by him *principle of virtual velocities*, a name that will used here through the text.

> If an arbitrary system of any number of bodies or mass points, each acted upon by arbitrary forces, is in equilibrium and if an infinitesimal displacement is given to this system, in which each mass point traverses an infinitesimal distance which expresses its virtual velocity, then

[308]p. 334.

> the sum of the forces, each multiplied by the distance that the individual mass point traverses in the direction of this force, will always be equal to zero. [114].[309] (C.68)

Lagrange maintained that any general principle of mechanics that could be found in future would differ from the principle of virtual velocities only for its formal expression [114]:[310] Because of the relevance of the principle, Lagrange attempted also a his own justification with more or less success; on this point see [26].

Another general formula used by Lagrange, is d'Alembert principle which he reformulated, or better chose among the various formulations given by d'Alembert. Lagrange distinguished between the accelerating forces with reversed sign:

$$-d^2x/dt^2, -d^2y/dt^2, -dz^2t^2$$

that actually moves a body (P, Q, R), from the external assigned accelerating forces. D'Alembert principle, in the version given it by Lagrange, affirms that the accelerating forces, both actual and assigned, multiplied each for the mass of the bodies on which they act, should equilibrate:

> If one imagines to impress to each body, the motion is will take, with sign reversed [$-ma$], it is clear that the system will be reduced at rest [115].[311] (C.69)

The success of the *Mechanique analitique*, besides the use of the generalized principle of virtual velocities, is also due to the massive introduction of the calculus of variations. A kind of calculus in which instead of the variation of ordinary variable, for instance time and position, there are variations of functions. The method has a long history and can be dated back to ancient Greeks; among its classical problem there is the search of the figure having maxima area for an assigned perimeter (isoperimetric problem). The name *calculus of variation* is due to Euler [85],[312] as equally to Euler should be attributed the solution of some important problems; but it was the young Lagrange who developed a formalism very efficient and closing related to that of the infinitesimal calculus.

Lagrange worked his ideas from 1755 until 1760 and published them in the 1760/61 issue of the Miscellanea Taurinensia [113]. He also published a number of other papers on the subject, which reached their climax at his hands in 1788 in his *Méchanique anlitique*, where he used the method of variation to solve many mechanical problems and as the natural language for the principle of virtual velocities.

The fundamental concept is that of *variation* expressed with the symbol δ. If $\varphi(x, y, z, t)$ is a generic function belonging to a some space of funtions, the variation with respect to a reference function $\varphi_0(x, y, z, t)$, is the function $\delta\varphi = \varphi(x, y, z, t) - \varphi_0(x, y, z, t)$. Generally it is assumed that φ and φ_0 are close, according to an appropriate metric.

---

[309] pp. 10–11.

[310] p. 12.

[311] p. 256.

[312] p. 54. Though the term variation is Lagrange's.

Combining the two general formulas and using the symbols of the calculus of variations, gives a single general formula, the symbolic equation of the whole mechanics (modern name). For a system of mass points it is:

$$\mathrm{S}\left(\frac{d^2x}{dt^2}\delta x+\frac{d^2y}{dt^2}\delta y+\frac{d^2z}{dt^2}\delta z+P\delta p+Q\delta q+R\delta r+\&\right)m=0$$

where P, Q, R &c, are the active accelerating forces (that is forces for unit of mass) on each mass point, which are assumed as assigned, and the accelerating forces [the acceleration, as a matter of fact]:

$$\frac{d^2x}{dt^2},\quad \frac{d^2y}{dt^2},\quad \frac{d^2z}{dt^2}$$

effectively acting on each point. The symbols preceded by δ, a variation indeed, indicate virtual displacements, respectively in the direction of the forces, (δp, δq, δr), and the global axes, δx, δy, δz. S means summation over the mass points. Sign are adjusted considering that forces P, Q, R and displacements p, q, r are positive in opposite directions.

Lagrange's general equation is not a true equation, for at least two fundamental reasons. First x, y, z are generally constrained, so that this equations should be flanked by equations expressing constraints. Lagrange did so using the so called method of Lagrange multipliers, which implies to add to the previous equations a term containing the equations of constraints [26].[313] Second, virtual displacements should be eliminated, because they are not definite magnitudes.[XI]

Lagrange's formalism applies very well also to the study of motion and equilibrium of fluids. No discussion is carried out of their internal structure. Only it is said that the particles which compose fluids are material and that for this reason the general laws of equilibrium are as applicable to them as they are to solid bodies. Indeed, the principal property of fluids and the only one which distinguishes them from solid bodies, according to Lagrange , is that all their parts have no resistance against the smallest force and can move among themselves with all possible facility. This property is easily modeled by Calculus and it follows that the laws of equilibrium and motion for fluids do not require a separate theory but that they are only a particular case of the general theory of mechanics. The symbolic equation of dynamics applied as well; the only particularity lies in the choice of the constraint equations. In the case of statics of incompressible fluids, for instance, the equation of conditions is derived by imposing the incompressibility at a local level. It requires that the volume of each particle be invariable. So, having expressed this volume by $(dxdydz)$, it must be $\delta(dxdydz)=0$. The symbolic equation of mechanics is thus:

$$\mathrm{S}\,(\Delta\ X\delta x+\Delta\ Y\delta y+\Delta\ Z\delta z+\lambda\delta(dxdydz))\,dxdydz=0$$

---

[313] pp. 240–274.

in which X, Y, Z, that replace P, Q, R, are the accelerating forces for each elementary mass and considering we are in statics, m is replaced by $dxdydz$; λ is the so called Lagrange multiplier.[XII]

3.7 Quotations

C.1 La portion infiniment petite dont une quantité variable augmente ou diminue continuellement, en est appellée la *Différence*. Soit par exemple une ligne courbe quelconque AMB, qui ait pour axe ou diametre la ligne AC, & pour une de ses appliquées la droite PM; & soit une autre appliquée *pm* infiniment proche de la premiere. Cela posé, si l'on mene MR parallele à AC; les cordes AM, A*m*; & qu'on décrive du centre A, de l'intervalle Am le petit arc de cercle MS-P*p* sera la difference de AP, R*m* celle de PM, S*m* celle de AM, & M*m* celle de l'arc AM. De meme le petit triangle MA*m* qui a pour base l'arc M*m*, sera la difference du segment AM; & le petit espace MP*pm*, celle de l'espace compris par les droites AP, PM, & par l'arc AM.

C.2 Definition I. La portion infiniment petite dont la différence d'une quantité variable augmente ou diminue continuellement, est appellée la *différence de la difference* de cette quantité, ou bien sa *différence seconde*.

C.3 Dans toutes sortes de mouvemens [...] soit que ces mouvements soient uniformes, ou accelerez, ou retardez [...] la somme des forces qui font le mouvement dans tous les instants de sa durée, est toujours proportionnelle à la somme des chemins ou des lignes que parcourent tous les points des corps mu.

C.4 Potentia indesinenter agens (P) ducta in elementum temporis (dt) aequipollet facto ex massa corporis (M) cui potentia applicata est, et elemento celeritatis (dc) quod tempusculo (dt) producit.

C.5 In Opere Hydrodynamico, quod non ita pridem in lucent edidit Filius meus, felicioribus auspiciis aggressus est materiam istam, sed fundamento nixus indirecto, conservatione scilicet virium vivarum, licet verissimo atque a me demonstrato, nondum tamen ab omnibus Philosophis recepto. Primus ego hanc hypothesin exhibui in Dynamicis solidorum [postquam Hugenius simili principia pro centro oscillationis determinando usus est] ostendique eandem constanter ex illa hypothesi solutionem elici, quam dant ordinaria principia dynamica ab omnibus Geometris admissa.

C.6 Mais a present que cette vérité est mise dans son jour & hors de toute atteinte, on a lieu d'admirer la parfait conformité qui regne entre les loix de la Nature, & celles de la Geometrie; conformité qu'elle observe si constamment & dans toutes les circonstances, qu'il semble que la Nature ait consulté la Géométrie, en etablissant les loix du Mouvement. Car s'il eut été possible que les forces des corps, qui sont en mouvement, n'eussent pas été en raison des produits des masses par les quarrez des vitesses, & que la Nature les eut faites en une autre raison; elle se seroit démentie, l'ordre de la Geometrie auroit été viole. La quantité des forces vives source unique de la continuation du mouvement dans l'Univers, ne se seroit pas conservée: plus d'égalité par consequent entre les causes. efficientes & leurs effets; en un mot, tout Ia Nature seroit tombée dans le desordre.

C.7 Dans chaque equilibre il ya une egalité d'energies des forces absolue, c'est a dire entre le produite des forces absolues par less vitesses virtuelles.

C.8 J'entends par vitesse virtuelle la seule disposition à se mouvoir que les forces ont dans un parfait equilibre, où elles ne se meuvent pas actuellement. C'est pourquoy eviter l'equivoque, an lieu de dire que leurs puissances ou les forces sont comme les produits des masses par leurs vitesse vous auriez peut-être mieux fait de vous exprimer ainsi, les energies des puissances ou des forces sont comme les produits de ces puissances ou de ces forces par les vitesses virtuelles.

C.9 Concevez plusieurs forces différentes qui agissent suivant différentes tendances ou directions pour tenir en équilibre un point, une ligne, une surface, ou un corps; concevez aussi que l'on imprime a tout le système de ces forces un petit mouvement, soit parallèle a soi-même suivant une direction quelconque, soit autour d'un point fixe quelconque: il vous sera aise de comprendre que par ce mouvement chacune de ces forces avancera on reculera dans sa direction, a moins que quelqu'une ou plusieurs des forces n'ayent leurs tendances perpendiculaires a la direction du petit mouvement; auquel cas cette force, ou ces forces, n'avanceroient ni ne reculeroient de rien; car ces avancemens ou reculemens, qui sont ce que j'appelle vitesses virtuelles, ne sont autre chose que ce dont chaque ligne de tendance augmente ou diminue par le petit mouvement; et ces augmentations ou diminutions se trouvent, si l'on tire une perpendiculaire a l'extremite de la ligne de tendance de quelque force, la quelle perpendiculaire retranchera de la même ligne de tendance, mise dans la situation voisine par le petit mouvement, une petite partie qui sera la mesure de la vitesse virtuelle de cette force [...] Tout cela étant bien entendu, se forme cette Proposition generale: *En tout équilibre de forces quelconques, en quelque maniere qu'elles soient appliquées, et suivant quelques directions qu'elles agissent les unes sur les autres, on mediatement, on immediatement, la somme des Energies affirmatives sera egale a la somme des Energies négatives prises affirmativement.*

C.10 Les Cartesiens conformément à la lettre que je viens de citer (art. 1.) de leur Maistre, avoient desja deduit de son principe la même egalité de Momens, ou d'energie, ou de quantités de mouvement, que vous employez pour deux puissances en equilibre sur les machines simples, & dans le fluides, par les commencemens de mouvement que M. Descartes prescrit dans cette lettre; mais vous etes le seul, que je sçache, qui ait étendu cette égalité d'energies à tant de puissances qu'on voudra supposer en équilibre entr'elles suivant des directions quelconques. Cette Remarque est fort belle; mais (comme j'ay desja dit) elle suppose équilibre entre des puissances donnée & de directions données sans le prouver.

C.11 Si les Sciences sont fondées sur certains principes simples & clairs dès le premier aspect, d'où dépendent toutes les vérités qui en sont l'objet, elles ont encore d'autres principes, moins simples à la vérité, & souvent difficiles à découvrir, mais qui étant une fois découverts sont d'une très-grande utilité. Ceux-ci sont en quelque façon les Loix que la Nature suit dans certaines combinations de circonstances, & nous apprennent ce qu'elle sera dans de semblables occasions. Les premiers principes n'ont guére besoin de Démonstration, par l' evidence dont ils font dès que l'esprit les examine; les derniers ne sauroient a voir de Démonstration physique a la rigueur, parce qu'il est impossible de parcourir généralement tous les cas on ils ont lieu. Tel est, par exemple, Ie principe si connu & si utile dans Ia Statique ordinaire; que *Dans tous les assemblages de corps, leur commun centre de gravité descend le plus bas qu'il est possible.* Tel est celui de la conservation des *Forces vives.* Jamais on n'a donne de Démonstration générale *à Ia rigueur*, de ces principes; mais jamais personne, accoûtumée à juger dans les Sciences, & qui connaîtra la force de l'induction, ne doutera de leur vérité. Quand on aura vi que dans mille occasions la Nature agit d'une certaine maniére, if n'y a point d'homme de bon sens qui croye que dans Ia mille-uniéme elle suivra d'autres loix.

Quant aux demonstrations *à priori* de ces sortes de principes, il ne paroît pas que la Physique les puisse donner; elles semblent appartenir a quelque science supérieure. Cependant leur certitude est si grande, que plusieurs Mathématiciens n'hésitent pas a en faire les fondements de leurs Theories, & s'en servent tous les jours pour résoudre des Problèmes, dont sa solution leur coûteroit sans eux beaucoup plus de peine. Notre esprit étant aussi peu étendu qu'il l'est, if y a souvent trop loin pour lui des premiers principes au point ou il veut arriver, & il se lasse ou s'écarte de sa route. Ces loix dont nous parlons, le dispensent d'une partie du chemin: il part de-la avec toutes ses forces, & souvent n'a plus que quelques pas a faire pour arriver la ou il desire.

C.12 Soit un système de corps qui pesent, ou qui sont tirés vers des centers par des Forces qui agissent chacune sur chacun, comme puissance n de leurs distances aux centres; pour que taus ces corps demeurent en repos, if faut que la somme des products de chaque Masse, par l'intensité de sa force, & par la puissance $n+1$ de sa distance au centre de fa force (qu'on peut appeler la somme des Forces du repos) fasse un Maximum ou un Minimum

C.13 PRINCIPE GENERAL. *Lors qu'il arrive quelque changement dans la Nature, la Quantité d'Action, nécessaire pour ce changement, est la plus petite qu'il soit possible.* La Quantité d'Action est le produit de la Masse des Corps, par leur vîtesse & par l'espace qu'ils parcourent. Lors qu'un Corps est transporté d'un lieu dans un autre, l'Action est d'autant plus grande, que la Masse est plus grosse; que la vîtesse est plus rapide; que l'espace, par lequel il est transporté, est plus long.

C.14 Eine flüssige Materie muss zu allererst diese Eigenschaft haben, dass ihre Teilchen nicht aneinander befestigt sind, so dass ein jegliches Teilchen ohne einigen Widerstand von den übrigen abgesondert und in Bewegung gesetzt werden kann

C.15 La matière, sur laquelle je voudrois à présent entretenir V.A. me fait presque peur. La variété en est surprenante, & le dénombrement des faits sert plutôt a nous éblouir qu'à nous éclairer. C'est de l'Electricité dont je parle, & qui depuis quelque tems en devenüe un article si important dans la Physique, qu'il n'est presque plus permis à personne d'en ignorer les effets.

C.16 Alles was undurchdringlich ist, gehört in das Geschlecht der Körper, und daher besteht das Wesen der Körper in der Undurchdringlichkeit, in welcher folglich alle, übrigen Eigenschaften ihren Grund haben müssen.

C.17 II y aura donc deux espéces de matiere, l'une qui fournit l'étoffe à tous les corps sensibles et dont toutes les particules ont la même densité, qui est très considerable et qui surpasse meme de plusieurs fois celle de l'or; l'autre espece de matiere sera celle dont ce fluide subtil qui cause la gravité est compose, et que nous nommons l'éther.

C.18 Die grobe Materie ist also an sick selbst keiner anderen Verönderung fähig als in Ansehung ihrer Figur, welche, wenn hinlöngliche Kröfte vorhanden, auf alle mögliche Arten veröndert werden kann.

[...] Dass die subtile Materie auch allezeit und allenthalben eine, beständige Dichtigkeit haben sollte, dergestalt, dass dieselbe durch keine Kröfte in einen kleineren Raum getrieben werden kannte, scheint der Wahrheit nicht gemöss zu sein. Vielmehr mochte auch hierin ein Hauptunterschied zwischen der groben und subtilen Materie bestehen, dass sick diese zusammendrücken liesse.

C.19 Ungeachtet wir aber hier stehn bleiben müssen und kaum hoffen können, jemals die wahre Ursache dieser Verminderung der elastischen Kraft des Aethers zu ergründen, so kann man sich doch damit leichter begnügen; als wenn man blosserdings vorgiebt, alle Körper seien von Natur mit einer Kraft begabt, einander anzuziehen. Denn da man sich von diesem Anziehn nicht einmal einen verstöndlichen Begriff mach en kann, so kann man im Gegentheil zum wenigsten überhaupt einsehn, wie es möglich sei, dass die elastische Kraft einer flüssigen Materie vermindert werde, und man begreift auch, dass dieses auf eine den Gesetzen der Natur gemösse Art geschehen könne.

C.20 Definitio 12. Quicquid statum corporum absolutum mutare valet, id vis vocatur; quae ergo, cum corpus ob causas internas in statu suo esset permansurum, pro causa externa est habenda.

C.21 Cette égalité des forces, d'ou dépend le grand principe de l'égalité entre l'action & réaction, est une suite nécessaire de la nature de la pénétration. Car, s'il étoit possible que le corps A pénétrât le corps B, le corps A seroit précisément autant pénétré par le corps B; donc, puisque le danger que ces corps se pénètrent, est égal de part & d'autre, il faut aussi que ces deux corps employent des forces égales pour resister à la penetration. Ainsi, autant que le corps B est sollicité par le corps A, précisément autant sera celui-cy sollicité par celui-là, l'un & l'autre déployant exactement autant de force qu'il faut pour prévenir la penetration. Or ces deux corps agissant l'un sur l'autre par une force quelconque, se trouveront dans le même état que s'ils étoient comprimés ensemble par Ia mê me force.

C.22 Idem omnino mihi, cum Neutoni Principia et Hermann Phoronomiam perlustrare coepissem, usu venit, ut, quamvis plurium problematum solutiones satis percepisse mihi viderere, tamen parum tantum discrepantia problemata resolvere non potuerim. Illo igitur iam tempore, quantum potui, conatus sum ex synthetica illa methodo elicere easdemque propositiones ad meam utilitatem analytice pertractare, quo negotio insigne cognitionis meae augmentum percepi. Simili deinde modo alia quoque passim dispersa ad hanc scientiam spectantia scripta sum persecutus, quae omnia ad meum usum methodo plana et aequabili exposui atque in ordinem idoneum digessi. Hoc in negotio occupatus non solum in plurimas antea nondum tractatas incidi quaestiones, quas feliciter solutas dedi: sed etiam complures peculiares methodos sum adeptus, quibus tam mechanica quam ipsa analysis non parum augmenti accepisse videantur.

C.23 Or on énonce communément ce principe par deux propositions, dont l'une porte, qu'un corps étant une fois en repos demeure éternellement en repos, à moins qu'il ne soit mis en mouvement par quelque cause externe ou étrangere. L'autre proposition porte, qu'un corps étant une fois en mouvement, conservera toujours éternellement ce mouvement avec la même direction & la meme vitesse, ou bien sera porté d'un mouvement uniforme suivant une ligne droite, à moins qu'il ne soit troublé par quelque cause externe ou étrangere.

C.24 Quoique les principe dont il s'agit ici soient nouveaux, entant qu'ils ne sont pas encore connus ou étalés par les Auteurs, qui on traité la Mécanique, on comprend néanmoins, que le fondement de ces principes ne saurait être nouveau, mais qu'il est absolument nécessaire, que ces principes soient déduits des première principes, ou plutôt des axiomes, sur le quels toute la doctrine du mouvement est établie.

C.25 Definitio I. Quemadmodum Quies est perpetua in eodem loco permanentia, ita Motus est continua loci mutatio. Corpus scilicet, quod semper in eodem loco haerere observatur, quiescere dicetur: quod autem labente tempore in alia atque alia loca succedit, id moveri dicitur.

C.26 Axiom 1. Omne corpus, etiam sine respectu ad alia corpora, vel quiescit vel movetur, hoc est vel absolute quiescit vel absolute movetur.

C.27 Axioma 2. Corpus, quod absolute quiescit, si nulli externae actioni fuerit subiectum, perpetuo in quiete perseverabit.

Axioma 3. Corpus, quod absolute movetur, si nulli externae actioni subiiciatur, secundum eandem directionem motu aequabili progredi perget.

C.28 Definitio 11. Proprietas illa corporum, quae rationem perseverationis in eodem statu in se continet, inertia appellatur, quandoque etiam vis inertiae.

C.29 Theorema 2a. Spatiola, per quae idem corpusuclum quiescens eodem tempusculo dt a diveris viribus promovetur, sunt ipsis viribus proportionalia.

C.30 Theorema 3. Si aequales vires corpuscula inaequalia quiescentia sollicitent, effectus eodem tempusculo producti erunt reciproce inertiae corpusculorum proportionales.

C.31 Definitio 15. Massa corporis vel quantitas materiae vocatur quantitas inertiae, quae in eo corpore inest, qua tam in statu suo perseverare quam omni mutationi reluctari conatur.

C.32 Theorema 4. Si corpuscula ratione massae inaequalia quiescant atque a viribus quibuscunque singula sollicitentur, erunt spatiola, per quae eodem tempusculo protrudentur, in ratione composita ex directa virium in inversa massarum.

C.33 Pro viribus ergo, quibus corpora iam mota sollicitantur, hanc dimetiendi rationem stabilimus, ut eas aequales iudicemus iis, quae in iisdem corporibus quiescentibus eodem tempore eundem effectum essent praestaturae. Haec autem ratio non indiget probatione, quia definitioni innititur nobisque adhuc liberum fuerat eam constituere. Si enim promotu quocunque spatiolas $s\sigma$ aequalia fuerint spatiola $S\sigma$, per quod idem corpusculum quiescens tempusculo eodem profertur a vi p, huic etiam illas vires aequales appellamus.

C.34 Quoniam autem immensi illius spatii cuiusque terminorum, [...] nullam nobis certam formare possimus ideam; loco huius immensi spatii eiusque terminorum considerare solemus spatium finitum, limitesque corporeos, ex quibus de corporum motu et quiete indicamus. Sic dicere solemus, corpus, quod respect horum limitum situm eundem conservat, quiescere, id vero, quod situm eodem respectu mutat, moveri.

C.35 Wenn der Zuschauer gleichgeschwind in einer graden Linie fortrückt und die Gegenden richtig, das ist, nach gleichlaufenden Linien schötzet, so werden zur Unterhaltung der scheinbaren Bewegung, wie sehr dieselbe auch von der wahren unterschieden sein mag, eben diejenigen Kröfte erfordert, als zur Unterhaltung der wahren Bewegung.

C.36 Wenn aber der Zuschauer sich nicht gleichförmig in einer graden Linie bewegt, dennoch aber die Gegenden richtig schötzet, so werden, um die scheinbare Bewegung aller Körper zu bewerkstelligen, noch ausser den Kröften, welche wirklich auf dieselben wirken, solche Kröfte erfordert, welche in einem jeden Körper alle Augenblicke eben die Verönderung hervorbringen, welche in dem Orte des Zuschauers vorgeht, aber nach einer umgekehrten Richtung.

C.37 Wenn aber der Zuschauer sich nicht gleichförmig in einer graden Linie bewegt, dennoch aber die Gegenden richtig schötzet, so werden um die scheinbare Bewegung aller Körper zu bewerkstelligen, noch ausser den Kröflen, welche wirklich auf dieselben wirken, solche Kröfte erfordert, welche in einem jeden Körper alle Augenblicke eben die Verönderung hervorbringen, welche in dem Orte des Zusehauers vorgeht, aber nach einer umgekehrten Richtung.

C.38 Ce qu'il y a de plus fâcheux pour moi, e'est que Ia Geometric est la seul occupation qui m'intéresse véritablement, sans qu'il me soit perm is de m'y livrer. Tout ce que je fais de littérature, quoique très-bénignement accueilli (à ce qu'il me semble) dans nos seances publiques de l'Académie Française, n'est pour moi que du remplissage et une espèce de pis-aller.

C.39 Mais si le nombre de nos connoissances certaines est fort petit, celui de nos connoissances directes l'est encore davantage. Nous ignorons, par rapport à un grand nombre d'objets, ce qu'ils sont & ce qu'ils ne sont pas; & nous n'avons sur beaucoup d'autres que des idées négatives y c'est-à-dire, nous savons ce qu'ils ne sont pas bien mieux que ce qu'ils sont; heureux encore dans notre indigence de posséder cette connoissance imparfaite & tronquée, qui n'est qu'une maniere un peu plus raisonnée & un peu plus douce d'être ignorans. Or dans tous ces cas on sera forcé d'avoir recours aux démonstrations indirectes. Les principales démonstrations de ce genre sont connues sous le nom de *réduction à l'absurde*; elles consistent à prouver une vérité par les absurdités qui s'ensuivroient si on ne l'admettoit pas. Dans cette classe doivent être placées toutes les démonstrations qui regardent les incommenssurables, c'est-à- dire, les grandeurs qui n'ont aucune commune mesure entr'elles. En effet l'idée de l'infini entre nécessairement dans celle de ces sortes de quantités; or nous n'avons de l'infini qu'une idée négative, puisque nous ne le concevons que par la négation du fini.

C.40 Dans l'art de conjecturer on peut distinguer trois branches. La premiere qui a été long-temps la seule; et qui n'a meme commence à être cultivée que depuis environ un siècle, est ce que les mathématiciens appellant l'analyse des probabilités dans les jeux de hasard [...] La seconde branche est l'extension [...] à differentes questions relatives à la vie commune; Comme celles qui ont rapport a la durée de la vie des hommes; au prix des rentes viagères, aux assurances maritimes à l'inoculation [...]. Cette troisième branche a pour objet les sciences dans lesquelles il est rare ou im possible de parvenir à la démonstration' et dans lesquelles cependant l' art de conjecturer est nécessaire.

C.41 Si nous savions *pourquoi il y a quelque chose*, nous serions vraisemblablement bien avancés pour résoudre la *question comment telle et telle chose existent-elle*? car vraisemblablement tous se tient dans l'univers plus intimement encore que nous ne pensons; et si nous savions ce premier *pourquoi*, ce *pourquoi* si embarrassant pour nous, nous tiendrons le bout du fil qui forme le système général des êtres, and nous n'aurions plus qu'à le developer.

C.42 Tous les êtres, & par conséquent tous les objets de nos connaissances, ont entr'eux une liaison qui nous échappe.

C.43 Toutes nos connaissance directes se réduisent à celles que nous recevons par les senses; d'ou il s'ensuit que c'est à nos sensations que nous devons toutes nos idées.

C.44 La premiere chose que nos sensations nous apprennent. et qui meme n'en est pas distinguée, c'est notre existence [...] La seconde connaissance que nous devons a nos sensations, est l'existence des objets extérieurs [...]. Ces objets innombrables produisent sur nous un effet si puissant, si continu, et qui nous unit tellement à eux, qu'après un premier instant ou nos idées réfléchies nous rappellent en nous-memes, nous sommes forces d'en sortir par les sensations qui nous assiègent de toutes parts, et qui nous arrachent a la solitude ou nous resterions sans elles [...]. Les affections involontaires qu'elles nous font éprouver, comparées avec la determination volontaire qui preside a nos idées réfléchies [...] tout cela forme en nous un penchant insurmontable à assurer l'existence des objets auxquels nous rapportons ces sensations, et qui nous paraissent en être la cause.

C.45 A foi et a serment, je ne trouve dans toutes les ténèbres métaphysiques de parti raisonnable que le scepticisme; je n'ai d'idée distincte, et encore d'idée complète, ni de la matière ni d'autre chose [...] je suis tenté de croire que tout ce que nous voyons n'est qu'un phénomène, qui n'a rien hors de nous de semblable à ce que nous imaginons, et j' en reviens toujours à la question du roi indicn: *Pourquoi y a-t-il quelque chose* car c'est la en effet le plus surprenant.

C.46 Dans la multitude des vérités que l'Encyclopédie embrasse, & qu'en vain on chercheroit à saisir toutes ensemble, il en est qui s'élèvent & qui dominent sur les autres, comme quelques pointes de rochers au milieu d'une mer immense. Ces vérités qu'il importe le plus de connoître, étant réunies & rapprochées dans des élémens de Philosophie qui serviroient à l'Encyclopédie comme d'introduction, l'utilité de ce grand Ouvrage en deviendrait sans doute plus générale & plus assurée.

C.47 Mais quelque différentes que ces Sciences soient entr'elles, soit par leur étendue, soit par leur nature, il est néanmoins des vues générales qu'on doit suivre dans la manière d'en traiter les élémens; il est ensuite des nuances différentes dans la manière d'appliquer ces vues générales aux élémens de chaque Science particulière; c'est ce qu'il faut développer.

C.48 Mais les principes de l'Algèbre ne portent que sur des notions purement intellectuelles, sur des idées que nous nous formons à nous-mêmes par abstraction en simplifiant & en généralisant des idées premières; ainsi ces principes ne contiennent proprement que ce que nous y avons mis, & ce qu'il y a de plus simple dans nos perceptions; ils sont en quelque façon notre ouvrage; comment peuvent-ils donc, par rapport à l'évidence, laisser encore quelque chose à desirer.

C.49 Les sciences abstraites ont occupé trop longtemps et avec trop peu de fruit les meilleurs esprits; ou l'on n'a point étudié ce qu'il importait de savoir, ou l'on n'a mis ni choix, ni vues, ni méthode dans ses études; les mots se sont multipliés sans fin, et la connaissance des choses est restée en arrière [...]. Les faits, de quelque nature qu'ils soient, sont la véritable richesse du philosophe. Mais un des préjugés de la philosophie rationnelle, c'est que celui qui ne saura pas nombrer ses écus, ne sera guère plus riche que celui qui n'aura qu'un écu.

C.50 Plus on peut tirer de l'application de la Geométrie à la Physique, plus on doit etre circonspect dan cette application. C'est à la simplicité de son object que la Géometrie est redevable de sa certitude; à al mesure que l'object devient plus composé la certitude s'obscurcit & s'éloigne. Il faut donc savoir s'arrêter sur ce qu'on ignore, ne pas croire que les mots & de Theoréme & de Corollaire, fassent par quelque vertu secrette l'essence d'une démonstration.

C.51 La seule utilité expérimentale que le Physicien puisse tirer des observations sur les lois de l'équilibre, sur celles du mouvement, & en général sur les affections primitives des corps, c'est d'examiner attentivement la différence entre le réésultat que donne la théorie & celui que fournit l'expérience; & d'employer cette différence avec adresse, pour déterminer, par exemple, dans les effets de l'impulsion, l'altération causée par la résistance de l'air; dans les effets des machines simples, l'altération occasionnée par le frottement & [...] car alors l'expérience ne servira plus simplement à confirmer la théorie y mais différant de la théorie sans l'ébranler, elle conduira à des vérités nouvelles auxquelles la théorie feule n'auroit pu atteindre.

C.52 La nature est une machine immense don les ressorts principaux nous son cachés; nous ne voyons meme cette machine qu'à travers un voile qui nous dérobe le jeu des parties les plus délicates. Entre les parties plus frappantes, & peut-être, si on ose le dire, plus grossieres, que ce voile nous permet d'entrevoir ou de découvrir, il est plusieurs qu'un meme ressort met en mouvement, & c'est là sur-tout ce que nous devons chercher à démêler.

C.53 Parmi les différentes suppositions que nous pouvons imaginer pour expliquer un effet, les seules dignes de notre examen sont celles qui par leur nature nous fournirent des moyens infaillibles de nous assuer si elles sont vraies. Le systême de la gravitation est de ce nombre, & mériteroit par cela seul l'attention des Philosophes. On n'a point à craindre ici cet abus du calcul & de la Géométrie, dans lequel les Physiciens ne sont que trop souvent tombés pour défendre ou pour combattre des hypotheses. Les planètes étant supposées le mouvoir, ou dans le vuide, ou au moins clans un espace non résistant, & les forces par lesquelles elles agissent les unes sur les autres étant connues, c'est un problème purement mathématique, que de déterminer les phénomènes qui en doivent naître; on a donc le rare avantage de pouvoir juger irrévocablement du systême Newtonien [...] il seroit à souhaiter que toutes les questions de la Physique pussent être aussi incontestablement décidées. Ainsi on ne pourra regarder comme vrai le syistême de la gravitation, qu'après s'être assuré par des calculs précis qu'il répond exactement aux phénomenes.

C.54 Je suppose seulement, ce que personne ne peut me contester, qu'un Fluid est un corps composé de particles très petites, détachées, & capable de se mouvoir librement [...] Car les Philosophes ne pouvant déduire immédiatement & directement de la nature des Fluides les loix de leur équilibre, ils les ont au moins reduites a un seul principe d'experience, *l'égalité de pression en tout sens* [...] En effet, condamnés comme nous le sommes à ignorer les premieres propriétés & la contexture intérieure des corps, la seule ressource qui reste à notre sagacité est de tacher au moins de saisir dans chaque matière l'analogie des Phenomenes, & de les rappeller tous à un petit nombre de faits primitifs & fondamentaux.

C.55 Ainsi dans la Méchanique ou science du mouvement des corps, on ne doit définir ni l'espace ni le tems, parce que ces mots ne renferment qu'une idée simple; mais on peut & on doit même définir le mouvement, quoique la notion en soit assez familiere à tout le monde, parce que l'idée de mouvement est une idée complexe qui en renferme deux simples, celle de l'espace parcouru, & celle du tems employé à le parcourir.

C.56 Force d'inertie, est la propriété qui est commune à tous les corps de rester dans leur état, soit de repos ou de mouvement, à moins que quelque cause étrangere ne les en fasse changer [...]. On l'appelle résistance, lorsqu'on veut parler de l'effort qu'un corps fait contre ce qui tend à changer son état; & on la nomme action, lorsqu'on veut exprimer l'effort que le même corps fait pour changer l'état de l'obstacle qui lui résiste.

C.57 Que ce que nous appellons causes, même de la premiere espece, n'est tel qu'improprement; ce sont des effets desquels il résulte d'autres effets. Un corps en pousse un autre, c'est-à-dire ce corps est en mouvement, il en rencontre un autre, il doit nécessairement arriver du changement à cette occasion dans l'état des deux corps, à cause de leur impénétrabilité; l'on détermine les lois de ce changement par des principes certains, & l'on regarde en conséquence le corps choquant comme la cause du mouvement du corps choqué. Mais cette façon de parler est impropre. La cause métaphysique, la vraie cause nous est inconnue.

C.58 Pourquoi donc aurions-nous recours à ce principe dont tout le monde fait usage aujourd'hui, que la force accélératrice ou retardatrice est proportionnelle à l'élément de la vitesse? [...]. Nous n'examinerons point si ce principe est de vérité nécessaire [...] pas non plus, avec quelques Géometres, comme de vérité purement contingent [...]: nous nous contenterons d'observer, que vrai ou douteux, clair ou obscur, il est inutile à la Méchanique, & que par conséquent il doit en être banni.

C.59 On vient de voir que de quelque maniere que le mouvement soit accéléré ou retardé, l'équation différentio-différentielle de la courbe sera toujours de cette forme $\pm dde = \varphi dt^2$. Or si on veut faire usage de cette équation, ainsi que des équations $\varphi dt = \pm du$ & $\varphi de = \pm udu$ pour déterminer dans un mouvement quelconque la relation entre u, t, e, il faut connoître φ & l'on pourrait penser que pour cet effet la connoissance de la cause qui accélere ou retarde le mouvement seroit nécessaire; l'objet de la Remarque est de faire voir que non, mais que φ est toujours donné par la définition même de l'espece de mouvement dont il est question; ainsi, conformément à cette même Remarque, quand on voudra faire usage des équations $\varphi dt^2 = \pm dde$, $\varphi dt = \pm du$, & $\varphi de = \pm udu$ pour déterminer la relation des espaces, des vitesses & des tems dans un mouvement dont la loi sera donnée, il suffira de substituer dans ces équations à la place de φ une quantité propre à exprimer la loi suivant laquelle on supposera que se font les augmentations ou diminutions de vitesse: quand on supposera, par exemple, que les diminutions instantanées de vitesse sont comme les quarrés de la vitesse, on écrira $gu^2dt = -du$ $gu^2de = -udu$ (g étant un coefficient constant), & ainsi du reste.

C.60 Il paroit d'abord que l'action du Createur sur les corps ne rend en aucune façon hipotétique l'execution des loix de la Méchanique dans l'univers, comme M. d' Alembert semble le supposer dans cet endroit; puisqu'il nous sera toujours permis de considerer cette action de Dieu comme une nouvelle force qui agit fur les corps: quels que soient alors les mouvemens qui en résultent, ils ne seroit jamais contraires aux principes de la Méchanique qui doivent immutables par eux mêmes.

C.61 Probléme I. Trouver la vitesse d'une verge CR fixe en C & chargée de tant de corps A, B, R qu'on voudra, en supposant que ces corps, si la verge ne les empêchoit, décrivissent dans des tems egaux les lignes infiniment petites AO, BQ, RT perpendiculaires à la verge.

C.62 Probléme V. Un fil CmM fixe en C, & chargé de deux poids m, M, étant infiniment peu éloigné de la verticale CO, trouver la durée des oscillations de ce fil.

C.63 Probléme IX. Un Corps dont la masse est m, & la vitesse u, se mouvant sur une même ligne avec un autre Corps dont la masse est M. SC la vitesse U, trouver la vitesse de ces Corps après le choc.

C.64 Soit un corps qui se meuve d'un mouvement quelconque, & dont toutes les parties aient chacune une vitesse différente représentée par l'indéterminée u dans un instant quelconque: soient aussi tant de forces accélératrices qu'on voudra, Ψ, Ψ', &c. qui agissent sur ce corps & en vertu desquelles la vitesse u que chaque partie a dans un instant quelconque, soit changée l'instant suivant en une autre vitesse u', différente pour chaque partie. Je dis que si on regarde la vitesse u comme composée de la vitesse u' & d'une autre vitesse u'', qui est infiniment petite; le système de toutes les parties du corps, animées chacune de la vitesse u'' doit être en équilibre avec les forces Ψ, Ψ', &c.

C.65 Soient en general les vitesses des différentes tranches du Fluide dans un meme instant, représentées par l'indéterminée v. Imaginons que dv soit l'increment de v dans l'instant suivant, cette quantité dv soit différente pour les différentes tranches, positive pour les unes, & negative pour les autres; en un mot, que $v \mp dv$, exprime la vitesse de chaque tranche lorsqu'elle prend la place de celle qui est immédiatement au-dessous; je dis que si chaque tranche étoit supposée tendre a se mouvoir avec la seule vitesse infiniment petite $\pm dv$, le Fluide resteroit en équilibre.

C.66 Supposons donc que le point A se meuve suivant AG; sur une courbe quelconque PAD, etant animé d'une force accélératrice réelle = π qu'en meme tems il soit sollicité de se mouvoir suivant AG par une force = F, qui par quelque raison que ce puisse être, se change en π; je dis que ce corps A, s'Il étoit sollicité suivant A P par une force = $F - \pi$ demeureroit en repos.

C.67 Egli, oltre a presentarcisi come uno dei pensatori più originali e profondi che si conoscano, ci appare eziandio come uno dei matematici più dotti a noi noti. L'ambito dei suoi studi si può ben dire abbracci tutta la matematica, pura ed applicata; aritmetica teorica e pratica; algebra e calcolo infinitesimale; meccanica e fisica matematica, con le principali applicazioni di tali discipline (p. es. la balistica); finalmente astronomia, con tutte le discipline ad essa collegate quali sarebbero l'orologeria e la navigazione. Ebbene per combattere e vincere in tutti questi ampli e svariatissimi campi Egli si armò di tutta la scienza del proprio secolo, rendendosi famigliare il maneggio di tutti, senza eccezione, i procedimenti logici e algoritmici, allora in uso: a provarlo basta rilevare che, benché vero e proprio analista per temperamento e per abitudini, Egli era però in grado di maneggiare con invidiabile disinvoltura i metodi cinematico-geometrici costantemente adoperati da Newton.

C.68 Si un système quelconque de tant de corps ou points que ton veut tirés chacun par des puissances quelconques est en équilibre & qu'on donne à ce système un petit mouvement quelconque, en vertu duquel chaque point parcoure un espace infiniment petit qui exprimera fa vitesse virtuelle; la somme des puissances multiplies chacune par l'espace que le point ou elle est appliquée parcourt suivant la direction de cette même puissance, sera toujours égale a zero.

C.69 Car, si l'on imagine qu'on imprime à chaque corps, en sens contraire, le mouvement qu'il doit prendre il est claire que le système sera réduit au repos.

Notes

[I]For instance, in the case of the expression $x^3 + y^3 = axy$ (AP = x, PM = y and AB = a), taking differentials gives:

$$3xxdx + 3yydy = axdy + aydx; \text{ and } dy = \frac{aydx - 3xxdx}{3yy - ax}$$

When the point P falls in the searched point E ($dy = 0$) it is: $y = 3xx/a$. "Substituting this value in place of y in the equation $x^3 + y^3 = axy$, one finds for AE the value $x = 1/3a\sqrt{2}$" [2], p. 43.

[II]Consider for instance the differential of the expression:

$$\frac{ydy}{dx}$$

L'Hôpital distinguished the two cases dx is constant or dy is constant. The first case corresponds, according to modern reading, to assume x as independent variable, the second to assume y. The resulting differentials of the previous expression in the two cases are respectively:

$$\frac{dy^2 + yddy}{dx}; \quad \frac{dxdy^2 - ydyddx}{dx^2}$$

[III]With Leibniz's symbols, the differential equation of motion of a planet moving with harmonic motion is given as:

$$ddr = bbaa\theta\theta - 2aaqr\theta\theta, : bbr^3 \tag{3.1}$$

where r is the distance from the center of the (harmonic) motion, θ the infinitesimal time, a the area described by the radius r in the unitary time, q the length of the major axis of the ellipse and $b = \sqrt{qq - ee}$ the length of the minor axis (the axis where there are no foci), being e the eccentricity; the symbol ", :" means that all the terms on the right side of the equation should be divided by bbr^3.

[IV]Varignon arrived to the expression [165], p. 228:

$$y = \frac{dsdds}{dxdt^2}$$

where ds is the space measured on the orbit (the curvilinear abscissa in modern terms), while dx is the variation of the distance from the center of the force.

[V]He probably referred to the memoir *Application de la regle generale des mouvements à toute les hypotheses possible d'accelerations ordonnes dans la chute des corps*. Here he had got for the motion on two inclined planes the relation: $ml^2H\pi\theta^2 = \mu\lambda^2hpt^2$, where h and H are the heights of the inclined planes of length

l and λ respectively (This is the relation reported in the present book, with different symbols). If the quantities ph/l and $\pi H/\lambda$, that are the component of weights p and π along the inclined planes, are assumed equal to the forces f and φ acting on the bodies, one obtains the relation referred to by Varignon.

[VI]Hermann presented results for different expressions of the force p; among which, $p=$ const, p varying linearly from a center and p varying arbitrarily. I refer only to the second situation. With reference to Fig. 3.4 the expression of the force is given by $p = \dfrac{ab - bS}{a}$, with b the value of the force at the beginning of the motion A (segment Aa in Fig. 3.4); S the space passed measured from A, and a the distance of A from the center G. After integration, the expression of the speed is given by $C = \sqrt{\dfrac{2abS - bSS}{a}}$, and time by $t = \dfrac{\text{angle AGH} \times \sqrt{AG}}{\sqrt{Aa}}$ [106], p. 144.

[VII]In the memoir of 1714, where he used the ordinary principles of mechanics, Bernoulli assumed that in a pendulum, a mass A located at a distance a from the fulcrum, can be replaced by a mass A^* located at a distance x, such that $A^*x = Aa$; that is by imposing the equivalence of the static moments (by then one of the ordinary principles of mechanics) of the masses-weights A and A^*. The same holds true for the masses B and C; the living forces of these masses is given by $A^*x^2 + B^*x^2 + C^*x^2$, which gives $xaA + xbB + xcC$ [9], pp. 215–217.

[VIII]Here with some details Euler's proof. Let assume three coordinate axes in the absolute space, OA, OB and OC and call the components of the motion (velocities) along the axes u, v, w. The motion of the observer is defined by velocities α, β, γ, that by assumption are constant. To the observer, the motion of the body as seen from the axis OA, for instance, will appear the smaller, the faster its own motions. The apparent motion in the direction of the axes OA, OB and OC will be $\underline{u} = u - \alpha$, $\underline{v} = v - \beta$, $\underline{w} = w - \gamma$ respectively. From the equations of motion, that Euler had already developed, it follows that for the maintenance of the true (absolute) motion one requires three forces, $MP = P$, $MQ = Q$, $MR = R$, so that:

$$P = \frac{Mdu}{ndt}; \quad Q = \frac{Mdv}{ndt}; \quad R = \frac{Mdw}{ndt}$$

where M is the mass and n a normalizing factor. Now replace u, v and w, by $\underline{u}$, $\underline{v}$, $\underline{w}$, in these equations. Since α, β and γ are constants, the differentials remain unchanged, so that it is apparent that for the maintenance of the apparent motion the same forces P, Q and R are required as for the true motion.

[IX]In this case also the proof is very simple. However arbitrarily the motion of the observer may change, with reference to the three assumed axes it can always be represented by the three motions (velocities) α, β and γ by taking these quantities to be variable. Now if the true movement requires the forces:

$$P = \frac{Mdu}{ndt}; \quad Q = \frac{Mdv}{ndt}; \quad Q = \frac{Mdw}{ndt}$$

By replacing u, v, w by $u-\alpha$, $v-\beta$, $w-\gamma$ the maintenance of the apparent motion will require the following three forces:

$$\begin{aligned}
&\text{Kraft} \quad MP = \frac{Mdu}{ndt} - \frac{Md\alpha}{ndt} = P - \frac{Md\alpha}{ndt}\\
&\text{Kraft} \quad MQ = \frac{Mdv}{ndt} - \frac{Md\beta}{ndt} = Q - \frac{Md\beta}{ndt}\\
&\text{Kraft} \quad MR = \frac{Mdw}{ndt} - \frac{Md\gamma}{ndt} = R - \frac{Md\alpha}{ndt}
\end{aligned}$$

[X]D'Alembert calculations are quite demanding; here for the sake of simplicity only the static moments of lost motions are considered, because more interesting. By renaming the coordinate axes as x, y, z, interpreting α'', β'', ω'' as the components of acceleration, or $\ddot{x}$, $\ddot{y}$, $\ddot{z}$ and indicating the element of mass as dM, the following relations are obtained:

$$\int_M (\ddot{x}z - x\ddot{z})dM; \int_M (\ddot{y}x - y\ddot{x})dM; \int_M (\ddot{y}z - y\ddot{z})dM$$

These expressions appear to a modern reader nearly the same as those found by Euler in his paper of 1750, *Decouverte d'un nouveau principe de mecanique* [29]. This fact is scarcely commented by modern historians who usually credit Euler to be the first to write the equations of motion of a rigid body rotating about a free axis [168]. What is sure is that d'Alembert's sent Euler his paper in 1749, shortly after its publication. Euler in his *Découvertee d'un noveaux principe de mécanique* of 1750, where he presented equations similar to those of d'Alembert, did not acknowledge d'Alembert's paper; d'Alembert was astonished and hurt by the fact and in 1752 published a comparison of his procedure with that of Euler concluding that "the method employed by this great geometer [Euler] is absolutely the same, fundamentally as my second method [...] of my work" [168], p. 237. In 1752 Euler published an *Avertissement* with an appreciation of what d'Alembert wrote in his *Researches sur la précessions des equinoxes* [78].

The similarity of relations and the above referred circumstances would lead to suppose that actually d'Alembert had preceded Euler in discovering the equations of motion of a rotating rigid body by using a 'lofty rule' very close to his. How much Euler was inspired by d'Alembert and derived his strategy from d'Alembert's is difficult to say, considering that d'Alembert himself was not completely original with his principle. For comments in this matter see [168].

[XI]An interesting example of Lagrange to reduce to a minimum the role of geometric intuition in his treatise, is shown clearly in the way he introduced the virtual displacements of the points of a rigid body, for which there is the constraints of the invariance of the distance of the various point. The traditional way was to start from the relations which expressed the change of coordinate of the various points, from a fixed coordinate system to a system of coordinate joined to the moving body; relations which are obtained using geometrical reasoning. Lagrange instead started directly from a local rigidity constraint, by imposing that the variation of distance of a given point and all points in its neighborhood be zero. In particular if (dx, dy, dz) express the difference of coordinates between a point of reference and its closest points, the conditions is expressed by Lagrange with the condition $\delta\sqrt{dx^2 + dy^2 + dz^2} = 0$. With a series of lengthy and not always simple passages, Lagrange arrived to an equation of the kind [114], p. 117:

$$\begin{aligned}\delta x &= \delta\lambda - y\delta N + z\delta M\\ \delta y &= \delta\mu + x\delta N - z\delta L\\ \delta z &= \delta\nu - x\delta M + y\delta L\end{aligned}$$

where δx, δy, δz are the sought virtual velocities and λ, μ, ν, L, M, N are six parameters independent of the coordinate of the points, function of the time only. Later on Lagrange will assign a meaning to these parameters, though not strictly necessary. The Greek letters denote translation, the Latin capital letters angle of rotation.

A classic example of the versatility of the calculus of variations can be found in the derivation of the principle of conservation of living forces, a concept to which Lagrange did not give any particular physical meaning. The starting point is the symbolic equation of motion. Here the virtual velocity-variations δx, δy, δz are free to assume any values, though compatible with constraints of the system, and thus they can also assume the values dx, dy, dz, that is the actual displacements, which for sure are compatible with constraints. So one can write:

$$S\left(\frac{d^2x}{dt^2}dx + \frac{d^2y}{dt^2}dy + \frac{d^2z}{dt^2}dz + Pdp + Qdq + Rdr + \&\right)m = 0$$

At this point Lagrange noticed that $Pdp + Qdq + Rdr + \&$ is 'usually' integrable, as in the case of central forces he assumed in the *Mechanique analitique*, and thus can be derived from a function Π of $p, q, r, \&$, so that $d\Pi = Pdx + Qdy + Rdz$, and rewrote the previous expression as:

$$S\left(\frac{d^2xdx + d^2ydy + d^2zdz}{dt^2} + d\Pi\right)m = 0$$

which integrated furnished the principle principle of conservation of living forces:

$$S\left(\frac{dx^2 + dy^2 + dz^2}{2dt^2} + \Pi\right)m = F$$

"where F designates an arbitrary constant of integration equal to the value of the first member of the previous equation in a given instant of time" and "$\frac{dx^2+dy^2+dz^2}{dt^2}$ the square of speed" [114], pp. 207–208.

[XII]Using the rules of the calculus of variations, after some lengthy calculations one obtains the relation of incompressibility in the form [114], p. 143:

$$\delta(dxdydz) = dxdydz\left(\frac{d\delta x}{dx} + \frac{d\delta y}{dy} + \frac{d\delta z}{dz}\right)$$

and the symbolic equation of equilibrium assumes the form:

$$\mathrm{S}\left(\Delta\ X\delta x + \Delta\ Y\delta y + \Delta\ Z\delta z + \lambda\frac{d\delta x}{dx} + \lambda\frac{d\delta y}{dy} + \lambda\frac{d\delta z}{dz}\right) dxdydz = 0.$$

References

1. Andersen K (1983) The mathematical technique in Fermat's deduction of the law of refraction. Hist Math 10(1):48–62
2. Anonymous (1696) Analyse des infiniment petits. Imprimerie Royale, Paris
3. Anonymous (1752) Élémens de musique, théorique et pratique suivant les principes de M. Rameau. Jombert and Bruyset, Paris-Lyon
4. Arnauld A, Nicole P (1775) La logique, ou l'art de penser, contenant, outre les regles communes, plusieurs observations nouvelles, propres a former le jugement. Humboldt, Paris
5. Bernoulli D (1728) Examen principiorum mechanicae, et demonstrationes geometricae de compositione et resolutione virium. Commentarii Academiae Scientiarum Petropolitanae 1:126–142
6. Bernoulli D (1738) Hydrodynamica, sive de viribus et motibus fluidorum commentarii. Dulsseker, Strasbourgh
7. Bernoulli D (1750) Remarques sur le principe de la conservation des forces vives pris dans un sens général. Mémoires de l'Académie Royale des Sciences et Belles Lettres de Berlin 4:356–364
8. Bernoulli J (1703) Démonstration générale du centre du balancement a toutes sortes de figure tirée de la nature du levier. Mémoires de l'Académie Royale des Sciences de Paris, pp 78–84
9. Bernoulli J (1714) Nouvelle theorie du centre d'oscillation. contenant une regle pour le déterminer dans les pendules composés & balançans non seulement dans le vuide, mais aussi dans les liqueurs; laquelle regle est appuyée sur un fondement plus sér qu'aucun qu'on ait publié jusqu'ici par rapport à cette matiere. Mémoires de l'Académie Royale des Sciences de Paris, pp 208–230
10. Bernoulli J (1727a) Du discours sur les loix de la communication du mouvement. Jombert, Paris
11. Bernoulli J (1727b) Theoremata selecta pro conservatione virium vivarum demonstranda et esperimenta confirmanda. In: Bernoulli J (ed) Opera omnia (1742, 4 vols), vol 3. Bousquet, Lausanne and Geneva, pp 124–130
12. Bernoulli J (1730) Nouvelle pensée sur le système de M. Descartes, & la maniere d'en deduire les orbites & les aphelies des planets. In: Bernoulli J (ed) Opera omnia (1742, 4 vols), vol 3. Bousquet, Lausanne and Geneva, pp 131–172
13. Bernoulli J (1735a) De vera notione virium vivarum, earunque usu in dynamicis. In: Bernoulli J (ed) Opera omnia (1742, 4 vols), vol 3. Bousquet, Lausanne and Geneva, pp 239–260
14. Bernoulli J (1735b) Essai d'une nouvelle physique céleste; servant à expliquer les principaux phénomènes du ciel, & en particulier la cause de l'inclination des orbites des planetes par

rapport au plan de l'equateur du soleil. In: Bernoulli J (ed) Opera omnia (1742, 4 vols), vol 3. Bousquet, Lausanne and Geneva, pp 261–364
15. Bernoulli J (1742a) Hydraulica. In: Bernoulli J (ed) Opera omnia (1742, 4 vols), vol 4. Bousquet, Lausanne and Geneva, pp 387–483
16. Bernoulli J (1742b) Opera omnia (4 vols). Bousquet, Lausanne and Geneva
17. Bernoulli J (1968) Hydrodynamics by Daniel Bernoulli and Hydraulica by Johann Bernoulli. Dover, New York, Translated into English by Carmody T, Kobus H
18. Bernoulli J, Maquet P, Ziggelaar A, Kardel T (eds) (1997) On the mechanics of effervescence and fermentation and on the mechanics of the movement of the muscles. Trans Am Philos Soc New Ser 87(3):i–v+1–158
19. Bernoulli J (2008) Die Werke von Johann I und Nicolaus II Bernoulli. In: Villaggio P (ed) Die gesammelten Werke der Mathematiker und Physiker der Familie Bernoulli (1955-), vol 6. Birkhäuser, Basel
20. Bernoulli J (2011) Unpublished correspondence, private communication by Radelet De Grave
21. Bertoloni M (1993) The emergence of reference frames and the transformation of mechanics in the Enlightenment. Hist Stud Phys Sci 23:301–335
22. Blackwell R (1977) Christiaan Huygens' The motion of colliding bodies. Isis 68(4):574–597
23. Boyer C (1949) The history of Calculus and its conceptual development. Dover, New York
24. Calinger R (2016) Leonhard Euler. Mathematical genius in the Enlightenment. Princeton University Press, Princeton
25. Cantor G, Hodge M (eds) (1981) Conceptions of ether. Studies in the history of the ether theories. Cambridge University Press, Cambridge
26. Capecchi D (2012) History of virtual work laws. Birchäuser, Milan
27. Capecchi D (2014) The problem of motion of bodies. Springer, Cham, Dordrecht
28. Capecchi D (2018) The path to post-Galilean epistemology. Springer, Cham, Dordrecht
29. Capecchi D (2020) Leonhard Euler between mathematics and natural philosophy: An introduction to natural science Anleitung zur Naturlehre. In: Sriraman B (ed) Handbook of the history and philosophy of mathematical practice. Forthcoming, Springer, Dordrecht
30. Capecchi D, Drago A (2005a) Lagrange e la storia della meccanica. Progedit, Bari
31. Capecchi D, Drago A (2005b) On Lagrange's history of mechanics. Meccanica 40:19–33
32. Casini P (1964) D'Alembert epistemologo. Rivista critica di storia della filosofia 19(1):28–53
33. Casini P (1970) Il problema d'Alembert. Rivista di. Filosofia 1(1):226–47
34. Casini P (1973) Introduzione all'Illuminismo. Laterza, Bari
35. Casini P (2008) Euler filosofo della natura e accademico. L'ipotesi dell'etere e la controversia antiwolffiana. Quaderni-Accademia delle scienze, pp 123–136
36. Cassirer E (1953) Substance and function and Einstein's theory of relativity. Dover, New York
37. Cernuschi A (1996) D'Alembert pris au jeu de la musique. Ses interventions musicographiques dans l'Encyclopédie. Recherches sur Diderot et sur l'Encyclopédie 21:145–161
38. Christensen T (1989) Music theory as scientific propaganda: The case of d'Alembert's Élémens de musique. J Hist Ideas 50(3):409–427
39. de Condillac EB (1746) Essai sur l'origine des connaissances humaines. Mortier, Amsterdam
40. Costabel P (1956) La 'loi admirable' de Christiaan Huygens. Revue d'Histoire des Sciences et de leurs Applications 9(3):208–220
41. Crépel P (2005) Jean Le Rond d'Alembert, Traité de dynamique (1743, 1758). In: Grattan-Guinness I, Cooke R, Corry L, Crépel P, Guicciardini N (eds) Landmark writings in western mathematics 1640–1940. Elsevier Science, Amsterdam
42. d'Alembert J (1743a) Essai d'une nouvelle théorie de la résistance des fluides. David, Paris
43. d'Alembert J (1743b) Recherches sur la precession des equinoxes et sur la nutation de l'axe de la terre. David, Paris
44. d'Alembert J (1743c) Traité de dynamique. David, Paris
45. d'Alembert J (1744) Traité de l'équilibre et du mouvement des fluides. David, Paris
46. d'Alembert J (1747) Réflexion sur la cause générale des vents. David, Paris
47. d'Alembert J (1753-1767) Melanges de littérature, d'histoire et de philosophie, vol 5 . Catelain, Amsterdam

48. d'Alembert J (1758) Traité de dynamique, 2nd edn. David, Paris
49. d'Alembert J (1761) Démonstration du principe de la composition des forces. Opuscules Mathématiques 1:169–179
50. d'Alembert J (1762) Élémens de musique, théorique et pratique suivant les principes de M. Rameau, éclaircis, développées et simplifiés, 2nd edn. Bruyset, Lyon
51. d'Alembert J (1767) Essai sur les elemens de la philosophie. In: d'Alembert J (ed) Melanges de littérature, d'histoire et de Philosophie (5 vol), 4th edn., vol 4. Catelain, Amsterdam
52. d'Alembert J (1805) Essai sur les elemens de la philosophie. In: d'Alembert (ed) Oeuvres philosophiques, historiques et littérariés de d'Alembert (18 vols), vol 2. Bastien, Paris
53. d'Alembert J (1821–1822) Oeuvres complètes (5 vols). Belin, Paris
54. d'Alembert J (1894) Discourse preliminaire de l'Encyclopédie, Edited by Picavet F. Colin, Paris
55. d'Alembert J (2018) http://numerix.univlyon1.fr/dalembert/d/indexracine.html
56. d'Elizagaray RB (1689) De la theorie de la manoeuvre des vaisseaux. Estienne Michallet, Paris
57. Descartes R (1657–1667) Lettres de M. Descartes. Edited by Charles Angot (3 vols). Angot, Paris
58. Descartes R (1668) Discourse de la method plus la dioptrique et les météores. Girard, Paris
59. Dhombres J, Radelet de Grave P (1991) Contingence et nécessité en mécanique. Etude de deux textes d'inédites de Jean d'Alembert. Physis 28(1):5–114
60. Diderot D (1875) Oeuvres complètes de Diderot (20 vols). Garnier, Paris
61. Drago A (1999) Il principio di d'Alembert non è un principio. Sua relazione col principio dei lavori virtuali. In: Tucci P (ed) XIX Congresso SISFA, Como, pp 185–209
62. Drago A (2003) La riforma della dinamica secondo G.W. Leibniz. Testi originali e loro interpretazione moderna. Hevelius, Benevento
63. Dugas R (1950) Histoire de la mécanique. Griffon, Neuchatel
64. Einstein A (1982) Ideas and opinions. Translated into English by Bargmann S. Three Rivers, New York
65. Encyclopédie (1751–1772) Encyclopédie ou dictionnaire raisonné des sciences, des arts et des métiers, par une societè de gens de lettres (17 vols). Briasson-David-Le Breton-Durand, Paris
66. Euler L (1736) Mechanica sive motus scientia analytice exposita (2 vols). Academiae Scientiarum, Saint Petersburg
67. Euler L (1738) De communicatione motus in collisione corporum. Commentarii Academiae Scientiarum Petropolitanae 5:159–168
68. Euler L (1739a) Explicatio phaenomenorum quae a motu successivo lucis oriuntur. Commentarii Academiae Scientiarum Petropolitanae 11:150–193
69. Euler L (1739b) Tentamen novae theoriae musicae ex certissismis harmoniae principiis dilucide expositae. Academiae Scientiarum, Saint Petersburg
70. Euler L (1740) De minimis oscillationibus corporum tam rigidorum quam flexibilium. methodus nova et facilis. Commentarii Academiae Scientiarum Petropolitanae 7:99–122
71. Euler L (1746a) De la force de percussion et de sa veritable measure. Mémoires de l'Académie des Sciences et Belles Lettres de Berlin 1:21–53
72. Euler L (1746b) Gedancken von den Elementen der Körper. In: Various (ed) Leonhardi Euleri Opera omnia (in progress), III, vol 2, Teubneri GB; [then] Fussli Turici; [then] Birkhäuser, Basel, pp 347–366
73. Euler L (1746c) Recherches physiques sur la nature des moindres parties de la matiere. Opuscula Varii Argumenti 1:287–300
74. Euler L (1748a) De observatione inclinationis magneticae. In: Leonhardi Euleri Opera omnia (in progress), III, vol 10, Teubneri GB; [then] Fussli Turici; [then] Birkhäuser, Basel, pp 110–137
75. Euler L (1748b) Dissertatio de magnete. In: Various (ed) Leonhardi Euleri Opera omnia (in progress), III, vol 10, Teubneri GB; [then] Fussli Turici; [then] Birkhäuser, Basel, pp 138–179

76. Euler L (1749) Recherches sur le mouvement des corps célestes en général. Mémoires de l'Académie des Sciences et Belles Lettres de Berlin 3:93–143
77. Euler L (1750) Reflexions sur l'espace et le tems. In: Various (ed) Leonhardi Euleri Opera omnia (in progress), III, vol 2, Teubneri GB; [then] Fussli Turici; [then] Birkhäuser, Basel, pp 376–383
78. Euler L (1752a) Avertissement au sujet des recherches sur la precession des equinoxes. Mémoires de l'Académie des Sciences et Belles Lettres de Berlin 6:412
79. Euler L (1752b) Dissertatio de igne, in qua ejus natura et proprietates explicantur. In: Recueil de piéces qui ont remporté le prix de l'Académie royale des sciences. Contenant les pieces depuis 1738 jusqu'en 1740, vol 4. Imprimerie Royale, Paris, pp 5– 21
80. Euler L (1752c) Recherches sur l'origine des forces. In: Various (ed) Leonhardi Euleri Opera omnia (in progress), II, vol 5, Teubneri GB; [then] Fussli Turici; [then] Birkhäuser, Basel, pp 109–131
81. Euler L (1756) Théorie plus complete des machines qui sont mises en mouvement par la reaction dell'eau. Mémoires de l'Académie des Sciences et Belles Lettres de Berlin 10:227–295
82. Euler L (1757) Principes generaux du mouvement des fluides. Mémoires de l'Académie des Sciences et Belles Lettres de Berlin 11:274–315
83. Euler L (1761) Comment to F.U.Th. Aepinus' Descriptio ac explicatio novorum quorundam experimentorum electricorum. In: Leonhardi Euleri Opera omnia (in progress), III, vol 10, Teubneri GB; [then] Fussli Turici; [then] Birkhäuser, Basel, pp CXXI–CXXIV
84. Euler L (1765) Theoria motus corporum solidorum seu rigidorum ex primis nostrae cognitionis principiis stabilita et ad omnes qui in huiusmodi corpora cadere possunt accommodata, presented 1765. Röse, Rostock
85. Euler L (1766) Elementa calculi variationum. Novi Commentarii Academiae Scientiarum Imperialis Petropolitanae 10:51–93
86. Euler L (1770-1774) Lettres à une princesse d'Allemagne sur divers sujets de physique & de philosophie (3 vols). Steidel & Compagne, Mietau, Leipzig
87. Euler L (1775) Lettres à une princesse d'Allemagne sur divers sujets de physique & de philosophie (3 vols). Societé Typographique, Berne
88. Euler L (1802) Letters of Euler on different subjects in physics and philosophy addressed to a German princess (2 vols). Translated into English by Hunter H, Murray et al, London
89. Euler L (1862) Anleitung zur Naturlehre. In: Fuss P and Fuss N (ed) Opera postuma: mathematica and physica (2 vols), III, vol 2. Eggers & Socios, Saint Petersburg, pp 449–560
90. Euler L (1911–2018) Leonhardi Euleri Opera omnia (in progress). Teubneri GB; [then] Fussli Turici; [then] Birkhäuser, Basel
91. Fellman E, Fleckenstein J (1970) Bernoulli, Johann (Jean) I. In: Gillispie CC (ed) Complete dictionary of scientific biography. Scribner, New York
92. Fermat P (1891–1912) Syntesis ad refractiones. In: Fermat P (ed) Oeuvres de Fermat (4 vols), vol 1. Gauthier-Villars, Paris, pp 73–179
93. de Foncenex DF (1760–1761) Sur les principes fondamentaux de la mécanique. Melanges de philosophie et de Mathématique de la Societé Royale de Turin Tomus alter, pp 299–322
94. Fraser CG (1983) J.L. Lagrange's early contribution to the principles and methods of mechanics. Arch Hist Exacts Sci 28(3):197–242
95. Fraser CG (1985a) d'Alembert's principle: The original formulation and application in Jean d'Alembert's Traité de dynamique (1743) (part one). Centaurus 28:31–61
96. Fraser CG (1985b) d'Alembert's principle: The original formulation and application in Jean d'Alembert's Traité de Dynamique (1743) (part two). Centaurus 28:145–159
97. Gaukroger S (1982) The metaphysics of impenetrability: Euler's conception of force. Br J Hist Sci 15(2):132–154
98. Glass B (1970) Maupertuis, Pierre Louis Moreau De. In: Gillispie CC (ed) Complete dictionary of scientific biography. Scribner, New York
99. Guicciardini N (1985) Johann Bernoulli, John Keill and the inverse problem of central forces. Ann Sci 52(6):537–575

100. Hakfoort C (1995) Optics in the age of Euler. Cambridge University Press, Cambridge
101. Hankins TL (1965) Eighteenth-century attempts to resolve the vis viva controversy. Isis 56(3):281–297
102. Hankins TL (1970) Jean d'Alembert. Science and Enlightenment, Clarendon, London
103. Hawkins T (1975) Cauchy and the origins of spectral theory. Hist Math 2(1):1–29
104. Heimann PM, McGuire JE (1971) Cavendish and the vis viva controversy: A Leibnizian postscript. Isis 62(2):225–227
105. Hermann J (1716) Phoronomia, sive de viribus et motibus corporum solidorum et fluidorum libri due. Wetstonios, Amsterdam
106. Hermann J (1727) Theoria motuum qui nascuntur a potentiis in corpora indesinenter agentibus. Commentarii Academiae Scientiarum Imperialis Petropolitanae 2:139–173
107. Hutton J (2019) Mactutor history of mathematics archive. http://www-history.mcs.st-and.ac.uk/index.html
108. Iltis C (1971) Leibniz and the vis viva controversy. Isis 62(1):21–35
109. Jammer M (1961) Concepts of mass in classical and modern physics. Harvard University Press, Cambridge, MA
110. Jourdain PE (1912) Maupertuis and the principle of least action. Monist 22(3):414–459
111. Koyré A (1965) Newtonian studies. Champan & Hall, London
112. Kuhn T (1962) The structure of scientific revolution. The University of Chicago Press, Chicago
113. Lagrange JL (1760) Essai sur une nouvelle méthode pour déterminer les *maxima* et les *minima* des formules intégrales indéfinies. In: Serret JA, [Darboux G] (1867–1892) (ed) Oeuvres de Lagrange (14 vols), vol 1. Gauthier-Villars, Paris, pp 333–362
114. Lagrange JL (1788) Méchanique analitique. Desaint, Paris
115. Lagrange JL (1811) Mécanique analytique. In: Serret JA, [Darboux G] (1867–1892) (ed) Oeuvres de Lagrange (14 vols), vol 11. Gauthier-Villars, Paris
116. Lagrange JL (1867–1892) Ouvres de Lagrange (14 vols). Serret JA [Darboux G] (eds). Gauthier-Villars, Paris
117. Lagrange JL (1997) Anal Mech. Translated into English by Boissonnade A and Vagliente VN, Springer, Dordrecht
118. Laudan LL (1968) The vis viva controversy, a post-mortem. Isis 59(2):130–143
119. Le Ru V (1994) Jean Le Rond d'Alembert philosophe. Vrin, Paris
120. Leibniz G (1686) Brevis demonstratio erroris memorabilis Cartesii et aliorum circa legem nature, secundum quam volunt a Deo eandem semper quantitatem motus conservari; qua et in re mechanica abutuntur. Acta Eruditorum 3:161–163
121. Leibniz GW (1689a) Dynamica de potentia at legibus naturae corporum. In: (1849–1863) GC (ed) Mathematische Schriften (7 vols), vol Band II. Reprographic reprint Hildesheim, Olms, pp 281–431
122. Leibniz GW (1689b) Tentamen de motuum coelestium causis. Acta Erud 2:82–96
123. Leibniz GW (1692) Essay de dynamique. In: Foucher de Careill A (ed) (1859–1875) Oeuvres de Leibniz (7 vols), vol 1. Didot, Paris, pp 470–473
124. Leibniz GW (1695) Specimen dynamicum. In: Gerhardt CI (ed) (1849–1863) Mathematische Schriften (7 vols), vol II. Asher A and Schmidt HV, Berlin, Halle, pp 234–246–Part II, pp. 246–254
125. Leibniz GW (1699) Essay de dynamique sur les loix du mouvement. In: Gerhardt CI (ed) (1849–1863) Mathematische Schriften (7 vols), vol II. Asher A and Schmidt HV, Berlin, Halle, pp 215–231
126. Leibniz GW (1849–1863) Mathematische Schriften (7 vols), Gerhardt CI (ed). Asher A and Schmidt HV, Berlin, Halle
127. Leibniz GW, Bernoulli Johann, Ficquet E (1745) Virorum celeber. Got. Gul. Leibnitii et Johann Bernoulli commercium philosophicum et mathematicum (2 vols). Bousquet, Lausanne & Geneve
128. de l'Hôpital GF, (2015) L'Hôpital analyses des infiniment petits. Birkhäuser, Dordrecht, Translated into English by Bradley RE, Petrilli SJ, Sandifer CE
129. Loria G (1913) Lagrange e la storia delle matematiche. Bibl Math 3(13):333–338

130. Lovejoy AO (1981) The great chain of being. Harvard University Press, Cambridge, MA
131. Mach E (1919) The science of mechanics: a critical and historical account of its development. Translated into English by McCormack TJ, Open Court, Chicago
132. Maclaurin C (1801) A treatise on fluxions (2 vols). Baynes et al, London
133. Maltese G (2000) On the relativity of motion in Leonhard Euler's science. Arch Hist Exact Sci 54(4):319–348
134. Maltese G (2003) The ancients' inferno: The slow and tortuous development of 'Newtonian' principles of motion in the eighteenth century. Essays on the history of mechanics in memory of Clifford Ambrose Truesdell and Edoardo Benvenuto. Springer, Basel, pp 199–222
135. Maupertuis PL (1740) La loi du repos des corps. Mémoires de l'Académie Royale des Sciences de Paris 1:170–176
136. Maupertuis PL (1744) Accord de différentes loix de la nature qui avoient jusqu'ici paru incompatibles. Mémoires de l'Académie Royale des Sciences de Paris, pp 417–426
137. Maupertuis PL (1746) Les loix du mouvement et du repos, déduites d'un principe de métaphysique. Histoire de l'Académie Royale des Sciences et Belles Lettres de Berlin, pp 267–294
138. Mouy P (1938) Malebranche et Newton. Revue de Métaphysique et de Morale 45(3):411–435
139. Nagel E (1961) The structure of science. Harcourt, Brace & World, New York
140. Newton I (1726) Philosophia naturalis principia mathematica. 3rd edn. Innys, London
141. Newton I (1822) Philosophiae naturalis principia mathematica, auctore Isaaco Newtono. Le Seur T, Jacquier F (eds). Duncan
142. Noll W (2004) Five contribution to natural philosophy, Carnegie Mellon University, Pittsburgh, PA
143. Panza M (2002) Mathematization of the science of motion and the birth of analytical mechanics: a historiographical note. In: Cerrai C, Freguglia P, Pellegrini P (eds) The application of mathematics to the sciences of nature. Springer, Dordrecht, pp 253–272
144. Paty M (2001a) D'Alembert, la science newtonienne et l'héritage cartésien. Revue de la Philosohie 38(2)
145. Paty M (2001b) Les recherches actuelles sur d'Alembert a propos de l'édition de ses oeuvres complètes. In: Michel A, Paty M (eds) Analyse et dynamique. Etudes sur l'oeuvre de d'Alembert, Presses de l'Université Laval, Québec, pp 25–94
146. Pourciau B (2016) Instantaneous impulse and continuous force: the foundations of Newton's Principia. In: Iliffe E, Smith GE (eds) The Cambridge companion to Newton. Cambridge University Press, Cambridge, pp 93–186
147. Radelet de Grave P (1998) La moindre action comme lien entre la philosophie naturelle et la mécanique analytique: Continuités d'un questionnement. Llull 21(41):439–484
148. Radelet de Grave P (2002) Déplacements, vitesses et travaux virtuels. Ph.D. seminar, Department of Ingegneria delle strutture, University of Pisa
149. Radelet de Grave P, Speiser D (2004) Introduction to volume III/10 of Leonhardi Euleri Opera omnia, electricity, heat and magnetism. In: Leonhardi Euleri Opera omnia (in progress), III, vol 10, Teubneri GB; [then] Fussli Turici; [then] Birkhäuser, Basel, pp XIII–CXXIV
150. Rameau JP (1750) Démonstration du principe de l'harmonie, servant de base à tout l'art musical théorique & pratique. Durand and Pissot, Paris
151. Robinet A (1960) Le groupe malebranchiste introducteur du calcul infinitésimal en France. Revue d'Histoire des Sciences 13(4):287–308
152. Schmit C (2013) Sur l'origine du principe general de Jean Le Rond d'Alembert. Ann Sci 70(4):493–530
153. Scott WL (1970) The conflict between atomism and conservation theory, 1644–1860. Elsevier, New York
154. Speiser D (2008) Euler, the principle of relativity and the fundamentals of classical mechanics. In: Fuss P, Fuss N (eds) Discovering the principles of mechanics (1600–1800). Birkhäuser, Basel, pp 179–224
155. Speiser D (2010) Il fisico Leonhard Euler. Il Club dei Volterriani 10:31–56
156. Suisky D (2009) Euler as physicist. Springer, Berlin

157. Terrall M (2002) The man who flattened the world: Maupertuis and the science in the Enlightment. The University of Chicago Press, Chicago
158. Truesdell CA (1954) Editor's introduction: rational fluid mechanics, 1687–1765. In: Leonhardi Euleri Opera omnia (1911–, in progress), 2, vol 12–13, Teubneri GB; [then] Fussli Turici; [then] Birkhäuser, Basel
159. Truesdell CA (1960) The rational mechanics of flexible or elastic bodies. In: Leonhardi Euleri Opera omnia (1911–, in progress), 2, vol 1, part 2, Teubneri GB; [then] Fussli Turici; [then] Birkhäuser, Basel
160. Truesdell CA (1968) Essays in the history of mechanics. Springer, New York
161. Varignon P (1666–1699a) Application de la regle generale des mouvements à toute les hypotheses possible d'accelerations ordonnes dans la chute des corps. Mémoires de l'Académie Royale des Sciences de Paris Printed in 1730 10:354–360
162. Varignon P (1666–1699b) Regles du mouvement accelerez suivant toutes les proportions imaginables d'acceleration ordonnes. Mémoires de l'Académie Royale des Sciences de Paris Printed in 1730 10:339–343
163. Varignon P (1666–1699c) Regles du mouvement en general. Mémoires de l'Académie Royale des Sciences de Paris Printed in 1730 10:225–233
164. Varignon P (1687) Projet d'une nouvelle mechanique avec un examen de l'opinion de M. Borelli, sur les propriétez des poids suspendus par des cordes. Martin & als, Paris
165. Varignon P (1700a) Des forces centrales ou des pesanteurs necessaires aux planetes pour leur faire decrire les orbes qu'on leur a supposez jusq'ici. Mémoires de l'Académie Royale des Sciences de Paris, pp 224–243
166. Varignon P (1700b) Maniere generale de determiner les forces, les vitesses, les espaces et les temps, une seule de ces quatre chose étant donné dans toutes sortes de mouvements rectilignes variés à discretions. Mémoires de l'Académie Royale des Sciences de Paris, pp 22–27
167. Varignon P (1725) La nouvelle mécanique ou statique (2 vols). Joubert, Paris
168. Wilson C (1987) D'Alembert versus Euler on the precession of the equinoxes and the mechanics of rigid bodies. Arch Hist Exact Sci 37(3):233–273

Chapter 4
Physics in General

Abstract D'Alembert counted as *physics in general* disciplines such optics, acoustics, positional astronomy, cosmology, magnetism and electricity. For the sake of space, this chapter deals with optics and electricity only. A good deal of the optical works, concerned the theories of propagation of light, with those of undulatory character that required complex mathematical treatments and the use of partial differential equations, becoming a fertile ground for mathematical physics. The experience with which the theories were compared was mostly based on experiments conducted in the 17th century by Newton and Huygens. Relevant new experimental work, on a quantitative basis, was carried out only relatively to what is today known as photometry with Bouguer and Lambert. The creation of the 18th century was the science of electricity. It assumes in the chapter the paradigmatic role of the development of the experimental sciences starting from the ascertainment of the phenomena at a qualitative level—remaining partially in the footsteps of the traditional natural philosophy—up to their quantification. The number and quality of experiments on electricity grew dramatically, especially after the 1750s when the discovery of the Leyden jar made it possible to accumulate large charges. After a brief mention to the situation in the 17th century, the chapter passes to the examination of the English experimenters and the continental ones to stop before Alessandro Volta's studies at the end of the century.

4.1 Theories of Light

The studies of optics in the 18th century related to those of the previous century with an important difference; there was little or no systematic experimental activity of the properties of light such as those carried out for example by Newton, Huygens and the followers of Descartes. The focus was on the theoretical studies of nature and propagation of light. Experimental activity was however not completely lacking but was focused on photometry studies.

As already commented in the previous sections, traditional optics was a complex conglomerate of arguments of mixed mathematics and natural philosophy, these latter

D. Capecchi, *Epistemology and Natural Philosophy in the 18th Century*, History of Mechanism and Machine Science 39,
https://doi.org/10.1007/978-3-030-52852-2_4

concerning the nature of light and the physiology of the eye. They derived from the Greek traditions, such as for example Aristotelian, Platonic, Epicurean and were generally devoid of quantitative considerations. In the 17th century and to a greater extent in the 18th century, these arguments were critically re-examined, especially by scholars with training in mathematics and experimental philosophy. Many of them did not have preconceived ideas, based on a priori principles of metaphysics or philosophy of nature (apart from the acceptance of mechanicism), or at least this was the image they intended to convey. The objective was to analyze the possible mechanisms of light transmission, defined by geometric models, on which the laws of mechanics could be applied and make calculations or develop reasonings to obtain results to compare with experiments for a feedback on the model.

From the point of view of the mechanical philosophy, dominant in the 18th century, the possible alternatives could be classified into two different types. This is what the French mathematician, physicist and astronomer Jean Jaques de Mairan (1678–1771) wrote on the purpose:

> All modern systems of light can be reduced to two. For, either the motions of the luminous body are transmitted to the eye only because they are communicated to the matter which is between the luminous body and us, as the vibrations of a sound body reach the tympanum of the ear, only because they excited in the air a similar movement; or the agitation of the luminous body produces in it an emission and a flow of corpuscles, which strike the organ of the eye in the same manner as the invisible parts which detach themselves from a flower. There is no alternative, either the luminous body sends to us particles of its substance, or it does not. It is necessary that light propagates according to one of these two ways, otherwise one should resort to occult qualities [147].[1] (D.1)

Actually the dichotomy suggested by Mairan, even if at first sight it seems well founded, does not allow to classify all the numerous approaches to the study of the propagation of light of the 18th century as the two categories sometimes overlap. The dichotomy is identified by various labels. The most common is particle (or corpuscle) and wave as in [1], but there are other such as Newtonian and Cartesian as in [80], emission and medium as in [116] and projectile and vibration as in [47]. In the following, instead to invent one my own I will adopt the labels suggested in [47]: projectile and vibration, because in my opinion it gives a more immediate idea. Vibration has been preferred to wave because that of wave has a modern taste which not always fit with the conceptions of the 18th century.

Even though today the dichotomy of the theories of light is considered a gross oversimplification, it was not considered such in the 18th century as: "to the supporters of one theory the other theory had faults so fundamental that no distinction between varieties of the same theory was sufficient to placate opposition to that theory. This meant that opponents of either the wave or the particulate theory seldom, in their attacks, distinguished between different varieties of either theory" [131].[2]

Table 4.1 lists the scholars who, for their own admission, supported of the different theories about light transmission. The supporters of vibration theories admitted that

[1] pp. 2–3.

[2] p. 47.

Table 4.1 Different theories of transmission of light and their supporters

	Greek	
	Effect from the object	*Effect from the eye*
Active medium	Propagation of the effect by the medium (Aristotle)	Ray with medium (Stoics)
No active medium	Emission of matter Atomists	Ray without medium (Euclid, Ptolemy)
	Projectile tradition	*Vibration tradition*
	17th century	
	Maignan	Descartes
	Newton	Hobbes Huygens
	18th century	
	's Gravesande	Johann Bernoulli II
	Musschenbroek	Euler
	Clairaut	Lambert

light was faster in less dense medium, the supporters of the projectile theory, to the contrary assumed that light was faster in the more dense medium. These assumptions did not derive from direct experimental measurements of light speed, what at the time was not possible, but where simple *ad hoc* hypotheses to justify some experimental results, for instance that passing from air to glass the refraction angle is less than the incidence angle. To find mechanical reasons to explain one theory or another was more a question or rhetoric than logic. Among the supporter of the projectile theory to signal the exception of Emmaunel Maignan (1601–1676) who as early as the 1648 in his *Perspectiva horaria sive de horographia gnomonica tum theoretica, tum practica*, assumed that light was faster in air than in glass, but simply because his mechanism of refraction derived from Hobbes's vibrational one [146].[3] For a discussion of Maignan's theory of light see [181].

In [47] it is suggested a third class of theories, that is somehow alien to mechanical philosophy: *fluid theories*. Fluid theorists claimed that light was a substance, usually consisting of small particles of matter, like the projectile theorists. However, they frequently denied most of the concepts related to the projectile theory. Central to their conception was some form of ubiquitous aetherial fluid to explain various phenomena, in particular heat and fire. Few fluid theorists were concerned with explaining the standard optical phenomena, most were concerned primarily with cosmological questions about the role of fire and light in the economy of nature. Thus instead of studying optics in a mechanist environment, the fluid theorists adopted a different style and approached theories of light via chemistry and theology.

[3]pp. 631–632.

In the 19th century fluid theories had become irrelevant in the study of optics but in 18th century they still had some appeal. One of the most important supporter of the fluid theory was the Dutch physician and chemist Herman Boherhaave (1668–1738) who formulated a theory of the nature of fire which became very influential; for instance he influenced's Gravesande and Musschenbroek. According to Boerhaave, the gross matter was inactive, lacking also the property of inertia. By contrast he attributed all activity and motion in the universe to *fire* which is described as the universal cause of all changes in nature. This is an active substance composed of very small smooth, round, impenetrable and indivisible particles; moreover, it was conserved and could not be created, destroyed or transmuted into particles of gross matter. Under normal conditions fire was uniformly distributed throughout the universe but some processes could increase its density in a limited region. Elementary fire was not directly observable but became apparent only through its effects; principal among these were heat and light. Thus light was a particular manifestation of fire and consisted of rectilinearly moving fire particles, whereas heat was produced by an excess of these particles [47].[4]

Another supporter of the fluid theories that spanned his activity through the whole 18th century was James Hutton (1726–1797). His role will be discussed below with some deepening, also because of his interesting epistemology of science. Hutton, according to the modern categories was a scientists; not a physicist but rather a natural scientist. According to the categories of the period he was a supporter of experimental philosophy and physico-mathematica, in the sense he used the strict way of reasoning of mathematicians in physical topics. Hutton had not a great mathematical background and made no substantial use of analysis and geometry in his researches. Here is how the web encyclopedia Mathematics Mactutor classifies him: "James Hutton was a chemist but is best known as a geologist. Why then include him in this archive? There are two reasons. First his remarkable theory of the age of the Earth was inspired by Newton's world view as presented in the teaching of Colin Maclaurin, and second that one of his main collaborators in his geological research was John Playfair (1747–1819) [a first rank mathematician]" [129]. Hutton is famous for his theory of uniformitarianism, that is the assumption that earth evolved from its original state and natural processes over geologic time explain the present features of the earth's crust. Hutton established geology as a science, and as a result he is usually considered the father of modern geology. He was born in a rather wealthy family and in 1740 entered the university of Edinburgh, where he was taught mathematics by Colin Maclaurin, in addition to logic and metaphysics by John Stevenson. He graduated in the spring of 1743, only seventeen. In 1747, Hutton went to Paris to perfect his medical studies, undertaken at Edinburgh, which he completed in Leiden later. However he never really intended to practice medicine. He moved to London after graduating and visited a friend of his, a certain James Davie, in Edinburgh to do a chemical works. When in London his interest in geology evolved, and in particular the *Discourse on earthquakes* by Hooke, the *New theory of the earth* by William Whiston, the *Protogaea* by Leibniz and the *Histoire naturelle* by Buffon, as well

[4] p. 95.

as Niels Steensen's treatise *Dissertation concerning a solid body enclosed by the process of nature within a solid*, caught his attention. He moved to Slighhouses near his father's farm in 1754, where he began to cultivate and work on his theories of geology. In 1767, he returned to Edinburgh and founded a laboratory to elaborate his theories on the history of the earth. When he realized that the soil was caused by rock erosion, Hutton also understood that there is another mechanism that creates rocks beneath the surface, which then form a new earth. Because he knew that this process was extremely slow, Hutton concluded that the earth must have been very old.

The Royal society of Edinburgh invited Hutton to give two lectures on his theory. The first conference took place in March 1785 and since Hutton was ill, was his friend Joseph Black to speak of the earth system. The second conference was delivered by Hutton himself a month later. Of course, many saw his conclusions as an attack on the Christian church and some strongly opposed his views. Hutton's reaction was what one would expect from an exceptional scientist: he made trips to see the rock formations to try to get more evidence that his theory was correct. Because Hutton's writing style was difficult, John Playfair rewrote *The theory of the earth*, Hutton's most known treatise. Playfair managed to explain Huttons' ideas in a simpler and clearer way and provided facts to support and arguments against in the *Illustrations of the Huttonian theory of earth* published in 1802 [168].

In his late treatise *Dissertation upon the philosophy of light, heat, and fire* of 1794, where a fluid theory of light is explored, Hutton introduced the difference between the man of science and the philosopher and exposed interesting epistemological considerations. Notice the introduction of locution *man of science* which predated the introduction of the term *scientist* usually attributed to William Whewell (1794–1866) in 1833. In the premise of his work he stressed the relevance for a man of science of the general considerations about the method of enquire. In his words: "The men of science may perhaps conceive that I have written the more general parts merely as introductory, or as subservient to those in which I give experiment and explanation of appearances". But, continued Hutton, "My view is very different; I wish to engage men of science, that is those who have sufficient knowledge of the particular branches of science, to employ their acquired talents in promoting general science, or the knowledge of that great system, where ends and means are wisely adjusted in the constitution of the material universe and I only, or chiefly, give those explanations of phenomena as an illustration of the doctrine, which upon all occasions I have held, with regard to the pursuit of science and philosophy" [128].[5]

According to Hutton a man of science and a philosopher are different things, although they both proceed in reasoning scientifically, after having distinguished their ideas. The purpose of the one, although occasionally extensive, is naturally limited, in having a subject which is particular; the purpose of the other again, though not perhaps unlimited, is universal and employs that general knowledge which science

[5] pp. IV–V.

had provided, in order to procure something still more interesting to the person who thus is made for a superior enjoyment [128].[6] A man of science should be a philosopher also.

The dominating inductivist approach of the experimental philosophy of the first half of the 18th century had already given up and a substantial role was given to the elaboration of theories. According to Hutton it is only general knowledge which is proper to direct the making of experiments. Thus, though natural philosophy requires to have matters of fact, yet, the indefinite collection of facts does not conduct to philosophy; and the properly disposing of one fact in the generalization of our knowledge, is more valuable than the indefinite progress of experiment made without a wise design [128].[7]

Hutton proposed a quite sophisticated version of the method for hypothesis. For him facts are fundamentals to validate a theory; but if a single fact contradict it, before leaving, an in depth analysis should be carried out. Indeed, asked Hutton, are we, upon the faith of a solitary experiment, to abandon our theory of heat and cold, which is the generalization of a broad experience? Are we to build new theory upon a single fact, before this fact is properly analyzed, in seeing every circumstance with which it is connected, and before comparing that analyzed fact with every other event which it should agree? [128].[8]

However Hutton could also be seen as an inductivist: "In the whole of Hutton's doctrine he rigorous guarded himself against principle which could not be founded on observation. He made no assumptions. Every step in his deductions was based upon actual fact, and the facts were so arranged as to yield naturally and inevitably the conclusion which he drew from them [102].[9]

Hutton was a supporter of the theory of phlogiston—assimilated by Hutton with fixed light—which postulates that a fire-like element called phlogiston is contained within combustible bodies and released during combustion. For him light and heat derive from phlogiston degradation and phlogiston dubbed the *solar substance*: "It is the light of the sun which is stored up in the substance of vegetable bodies, as fixed light, or phlogiston" [8, 128].[10] Light and heat are not the same but light and radiating heat cannot be distinguished: "What dilates the thermometer, is surely heat; and, when that substance quits the contracting thermometer, in order to dilate the contiguous atmosphere, it is also heat, because it then observes the laws of heat, and produces its effects. But, when a substance is irradiated from a warm body, when it traverses with the velocity of light a transparent medium is reflected by one metallic speculum to be again concentrated by another; and when it then heats a black body more easily than a bright one, we find a substance acting the part of light, in all

[6] pp. VII–VIII.

[7] pp. VII–XII.

[8] p. 4.

[9] p. 315.

[10] p. 62, p. 323.

respects, and not that of any species of heat; therefore, while this substance must be properly named light, it cannot, without absurdity, without committing a flagrant transgression in our science to be named heat" [128].[11]

In the following a few theories, vibration and projectile, are discussed to show the way mathematicians addressed this subject of natural philosophy. A greater space is devoted to vibration theories for two reasons. First the projectile theory has largely been discussed in Chap. 1 in the form given to it by Newton; second vibration theories are more interesting for the present book, because in it the role of mathematicians was more relevant.

4.1.1 Projectile Theories

Newton is considered the champion of projectile theory, though most historians attribute him a weak conception of it. He indeed left open some other ways of transmission of light, even those of not mechanicistic mould. He was however read by his contemporaries as a strong sustainer of the projectile theory. And a modern by reading his writing, without the warning of historians, is tempted to see Newton as clearly sided toward projectile theory.

After the various editions of the *Opticks*, before the half of the 18th century, in England the projectile theory became shaped into a system [47].[12] In the Continent it found supporter in the Netherlands by 's Gravesande and Musschenbroek, who however made some changes, influenced by Boherhaave's concept of light as projected fire [116].[13] In France the projectile theory did not find place at the beginning, because of the dominant Cartesian theory of light. Things changed after Voltaire, Clairaut and Gaspard Le Compasseur de Créqui-Montfort, marquis de Courtivron spread Newton's ideas. Important was the role of Courtivron who in 1752 published a Newtonian treatise of opticks, the *Taité de optique* [142], which had a great diffusion.

However among the Cartesians, or if one prefers the Neo-Cartesians, there was an important scholar, the already cited Jean Jacques Dortous de Mairan (1678–1771), who was president of the l'Académie des sciences de Paris for some period and before Voltaire and other Newtonians supported the projectile theory. As Descartes he imagined the universe filled with elements of the second type according to Cartesian classification. But in this plenum, differently from Descartes, he assumed that light was carried by small corpuscles and that such a transmission occurred in a finite time.

Mairan afforded the theme of light transmission in a early paper of 1717, the *Dissertation sur la cause de la lumiere des phosphores et des noctiluques* [147], which won prize of the Académie de Bordeaux; a dissertation on what is now called luminescence. Though Mairan had a good mathematical background, he was part of

[11] pp. 55–56.

[12] p. 25.

[13] p. 42.

the Malebranchian circle, Calculus and geometry do not transpire from his paper, which could be classified as an experimental philosophy work. Here empiric data are relevant and, at least in principle, they are used to accept or refute the various theories of light. But differently from most experimental philosophers, he did not follow an inductive approach and, according to his Cartesian education, made extensive use of mechanical models, that were considered consistent not only against the experimental data but also against a pre-established system of natural philosophy.

In his dissertation Mairan first contrasted the vibration view, with a thesis very common among the supporters of projectile theories, that it could not explain the rectilinear propagation of light. Then he passed to present his own theory, according to which light is nothing but a "successive flux" of some subtle matter caused by the agitation in the luminous bodies, such as the sun for instance. The subtle matter of light would propagate in the aether with a very high speed; it covers the distance from sun in seven minutes [147].[14] According to Mairan there is no difficulty to admit that the aether has a very very low density so it does not oppose any substantial resistance to the motion of the corpuscles of light. With regard to the objection commonly addressed against the supporter of projectile optics, according to which the sun, the classic luminous body, if it were to actually issue corpuscles, after so many centuries would be exhausted, Mairan proposed two solutions, which could coexist; (1) either light is formed by corpuscles of negligible mass, though much larger than the particles of aether, that not even in many centuries could be perceived the consumption, (2) or light coming from the other stars supplies the consumption.

To luminous corpuscles Mairan attributes not only motion and extension, but also a quality, entering into the world of chemistry. The entrance of quality in his mechanist vision is that of Robert Boyle. All the bodies of the world differ among them only by the figure and motion of the invisible parts which compose them (first principles), or in other words, by the different size and by the various arrangements of parts. This idea so simple in appearance, however, are the source of a prodigious variety. But, said Mairan, the industry of men, in the search for the intrinsic nature of body, is unable to reach these last parts, owing to their smallness, thus "They [men] have given the name of (second) principles of the bodies to the simplest parts, in which the art can reduce them. These (second) principles are the so-called chemicals. They are small masses of matter, corpuscles whose insensitive parts have received certain configurations, or some motions and which are so closely related to each other, or of such great subtlety, that the analysis has not been able to separate them so far. It is one of those principles that is the matter of the light. This matter must be by all the same and that of a luminous body differs from that of another luminous body but for the assemblages of the different corpuscles" [147].[15] Mairan named *sulfur* the principle of light.

[14]p. 11.
[15]pp. 21–22.

So I think the *luminous material consists of a very subtle, very delicate sulfur* [emphasis added] and that it is nothing but the active ingredient of Chemists, so named, because it acts alone and makes others act. Indeed of the five principles of Chemistry, there is only the sulfur, which has this property: among the other four, one counts as purely passive, the Earth, which only serves as a receptacle or to the others, & three means, which are Salt, Water & Mercury, which become able to act when are joined to the sulfur [147].[16] (D.2)

Mairan insisted that it was necessary to distinguish the common sulphur, or the fat, oily, sulfurous matters, from the sulfur which is in question here; for whatever they abound in this principle, nevertheless, as they can be decomposed, reduced by analysis into simpler components, they could only be classified as mixed. He maintained, with Newton, that colors are not modification of light, but that the light contains in itself all the colors, independently of the internal or external configurations of the bodies through which it passes, or on which it is reflected. That is, a ray of the sun, for instance, is composed of particles of different species, each of them with the property of exciting in the soul, by means of the organ of sight, the particular feeling of a color, without any reflection or refraction ever changing it.

White is not a color, but a compound of all colors and black, on the contrary, is the negation of all colors, or light.

Each color of bodies does not consist in the figure, and in the particular arrangement of the parts which compose them if not that they are better adapted to absorb in their pores the light of a certain color and to reflect that of another color. Thus the carmine, for example, is very red, because it reflects only the red light, and all the other species of light break up and lose themselves in its pores without reflecting [147].[17] (D.3)

Finally, each species of light has its determined index of refraction, that is, that each color passing obliquely from one medium to another, air, for example, in the crystal, breaks up to meet with a its characteristic angle. "This is what Mr. Newton, author of this discovery, calls the different refrangibility of the colors of the light. It is principally by this property that he recognized all the other properties; and his clever experiments, would alone be sufficient to immortalize a name less famous than his" [147].[18]

Mairan returned to the subject of light, in four lengthy memoirs with the main purpose to give a Neo-Cartesian derivation of the laws of refraction (and reflection as well) in 1722, 1723, 1738 and 1740 [148, 149, 151, 152, 167]. In these memoirs he gave more prominence to the vibrational theory of light than he had done in the past. He repeated that reflection and refraction were equally explained by both vibrational and projectile theories: "I do not want here to exclude any physical system about light unless it goes outside mechanic philosophy or is manifestly contrary to experiments" [151].[19]

In these memoirs the chemical considerations about the particles of light are no longer resumed and a fully mechanical approach is followed. The difference in color

---

[16] p. 22.

[17] pp. 47–48.

[18] p. 48.

[19] p. 4.

is explained with the different speed of the particles of light, and the refraction is made to depend on the speed of light. Because white light is a mixing of particles with different speed, they are separated during refraction, and Newton experimental findings can be justified.

> Corol. 39. Thus the different velocities, the different degrees of refrangibility and the different colors of light are in themselves and independent of us, one and the same property which only expresses a graduation of the effects due to one single cause [151].[20] (D.4)

Of some interest is also Mairan memoirs on sound, *Sur la propagation du son dans les differens tons qui le modifient* of 1637 [150], because here the transmission of sound is compared with that of light and some considerations on the possibility that light transmits by vibrations are reported. Here he assumed that air, as a vehicle of sound, is an assemblage of an infinity of particles of different elasticity, the vibrations of which are analogous in duration to those of the different tones of sound bodies. Moreover between all these particles only those of the same species, of the same duration of vibration and in unison with the sound body, can retain the vibrations of this body and transmit them to the ear. "That the smallest mass of sensible air contains several of these particles of all kinds, and that all their vibrations, or the trembling of the mass in all its parts, can only produce sound or noise" [150].[21]

4.1.2 Vibration Theories

For what the vibration theories is concerned, Descartes was considered by nearly all the scholars of the period, the starting point even though no vibrations exist in his system. More complex is the evaluation of the role attributed to Huygens, already discussed in Chap. 1, who was instead a true former supporter of a vibrational optics. For a discussion on the acceptance of the wave theories before and after Newton see [12].

A scholar which was cited often about light transmission in the treatises of the 18th century is the 'Cartesian' Nicolas Malebranche (1638–1715). He not only contributed, at the foundation level with metaphysical considerations [117], but also with technical contributions [143]. One of the most well-known results of his scientific activity relate to the mechanism of light transmission; his considerations are mainly reported in *Reflexion sur la lumiere et les couleurs, et la generation du feu* of 1699 [154] and in the *Eclaircissements* of *De la recherche de la vérité* of 1712 version [153]. This last writing included the paper of 1699, published in the Mémoires de l'Académie des science de Paris, with few but important changes, that will be commented later, and some details about the property of the medium that transmits light.

[20] p. 27.
[21] p. 3.

In the paper of 1699 Malebranche expressed his criticisms to Descartes's theory of light, because for instance it made impossible for rays of light of different colors to cross without interacting and losing their color. Indeed for Descartes color was associated to a certain value of the spin (angular moment, rotatory speed) of the hard spheres of subtle matter (aether) and, when the rays meet, for their colors do not change it is necessary that the spheres had two distinct spins, which is impossible. For this reason he removed the assumption of the perfect rigidity of the assembly of aetherial particles. One more reason is that the Descartes's theory implied an infinite speed of light, while recent measurement indicated that it was finite. However in this point Malebranche remained ambiguous and, though he seemed to accept the finite value of the speed of light, he left some space to the possibility it was infinite.

Malebranche's aether is a unique primary substance that, forced to move at high speed in a closed and plenum world, is obliged to whirl in vortices whose dimensions can decrease without limit. Matter, considered essentially as a continuum, can be divided with no effort into particles as small as you like: "matter can be divided *à l'infini* and each part makes no resistance at all to be divided" [153].[22] In this divisible matter the vortices of tiny particles are characterized by a very hight speed. The centrifugal force then requires that these small vortices are capable of releasing a "fearful" force upon breaking up. Thus Malebranche's aether is elastic; however its elasticity does not come from the elasticity of the particles in themselves but is associated to the dynamics of rigid particles. Under normal conditions the forces or pressures exchanged among the vortices are balanced: "the vortices counterbalance each other" [153].[23] The balance is disturbed by the action of the luminous bodies which determine pulses of pressure in the elastic medium.

In the work of 1699, however, Malebranche did not insist on the discussion of the nature of the aether and took as an example of propagation of pulses of pressure what would happen in a spherical vessel filled with a fluid compressible but very stiff.

Suppose, said Malebranche, to make in the balloon of Fig. 4.1 filled with a compressible and compressed fluid, a small hole as in A. All parts of the fluid, like those, for example, which are in R, S, T, will tend towards A, by the straight lines RA, SA, etc., because all the parts which were equally pressed, ceasing to be such on the side opposite to the hole A must tend to A since all pressed body must tend to move toward the side where they find less resistance. Equally, if a piston operates at the opening A and pushes the fluid promptly forward, the same parts R, S, T, V, etc., will all tend to get away from the hole by the same straight lines AR, AS, etc., because they are more pressed by the side that respond to A directly than any other. Finally, if the piston moves forward and backwards promptly, all the parts of the fluid matter that fills the ballon will receive an infinite number of pulses that can be called *vibrations of pressure*.

Let's apply this reasoning to light and colors, said Malebranche. Everything around us is full and our eyes are currently compressed. But this compression of

---

[22] Vol. 2, p. 495.

[23] Vol. 2, p. 496.

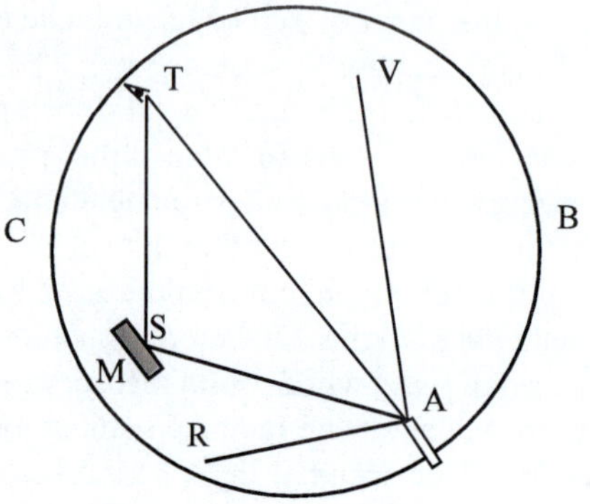

Fig. 4.1 A ballon filled with an elastic fluid. Redrawn from [154], p. 23

the optic nerve does not excite any sensation of color, because this nerve is always in the same state; by the same reason that we do not feel the weight of the air that surrounds us, though as heavy as 28 in. (∼760 cm) of mercury. But imagine a T-shaped eye, or everything else turned towards a flame A of a torch (Fig. 4.1), the parts of the flame being in a continual motion, will press the subtle matter harder than in the darkness by vibrating very promptly pressing the bottom of the eye; thus the optic nerve more compressed than usual and shaken by the vibrations, will excite in the soul a sensation of brightness. If a black body M is assumed at S, the subtle matter not being reflected to the eye and not shaking at all the optic nerve, it will be seen as black. If the body M is such that the subtle matter is reflected from this body to the eye and produce vibrations equally prompt, this body will appear white. It will even appear very bright as the flame of the torch, if the body M is polished and the rays are reflected mostly in the same order, because the brightness comes from the force of the vibrations, and the color of their promptitude.

> But if the body M is such that the subtle reflected matter excites in the eye more or less rapid vibrations in certain degrees, that I do not think we can determine exactly, we will have some of the colors we call primitive; yellow, red, blue, if any part of the body M reduces equally the vibrations caused by the flame in the subtle matter. And we will see all the others colors that are made by the mixture of the primitive ones when the parts of the body M will decrease unequally the promptness of the vibrations of the light [154].[24] (D.5)

That is what I meant when I went ahead in some of my books [*Recherches de la verité* of 1674–1675 and the *Entretiens sur la métaphysique et sur la religion* of 1688], said Malebranche, that light and colors consist nothing but that in various shakes or vibrations of aethereal matter, or vibrations of pressure more or less prompt *[that is with low or high frequency]*, that the subtle matter produces on the retina [154].[25]

In the previous passage there is a summary of Malebranche's wave theory of light. He tended to justify his theory by making analogies between the propagation of light and that of sound; a phenomenon this last that at the time was quite well studied, even if some problems remained to be solved. An analogy which was the inverse of

[24] p. 24.

[25] pp. 24–25.

that made by Mersenne, who in the *Harmonie universelle* studied the properties of sound starting from those of light [159].[26] Essentially Malebranche believed he had proved that the different colors consist only of the different frequencies of vibration of the pressure that propagates in the aether, in analogy to the different musical tones that derive from the different frequency of vibration in the air [154].[27]

In the *Reflexion sur la lumiere et le couleurs, et la generation du feu* as reported in the *Recherche de la verité* of 1712, Malebranche made two important changes in the text; one concerning the theory, another concerning the speed of propagation of light. In 1699, Malebranche did not know Newton's color theory and believed that color was a modification of a uniform light caused by interaction with colored bodies. In 1712, after reading Newton's *Optics*, Malebranche, expressed the following, apparently severe judgment: "Although Mr. Newton is not a physicist, his book is very curious and very useful to those who have good principles in physics. Besides, he is an excellent geometer. Everything I think about the properties of light can be adjusted to his experiment" [155].[28]

Malebranche agreed with Newton's phenomenological approach and accepted Newton's conclusions and admitted that the colors are contained in the light from the beginning and derive from the absorption of a part of the colors contained in the white light that hits the colored body under observation. And the red, the orange, the yellow, the green, the blue, the indigo or dark blue and the violet, do not change their color or the amplitude of their vibrations, as Descartes said, and that their refractions always have the same relation with each other, "which Mr. Newton has proved by several decisive experiences" [153].[29]

> But if the body M is such that the subtle reflected matter excites in the eye more or less rapid vibrations to certain degrees, that I do not think we can determine exactly, we will have some of the simple homogeneous or primitive colors, like red, yellow, bleu, etc., and we will have the other compounded colors, and even the whiteness that is the more composed of all, according to various mixtures of rays whose vibrations will have various promptitudes [frequencies]. I say that whiteness is the most composed of all because it is composed by the assembly of vibrations of different promptness, that each small part of the flame produces in the subtle matter. Since everything is full and infinitely compressed, each ray retains in all its length the same promptitude of vibration that has the small part of the flame that produces it. And because the parts of the flame have a movement varies, the rays of colors necessarily have vibrations that make different refractions. But we should see on that the experiences that will be found in the excellent work of Newton [153].[30] (D.6)

According to Malebranche, one must not judge the cause by the sensation it has produced. It is easy to conceive of two or more unequal vibrations, pressing together on the same fiber of the optic nerve, which can shake the main part of the brain in the same manner as a mean vibrations. Any white ray is a mean of all simple rays, red, yellow, blue, etc. which all have different vibrations and refractions; and all the

[26]Vol. 1, Livre premiere des movemens, p. 14.

[27]p. 12.

[28]p. 771–772.

[29]Vol. 2, p. 525.

[30]Vol. 2, pp. 483–484.

different colors come only from the various mixtures of simple rays, or transmitted or reflected from the small transparent parts of opaque bodies.

Regarding the speed of light, Malebranche questioned the results reported by Huygens—that witnessed a finite speed of light—basing on alternative results obtained by Giovanni Domenico Cassini (1625–1712); "because Mr. Cassini has observed eclipses of the satellites of Jupiter in different distances from the earth, which do not agree with the conclusion of M. Huygens" [153],[31] and this seems to reinforce his doubt on the possibility that the speed of light could be finite, even if there is no precise judgment on the point.

To decide about the originality and relevance of Malebranche's contribution to the theory of light propagation is not simple. He was considered by his contemporaries as a forerunner for the hypotheses that light propagates as periodic pulses and each color is associate to a particular frequency of vibration. Two were the assumptions that allowed this result: (1) The aether is an elastic medium; (2) Light propagates likewise sound. The first assumption is the most interesting and original one because Malebranche justified it inside Cartesian physics, with a space filled with rigid and not elastic matter. The second assumption is not new; there has always been association between sound and light. Malebranche was in an advantageous position to start from a theory of sound well defined in which the periodic nature of sound waves was generally accepted.

From the point of view of the present book, Malebranche's contribution is relevant because it shows the way a scholar strongly oriented toward metaphysics, but who also had a good mathematical background, afforded topics of natural philosophy. Differently from scholars more mathematically oriented, as Huygens and Euler for instance, Malebranche inserted the explanation of the nature of light in a coherent system, by integrating it with his metaphysical conceptions.

An important contribution to the study of nature and light propagation is that of Johann II Bernoulli (1710–1790), the youngest son of Johann I. His contribution, not much cited by modern historians, is particularly interesting because it presents the right balance, for the times, among the considerations of natural philosophy of a Cartesian qualitative nature and their quantification using refined instruments as Newtonian equations of motion and mathematics.

Bernoulli's work commented here is the *Recherches physiques et géométriques sur la question: Comment se fait la propagation de la lumière* of 1736. Here Bernoulli referred to the astronomical measurements reporting the value provided by Rømer according to which the light would take 11 min to reach the earth, and that provided by Newton, according to which it would be sufficient 7–8 min. Thus, even though the speed of light is not infinite; it is however very high and Bernoulli tried to convince the reader that this is not so unlikely. And for this he referred to the second 'Newtonian' law of motion written in a quite modern notation as $f = ma$, in which "f is the motive force, m the mass and a the accelerating force" [35].[32] For the same value of f the acceleration, which is considered to be representative of the variation of speed,

[31] Vol. 2, p. 486.

[32] p. 6.

or of the speed tout court, is the greater the smaller the mass. And therefore for a very small mass and for sufficiently large forces such as that of the aether particles one will have a very great speed. It it is noteworthy that Bernoulli's text is contemporary (same year) to the *Mechanica* of Euler (a friend of his), in which the second law of motion is expressed, using a modern notation, as $dv = mf\,dt$, an expression that in the 18th century was preferred to $f = ma$.

The 'Cartesian' Bernoulli challenged the Newtonians, by asserting that "if it is allowed to Messrs the Newtonians to suppose an universal gravitation among bodies, even though no physical explanation is furnished for this, with a greater reason it could be allowed to imagine a pressure that act on the subtle matter which fills the vast spaces of the world" [35].[33]

At this point Bernoulli introduced the first important hypothesis of his theory: the aether is endowed with elasticity. And unlike Newton and Huygens which according to him had introduced the elasticity of the aether without any explanation [35],[34] he will give a physical explanation of it. About Newton Bernoulli probably referred to the considerations he made in the Queries of *Optics*; while the accusation to Huygens seems to be improper because he gave a physical justification, albeit a summary one, of the elasticity of the aether; see Sect. 1.1.3.1.

Bernoulli's explanation is that of Malebranche, who is explicitly named: "I do not find anything more appropriate for my project than the small vortices of P. Malebranche" [35].[35] At this point Bernoulli can present his theory on light propagation:

> I imagine that all this cluster of small whirlpools that fill the vast spaces of the world is mingled with very subtle, hard or solid corpuscles, leaving among them spaces, if you like, a thousand times longer than the diameter of one of these corpuscles, I do not need to determine the length of them, it is sufficient that I understand very clearly that each straight line drawn from one point to another, will contain an infinity of these small corpuscles, of which I can suppose the intervals nearly equal, since these corpuscles are uniformly dispersed among the small whirlwinds, though the corpuscles themselves may be of different magnitudes [35].[36] (D.7)

That is, the medium through which light propagates is made of infinitely many aetherial elastic whirlpools mingled with infinitely many hard particles of luminous matter.

According to Bernoulli, each point of a source of light can excite, in the universe conceived by him, an infinity of rays of different direction, as many as there are straight lines coming out of this point. Each of these straight lines, filled with small corpuscles (and vortices) distanced of many units from each other, receives from the (point-wise) luminous source a violent shock which compresses the closest vortices displacing the corpuscles of luminous matter from their positions of equilibrium, by producing vibrations along each straight line.

[33]p. 6.

[34]p. 8.

[35]p. 9.

[36]p. 10.

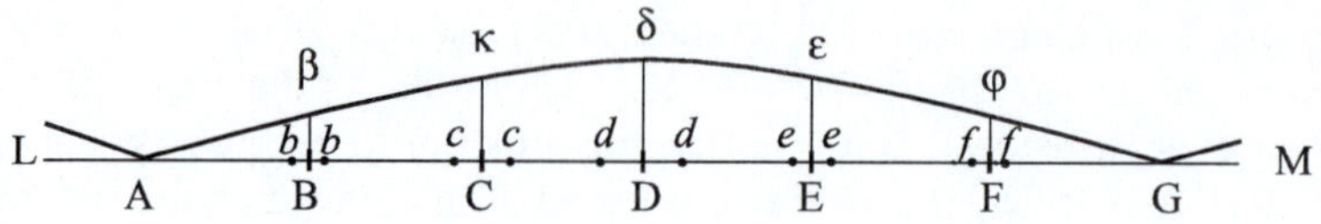

Fig. 4.2 A luminous filament. Redrawn from [35], figure III of the table of figures

After the exposition of the basic aspects of his optics, Bernoulli introduced a key concept, that of *luminous fiber* (fibre luminous) as the segment that connect the beginning of the perturbation of the aether to its end along a given line [35];[37] a *luminous ray* is defined as a chain composed by a great number of luminous fibers aligned along a straight line, at least when light propagates in a homogeneous medium [35].[38]

According to Bernoulli, the first hard luminous particle of the medium displaced by the impulse received by the luminous source (particle D of Fig. 4.2) causes a compression of the aether comprised between the first and second particle C (interval DC). This causes the displacement of the second particle C which compresses the space CB between the second and the third particle, and in sequence the third and the forth with an effect that tends to decreases with the distance from D, until the last particle which is moved, A. The length of the interval between the beginning of the perturbation (D) and its end (A) is the luminous fiber: indeed it is only one half, as Bernoulli will clarify later; the full length being AG [35].[39] It is natural for a modern reader to see in the length of the luminous fiber the wave length of light, even though with some perplexity.

Once that the interval DC (transformed in *cd*) between the first and second particle has reached the maximum compression, the particle D inverts its motion and displaces the particle E, then the particle F until in G the motion is exhausted. The cycle continues for a great number of oscillations very small but very speedy.

Once the nature of light propagation has been described in a general way, Bernoulli began to make quantitative considerations, using the then known (Newtonian) laws of dynamics. The first result obtained is that the oscillations of the luminous particles are synchronous. In other words, the period of vibration is always the same regardless of the amplitude of the oscillation. It is a question of proving the following general proposition, not new at the times:

> A body placed in a center of equilibrated forces, if it is displaced by whatever cause, up to a small interval in the direction of the two springs or motive forces, will return in its initial position, and will make vibrations in equal time in the form of tautochronous oscillations [35].[40] (D.8)

[37] p. 12.
[38] p. 12.
[39] pp. 11, 26.
[40] p. 17.

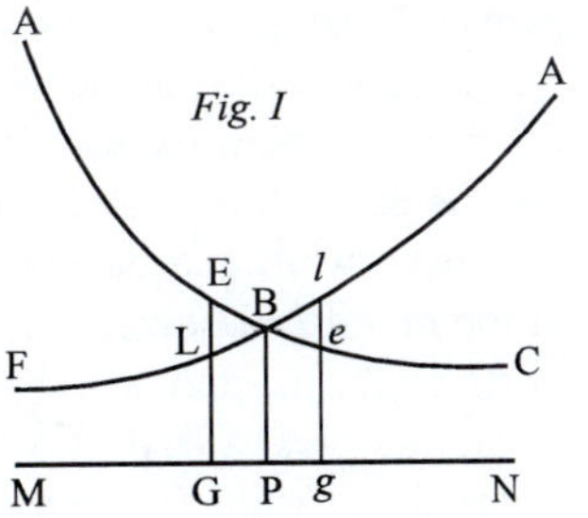

Fig. 4.3 Elastic forces. Redrawn from [35], figures I of the table of figures

The proposition is demonstrated by showing that when considering small oscillations the oscillating body is subjected to a force that varies linearly with the displacement, "according to the known properties of these forces, the corpuscle will do isochrones vibrations of equal or unequal excursions" [35].[41]

The luminous particles are of different sizes and masses, randomly distributed, so it should turn out that the luminous fibers are formed by different kind of particles. Actually, according to Bernoulli, this does not occur because a resonance phenomenon triggered, similar to those that occur between vibrating strings and each luminous fiber is made up only of particles of equal mass. In essence, the structure of a light fiber is determined by the first particle that is excited by the light source; it begins to vibrate and resonates only with the contiguous particles of the same mass. Particles of different mass are expelled and become part of other luminous fibers.[I]

This way to communicate motions in the bodies with the same disposition to the motion, is something very ordinary to nature, said Bernoulli; one sees for example that several strings of music, located close together each others, some of which are at the unison. If one of these is pinched it will urge all those that are strained in the same tone and will live all the others at rest, even the nearest, if it is not tuned to the octave or the fifth, which will also receive some small, sensible impression; but in general the strings that give high dissonant tones receive no impression [35].[42]

Bernoulli's explanation of the formation of luminous fibers, although ingenious, appears inconsistent, and not only if considered from a modern point of view. The idea that the impulse given to the first particle of the fiber is attenuated with distance until it is exhausted, even if intuitive, is not consistent with the laws of mechanics then available. The attenuation phenomenon that Bernoulli suggests should be seen rather as a delay in the time of transmission of the impulse. As the first particle is stimulated, the impulse it receives is transmitted to the adjacent luminous particles with a certain finite velocity and reaches them after a certain interval of time, until the motion of the first particle has breached its maximum.

Still less convincing, and apparently contradictory, is the explanation of the transmission of the movement of the light fiber to particles external to it to form other fibers, which are called *secondary fibers*, leaving the title of *primary fiber* to that

[41]p. 18.

[42]p. 15.

in which the movement originated of the disorder caused by the luminous body. According to Bernoulli, when the particles B, C, D, etc. of Fig. 4.2 have reached the limits of their excursions in b, c, d, etc. the aether or elastic material around A will be accumulated and condensed very strongly; therefore making an effort to to restore forward and backward, as do all springs, it will not only repel the particles of the first fiber but also spreading it on the opposite side will stir the particles found in its region L, and will produce a new fiber that will be called secondary [35].[43] What does not work in this reasoning is that there is no reason that the aether should accumulate around A, because as Bernoulli explained, the movement transmitted by the first particle has been exhausted and therefore around A the aether should be substantially undisturbed.

According to Bernoulli, sound and light have a great affinity. Sound as light takes its origin with the production of fibers that can be called *sound fibers*, which extend to more or less large distances, depending on the greater or lesser force with which the sonorous body has excited the air. Luminous fibers and sound fibers have the same nature, because both require an elastic medium in a state of compression. The luminous fibers are generated in an extremely thin aether, composed of vortices of an unimaginable smallness that determine an enormous elasticity and pressure. Air particles are much heavier than those of the aether so that the frequencies of vibration of sound are much lower than those of light. The light rays travel in a straight line because the solid corpuscles are incompressible and cannot expand transversely, while the air particles, which in the sound fibers take the place of the luminous solid particles, being deformable in all directions when they are compressed, tend to deform laterally, which causes the formation of new transverse fibers as branches, which bear the sound, albeit less strongly, through oblique directions [35].[44]

After defining the mathematical model of light fiber, Bernoulli moved on to quantitative determinations concerning the shape and frequency of vibrations and the speed of light. I do not enter into the merits of evaluating the correctness of the results, which is not an easy task given the difference in notation between the analysis of the 18th century and the modern one, but above all because rigor was then more an exception than a rule. I will therefore limit myself to describing the procedure in a very general way. The first thing Bernoulli did, in the section entitled *Calcul*, is to define what, in a quite improper way, could be called waveform. In particular he wanted to determine the equation of the curve $A\beta\kappa\gamma\delta\epsilon\varphi G$ of Fig. 4.2, which has as abscissas the position of the luminous corpuscles and in the ordinates the displacements (Dd, Cc, Bb, etc.) of these points when the disturbance of the point D is at its maximum. Note that in the curve there is no difference if the displacements are to the left (like those of the stretch AD) or to the right (like those of the stretch DG). After a series of more or less conclusive arguments Bernoulli came to the conclusion

[43] p. 27.
[44] p. 25.

that it is a cycloid: "It is what is proper to the Roulette or the Cycloide" [35].[45] For comments on Bernoulli's calculations, see [44].[46]

At this point Bernoulli pointed out that the same curve was obtained by his father Johann I for the vibration of a taut string, reported in the acts of the Petersburg academy of 1728 in a memoir entitled *Meditationes de chordis vibrantibus* [34], although his are longitudinal oscillations (*longitudinales*) and those of the father transversal oscillations (*latitudinales*). The memoir of 1728 shows that a string, homogeneous and uniform, vibrates with a shape which "is an elongated companion of the trochoid" [34].[47] The work continues by demonstrating that when the sine of contact angle is confused with the angle itself a sinusoidal shape is obtained [187].[48]

The coincidence of the shape of the longitudinal oscillations of the luminous particles and the transversal oscillations of the vibrating string is sufficient for Bernoulli to consider valid also for the light the equation, found by his father, that provides the frequency of vibration in the form:

$$\frac{p\sqrt{D \times A}}{AG}$$

Where $p = 2\pi$, AG the length of the luminous fiber of Fig. 4.2, A is representative of the pressure of the aether, and D the arbitrary length of a simple pendulum; the above relation furnishes the number of vibrations of the sound fiber in the same time a pendulum of length D makes an oscillation. If D represents the length of a pendulum with a period of one second, the previous relation furnishes what presently is called frequency, that is the number of vibrations in one second.

To obtain the speed of light Bernoulli made the assumption that the number of the vibrations of the luminous body in the primary fiber equals the number of secondary fibers that are activated. This means that by multiplying the length AG of the fiber by the number of the cycle in a period the distance covered in that period is obtained, which is given by:

$$p\sqrt{D \times A}$$

Bernoulli used this formula for the speed of sound and compare it with the values obtained by Newton in the second book of the *Principia*, and found the value of 979 *pieds d'Angleterre, moins a demi pouce* per second [35].[49] Comparing the values of

[45] p. 35.

[46] pp. 77–80.

[47] p. 26. A trochoid is the curve described by a point linked to a disk. If the distance of the point is equal to (lower than) the radius of the disk, the curve is also called (curtate) cycloid and looks like a sinusoid.

[48] p. 135.

[49] Corresponding to 300 m/s. The value obtained by Newton was distorted downwards. Later, Pierre-Simon Laplace saw the flaw and ultimately corrected Newton's result. He assumed that the process of sound transmission was not isothermal as Newton had thought, but adiabatic. Bernoulli suggested his own reasons why the actual speed, then estimated in 1080 *pieds d'Angleterre*, was grater than that found by Newton [35], pp. 38–39.

the experimental speeds of light and sound then available, Bernoulli found that the value of A in the previous relation in the case of light should be 4.9×10^{11}, whose root 7×10^5 represents the ratio between the speed of light and that of sound.

Bernoulli went on to study reflection and refraction of light. Regarding refraction, he started from the premise that the speed of light is greater in denser than in rarer bodies. Very simply the reason is to be found in the fact that in denser materials the pores are smaller than in the less dense ones and therefore the vortices of ether have to move in narrower spaces with a consequent decrease in their diameter and therefore an increase in the centrifugal force and therefore of compression. And where the compression is greater, for the formula Bernoulli found, the speed is greater [35].[50]

Bernoulli referred back to the memory of his father of 1701 in the Acta Eruditorum, *Disquisitio catoptrico-dioptrica exhibens reflexionis & refractionis naturam* [33], where refraction was explained by reducing it to the equilibrium of three forces, which act on a mobile point in different directions. For the laws of statics they will produce equilibrium only when two any of them are inversely proportional to the sinus of the angles that their directions make with the third force [35].[51] Bernoulli's task was to determine the origin and values of the forces of which the father spook. The first force is given by the effort with which the incident ray of the first medium enters obliquely in the second medium, the second force is the more or less large resistance encountered in the second medium, the third force is simply passive and is such that the point of incidence (assumed as a material body) stressed by the two other forces is prevented from slipping on the surface of separation between the two mediums.

With reference to Fig. 4.4, consider the fiber AEB oblique to the surface of separation between air and glass CD. The two ends A and B of AEB play the role of fixed supports. It is clear that the particles of the fiber AE will not be able to continue their motion along AEF, because the pressure in the aether of the glass is greater than that in the aether of the air and therefore there would be an imbalance of forces in the direction of propagation. Thus there should be a deviation of the luminous ray; because of this deviation the corpuscle in E would be stressed by forces of different intensity and direction, with the pressure of the part EB that would tend to push the particle E towards the left. Bernoulli did not make explicit that the imbalance that would occur between the two forces acting on different lines in the particle E is compensated by a reactive force, provided by the mediums.

The dispersion of a light bundle, formed by different fibers each characterized by corpuscles of the same mass, is explained by Bernoulli, admitting that the hard particles that transmit the light in the medium of the incident ray, have mass different from those that are in the medium of the refracted ray and that the angle of refraction of the luminous fibers "contained in the same bundle will suffer larger or smaller refractions, as much as the accelerating forces of the corpuscles of the media through

[50] p. 46.

[51] p. 44.

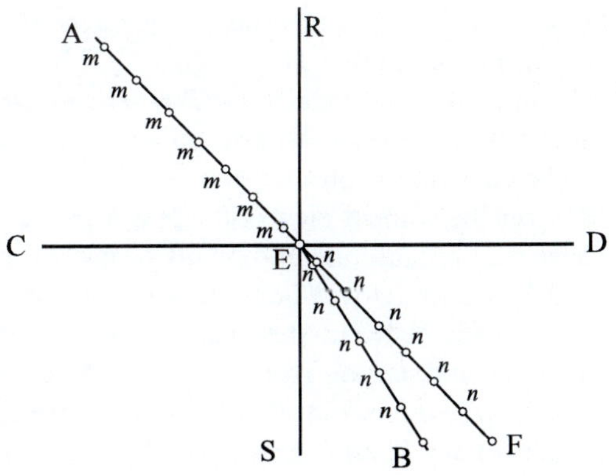

Fig. 4.4 Refraction of a luminous filament. Redrawn from [35], figure IV of the table of figures

which the light must pass are different. So the compound ray entering there must disperse in single rays, and be seen each the color that suits it" [35].[52]

The vibratory theory of light reached its peak in the middle of the 18th century with Euler, who developed his considerations on the subject in the small treatise *Nova theoria lucis et colorum* [89], less than 80 pages, presented at the Académie des Sciences et belles lettres de Berlin in 1744. A summary appeared in the Histoire of the academy for the same year (published however in 1746) [91], but this text was never published in full; it survives, as a manuscript with the title *Pensées sur la lumiére et les couleurs*, in the archives in Saint Petersburg [124].[53]

Euler knew the work of Johann Bernoulli II because, in 1737, Johann I sent him a copy of it and asked for an opinion. Euler replied by praising the work; he was only critical toward the derivation of the velocity of sound, which reproduced Newton's formula and contradicted his own [80][54] And he was aware of Malebranche's color theory, who is cited in his text, as also mentioned were Mairan and Descartes. He was also inspired by Newton's sound theory as reported in the second book of the *Principia*. Quite strange for a modern, he did not cite Huygens, of whom he should know about. The reasons are not clear; certainly the theory of Euler was profoundly different from that of Huygens; Euler did not accept for example the so called Huygens principle, whereby every point struck by a luminous impulse in turn becomes a source. It should however be said that at the time of Euler the theory of Huygens, had already been set aside by opticians and Euler could very simply have followed the others in ignoring it [80].[55]

Euler's light transmission theory has been the subject of a number of studies, but less than those that have been devoted to his mechanical or mathematical writings. Probably Euler's contribution to optics most valued by historians is his discovery of achromatic lenses. This work falls within the field of geometrical optics, a subject

[52]p. 57.

[53]p. 529.

[54]p.149.

[55]p.75.

filling four volumes of Euler's *Opera omnia*, two more volumes are occupied by the *Dioptrica*, published in 1769–1771. In contrast, the articles on physical optics, including the *Nova theoria the lucis et colorum*, were contained in one volume only. However, Euler gave just as high a value, or a higher, to his *Nova theoria* as he did to his works on dioptrics.

From the point of view of the present book, Euler's work has a relatively marginal relevance, because the arguments of natural philosophy are all considered few and large space is left to mathematical developments. Euler could act with optics as a pure mixed mathematicians because he had already developed his own physical theory of the aether starting from the *Dissertatio de igne, in qua ejus natura et proprietates explicantur* written in 1737, developed in an extended form in the *Recherches physiques sur la nature des moindres parties de la matiere* of 1744, practically at the same time of the *Nova theoria the lucis et colorum*, and then in the *Anleitung*. The aether was treated as an absolutely continuous, elastic medium with highly compressed in its equilibrium configuration.

Euler devoted a long introductory part to refutation of Newton's projectile theory of light and to support the vibration theory. He believed that the propagation of light in its medium was a phenomenon completely analogous to that of the propagation of sound that he had already studied in a short dissertation *Dissertatio physica de sono* written in 1727 as part of applying to the physics chair of the university of Basel at the age of 20. This work was somewhat expanded in the first chapter of his treatise *Tentamen novae theoriae musicae ex certissismis harmoniae*, published only in 1739, although written around 1730 [88].

Euler conceived of air as made up of infinitely small but nevertheless real and springy particles that are compressed together. Euler had the qualitative picture of sound as being a pressure vibrations of the air induced by a sound source. These vibrations, like those of pendulums, are harmonic, that is sinusoidal, with pitches described by their frequency of vibration in the case of simple sounds. Composite sounds are made up of simple sounds, sounding consonant when the frequencies have a simple ratio.

Regarding the transmission of light, since the medium in which it is propagated, the aether, is continuous and therefore finite particles of it do not exist, as it is for air, Euler had to consider the vibration of an infinitesimal element of aether, to be treated as a purely geometric entity, as Newton had done in *Principia* in his study of the propagation of sound in a fluid.

Euler believed that light propagation were essentially a mono-dimensional phenomenon, although light rays could propagate in all directions and a spherical front could be determined.

> Once a pulse has been formed it moves forward in a straight line if the medium is uniform, as it is assumed for the rays of light. The transmission of pulses arises from an agitation of the particles of the elastic medium in which the pulse is situated. Since the agitation at all times tends in a definite direction, it gives the pulse a motion towards that same region [89].[56] (D.9)

[56] p. 198. Translation adapted from [124].

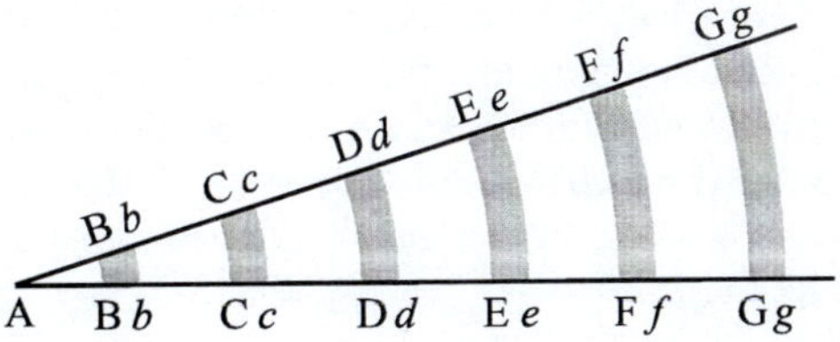

Fig. 4.5 Spreading of pulses of light. Redrawn from [89], Fig. 3 of Table V

Figure 4.5 illustrates the propagation of light as conceived by Euler. A source in A produces pulses in all directions; in this way spherical front waves such as GG, FF, EE, DD, CC, BB are generated. If the pulse at G affects the organ of sight, the ray of light is judged to have arrived along the direction normal to the pulse GG. Therefore rays exist in the aether only in as far as straight lines normal to the pulses are conceived there [89].[57]

In Chap. 1 of the *Nova theoria lucis et colorum* Euler studied the mechanism of propagation of a pulse. He used Newton's treatment of pulses in elastic media as developed in the *Principia* by reformulating the geometric treatment in algebraic form. Euler's ultimate aim was the deduction of the equation for the speed of light confining his attention to the propagation of a single pulse. The basic hypotheses are (1) that the change in pressure of the aether is proportional to the variation of its density, or if desired of its volume, (2) that the restoring force is linear elastic and that the impulse propagates according to a harmonic law, that is sinusoidal. Under these conditions Euler obtains the following formula for the speed of sound [89]:[58]

$$\frac{a}{T} = \sqrt{\frac{k}{2}}$$

where k is a constant that multiplied by the specific weight D (weight density) gives the elastic force of the medium, a is the space passed in the time T.[59] The formula, Euler pointed out, is the one obtained by Newton. Replacing in this formula the value $k = 28678$, as made by Newton, one gets for the sound the speed of 975 English feet per second [89].[60] The speed of light is obtained by assuming for k a value of 3.8510^{11} greater than that of the sound. In [116][61] comments to the passages leading to this value are reported in detail.

---

[57] p. 205.

[58] p. 192.

[59] Following the standard symbols, the speed v of sound according to the modern theories is given by the relation $v = \sqrt{E/\rho}$, where ρ is the mass density of the medium, not to be confused with the weight density D, and E its modulus of longitudinal elasticity, also known as Young modulus, which stands for the elastic force. Factor 2 in Euler's formula derives from the values he assumed for the acceleration of gravity g: $D = \rho g = \rho \times 2$.

[60] p. 193.

[61] pp. 91–97.

In Chap. 3 Euler examined the propagation of successive pulses. Here he introduced a magnitude, the distance c between two successive pulses, which is to be compared with the modern concept of wavelength, and derived the well-known formula of wave theories [89][62]:

$$c = \frac{n}{i}$$

where c is the distance between two consecutive pulses, n the speed of light and i the frequency of pulsation, valid in the case of isochronous vibrations.

Euler believed that the pulses within a medium different from the aether, water for instance, propagated exclusively in the water particles, rather than in the aether filling the voids left by water particles. If one used the standard formula for the speed within a medium—air or water—for calculating the speed of light, the resulting value would be much lesser than the actual one as water density and consequently the coefficient k is much lower, less than $1/(3.8510^{11})$. Euler thought however it was incorrect to employ the equation he had found for the aether, because the propagation through water is entirely different from the propagation trough aether, since the contact of the parts of the water causes the pulses to be propagated nearly as in an instant (ut quasi in istanti impressiones receptae transferuntur) and the speed in water is not very different from the speed in aether [89].[63] This last statement is stated without Euler commented the rather strange circumstance that two entirely different propagation mechanisms result in virtually the same speeds. Neither he was disturbed by the fact that various differing materials, such as water, glass and diamond, demonstrated almost the same propagation speed.

Euler divided the rays of light into two types. The first type is constituted by *simple rays*, in which the pulses follow each other at a constant distance (in time and in space); they are identified with the homogeneous rays of Newton and as in Newton's theory the perception of a color is connected to a particular simple ray, characterized by a certain distance of pulses and therefore frequency. To make this idea acceptable, he affirmed that the organ of sight is able to distinguish the number of impulses it receives in an assigned time as the perception of the high and low tones in the sound by the ear is based on a similar process. The second type of rays is constituted by *composite rays*, in which the pulses follow each other at irregular intervals. Euler recognized that the terminology was improper because, at least for him a "composite ray" was a single ray produced by a non-isochronous vibration and not a mixture of different rays; this is the case of solar rays. Even so, he preferred not to deviate from the established terminology derived from Newton [89].[64]

The refraction of a simple ray of light is explained by Euler with reference to Fig. 4.6. Consider the pulse Pp with P touching the transition surface from a less dense medium ACB (air for instance) to the denser medium ADB (glass for instance). The part of the pulse indicated by P moves in the denser medium with a speed v_2 less

---

[62] p. 192.

[63] p. 229.

[64] p. 208.

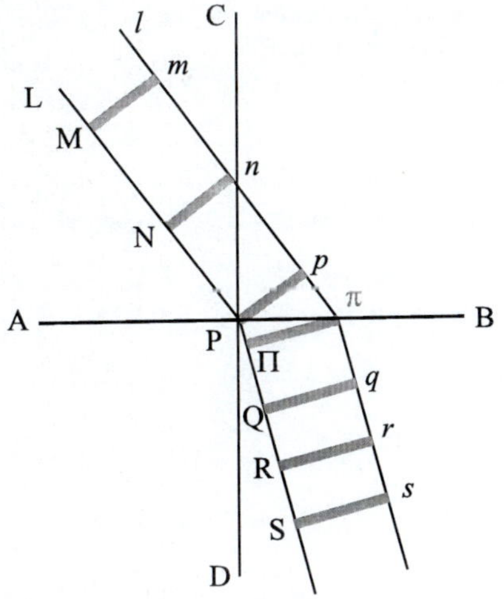

Fig. 4.6 Refraction of rays of light. Redrawn from [89], Fig. 5 of Table VI

of the speed v_1 of the part p which remains in the less dense medium ACB, until it reaches the surface AB at π. Meantime the part P of the pulse has reached the point Π, with PΠ less than $p\pi$ in the same ratio of v_1/v_2. Euler noticed that the pulses Pp and $\Pi\pi$ cannot be parallel and because the rays are perpendicular to pulses, by definition, the direction of the ray should change, as shown in the figure and the ratio of the sinus of the angle of incidence and that of refraction equals the ratio v_1/v_2.

Euler's reasoning is not very stringent however, because based on two not proved assumptions. First, that the pulses propagated less rapidly in a denser medium than in a less dense medium [89].[65] A quite common assumption for the vibration theories of 18th century. Only later in the treatise he suggested an explanation. Second, that the pulses of light propagates as parallel lines after the refraction. This proposition can be found in Huygens also and had become background knowledge of the 18th century [181].[66]

To explain the phenomenon of dispersion of light, Euler had to postulate that the light rays of different colors move at different speeds. This is in contrast with the theory developed in Chap. 2 of his treatise, in which it is found that the speed of light is constant regardless of the frequency of the pulses. Euler suggested, probably an a hoc hypothesis, that uniform speed only operates for isolated pulses. In presence of a sequence of pulses they interact, which results in an increase in speed. The increase in speed is not constant for each color, but it depends on the frequency associated to the color. According to Euler the ratio of speed passing from a less denser to a denser medium is the greater the greater the frequency and consequently the pulse with lower frequency has the greater refraction.This means that violet that is more refracted among the solar rays should have the lowest frequency while the red that

[65]p. 228.

[66]p. 255.

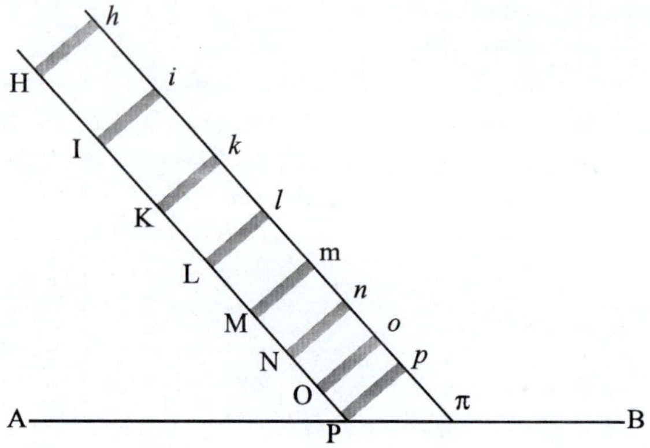

Fig. 4.7 Dispersion of white solar rays of light. Redrawn from [89], Fig. 6 of Table VI

is less refracted should have the highest frequency; which is in contrast with modern theories of light, that assume that red has the lowest frequency [89].[67]

This type of explanation is good for simple rays of light and reflects Newton's explanation. As far as sunlight is concerned, Euler's ideas are different; he believed that rays coming out of the sun cannot be synchronous and therefore they were composite rays. Indeed, because the sun's particles are in continual motion (as a result of the high temperatures), at some point in time one frequency will be the prevalent one, and at a later time, another. Consequently the composition of sunlight would not be constant; its pulse frequency would continuously change [89].[68]

The dispersion of sun white light can be explained with reference to Fig. 4.7.

In the diagram the intervals [HI, IK, [...], OP, of pulses of a non synchronous ray are drawn different from each other to represent pulse with different distances. There are seven intervals with an implicit reference to Newton's seven main colors. Pulses as H*h* can be imagined as parts of a sequence of isochronous pulse having HI as wave length. These pulses (violet pulses, for instance) having the greatest distance have the lowest frequency ant thus are refracted more than the others; pulses as P*p* have instead the greatest frequency (red pulses), so they are refracted the least.

Possibly the idea of non synchronous rays is one of the most interesting of Euler's work, because it explains the dispersion without the recourse to the idea of superposition of synchronous rays, which as will be explained below has not a clear physical meaning. While for the supporters of the projectile theory it is easy to imagine different rays made of different particles corresponding to given colors that remain distinct in a bundle of rays, it is more difficult for the supporters of vibration theory to imagine that pulses of pressure with different frequencies could maintain their individuality, especially in situation like that suggested by Euler of a homogeneous medium.

[67]pp. 217–218.

[68]pp. 219–220.

The idea of a light beam composed of several light rays of different frequencies that remain distinct is suggestive but groundless; this notwithstanding it is an idea commonly conveyed today in textbooks of physics, used to an abstract approach to physics. Using the modern mathematical language of wave theory one can say that the physical superposition of several waves of light with different frequencies, each of which has a sinusoidal shape, gives raise to a single wave of complex shape, no longer sinusoidal and in general not even strictly periodic, albeit fluctuating. At the mathematical level one can still imagine this wave composed of simple waves, which can be visualized using appropriate mathematical algorithms such as the Fourier transform, but it is only a mathematical and not a physical decomposition.

Assuming this point of view, it can be said that the assumption for which colors are contained in sun, white light, as Newton did, does not make much sense. According to Euler, when the white light falls on an opaque body it is not reflected as an elastic ball in motion is reflected by a wall. What happens is that the particles on the surface of the body begin to vibrate with a their proper frequency and emit rays with color corresponding to these frequencies. "Opaque bodies, as long as they are not illuminated, must be compared to musical instruments not in use, or, is you will, to strings which emit no sound till they are touched" [93].[69] This kind of reasoning works well not only if one imagines light as formed by separated color rays but also as composite rays. Indeed the equation of motion of vibrating bodies make clear that even though waves that hit the body is not considered as a mix of separate rays, the bodies gives back a vibration corresponding to its natural frequency.

One more very interesting Euler's idea is that of *colores derivativi* [89],[70] that in modern term can be translated as over and under colors [116],[71] roughly representing our infrared and ultraviolet rays, that is rays outside the visual spectrum. These rays are associated with the ordinary rays, using an analogy from sound. Just as in music, sounds with vibrations, that bear a double, quadruple, eight-fold (etc.) ratio to the main tone are considered similar, so simple rays of light containing in a given time, two times, four times, eight times the pulses of red, for instance, should be considered as variation of red even though they cannot be seen [89].[72]

> This holds true for pure or high colors, such as the rainbow and prism present them to us. The other colors, lower colors, differ only as tones of various octaves. So in the case that a red ray makes 10,000 vibrations per second, rays which make 5000, or 2500, or 1250, or 625 vibrations in the same time, will also produce a red color, but less high than the first. Therefore, there will be several differences of color for each name, as we have in a harpsichord several tones expressed by the same letter [90].[73] (D.10)

[69] Vol. 1, pp. 106–107, letter 26. See also [92], pp. 234–235.

[70] p. 239.

[71] p. 9.

[72] pp. 237–239.

[73] p. 23.

According to Euler the idea of *colores derivativi* could be applied to the explanation of Newton's rings [116].[74] Euler was explicit on this point in his *Essai d'une explication physique des couleurs engendrees sur des surfaces extremement minces* published in 1754 [92].[75]

Vibration theories enjoyed limited acceptance during the 18th century; the dominant theory remained, the projectile theory both because of the prestige of Newton and because of they were able to explain in a satisfactory way the rectilinear propagation of light and a the complex and somehow reflection, refraction and dispersion. Euler's theory was accepted in the 18th century only to himself and a few others, especially in Germany [116]; it did not, however, arouse much interest even for later proponents of wave theories, such as Thomas Young and Augustin-Jean Fresnel, if not for the fact that a great mathematician like Euler had previously believed that light was undulatory and colors depended on the frequency of vibration. Young, for instance was quite well aware of the adverse criticism that Euler's theory had received and was cautious, in his optical writings at least, to point out where he thought Euler had made mistakes [131].[76] To note that Ernst Mach in his *The principles of physical optics* cited Euler only incidentally and not for his pulse theory [144].

4.2 Photometry, a New Field of Optics

Today the term *photometry* refers to the measurements of the power of light. In the 18th century the term *photometry* had a broad meaning and also comprehended illuminating and optical engineering, astro photometry, etc. Elements of photometry can be traced back to ancient optics; but it is only in the 18th century that it saw a new course. A brief history can be found in [141]. From the Middle Ages to Maurolico, to Kepler to get to François Aguilon (1567–1617) who is credited with having conducted the first modern photometric experiment in his *Opticorum libri sex philosophis iuxta ac mathematicis utilis* of 1613, documented by one of the splendid seven engravings by Pieter Paul Rubens (1577–1640), reported in Fig. 4.8.

Aguilon described his experiment as follows:

> Now we explain in what manner of converse reckoning the same things arises. Fixed at one end of the table is the board behind which is arranged the card which intercepts the rays of both lights. Then the lights are moved and yet the distance from the card is unequal, so that the interval of the twin lamp is double of that by which the single lamp is distant. Plainly therefore, by considering the card, the light of the twin lamp appears brighter than the single; and so that the lights observed on the card become equal, it is necessary for the twin lamp to be moved further, because the decrement of light brought about by a double space is less than double [6].[77] (D.11)

[74] p. 113.

[75] p. 278.

[76] p. 50.

[77] p. 378. English translation in [141].

Fig. 4.8 Aguilon's photometric experiments [6], front cover of book V, p. 356. Reproduced with the permission of ETH-Bibliothek Zürich, Alte und Seltene Drucke

Here Aguilon stated that two equal bright lamps located at double distance than a single lamp produces less light. According to [141],[78] Aguilon's more than a test of photometry was simply a test of the effect of illumination with the distance of the lamp.

Mersenne, Huygens, Anders Celsius (1701–1744) and Euler also contributed to photometry, but the one who often is named alongside Lambert as the father of photometry was Pierre Bouguer (1698–1758). In 1729 Bouguer published his *Essai d'optique sur la gradation the la lumiere* [40] and posthumously in 1760 an augmented version, the *Traité d'optique sur la gradation the la lumiere* [41]. Below only some hints of the *Essai* are given for the sake of space.

The *Essai d'optique sur la gradation the la lumiere* is the first extensive work devoted exclusively to photometry. It is divided into two parts. The first part is mainly experimental and gives fundamental procedures. Here Bouguer exposed clearly the way to compare sources of different luminance. Bouguer compared the brightness produced by two or more equal lamps—joined to give one single source—with a source made of a single lamp, changing the distances from the sources to an illuminated screen until the brightnesses are equal for the two sources together and the one alone. A general law is obtained: The ratio of the square of the distances gives the ratio of number of lamps:

[78] p. liv.

> In fact, it is only necessary to take a torch, to move it away or to approach it until it lights up successively in the same manner as the two lights that one wants to compare; and if we take the squares of the two different distances, these squares, placed in an inverted order, will express the ratio of the light [40].[79] (D.12)

The second part of Bouguer's treatise is devoted to the decrease of light passing though translucent media.

4.2.1 Lambert's Contribution. A Philosopher and a Physicist

The most important work on photometry of the 18th century was the *Photometria, sive, De mensura et gradibus luminis, colorum et umbrae* of 1760 (herein after simply *Photometria.* Lambert was not the first to introduce elements of photometry, nor was he the one who introduced almost all the elements, but he was the first to give it a definite structure.

Johann Heinrich Lambert (1728–1777) was born in Müllhausen, a small town of Alsace, then part of the old Swiss confederation of 1536–1798. Lambert's formal education at an elementary school ceased when he was only twelve. Since then he started his life program of self-education. In 1745 Lambert became a secretary to Jean Rodolph Iseline (1705–1779), a professor at the Institute of Basel. Later on he got the position as private tutor for the children of count Peter von Salis and was in the position to become acquainted with Swiss scholars. In 1755 he published his first scientific memoir on heat. In 1756 he left Switzerland for a tour through Europe. In Göttingen met the mathematician Abraham Kästner and the astronomer Tobias Mayer. Reading Kästner's book on opticks was very important for him. The French conquer of Göttingen pushed Lambert to Utrecht where he remained for two years. In this period he met Musschenbroek. After a trip to Paris, where he got in touch with d'Alembert and Messier, Lambert continued his travel through Marseille, Nice and Milan for eventually returning home at the end of 1758.

After this he started a new and longer trip which lasted five years, through Switzerland and Germany. During this period he wrote continuously. In 1760 in Augsburg he published his *Photometria* and in 1761 two important books, the *Insigniores orbitae Cometarum proprietates* where he introduced the method of determining the orbit of comets for the case of parabolic orbits based on three astronomical observations and *Cosmologische Briefe fiber die Einrichtung des Weltbaues* where he described the structure and the motion of the universe and expressed the opinion that the Milky way might be visual effect of a lens shaped universe [192]. In 1765 Lambert obtained a place in the Académie royale des sciences et belles-lettres de Berlin, where he remained until his death.

---

[79]pp. 6–7.

A true polymath, he was called the Newton or the Leibniz of Alsace, Lambert's interests ranged from theoretical subjects, such as philosophy, logic, mathematics, natural philosophy to instrument designs, practical activity of surveying and so on. He not only proposed a his own philosophy and epistemology, but also suggested a methodology of science and was a practicing great theoretical and experimental 'scientist'.

Lambert was a fairly prolific writer; for instance in the twelve years he stayed at the academy of Berlin he wrote more than 160 works, not all published during his life. His writings today are classified either as scientific or philosophical, following a dichotomy typical of modern highly specialized organization of studies. The scientific writings are mainly in French and Latin in the juvenile period and in German late, as it is the case for the *Pyrometrie oder vom Maasse des Feuers und der Wärme*, published posthumous in 1779 [139], where Lambert gave a comprehensive, for the time, theory of heat in the style of the *Photometeria*. He dealt not only with radiation but also with reflection of heat, though this effect could not be proved and his results could have been only speculative. The philosophical writings, with some exceptions, are in German; among them should be cited his probably two main treatises: the *Neues Organon* of 1764 and the *Adage zur Architectonik* of 1771 and some texts on logic, whose appreciation is controversial. The source of bibliography on Lambert and of Lambert is still represented by [183]; interesting updating can be found in [178], from which most of biographic information is drawn, and in [141].

Lambert had not received much attention by philosophers and scientists of his period. Philosophers considered him an epigone of Locke and Wolff and overshadowed by the rise of his contemporary Immanuel Kant (1724–1804). A certain revival of interest occurred towards the end of the 19th century, determined by the outcome of Neo-Kantianism. He had no better luck as experimental philosopher and mixed mathematicians, because his interests laid mainly outside the mainstream of 18th century science. For instance he did not devoted much effort to electricity or to rational mechanics, the two dominating topics of the science of the period. Modern historians of science are however interested in his production.

From the studies of the last century by philosophers and historians, Lambert emerges as an important protagonist of the German Aufklärung. A philosopher and a scientist who based knowledge on few fundamental ideas and axioms, but differently from the German school then represented by the Leibnizian Wolff, he considered that experimental evidence has not only the role of giving a value of truth to an hypothesis, but also it gives guidelines for the process of knowledge itself; in other words analysis and synthesis cannot be separated. He was however conscious that the experimental evidence may be misleading and a great care must be devoted to experimentations. He strongly insisted on the quantitative nature of science and saw mathematical laws at the basis of scientific knowledge, while mechanicistic explanations may be useful to suggest hypothesis but in the end they should have no role in final account of phenomena [81].

4.2.1.1 Photometria

Lambert's treatise, the *Photometria*, is an impressive work divided into six parts; below the title of each of them.

Part I In which are set forth assessments and degrees of direct light its brightness and illuminating power.
Part II In which the assessments of light from transparent bodies, chiefly from glass, are subjected to experiments and calculation.
Part III In which assessments of light depending on opaque bodies are surveyed by experiments and calculation.
Part IV In which the sense of light and its apparent brightness is defined by calculations and experiments.
Part V In which the dispersion of light in a transparent medium is investigated, chiefly traveling in the atmosphere of the earth.
Part VI In which the illumination of the planetary system is subjected to calculation [137].[80]

In the following I will refer with some details on part I only. It is divided into three chapters. Chapter 1 exposes the principles of photometry, or better its main principles. Other assumptions are introduced to afford complex phenomena in the more applicative parts of the treatise. Chapter 2 concerns the transmission of light from extended sources. Chapter 3, comes back to the principle of photometry and reports a series of experimental data which confirm them by the use of a statistical analysis.

In the preface to *Photometria* Lambert defined the object: "it concerns itself with the brightness of light, its density, its illuminating force, its modification in color and shade, its degrees, its increase and decrease which it undergoes in all cases" [137][81] and the epistemology:

> Between hypotheses which I will call *mathematical* and those which are *physical* there is the greatest difference. The physical are frequently assumed, so that it is not known where in a matter they may stray from the truth; whence it happens that in turn, each may be rejected again, only as their aberration from the truth is revealed with the advance of time. In the mathematical it is almost always known not only in what part they recede from the truth, but also it is possible in many cases to define in advance the importance of the aberration [137].[82] (D.13)

While it is clear what Lambert could mean for a physical hypothesis, it is a mechanist explanation of Cartesian mould, less clear is the meaning of mathematical hypothesis. He seems to refer to mathematical function assumed to represent the quantitative aspect of a phenomenon, whose shape can be determined by means a best fitting procedure (modern meaning). But it is possible Lambert used the term in a more general way, for instance as a mathematical model or algorithm. The role of hypotheses in

---

80 Index capitum. English translation in [141].

81 Preface.

82 Preface. English translation in [141].

Lambert's epistemology is further specified some pages below. The most certain criteria for a hypothesis arriving at truth is that when one is able to predict the outcome of new phenomena from the hypothesis, thence to deduce propositions which experiments, established for this end, support [137].[83]

Optics, classical geometrical optics, has been framed among the mixed mathematics in which considerations of traditional natural philosophy, nature of light, physiology of eye for instance, were present and important. But, as optics was also the mixed mathematics closer to pure mathematics (geometry) itself, it was natural for Lambert, once he had framed his *Photometria* in optics, to proceed as a mathematician with the renounce to the search for the ultimate causes. Even though he was dealing with an aspect of optics that in principle was far from being caught by geometry. Lambert had a good knowledge of traditional natural philosophy; he wrote for instance a treatise on cosmology and physical astronomy, the *Système du monde*, first printed in 1770 [140], but in the *Phoronomia* he acted as a mathematician and as an epistemologist, subtracting a part of natural philosophy to the canonical philosophers.

In Chap. 1 the main assumptions of photometry, which he referred to as laws are presented. The chapter opens in a very similar way Aristotle opened his *Physica* [11]:[84]

> It appears to be the common fate of human knowledge that those things which are most apparent to the senses, can be more remote from understanding. Certainly the theory of light sets a clear example of this declaration for us. For many very serious difficulties oppose fully exploring its very power and nature, and they can hardly be overcome; so that it is remarkable how in this matter-which is the very fount of clarity-our understanding is still enveloped in so much darkness, and such great shadows remain on light itself [137].[85] (D.14)

Lambert stated that an acceptable physical theory of light is lacking, and both the Eulerian (vibration theory) and Newtonian (projectile theory) are unsatisfactory, even though he expressed his preference for Euler: "the Eulerian seems more consistent with the proper nature of light" [137].[86] He admitted that photometry is in the same position it was the thermology before the invention of the thermometer and expressed the hope that a photometer could be devised, which exposed to a light indicated its intensity and brightness. At the moment there is the eye only, which is a precise enough instrument, but which does not furnishes an absolute measure of brightness, so that only comparative results of equal or more or less brightness are possible.

Lambert built a vocabulary of the magnitudes and concepts he introduced; in this he advanced previous authors, Bouguer included. Table 4.2 refers some of Lambert's definitions compared with modern ones. It must be noticed that however not all the concepts of the table have precise modern correspondents and that Lambert himself was not completely consistent, because the same concept is sometimes expressed

---

[83] p. 3.

[84] 184a.

[85] pp. 1–2. English translation in [141].

[86] p. 3.

Table 4.2 Lambert's nomenclatura in the *Photometria*. Main concepts. Adapted from [141], p. cv

Lambert	Modern	
Radiuos luminous (Ray of light)	The fundamental unity. Newtonian ray	Section 42
Punctum radiante (Radiant point)	Infinitesimal element of light source	Section 48
Quantitas radiorum (Quantity of rays)	Luminous flux	Section 43
Densitas radiorom (Density of rays)	Luminous flux per unit of area or solid angle: intensity or illuminance	Section 44
Intensitas (Luminous intensity)	Flux for solid angle. Luminous intensity	Section 39
Claritas (Brightness)	Illuminance. Flux for unit of illuminated surface	Section 98
Illuminatio (Brightness)	Illuminance. Flux for unit of illuminated surface	Section 98
Claritas visa (Perceived brightness)	Brightness	Section 37
Splendor (Splendor)	Luminance. Flux for unit surface of luminous source	Section 98

with different terms. Notice that, in the column of modern terms, luminous flux (a measure of the total amount of light a source of light puts out) is used to define most magnitudes.

The introduction of precise mathematical laws is preceded by the exposition of semiquantitative empirical laws, which are known from everyday experience:

1. Two or more candles illuminate more than one.
2. An object becomes brighter as it is moved closer to a light.
3. Light incident obliquely on a surface illuminates it less [137].[87]

They are then specified by assigning mathematical functions, as follows:

1. Illumination [the brightness] is greater in proportion with the number of [equal] candles, or lights, or radiant points [surface] by which a page of paper or a plane exposed to these is illuminated.
2. It [The brightness] is less to the degree that the square of the distance of the illuminated plane to the luminous body is greater.
3. It [The brightness] decreases in proportion with the sine of the angle of incidence [137].[88]

These laws can be proved a priori, if one accepts some principles of natural philosophy. The first law is suggested by the evidence that light does not impede light when it traverses the same space. If one takes the candles to enjoy equal brightness, equal distance from a screen, and finally equal size, since one light does not block

[87] Section 46, p. 24. English translation in [141].

[88] Section 226, p. 105. English translation in [141].

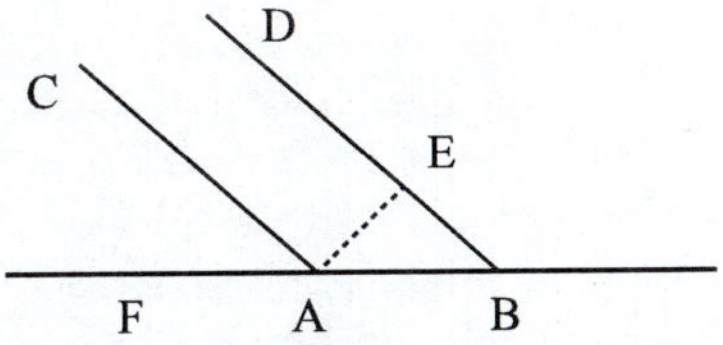

Fig. 4.9 Oblique incidence of rays of light. Redrawn from [137], Fig. 1 of Table I

another, it is clear that for any new candle an equal degree of brightness is added to the paper. The same hold good if one substitutes another light in place of the candles, equally bright. The second law could be derived by assuming that light propagates in spherical surfaces. It is evident that the same number of rays traverse each spherical surface. But on the larger surface the rays are diffused over a larger space, whence the density of rays is less and the brightness is lower. Equally the third law is easily proved by noticing that the number of rays is less when they strike the same surface obliquely, whence they cannot be but more rare.

That the brightness decreases in the same ratio by which the sine of the angle of incidence decreases, is demonstrated by a simple geometrical argument. With reference to Fig. 4.9, on the plane AB, parallel rays are incident between CA and DB. Since the brightness is associated to the number of rays divided by the surface on which they are incident, the same number of rays must be divided in a case (normal incidence) by AE and in another case by AB (oblique incidence), and therefore the density on AE will be to the density on AB reciprocally as the lines themselves, or directly as AB to AE, thus the third law.

But Lambert believed that these a priori arguments were not completely stringent, also because he saw a certain circularity in these arguments as they partly take for granted what one wants to prove and he suggested that everything must rest with the experience; not only to verify the three laws separately but also and especially the consequences that derive from them. But also an experimental verification maintains a circularity. The three laws cannot be independently proved, even empirically, but one of them must be given by definition or rather as a hypothesis that cannot be directly verified. This depends on the fact that there is not a photometer that allows a direct measure. Lambert was not explicit, but he assumed the first law as a hypothesis. He said that it is clear for instance that two candles produce a greater brightness than one, but this is all the eye can say because it has no criteria to establish that the brightness is doubled. It is however natural assume that if the intensity of light doubles also the sensation doubles; even though this is not certain, and indeed in the case of sound for instance, the sensation increases with a logarithmic law. Similarly one can see a sheet of paper more remote from the candle to be dimmer than another which is closer, but how much the eye cannot say.

Of the experiments Lambert carried out to prove his laws, only one, named Experiment 1, is referred to here. It is the simplest but contains the essence of photometry. With reference to Fig. 4.10, let consider the horizontal plane ABC with two equally bright candles placed at A, a white screen BE is set up so that the rays from A are incident normally on it. At HL it is placed another smaller screen, so that the shadow

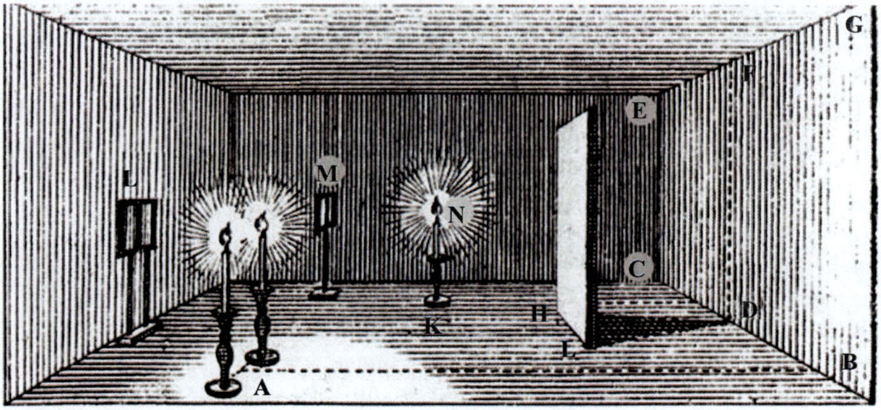

Fig. 4.10 An experiment of photometry, redrawn from [137], Fig. 2 of Table I

proceeding from each candle at A covers the part DFEC of the bottom screen. A third candle equally bright as the preceding ones, is placed at K, so that only the left part of the bottom screen, DFEC is illuminated by it. While maintaining this condition, candle K is moved toward the plane BE or moved away from it, until both part DG and DE appear equally illuminated.

Having done this, the distance of the candles from plane BD is measured; it is found that AB is to KC as $\sqrt{2}$ is to 1; or, the square of the distance AB is to the square of the distance KC as 2 to 1, or more generally the square of the distance AB is to the square of the distance KC as the candles at A are to the candles at K. The experiment will be more accurate when the size and brightness of individual candles are more nearly equal. Notice that in this experiment only mathematical laws 1 and 2 are considered, because only orthogonal incidence is assumed [137].[89]

The three laws of photometry are reexamined in depth in Chap. 3, by rediscussing their analytic structure. Lambert did so by starting with general analytical expressions for these laws, that for him could always be represented by a series expansion. What he gave for granted were only the three semiquantitative laws already considered in Chap. 1.

At the end of lengthy reasonings of which I do not intend to discuss here the legitimacy, Lambert obtained the expression of the brightness η as a function of the distance x, the number of candles z, the sine of the angle of incidence s and and inessential constant of proportionality A, in the form:

$$\eta = A\left(\frac{zs}{xx}\right)^m$$

[89]Section 58, pp. 29–30.

To conclude later that $m = 1$ should be the correct choice not to go against common experience [137].[90] This equation resumes all the three Lambert's experimental laws.

To be sure of the correctness of his law Lambert performed several experiments similar to those referred to in Chap. 1. However, they were more accurate because repeated more than once and the estimated results were obtained through averages. For instance Lambert instead of the value $\sqrt{2} = 0.707$ for the dependence of the distance obtained 0.714 [137].[91] This difference is "from the errors which can creep into the judgement of the eye" [137].[92]

In the elaboration of the experimental data Lambert introduced a statistical analysis, which did not reduce to the mean only, and if even based on simple criteria, it represented the first effort for a critical analysis of a series of experimental data [108].[93] Indeed Lambert should be credited as the main predecessor of Gauss for the theory of errors. Referring to Jakob Bernoulli *Ars conjectandi*, published in 1713, Lambert observed that:

Section 272 If some law is tested, it is enough that the residues be less than the maximum error the eye can make.

Section 273 If the experimental conditions are varied and the residues vary in proportion to the changes made, then the law is untrue.

Section 274 If the law is true and universal, positive and negative errors should occur indifferently.

Section 275 If either positive or negative errors predominate, then either the assumed law is not true in all its details, or there is some systematic instrumental error.

Section 279 If the same experiment is taken to have been repeated infinitely, it is correct to assume that the mean value among all does not differ from the truth.

He then classified errors according to their origin (Section 282), proved the necessity of rejecting the extreme observation (Section 287–291), estimated the precision of observations (Section 294) using the divergence of the arithmetic mean of all observations from the arithmetic mean obtained after having rejected the extreme observations. He raised also the problem of introducing a statistical analysis based on the maximum probability that the measured magnitude would differ the less from the true value (Section 295), he considered a continuous frequency curve (Section 296), formulated the principle of maximum likelihood (Section 303) and deduced the likelihood equation, stating, however, that in most cases the maximum likelihood estimate will not differ from the arithmetic mean (Section 306) [182].[94]

In Chap. 1 Lambert had argued that his three laws of photometry cannot be verified experimentally separately, but that one of them must be considered true by hypothesis, essentially because there is not a photometer that gives an absolute measure of the

[90] Section 251, p. 117.

[91] Section 257, p. 122.

[92] Section 269, p. 128.

[93] p. 24.

[94] p. 250.

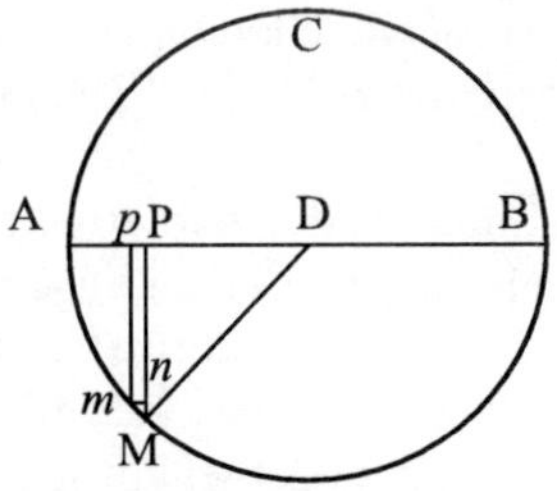

Fig. 4.11 Illumination from elements with different inclinations, redrawn from [137], Fig. 5 of Table I

amount of light, or rather illumination. The treatment carried out in Chap. 3 allows us to see these statements in a different light. In effect empirical experience shows that the three laws, even if they cannot be verified separately, can in some way be verified simultaneously. Or rather what can be experimentally verified is that if two sets of quantities (z_1, s_1, x_1) and (z_2, s_2, x_2) give the same value of the function $\eta = zs/xx$, then the brightness found is the same. At this point two things are possible. One can stipulate that η is the measure of brightness, intended as the perceived light; or one can say that η actually measures a physical magnitude, in particular the illumination, or the luminous flux per unit area; in such a case it is only assumed that the brightness, depends on the illumination, in the sense that two brightnesses are equal when their illumination is equal. Lambert seems to opt for this second possibility.

In Chap. 2, Lambert considered the illumination due to an extended sources. Reference is made to the sun surface as a practical example. The question is posed: does elementary surfaces seen with different angles produce the same illumination? In particular does a particle located at the center of the sun disk produce the same illumination of a particle located at the borders? The response is simple and furnished soon by the examination of Fig. 4.11 which is assumed to represent the sun disk, where the diameter AB is orthogonal to rays converging toward the eye of the observer located with his feet on the plane of the figure. From the figure one can see that the small surface *m*M on the boundary of the disk has an apparent surface *mn* (or *p*P), thus: "the increment of illumination proceeding from any element of surface, *m*M, is not as the true area of the particle *m*M, but rather as the apparent, *p*P, which covers the solar disk" [137].[95]

Thus, if I is the luminance (light flux per unit of emitting surface), representative of the number of candles per unit of surface, in the orthogonal direction, the effective luminance is $I d\omega \sin\varphi$, where φ is the angle ADM. The brightness/illumination does not depend only on the inclination (φ) of the radiating surface, but also by the inclination of the receiving surface (ψ). With reference to Fig. 4.12, the luminous flux on the elementary surface dA, or the brightness, due to the elementary illuminating surface $d\omega$ (representative of the area M*m*), is given as:

$$dd\eta = \frac{I}{D^2} d\omega \sin\varphi dA \sin\psi$$

[95] Section 79, p. 40.

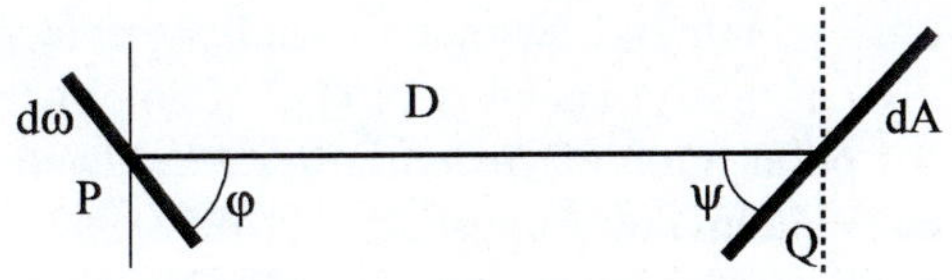

Fig. 4.12 Angles of emission and incidences

Lambert did not furnish an explicit expression like this; but expressed it rhetorically and used as a matter of fact this expression in some numerical applications he carried out. It should be noticed that the above relation is a local one, and in principle it allows to evaluate the total flux that a luminous source of any shape transfers to a receiving body of any shape, on condition that one knows the distribution of I and is able to solve a double integral.

Lambert Versus Brouguer

In 1758 Lambert had bought a copy of Pierre Bouguer's *Essai d'optique*. He acknowledged his debt to Bouguer's earlier work as well as to that of Euler and others, but he had not seen Bouguer's *Traité* before his own *Photometria* was published. Lambert and Bouguer covered much the same ground, but their approaches were very different in nature. Both dealt with photometric measurement of direct, reflected and transmitted light, developed formulae for the attenuation and dispersion of light in transparent media and applied in the investigation of light from the sun, planets and stars. Apart from his cosine law for the intensity of illumination of a surface by an oblique beam of light, there was little in Lambert's work that had not also appeared in Bouguer's *Essai d'optique*. There is no doubt that Bouguer was by far the more inspired experimenter of the two and the originator of many results that Lambert merely copied from his *Essai*. Both were adamant about the primacy of experimental data; they differed in that Bouguer was vague about his definitions of photometric magnitudes and in places relied on deduction from a mechanical model in order to predict his experimental results, whereas Lambert's theoretical approach was purely phenomenological and mathematical in nature. Both however, like so many of their contemporaries, relied unconsciously on an assumed simplicity in the laws of nature.

The pervasive use of mathematics, and of Calculus in particular, is one of *Photometria* characteristic. It was not uncommon however in optical works, and Bouguer also made use of extensive use of mathematics, of Calculus too, though to a very little extent. Lambert's approach caused that criticisms were raised by experimental physicists of the time who accused him of being more interested in mathematics than in photometry [141].[96] This is however a criticism that could be extended to all modern physics where the language of mathematics has replaced that of traditional experimental philosophers.

There were few, if any, use and appraisal of *Photometria* for almost a hundred years from its publication. Only Lambert returned to photometrical studies twice while he was at Berlin academy of science. Particularly interesting is what he wrote in 1768 on photometric aspects of paintings [138]. At the moment there is not a

---

[96]pp. xcviii-xcix.

clear picture of how much *Photometria* influenced studies carried out by the physicists of the 19th century [141].[97] Photometry was taken up again only in 1854 with the publication of *Grundiss des Photometrishen Calcüles* (Outline of photometric calculations) by August Beer (1825–1863) [32]. In his introduction Beer stated that nothing substantial had been added to photometry since Lambert's treatise.

4.3 Electricity as a Paradigm of Experimental Sciences

The history of studies on electricity is particularly interesting because it shows the passage that took place in a single century, the 18th century, from the conception of traditional natural philosophy, in which electricity was little more than a curiosity—rubbed amber attracted small pieces of paper or other—to a science and a technology based on precise mathematical laws that showed the potential of this phenomenon, fully developed only in the 19th century. As a matter of fact if the 18th century from a philosophical or political point of view is considered as the century of the Enlightenment, from the point of view of science it should be considered the century of electricity.

Of course, even though the 18th century saw a rapid development of the knowledge of electricity, this did not fall from the above but was a consequence of a long incubation period, in which though no new phenomena were discovered, electricity was at the center of the attention of natural philosophers, experimental philosophers and mathematicians. Beginning in the second half of the 17th century, new scholars of the philosophy of nature appeared on the scene who did not come from universities or academies but were self-taught or had followed a short period of scientific literacy by attending numerous seminars then in fashion, or private lessons. Electricity was an extremely interesting topic for these groups. Although no one caught a glimpse of technological use, as was the case for other sectors of physics such as mechanics, hydraulics and optics, and even terminology, electricity began to present a series of very interesting phenomena, which sometimes also took a turn that we could define parlor.

It should not be ignored indeed that a substantial contribution to the positive results in the study of electricity had come from the pleasant form that the scientific disclosure had taken, particularly congenial to the spirit of the times, especially in the 18th century. To cultivate physical and in particular electrical experiences, became a fashion soon widespread in all elegant European environments to the point of infecting the same French court. Scholars of the time indulged with frivolous but spectacular experiments: for example tables electrified secretly, which made sparks emerge from the guests' forks; or the electric Venus, the kiss of a lady, electrified and isolated, from whose lips an electric shock was received. With the use of the Leiden jar, a condenser which allowed a more violent discharge, it also leaded to mass

[97] pp. clxvi-clxxxiii.

shocks: in the real presence, the academician Louis-Guillaume Le Monnier (1717–1799) electrified 140 courtiers, overtaken by Antoine Nollet with 180 gendarmes in the royal gardens, and 200 Cistercians in their convent.

If these behaviors arouse our perplexity it should be remembered that the spirit of the time required that even the most serious arguments were treated with finesse and elegance. At that time the scientific cabinets were not far from the living room and even very strict scholars felt he need to address their readers by introducing scientific arguments with gallant tales. After all the results encouraged them, as testified by the Italian and European fortune of the *Il newtonianismo per le dame* of 1737 [7], the original work by Francesco Algarotti (1712–1764) on Newtonian optics, in which the author made rococo lightness coexist with the anxiety of philosophical renewal.

4.3.1 Early Theories About Electricity

In the following few sections I will try to summarize the main contributions of the 17th century to electricity which were very numerous. Every philosopher of nature who dealt with the problem proposed his own explanation. The purpose of this section cannot be that of referring a complete account of the various approaches, for which one can make reference to excellent past and present histories, such as for instance [122, 171]. It is rather to give a general idea of the way the study of electricity was subtracted to traditional natural philosophers to pass into the hand of experienced experimenters with a mathematical background more or less profound who mostly avoided metaphysical considerations. To this end I will briefly summarize the ideas of some of the protagonists. These were characters that according to modern categories are classifiable as philosophers of nature but not of scholastic style and in some way heretics: William Gilbert, Niccoló Cabeo and Descartes. All three followed in some way a natural philosophy of Aristotelian mould. Gilbert maintained the ideas of form and substance, but made an extensive recourse to experience. Cabeo approached a substantially mechanicistic conception of nature and made an important use of mathematics. Descartes, who officially declared himself an enemy of the philosophical approach of schools, still moved within the traditional framework of the philosophy of nature with the fundamental objective of the search for (mechanical) causes.

4.3.1.1 Mechanical Effluvium Theories

William Gilbert
The official entrance of electricity into natural philosophy as a subject worthy of interest by scholars can be rooted in the *De magnete* by William Gilbert published just at the beginning of the 17th century [103]. Gilbert's argumentations can today be classified as natural philosophy, although the importance given to experience, also to contrived experiments, was fundamental. Independently of the fact one wanted

to consider Gilbert as a "revolutionary hero" or a "moderate peripatetic and not above plajarizing those he critics", the *De magnete* is a milestone and affected either positively or negatively the ideas of the 17th century in electricity and magnetism.

Electricity is dealt with in Chap. II of Book I of the *De magnete*, devoted in particular to the behavior of amber. Here it is distinguished from magnetism at the ontological level:

> In all bodies everywhere are presented two causes or principles whereby the bodies are produced, to wit, *matter and form* [emphasis added]. Electrical movements come from the *materia*, but magnetic from the prime *forma*; and these two differ widely from each other and become unlike, the one ennobled by many virtues, and prepotent; the other lowly, of less potency, and confined in certain prisons, as it were [103].[98] (D.15)

Gilbert had to differentiated sharply the pure bond of sympathy uniting iron and lodestones from the promiscuous behavior of amber. The distinction was crucial; a sympathy should be mutual and shared between two similar objects, while the attraction of amber is toward many objects of different kind. According to the principle that no action can be executed by matter except by contact—a part from sympathy—a principle accepted both by corpuscularians and scholastics, Gilbert deduced that a material bond should exist between the excited electric body and the attracted chaff.

Gilbert claimed that the attractive property of amber belongs to a wide variety of different substances such as common glass, sulfur, resin, precious stones and wax; and this was an important achievement of his. He observed that all these substances were almost hard and transparent and therefore, for the ideas of that time, formed by consolidation of watery liquids. Thus, he concluded that the common menstrum of these liquids must be a particular humor, whose possession explained the electrical properties.

Friction might be supposed to warm or otherwise excite and liberate the humor, which would then issue from the body as an effluvium and form an atmosphere around it. The effluvium, meeting light objects produces the union commonly (but improperly) known as an electrical attraction. The 'attraction' is explained by an analogy; all bodies, said Gilbert, are cemented together by moisture; wet bodies on the surface of water attract one another when sufficiently close, and drops of water on a dry surface unite when contiguous, thus the electric humor acts similarly to water by joining the various parts of chaff among them and to the electric body.

Below the description of the causal explanation of the electric attraction:

> Therefore the effluvium called forth by a friction that does not clog the surface—an effluvium not altered by heat, but which is the natural product of the electric body—causes unition and cohesion, seizure of the other body, and its confluence to the electrical source [...]. The effluvia spread in all directions: they are specific and peculiar, and *sui generis*, different from the common air; generated from humor; called forth by calorific motion and rubbing, and attenuation; they are as it were material rods—hold and take up straws, chaff, twigs, till their force is spent or vanishes; and then these small bodies, being set free again, are attracted by the earth itself and fall to the ground [103].[99] (D.16)

---

[98] pp. 52–53. Translation in [104].

[99] pp. 59–60. Translation in [104].

Niccoló Cabeo

Nicoló Cabeo (1586–1650) was a disciple of Giuseppe Biancani (1566–624) and professor at Parma and Genoa. His most successful treatises were *Philosophia magnetica, in qua magnetis natura penitus explicatur, et omnium quae hoc lapide cernuntur, causae propriae afferuntur* of 1629 [42] (hereinafter *Philosophia magnetica*) and a commentary on Aristotle's *Meteorologica*, *Commentaria in libros meteorologicorum* in 1646 [43] (herein after *Commenatria*) which was reedited posthumous in 1686 with the title *Philosophia experimentalis*.

The *Commentaria* was actually an original presentation of the physics of terrestrial phenomena based on observations and experiments. As in other works of this type by Jesuit authors, elements of experimental physics were integrated along with Aristotelian doctrine. Cabeo separated *physica*, as he called natural philosophy, from metaphysics and mathematics, both of which he considered speculative. For him physica is concerned with the sensible only.

Substantial forms were the main target of Cabeo's attack on metaphysics. The common conception of forms as essences was mistaken, according to Cabeo; for him forms were real, physical, material entities, namely spirits and vapors with powers and virtues. What Aristotle called form, and what some considered metaphysical, is in fact a specific type of body. It is a spirit, a vapor that consists of small particles of matter and contains active forces that order the world [156].

Cabeo had referred about electricity in his *Philosophia magnetica*, a text that though modeled on Gilbert's *De magnete*, contained many and interesting new experiments. According to him for the earlier opinions about the cause of electrical attraction, only Gilbert's, was found worthy of examination, though he also was largely unsatisfactory. Whatever attraction is, it does not involve a humor that cements all things: "These are words introduced for eloquence, not for explaining the cause and method of attraction" [42].[100] Experiment shows that some things concreted of humor—metals and certain gems, for example—do not attract; and that others, which contain no more humor—like glass and other gems—do. In any case, for Cabeo, Gilbert's watery effluvia cannot act as put across. Fluids adhere in proportion to their viscosity; since cohering glass plates separate more easily in air than under water, they should separate still more easily in a subtle humor [42].[101] What then causes the attraction between floating sticks that Gilbert saw as the prototype of electrical interactions?

Cabeo distinguished four species of attractions. First was the standard sympathy which he accepted as the cause of magnetic action. Then came attraction by gravity or levity, through which a body tends towards its place by a native quality. The third attraction operated when bodies moved to fill a vacated place, or rarefied air condenses. Finally there was proper attraction, by which one body draws a second through others conjoined, as in pulling a boat with a rope [42].[102] It is the third kind of attraction which was implicated in electrical phenomena.

[100] p. 183.

[101] p. 185.

[102] p. 192.

According to Cabeo, friction opens pores of the electric body to streams of subtle effluvia, which beat back and rarefy the surrounding air; the air returns to re-establish its former density, driving light bodies before it. This resemble Plutarch's theory toward which Galileo also inclined "Amber, diamond, other joys and very dense materials, heated, attract light corpuscles, and this is because they attract the air in cooling down, and the air blows to the corpuscles" [101].[103]

Cabeo observed that chaff attracted by a large flat electric body tends to go to the edges, a consequence, he said, of the concentration there of the effluvia projected from the center of the body. In other experiments attracted sawdust particles adhered to one another to form a thread. Cabeo thought that the wild fluctuations of the far ends of these threads, from which particles occasionally flew off, were an ocular demonstration of the suppositious aerial motions [122].[104]

Moreover, Cabeo observed that a strongly electrified body sometimes drew scraps of iron or wood with such force that they rebounded to a distance of three or four inches [42].[105] On the strength of this remark he has been sponsored as the discoverer of electrostatic repulsion. He did not however consider that he had found anything novel; on the contrary, he saw in this 'repulsion' confirmation of his mechanism of attraction, indeed were sticky effluvia (as supposed by Gilbert) the agent of electricity, drawn bodies could never rebound, but once arrived must remain attached as Gilbert wrongly said they did.

René Descartes and the Cartesians

Descartes considered a parochial argument that related to electricity and treated it very shortly in his *Principia*; just after a long discussion on magnetism. Descartes explanation was not very different from those of Gilbert and Cabeo. But while the last two did not explain in detail the mechanism of attraction, Descartes did, even though his explanation may baffle a modern reader.

According to Descartes the electric matters such as amber, wax, resin, glass etc, have all the same property of glass; a product of fire which makes the constituent third-element particles, smooth and plane so that they can adhere. There are of course intervals between parts of the third-element, which are mostly long, so that only the middle of these intervals is wide enough to give passage to parts of the second element which make transparent the glass. What remains of these intervals are small slots, so narrow that there is nothing but the first element that can occupy them. It is to notice, concerning this first element, whose property is to take the figure of the places where it is, that while it flows by these small slots, the less agitated parts of it join and make up strips that are very thin, but which have a bit of breadth and come spinning around all the parts of the glass, without ever to get away from it, because the passages they find in the air, or in the other bodies that surround it, are not adjusted to their measure, to receive them [83].[106] When one rubs the glass strongly enough, so that it warms

[103] Vol. 3, p. 399.

[104] pp. 192–194.

[105] p. 194.

[106] p. 496.

up a little, these strips are driven out of their pores and are forced to go to the air and other bodies around, where not finding pores so fit to receive them they go back to the glass, and bring with them feathers or others small bodies [83].[107]

Descartes's disciples followed his approach by giving electricity scarce relevance and repeating substantially master's arguments. This is also true about Jacques Rohault in his physical treatises *Traité de physique* of 1671 and *System of natural philosophy* of 1723, the famous English translation of the *Traité* with notes by Samuel Clark. Below as Rohault explained the reason of electricity; he did not use the same terminology as Descartes and his explanation is more didactical, but the main assumptions are the same.

Suppose, said Rohault, that there is a certain matter, which is, very subtle continually moving in the smallest pores of the electric bodies and that it comes from the center to the surface where it is reflected inwards by the resistance of the air which it then meets with. Now when these bodies are rubbed this gives a sufficient force to the subtle matter contained in them, to overcome the resistance of the air and to extend itself to a little distance all round them; but because the subtle matter cannot go very far without losing some of its force, the agitation and circulation of the air will drive it back and force it to return and enter into some of the pores which it came out of, and where other matter cannot so conveniently enter, because it is not so well proportioned to the bigness and figure of those pores. Thus in amber, for example, that has been rubbed, a great number of the particles of the subtle matter, like so many fine threads, too small to be seen, come out of it and dart themselves in to the air, where meeting with small bodies, they get in to the pores of them and then return back into the amber; at the same time, the air continually repelling these small threads, and forcing them to contract themselves into less and less compass, presses likewise in the same manner upon the light bodies into the pores of which these small threads have thrust themselves; so that in returning back to the amber they carry small straws, in whose pores they are engaged along with them [176].[108]

Neither Huygens devoted much attention to electricity; he started to have some interest only since 1672, under the inspiration of Otto von Guericke's (1602–1686) work with the sulphur sphere. Some notes of Huygens' studies of 1692-93 are collected in his *Oeuvres*. Below an explanation of the attraction of a flock of wool by an electrified amber (which had replaced Guericke's sulfur) sphere, based on the assumption of vortices:

> Evidently a certain vortex of invisible matter surrounds the flock of wool, which vortex originates in and is transmitted by that set up about the sphere [of amber] by rubbing. The vortices impede and prohibit the flock's motion towards the sphere. But once the finger is removed or because of some other disturbance, the flock accesses the sphere [130].[109] (D.17)

Huygens also recognized electrostatic repulsion; for him two flocks each having acquired a vortex from the sphere will repel each other without a previous attraction.

[107] pp. 496–497.

[108] pp. 186–187.

[109] Vol. 19, p. 615.

4.3.1.2 Fluid Theories

Gilbert, Cabeo and Descartes assumed a similar mechanism: by rubbing an electric body an effluvium comes out and acts on light bodies so that they tend to join the electrified body; the effluvium then reenters the electrified body.

Other explanations of electricity were however possible. Of particular relevance are those suggested by Newton. He assumed the existence of an electric spirit which, for reason unknown to us, is endowed with attractive forces. Newton's ideas on the matter are spread in many writings, drafts of the *Principia* and *Optics*, glosses and in the alchemical writings, largely unexplored, as it has finally become clear to the 20th century historians. Here there is no room to discuss Newton's opinions on electricity, that can be found in recent literature [118, 119, 125, 166, 195] and only some quotations where he described the properties of his electric spirit are referred to. Instead of the well known quotations from the Queries of the *Opticks* and the general scholium of the *Principia*, less known but very explicit writings are referred to.

Below an interesting comment from Newton's *Second paper on colors and light* of 1675 (read at meetings of the Royal society, but never printed in the Philosophical Transactions):

> And as this condensed matter by rarefaction into an aetherial wind (for by its easy penetrating and circulating through glass I esteem it aethereal) may cause these odd motions, and by condensing cause electrical attraction with its returning to the glass to succeed in the place of what is there continually recondensed; so may the gravitating attraction of the earth be caused by the continual condensation other such like aethereal spirit, not of the main body of phlegmatic aether, but of something very thinly and subtilly diffused through it, perhaps of an unctuous or gummy, tenacious, and springy nature, and bearing much the same relation to aether, which the vital aereal spirit, requisite for the conservation of flame and vital motions, does to air [36].[110]

So by rubbing an electric body the aetherial (electric) spirit is issued from it as a wind, which by reaching nearby bodies makes them to be moved toward the body by its condensation and its unctuosity. A not dissimilar picture from Gilbert's, of whom Newton knew the *De magnete* [195].

The possibility for an electric aether and its properties, are better specified in the following quotation, drawn from a draft of Query 31 of the *Opticks*:

> There are therefore Agents in Nature able to make the particles of bodies attract one another very strongly & to stick together strongly by those attractions. One of those Agents may be the Aether above mentioned whereby light is refracted. Another may be the Agent or Spirit which causes electrical attraction. For tho this Agent acts not at great distances except when it is excited by the friction of electrick bodies: yet it may act perpetually at very small distances without friction, & that not only in bodies accounted electrick but also in some others. And as there are still other mediums which may cause attractions, (such as are the Magnetick effluvia); it is the business of experimental Philosophy to find out all these Mediums with their properties [119].[111]

[110] Vol. 3, pp. 250-251.

[111] p. 102.

The idea of an electric fluid (that is a not inert fluid such as water, but endowed with a power) was pursued further after Newton death, framed in a more general context of ancient philosophy and alchemy for which many physical phenomena were attributed to some fluids. An important representative of them was heat, conceived by many physicists and chemists of the 18th century as a fluid substance and no longer associate to the agitation of elementary particles (return to Aristotle?). After Boherhaave studies of 1720s of chemical nature, in particular after the publication of his *Elementa Chemie* in 1732, where heat was understood in terms of a fluid (fire), even electricity started to be associated to fluids. Two main theories should be evidenced, that of two fluids, published by Robert Symmer in 1759 and that one fluid developed first by William Watson and then by Benjamin Franklin in 1746–1747.

The theories of two fluids is well summarized by Priestley in his *The history and present state of electricity, with original experiments* of 1771. Symmer assumed that there were two different electric fluids, or emanations of two electric powers, essentially different from each other; that electricity does not consists in the afflux and efflux of these fluids, but in the accumulation of the one or the other of them in electrified bodies; or, in other words, it consists in the possession of a larger portion of one or the other power, what is requisite to maintain a balance within the body; and lastly, that, according as the one or the other power prevails, the body is electrified in the one or the other manner. According to Symmer, the principle of two distinct electrical powers, upon due consideration, do not disagree with the general laws of nature. It is one of the fundamental laws of nature, that action and reaction are inseparable and equal; and, when one looks round, he finds that every power which is exerted in the material world meets with a counteracting power, which controls and regulates its effects, so as to answer "the wise purposes of providence" [171].[112]

Let us suppose that the friction of any electric body produces a separation of these two fluids, causing, in the usual method of electrifying, the vitreous electricity of the rubber to be conveyed to the hand and the resinous electricity of the hand to be conveyed to the rubber. The rubber will then have a double share of the resinous electricity, and the conductor a double share of the vitreous; so that no substance whatever can have a greater or less quantity of electric fluid at different times. The quality of it only can be changed [171].[113]

Franklin, following Watson, assumed electricity due to one fluid only. He recognized however that there were two different kinds of electrification, the one associated to an excess of electric fluid and the other associated to a defect of electric fluid, to which Franklin referred to as positive or negative electrification respectively. Below there is the description that Franklin himself gave of his theory.

1. The electrical matter consists of particles extremely subtle, since it can permeate common matter, even the densest metals, with such ease and freedom, as not to receive any perceptible resistance.

---

[112]p. 255.

[113]p. 432.

2. Electrical matter differs from common matter in this, that the parts of the latter mutually attracts, those of the former mutually repel, each other. Hence the appearing divergency in a stream of electrified effluvia.
3. But though the particles of electrical matter do repel each other, they are strongly attracted by all other matter.
4. In common matter there is (generally) as much of the electrical, as it will contain within its substance. If more is added, it lies without upon the surface, and forms what we call an electrical atmosphere: and then the body is said to be electrified [94].[114]

Fluids have the properties of conserving; thus electricity or charge (Franklin term) conserves. When a piece of resinous body, such as amber, is electrified by rubbing it at one extremity, the electric fluid which is subtracted from this extremity is restored from the other extremity by the hand of the man rubbing the amber. Indeed being the hand and the body of the man a conductor they can receive electric fluid from the ground which contains an enormous amount of it. This is what is called pumping effect.

A modern reader probably could not pronounce about the theories of one electric fluid or two electric fluids, because the present conception is a merging of the two. Today it is thought that there are two different kinds of electricity, the positive one associated to protons and the negative one associated to electrons, and this is in accord with the theory of two fluids. Only electrons can however move, and this is in accord with the theory of one fluid, whose particle are the electrons. Notice that there is an inversion in the nomenclature. While according to Franklin an excess of electric fluid gives a positively charged body; today a body with an excess of electrons is named negatively charged; this because the resinous electricity, that today is characterized by an excess of electrons, was in the past considered deprived of fluid electric and then negatively electric.

4.3.2 Some Elements in the History of Electricity in the 18th Century

The present section does not want to report a history of electricity, even though largely summary. The goal is only to provide sufficient elements for the understanding of the following sections in which the relationship between natural-philosophy-epistemology-mathematics of some of the protagonists of the electricity history is explored.

The development of electricity in the 18th century is not only characterized by its rapid growth but also by its modalities in particular:

1. Many of the contributors did not have an academic background.

[114]pp. 52–53.

Table 4.3 Electricians by professions 1700–1790. Taken from [122], p. 99

		1700	00–39	40–49	50–59	60–69	70–79	80–89
A.	Jesuits	7		6	6	6	6	1
	Univ. profs.	4		4	1	1	4	1
	College profs.	1		2	4	4	2	
B.	Academicians[a]	3	1	4	10	4	5	10
	Big three[b]	1	1	4	10	3	3	5
	Others	2				1	2	5
C.	Professors[c]	4	10	31	22	16	29	25
	Universities	4	9	24	18	11	23	15
	Colleges		1	7	4	5	6	10
D.	Lecturers	3	4	4	2	2	3	3
E.	Others	10	5	43	32	19	28	40
	Britain	5	5	22	14	11	15	13
	Elsewhere	5	21	18	8	13	27	
	TOTALS	27	20	88	72	47	71	79

a Salaried only, except for associate at AS
b Paris, Berlin, Petersburg
c Exclusive of Jesuits

2. This fact was made possible by the affirmation of the experimental philosophy that had almost entirely replaced the traditional natural philosophy.
3. By the development of a bourgeoisie that had enough spare time and also an entrepreneurial mentality.
4. By the diffusion of the experimental philosophy outside the academic sphere with cycles of lectures.

Below I will provide a brief comment on these points. The following Table 4.3 lists the electricians of the 18th century according to their profession during the years 1700–1790; from it is clear the relevance of not professional researchers.

According to Priestley, the history of electricity before Coulomb, can be divided into 10 periods [171]:

I Before the first important discoveries of the 18th century.
II Experiments of Francis Hauksbee, Isaac Newton's lab assistant.
III The experiments of Stephen Gray who discovered the transmission of electricity at distance.
IV Experiments and discoveries of Charles François de Cisternay Dufay who discovered that electricity comes in two forms which he called resinous and vitreous.
V The continuation, and conclusion of the experiments of Stephen Gray.
VI The experiments of John Theophilus Desaguliers.
VII The experiments of the Germans and of William Watson, before the discovery of the Leyden jar in the years 1745–1746.

VIII From the discovery of the Leyden jar in the 1745 and 1746, to Benjamin Franklin discoveries. The methods used by the French and English physicists, to measure the distance to which the electric shock can be carried, and the velocity with which it passes. Experiments on animal and other organized bodies in this period and other experiments connected with them, made chiefly by Antoine Nollet. The history of the medicated tubes, and other communications of medicinal virtues by electricity.

IX The experiments and discoveries of Franklin himself. The discoveries concerning the Leyden jar, and others connected with them.

X From the time that Franklin made his experiments in America, till the year 1766.

4.3.2.1 Time Table of the Main Achievements on Electricity

1600 William Gilbert (1544–1603) first diffused the term electricity (electrica) and electric from the Latin word electrum for amber (electricus, 'of amber') and Greek (ἤλεκτρον).

1660 Otto von Guericke (1602–1686) invented a machine, a rotating sulphur sphere, that produced static electricity.

1675 Robert Boyle (1627–1691) discovered that electric force could be transmitted through a vacuum and observed attraction and repulsion.

1705 Francis Hauksbee (1660–1713) discovered that rubbing a glass ball, evacuated of air in order to build up a charge, a glow was visible at distance if he placed his hand on the outside of the ball. This effect later became the basis of the gas-discharge lamp, which led to neon lighting and mercury vapor lamps. Hauksbee clearly showed that the bodies previously attracted could then be rejected, but he did not associate the fact with repulsion of similarly charged bodies.

1729 Stephen Gray (1666–1736) discovered the transmission of electricity at distance and existence of two kinds of matter, today known as insulator and conductor, terms introduced by Desaguliers in 1742, though in a not technical way. He also enlarged the list of material that could be electrified [82].[115]

1733 Charles François de Cisternay Dufay (or Du Fay) (1698–1739), a polymath member of the Académie des sciences de Paris, discovered that electricity comes into two forms which he called resinous and vitreous. Similar electricities repel, different electricities attract. What now seems a simple fact confirmed by experience was not yet well recognized. Until then there was a shared idea that electrified bodies attract small pieces of matter and the repulsion which was sometimes observed could always traced back to attraction. Dufay also found that every body can acquire electrification, except metals and substances too soft or fluid to rub. Differently from the English electricians, Dufay was an educated man. Fontenelle said that he was the only man of his time to submit contributions to the Académie in any field then object of study; anatomy, astronomy, botany, chemistry, geometry, mechanics and general physics. According to Heilbron, Dufay's substantive

[115] pp. 2, 17.

discoveries: the sequence attraction-contact-repulsion, the two kind of electricities, shocks and sparking, are but one aspect and perhaps not the most significant, of his achievement. His insistence on the universal character of electricity, on the necessity of organizing known facts before grasping for new ones, all helped to introduce order at precisely the moment when the accumulation of data about electricity began to require them. He found the subject a record of often capricious, disconnected phenomena, the domain of polymaths, textbook writer, and professional lecturers, and left it a body of knowledge that invited and rewarded prolonged scrutiny from serious professional physicists [122].[116]

1745-6 Ewald Jürgen Georg von Kleist (1700–1748) and Pieter van Musschenbroek (1692–1761) with Andrea Cunaeus 'invented' the *Leyden jar* the first electrical capacitor.

1746 William Watson (1715–1787) proposed one fluid theory. He likened the agent of electricity to an aether whose particles act upon one another over microscopic distances, and upon ponderable bodies by impact. He presented his theory in the *A sequel to the experiments and observations tending to illustrate the nature and properties of electricity* of 1746 [194]. Watson concluded his treatise in the style of Newton's *Opticks* with a series of *queries* (ten); some of them are summarized below: (1) Wether or not electrical attraction and repulsion be attributed to the flux of an electrical aether? (2) Wether or not electricity or electrical aether may be elementary fire? (3) Wether or not this elementary fire may appear in different forms; as air when diffused over a large surface, as lambent flame when brought towards a point. Wether or not does it explode and become the object of feeling as well of our hearing? (4) Wether or not is this fire always intimately connected with all bodies at all times? (6) May its elasticity be inferred from empirical observations? (7) May an electrical machine be called a fire-pump in the same sense that the instrument of Otto Guericke and Mr. Boyle is called an air-pump and the aether is not generated by the machines but pumped from the ground? (8) Does the separation of fire from bodies by motion, and its restoration to them after that motion has ceased, causes us to incline that fire is an original, a distinct, principle, a substance, formed by the Creator himself, as s'Gravesande and Boherhaave believed, rather than mechanically producible from other bodies, as believed by Boyle and Newton besides others scholars? [194].[117]

1747 Benjamin Franklin (1706–1790), an American polymath and one of the founding fathers of the United States. A leading author, printer, political theorist, politician, freemason, postmaster, scientist, inventor, humorist, civic activist, statesman, and diplomat. As a scientist, he was a major figure in the American Enlightenment and the history of physics for his discoveries and theories regarding electricity. To be precise he was considered as a great natural philosopher in Europe, while in America he was mostly famous as a state man and an inventor. Franklin was not the uneducated self made man as often he is portrayed. Youth he

---

[116] p. 260.

[117] pp. 71–72.

received a good scientific training and surely read Newton's *Opticks* [66].[118] Eighteenth he went to England and made the acquaintance of a number of 'scientists', among them Henry Pemberton (1694–1771), who was then preparing—under Newton's direction—the third edition of the *Principia*.

His first studies in electricity are documented in the *Experiments and observations in electricity made at Philadelphia in America*, published for the first time in 1751 [94]. It consists in a series of letters, mostly to Peter Collison (1694–1768) a botanist fellow of the Royal society. There were five English edition of the letters. The fourth one, of 1769, is the first complete editions and also contains some other writings by Franklin [97]. There were also three French editions, and one in German and Italian as well.

Franklin took inspiration from Watson for his general one fluid theory; however even Newton could have been of inspiration. Below an interesting comparison between the properties of Franklin fluid and Newton's aether

> It is a little remarkable, that the electric fluid, in this, and in every other hypothesis, should so much resemble the ether of Sir Isaac Newton in some respects, and yet differ from it so essentially in others. The electric fluid is supposed to be, like ether, extremely subtle and elastic, that is, repulsive of itself; but, instead of being, like the ether, repelled by all other matter, it is strongly attracted by it; so that, far from being, like the ether, rarer in the small than in the large pores of bodies, rarer within the bodies than at their surfaces, and rarer at their surfaces than at any distance from them; it must be denser in small than in large pores, denser within the substance of bodies than at their surfaces, and denser at their surfaces than at a distance from them. But no other property can account for the extraordinary quantity of this fluid contained within the substance of electrics *per se*, or for the common atmospheres of all excited and electrified bodies [171].[119]

One more idea Franklin took from Watson was that of electric pump. That is that electricity produced by an electric machine, for instance, is not created but driven from ground.

An example of esteem enjoyed by Franklin is given by the great Swedish electrician Johan Carl Wilcke (1732–1796) wrote in his preface to his German translation of Franklin's text in 1759. According to Wilcke, the elaboration of the theory of electricity is the chief part of Franklin's work even though its principles were not the invention of Franklin. Indeed all these may already be discovered in other works written earlier and with which, according to Franklin's own statement, he was familiar. Still, continued Wilcke, great credit must be given to Franklin. For not only did he clarify them, but he also applied them with ingenuity to the discharging-or shock experiments which are generally associated with the names of Leyden and of Musschenbroek. These hitherto had no natural explanation which could be considered as a proof. If he advanced thereto by means of a trifle too artificial, still his explanations are less artificial than others still more incredible [66, 96].[120]

---

[118] pp. 206–209.

[119] p. 424.

[120] Preface, pp. 11–12 (not numbered pages); 595, 597.

A comparison between Newton and Franklin was not uncommon, and is still such. Franklin was considered by many the Newton of electricity. On the similarity between Newton and Franklin, are of the interest the considerations by I. Bernard Cohen [66]. In recognition of his work with electricity, Franklin received the Royal society's Copley Medal in 1753 and in 1756 he became one of the few 18th-century Americans elected as a Fellow of the society. He received honorary degrees from Harvard and Yale universities.

1752 Benjamin Franklin 'invented' the lightning rod and demonstrated that lightning was electricity. He proposed an experiment to prove that lightning is electricity by flying a kite in a storm that appeared capable of becoming a lightning storm. On May 10, 1752, Thomas-François Dalibard (1709–1778) of France conducted Franklin's experiment using a 40-foot tall (∼12 m) iron rod instead of a kite, and he extracted electrical sparks from a cloud. On June 15, 1752, Franklin may possibly have conducted his well-known kite experiment in Philadelphia, successfully extracting sparks from a cloud. This account was read at the Royal society and printed as such in the Philosophical Transactions [95]. Franklin was careful to stand on an insulator, keeping dry under a roof to avoid the danger of electric shock [57]. Others, such as Georg Wilhelm Richmann (1711–1753) in Russia, were indeed electrocuted in performing lightning experiments during the months immediately following Franklin's experiment.

Franklin's electrical experiments led to his invention of the lightning rod, for which he is famous, even though as often occurs in history of inventions there is the possibility Franklin was not the first to actually introduce the lightning rod. The exact non-Franklin origins of the lightning rod are hotly debated, see for instance [127].

1753 John Canton (1718–1772), a London schoolmaster, was to become the leading English Franklinists of the 1750s. He discovered that clouds could become electrified positive and negative independently of Beccaria and Franklin. In 1753 he published in the Philosophical Transactions of the Royal Society a seminal paper, *Electrical experiments, with an attempt to account for their several phaenomena, together with observations on thunder clouds* with the declared purpose to study the nature of electricity in the clouds [45]. Here he presented a series of experiments (9), that are now considered the proof of the electrostatic induction (modern term). Today we refer to electrostatic induction, as a modification in the distribution of electric charge on one material, typically a conducting one, under the influence of nearby objects that have electric charge. Thus, a negatively (positively) electrified body A brought near an electrically neutral body B induces a positive (negative) electricity on the side in front to it and a negative (positive) one on the far side. B, furthermore, becomes charged positively (negatively) in all its parts if its negative (positive) side is grounded. With the phenomenon of induction nearly any behavior of electrostatics then known could be accounted for: the Leyden jar, the condenser, the electrophore and so on.

But for Canton and his contemporaries, the result of the experiments he presented was seen simply as a new and curious phenomenon: bodies could be electrified without getting in touch. It will be only Aepinus in 1759 who explained the phenomenon using the same principles adopted today, but his results were not accepted at the moment, and various and not always convincing explanations were given for many years.

Below only the first two of Canton's experiments are discussed, which though simple, represent the essence of the phenomenon of induction. In the first experiment, two very tiny balls of cork or brass were suspended from the ceiling by conducting threads so that they were close to each other. Canton found that bringing an electrified (positively) glass tube three or four feet (1.22 m) below the balls caused a separation. Bringing the glass tube closer caused them to separate further. Finally, on removing the tube altogether, the cork balls come together again. In the second experiment the two balls were suspended by insulating threads of silk. Canton found that in this case the electrified glass tube had to be brought much closer within eighteen inches (50 cm) from the balls before they repelled each other. Furthermore, the repulsion continued for some time after the tube had been taken away [45].[121] Similar phenomena occur if a wax tube (negatively electrified) is used.

Canton explanation derived from Franklin's theory and is only partially in agreement with modern explanations. It is indeed such, or not very different, for the case of electrified wax. In such a case, according to Canton, when the excited stick of wax is brought near the balls suspended by conducting threads, the electric fluid comes from the ceiling through the threads into the balls, and condensed there, because it is attracted by the negatively charged wax which wants electric fluid. When the wax is withdrawn, the balls, being grounded, return to their natural (that is, neutral) state. For a glass electrified tube, the explanation of Canton is far from that given today, which would be specular to the previous one. Canton assumed than when the atmosphere of the positively electrified glass comes close to the cork balls, some electric fluid is assumed by the balls that becomes electrified positively and thus repels. The reason of this asymmetrical explanation is not given by Canton, but it must be probably searched in the difficulty existing in Franklin's theory to explain the repulsion of negatively electrified bodies.

1759 Robert Symmer (1707–1763), educated at the university of Edinburgh, where he took a belated M.A. in 1735, proposed the theory of two electric fluids: "I confess it was unlucky that I felt myself obliged to use, in some respect, the same terms that Mr. Franklin and others, who follow his system, make use of, while there is an essential difference in the things meant by them and by me. By the terms positive and negative, they mean, as in algebra, simply plus and minus. By the same terms I mean two distinct powers (both of them in reality positive) but acting in contrary directions, or counteracting one another" [121].[122]

[121] p. 350.

[122] Letter of Symmer to Michell, July 10. p.17.

1759 Franz Ulrich Theodor Aepinus published his *Tentamen theoriae electricitatis et magnetismi*, the first mathematical and rigorous treatment of static electricity.
1785 Charles Augustin Coulomb (1736–1806) published his paper on the inverse square law for electric charges.
1786 Luigi Galvani (1737–1798) demonstrated what is now understood to be the electrical basis of nerve impulses when he made frog muscles twitch by jolting them with a spark from an electrostatic machine.
1800 Alessandro Volta (1745–1827) invented the first electric battery. Volta proved that electricity could travel over wires.

In the previous time table two important characters of the science of electricity in the 18th century are missing: Antoine Nollet and Henry Cavendish. The former because, though one of the most famous electrician of the 18th century, he was more a science communicator than a creative person, this is at least the shared view. The latter because, notwithstanding he was a brilliant mathematician and experimenter in many field of physics, most of his writings remained hidden in his drawer, and if published were not noticed as due. Cavendish's writings were rediscovered and printed by Maxwell a century later; also his influence on the contemporary is reevaluated.

Henry Cavendish, (1731–1810), considered today the greatest experimental and theoretical English chemist and physicist of his age, was also famous as a wealthy man. Cavendish was distinguished for great accuracy and precision in researches and measurements into the composition of atmospheric air, the properties of different gases, the synthesis of water, the law governing electrical attraction and repulsion, a mechanical theory of heat, etc. Below an efficient portrait

> Among eighteenth-century British natural philosophers, Cavendish stands out for the sustained intensity of his inquiry into the workings of nature. Simply put, his life was about natural philosophy. First and foremost, natural philosophy was his work. Not ordinarily thought of as an occupation, natural philosophy offered him an activity of a kind that was compatible with his aristocratic position in the wider society. It opened for him a career of public service fully as absorbing as traditional careers in politics and government, the military, religion, law, and medicine. With his career came fellowship. Inordinately shy in public, Cavendish came together with a limited society with which he could make human contact [158].[123]

Of him only two papers on electricity were published, in 1771, the *An attempt to explain some of the principal phaenomena of electricity, by means of an elastic fluid* [55] and in 1776, the *Attempts to imitate the effects of the torpedo* [56]. But they did not fail to impress those who read them, with his great mastery of theory and experiments. Of the 20 parcels of (unpublished) papers on electricity 18 belong to the years 1771, 1772, 1773; the remaining are dated 1775, 1776 [85].[124]

To note that in fifty years Cavendish devoted to sciences, in only five was he occupied with electricity. His tastes and interests were extraordinarily varied; apart from his contribution to electricity he made important contributions to chemistry, gravitation, heat, magnetism and meteorology. His most known result is the determination

[123] p. 10.
[124] p. 10.

of the weight of the earth or, equivalently, the value of the universal gravitational constant, published in the memoir *Experiments to determine the density of the earth* of the Philosophical Transactions of 1798 [57]. His experiment to weigh the earth has come to be known as the *Cavendish experiment*.

In the paper of 1771, the only to be commented here, Cavendish following the same line of research traced by Aepinus, but "as I have carried the theory much farther than he has done, and have considered the subject in a different, and, I flatter myself, in a more accurate manner, I hope the society will not think this paper unworthy their acceptance" [55].[125]

In his paper Cavendish assumed as a principle (hypothesis) of his theory that attractions-repulsions between particles of electric fluid and between particles of ordinary matter vary according to the same function of the distance.

> Or, to express it more concisely, if you look upon the electric fluid as matter of contrary kind to other matter, the particles of all matter, both those of the electric fluid and of other matter, repel particles of the same kind, and attract those of a contrary kind, with a force *inversely as some less power of the cube* [emphasis added] [55].[126]

Notice that the word hypothesis is emphasized as the title of a paragraph. Thus Cavendish recognized, very honestly, that his theory is based on an assertion non-verifiable directly through experiments; only the consequences can. However, the acceptance of a not directly verifiable assumption is common to all the theories of electricity of the time: from Franklin to Aepinus to Nollet; only they were not explicit on the fact. One more thing to notice is that Cavendish did not make the hypothesis that the force varies as the inverse square but a more general hypothesis (inversely as some less power of the cube). This in the footprint of Newtons that in the *Principia* studied different forms for the laws for the centripetal force.

In Part 1 of his paper Cavendish laid down the theory of the electric fluid; in Part 2 he compared the propositions he found with known experiments: "I have now considered all the principal or fundamental cases of electric attractions and repulsions which I can think of; all of which appear to agree perfectly with the theory" [55].[127]

In Proposition 5 (Problem 1) of Part 1 he assumed as a special case the inverse square law for the electric force. The proposition proofs that in a conducting sphere the electric fluid at the equilibrium should be located on the surface, or equivalently that no electric fluid could be inside [55].[128] In Proposition 6 (Problem 2) Cavendish assumed the case of electric force varying with a law different from the inverse square. He confessed not to be able to furnish a general solution, but suggested that in such a case there could be electric fluid inside the sphere.

Cavendish extended these results to the case of conducting bodies with any shape, even though he stated his result in a dubitative form by asserting:

---

[125] p. 584.

[126] p. 585.

[127] p. 659.

[128] p. 593.

> From the four foregoing problems it seems likely, that if the electric attraction or repulsion is inversely as the square of the distance, almost all the redundant fluid in the body will be lodged close to the surface, and there pressed close together, and the rest of the body will be saturated. If the repulsion is inversely as some power of the distance between the square and the cube, it is likely that all parts of the body will be overcharged: and if it is inversely as some less power than the square, it is likely that all parts of the body, except those near the surface, will be undercharged [55].[129]

These propositions were not considered as particularly interesting by the readers of the 1771 paper. A different reaction would have been obtained if the same readers could have read further papers, where Cavendish returned to his propositions and made accurate measurements of electricity inside an electrified spherical surface, by finding a vanishing electric fluid, a part from experimental errors and thus giving an indirect proof that electric attraction is inversely as the square of the distance. Cavendish results's can be found in his writing collected by Maxwell in 1879 [58, 59].

To carry out his experiment Cavendish took the globe G of Fig. 4.13, 12.1 (30 cm) inches in diameter,[130] and suspended it by a solid stick *ss* of glass run through the middle of it as an axis, and covered with sealing-wax to make it a more perfect non-conductor of electricity. He then enclosed this globe between two hollow pasteboard hemispheres (H and *h*), 13.3 (34 cm) inches in diameter, and about 1/20 of an inch (1 mm) thick, in such a manner that there could hardly be less than 4/10 (1 cm) of an inch distance between the globe and the inner surface of the hemispheres in any part, the two hemispheres being applied to each other so as to form a complete sphere. The inner globe and the hemispheres were both coated with tinfoil to make them the more perfect conductors of electricity.

A communication was made by a piece of a conducting wire run through the hemispheres and touching the inner globe, a piece of silk string being fastened to the end of the wire, by which it could be drawn out at pleasure. The hemispheres were electrified by the positive side of a Leyden jar, and then immediately the wire which made communication between the inner and the outer globes was drew out by the silk string. Then the two hemispheres were instantly separated. The result was, that though the experiment was repeated several times, neither the pith balls of the electroscope T*t* (see Fig. 4.13, bottom) separate nor any signs of electricity is shown in the globe G [58].[131]

These are Cavendish conclusions:

> The 1st experiment shews that when a globe is electrified the whole redundant fluid therein is lodged in or near its surface, and that the interior parts are entirely, or at least extremely nearly, saturated, and consequently that the electric attraction and repulsion is inversely as the square of the distance, or to speak more properly, that the theory will not agree with experiment on the supposition that it varies according to any other law [59].[132]

---

[129] pp. 607-608.

[130] All the following measurements in cm in the following are approximated.

[131] pp. 105–108.

[132] p. 140.

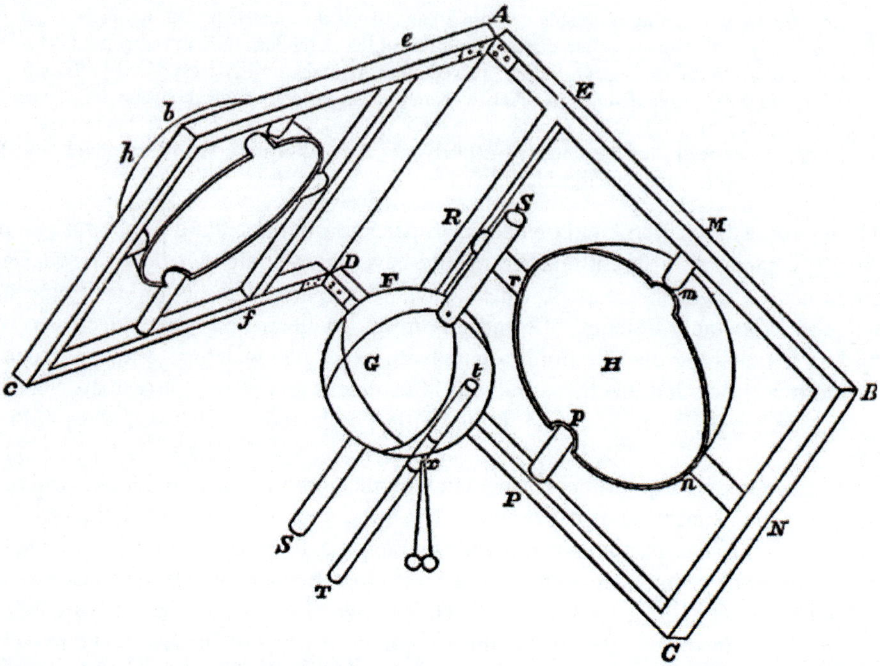

Fig. 4.13 Cavendish's apparatus to prove that electricity is logged on the surface of a conducting body [58], p. 106

In his writings Cavendish also gave an estimate of the experimental errors he could have done, so giving credence to his statement. By referring to his sophisticated experimental apparatus and the precision with which he could measure the absence of electric fluid inside a conductor, he could assert: "We must therefore conclude that the electric attraction and repulsion must be inversely as some power of the distance between that of the $2 + 1/50$th and that of $2 - 1/50$th" [58].[133]

To be honest, what Cavendish was authorized to assert is that if the electric force varies as r^{-n}, the exponent n should be close to 2; but the possibility remains that the force could vary in another arbitrary way with r, for instance according to a not algebraic function.

Jean Antoine Nollet (1700–1770) one of the great popularizers of the new electrical science in the salons and at the court of 18th-century France, studied humanities at the Collège de Clermont in Beauvais, starting in 1715. He completed a master's degree in the faculty of theology at the university of Paris in 1724. Nollet was particularly interested in the new science of electricity. He assisted Dufay, especially with electrical experiments and travelled with him in 1734 to meet physicists in England and in 1736 in the Netherlands. He was a member of the Royal society of London from 1735 (as resulting from the official list of fellows of the Royal society). In 1739

[133] pp. 111–112. The mathematical proof of this assertion is given at pp. 112–113.

he became *adjoint mécanicien* of the Académie des sciences de Paris and in 1742 *associé mécanicien*. From at least 1743, this academy identified Nollet as the person who was particularly in charge of research about electricity. In 1753 he became the first professor of experimental physics in France.

But whatever low the merits of Nollet as an electrician could have been, the world of science owes him a debt of gratitude. He was a master showman and the success of his public demonstrations of electrical phenomena stimulated interest in the subject. He was particularly well known in the world of *physique du salon*, the presentation of physics experimenting with elegant ladies of the French court.

Nollet was the great contender of Franklin, not so much at personal level but at a philosophical one. There was a profound clash between the theories of the two antagonists: the Cartesian (Nollet) and the Newtonian (Franklin), intending these labels mainly as representing mechanicistic explanation (ultimate causes) versus force explanation (proximate causes). A prolific writer, Nollet published many papers on electricity, some of them were devoted to a frontal attack to Franklin. The sixth volume of his famous *Leçons de physique expérimentale*, published in six volumes between 1743 and 1748 and reprinted many times [163], contains a vast accumulation of curious and ingenious experiments. Most of them were spectacular and they could entertain as well as instruct a courtly audience (Fig. 4.14).

Nollet theory of electricity is expressed in eighteen fundamental propositions in Chap. 21 of the sixth volumes of the *Leçons de physique expérimentale*. The propositions postulate on the existence of a subtle matter which penetrates all bodies, with different degrees of ease. It is a fluid which can be compared with elementary fire and light. An electrified body emits rectilinear divergent rays of electric fluid but simultaneously receives convergent rays of the same electric fluid, these are two characterizing principles: effluence and affluence. Bodies have two typologies of pores one for the emission of electric fluid, the other for receiving it, however the returning fluid was not necessarily that very matter fluid had left the electrified body. Yet, independently of the number of pores, some bodies at some times have a stronger current of efflux than of afflux, and hence there may be both resinous and vitreous electrification [163].[134]

As clear from previous exposition, Nollet's theory is Cartesian rather than Newtonian; indeed the electric fluid is not elastic and repelling in the sense of Newton's aether, but was more like the inert elementary matter of Descartes (Fig. 4.15).

Nollet's theory was considered very tortuous by Priestley, who commented it as follows: "A man of less ingenuity than the Abbé could not have maintained himself in such a theory as this; but, with his fund of invention, he was never at a loss for resources upon all emergencies, and in his last publication appears to be as zealous for this *strange hypothesis* [emphasis added] as at the first" [171].[135]

[134] Vol. 6, pp. 407–410.

[135] p. 416.

Fig. 4.14 Electricity for the dames [162], Frontpage. Reproduced with the permission of ETH-Bibliothek Zürich, Alte und Seltene Drucke

Fig. 4.15 Electricity for the dames [162], pl. 4, p. 216. Reproduced with the permission of ETH-Bibliothek Zürich, Alte und Seltene Drucke

4.3.2.2 Memoirs on Electricity Published in the Philosophical Transactions of the Royal Society of London

A large body of the experimental (and theoretical) researches on electricity of the 18th century was reversed on the Philosophical Transactions. Table 4.4 refers to all papers on electricity published in the first half of the century; Table 4.5 refers instead to a selection of papers of the second half of the century, when the researches on electricity had growth enormously.

4.3.3 A Representative of British Electricity. Stephen Gray

British contribution to electricity, especially in the experimentation, was fundamental in the 18th century. The Philosophical Transactions published a huge number of papers on electricity, attracting also the contributions from the Continent. A short list of British electricians helps to understand the scale of the phenomenon: Francis Hauksbee (1660–1713), Stephen Gray (1666–1736), Robert Symmer (1707–1763), Benjamin Wilson (1712–1788), William Watson (1715–1787), John Canton (1718–1772), Henry Cavendish (1731–1810), Joseph Priestley (1733–1804); to them one can add the American Benjamin Franklin (1706–1790). Below I will discuss only the contribution of Stephen Gray, which is probably the most important because he opened the main stream.

Stephen Gray (1666–1736) was an intriguing figure in the history of physical science in the 18th century. He is given great credit for his discovery that electrical effects can travel long distances over suitable 'lines' (which gives him the best of all claims to being the father of electrical communication). The content and quality of his other scientific works have remained substantially unknown; moreover only a few hints about his biography is available; below I refer a summary of what is referred to in [61, 65, 122]. The son of a dyer and brother to a dyer, carpenter and grocer, he was born in Canterbury where he lived until 1706. He had a fair education, including enough Latin to puzzle out Scheiner's *Rosa ursina*; it is not known at what extent he knew about mathematics, in particular of Calculus. He may have studied in London or Greenwich for a time, perhaps under the astronomer John Flamsteed, a friend of his. Gray belonged to the category of amateur, in the sense that he pursued his researches in his spare time and without earning a living with them; in any case he was from many points of view an educated man in scientific matter. From the beginning Gray's scientific work was characterized by astronomical observations under Flamsteed and won him the regard of the scrupulous. Much earlier, a few (11) consecutive papers, some of them very short, mainly on optical and astronomical topics were published on the Philosophical Transaction of the Royal Society of London between 1696 and 1706 [65].

Meantime his trade had become too strenuous and Gray managed for the admission to the Charterhouse, a foundation established as a day school for poor boys and a home for eighty gentlemen pensioners; his request was eventually accepted in 1719.

Table 4.4 Papers on electricity of the first half of the 18th century (52 papers over 300). The dates are according to the calendar adopted in England at the times

1708 Hauksbee	An account of some experiments, touching the electricity and light producible on the attrition of several bodies
1720 Gray	An account of some new electrical experiments
1731 Gray	A letter to Cromwell Mortimer, containing several experiments concerning electricity
	A letter concerning the electricity of water, from Mr. Stephen Gray to Cromwell Mortimer
	A letter from Mr. Stephen Gray to Dr. Mortimer, containing a farther account of his experiments concerning electricity
	Two letters from Mr. Stephen Gray, to Dr. Mortimer, containing farther accounts of his experiments concerning electricity
1733 Dufay	A letter from Mons. Du Fay to his grace Charles duke of Richmond and Lenox, concerning electricity
1735 Gray	Experiments and observations upon the light that is produced by communicating electrical attraction to animal or inanimate bodies, together with some of its most surprising effects
	A letter from Stephen Gray to Dr. Mortimer, containing some experiments relating to electricity
	An account of some electrical experiments intended to be communicated to the Royal society by Mr. Stephen Gray, taken from his mouth by Cromwell Mortimer the day before he died
1739 Desaguliers	Some things concerning electricity
	An account of some electrical experiments made before the Royal society
	Electrical experiments made before the Royal society
	Several electrical experiments, made at various times, before the Royal society
1739 Wheler	Some electrical experiments, chiefly regarding the repulsive force of electrical bodies
1739 Wheler	An account of some of the electrical experiments made by Granvile Wheler, at the Royal society's house, on May 11 1737
1739 Wheler	A letter from Granvile Wheler to Dr. Mortimer, containing some remarks on the late Stephen Gray. His electrical circular experiment
1739 Desaguliers	Some thoughts and experiments concerning electricity
	An account of some electrical experiments made before the Royal society on thursday the 16th of February 1737–8
	An account of some electrical experiments made at his Royal highness the prince of Wales's house at Cliefden, on Tuesday the 15th of April 1738. Where the electricity was conveyed 420 feet in a direct line
1742 Desaguliers	Some further observations concerning electricity
	Some conjectures concerning electricity, and the rise of vapours
1744 Wintler	Abstract of what is contained in a book concerning Electricity

(continued)

Table 4.4 (continued)

1744 Miles	A letter from the reverend Henry Miles, to Mr. Henry Baker, of firing phosphorus by electricity
	A letter from the reverend Henry Miles to the president; containing observations of luminous emanations from human bodies, and from brutes; with some remarks on electricity
1744 Watson	Experiments and observations, tending to illustrate the nature and properties of electricity.
1746 Watson	A continuation of a paper concerning electricity
1746 Watson	A sequel to the experiments and observations tending to illustrate the nature and properties of electricity; in a letter to the Royal society from the same
1746 Miles	Extracts of two letters from the rev. Henry Miles to Mr. Henry Baker, concerning the effects of a cane of black sealing-wax, and a cane of brimstone, in electrical experiments
1746 Watson	Further experiments and observations, tending to illustrate the nature and properties of electricity
1746 Miles	Extracts of two letters from Henry Miles, to Mr. Henry Baker containing several electrical experiments
1746 Trembley	Part of a letter from Mr. Trembley, to Martin Folkes, concerning the light caused by quicksilver shaken in a glass tube, proceeding from electricity
1746 Miles	Part of a letter from Dr. Miles, to Mr. Henry Baker concerning electrical fire
	A letter from Dr. Miles to Mr. Baker concerning the electricity of water
	A letter from Dr. Miles to Mr. John Ellicot of weighing the strength of electrical effluvia
	Part of two letters from Henry Miles, to Mr. Henry Baker, containing some electrical observations
1746 Winkler	An extract of a letter from Mr. John Henry Winkler to a friend in London; concerning the effects of electricity upon himself and his wife
1746 Robins	A letter to Mr. Benj. Robins, shewing that the electricity of glass disturbs the mariners compass, and also nice balances
1746 Needham	Of a letter from Mr. Turbervill Needham to Martin Folkes, concerning some new electrical experiments lately made at Paris
1746 Le Monnier	Of a memoir concerning the communication of electricity; read at the public meeting of the Royal academy of sciences at Paris, Nov. 12. 1746. By Monsieur le Monnier the younger, M.D. of that academy, and F.R.S. Communicated by the author to the president of the Royal society
1746 Browning	Part of a Letter from Mr. John Browning, of Bristol, to Mr. Henry Baker, dated Dec. 11. 1746. Concerning the Effect of Electricity on Vegetables

(continued)

Table 4.4 (continued)

1746 Watson	Observations upon so much of Monsieur Le Monnier the younger's memoir, lately presented to the Royal society, as relates to the communicating the electric virtue to non-electrics
1748 Watson	A collection of the electrical experiments communicated to the Royal society by Dr. Watson, read at several meetings between October 29 1747 and Jan. 21
1748 Nollet	Part of a letter from abbé Nollet, of the Royal academy of sciences at Paris to Martin Folkes, concerning electricity
1748 Ellicott	Several essays towards discovering the laws of electricity, communicated to the Royal society by Mr. John Ellicott and read on the 25th of Feb. 1747. And at two meetings soon after
1748 Baker	A letter from Mr. Henry Baker F. R. S, to the president, concerning several medical experiments of electricity
1748 Roche	A Letter from Mr. Robert Roche to the president, of a fustian frock being set on fire by electricity
1748 Hales	Of a letter from the Dr. Stephen Hales to Mr. Westly Hall, concerning some electrical experiments
1748 Watson	An account of the experiments made by some gentlemen of the Royal society, in order to measure the absolute velocity of electricity; communicated to the Royal society
1749 Bose	Of a letter from Mr. Matthias Bose, of Wittemberg, to Mr. W. Watson on the electricity of glass, that has been exposed to strong fires
1749 Watson	A letter from Mr. Watson, F. R. S. to the Royal society, declaring that he as well as many others have not been able to make odours pass thro' glass by means of electricity; and giving a particular account of professor Bose at Wittemberg his experiment of beatification, or causing a glory to appear round a man's head by electricity
1749 Nollet	Of a letter from the Abbe Nollet, to Charles Duke of Richmond accompanying an examination of certain phaenomena in electricity, published in Italy, by the same

As an inmate of the Charterhouse he would have time to pursue his inquiries relating to astronomy and navigation and might happily find out something that might be of use. Shortly after his admission to the Charterhouse Gray showed the Royal society his first paper on electricity [109]. The paper was the only communications by Gray printed during Newton's presidency of the Royal society. The coincidence of Gray's silence and Newton's tenure has prompted the suggestion that Newton, who disliked Flamsteed, deliberately muzzled his protege, who however was appreciated by Newtonians like John Keill and Brook Taylor.

Table 4.5 Main papers on electricity of the second half of the 18th century (45 papers over 300)

1752 Watson	An account of Mr. Benjamin Franklin's treatise, lately published, intituled, Experiments and observations on electricity, made at Philadelphia in America
1752 Watson	An account of Dr. Bianchini's Recueil d'experiences faites à Venise sur le medicine electrique
1752 Watson	An account of professor Winkler's experiments relating to odours passing through electrised globes and tubes
1752 Nollet	Two letters of the abbé Nollet to Mr. William Watson, relating to the extracting electricity from the clouds
1752 Watson	An account of the phaenomena of electricity in vacuo, with some observations thereupon.
1752 Franklin	A letter of Benjamin Franklin; to Mr. Peter Collinson, concerning an electrical kite
1753 Nollet	An account of a treatise, presented to the Royal society, intituled, "Letters concerning electricity; in which the latest discoveries upon this subject, and to the consequences which may be deduced from them, are examined
1753 Wilson	A letter to the right honourable the earl of Macclesfield, president of the Royal society, from Mr. Benjamin Wilson, concerning some electrical experiments, made at Paris
1753 Canton	Electrical experiments, with an attempt to account for their several phaenomena; together with some observations on thunder-clouds
1753 Winkler	A letter from John Henry Winkler, professor of natural philosophy at Leipsic, and fellow of the Royal society, to Thomas Birch, secretary of the Royal society, relating to two electrical experiments
1753 Canton	A letter to the right honourable the earl of Macclesfield, president of the Royal society, concerning some new electrical experiments, by John Canton
1753 Nollet	An account of a treatise, presented to the Royal society, intituled, Letters concerning electricity; in which the latest discoveries upon this subject, and to the consequences which may be deduced from them, are examined
1753 Winkler	A letter from John Henry Winkler, professor of natural philosophy at Leipsic, and fellow of the Royal society, to Thomas Birch, secretary of the Royal society, relating to two electrical experiments
1755 Franklin	Electrical experiments, made in pursuance of those by Mr. Canton, dated Decem. 3, 1753
1755 Franklin	Extract of a letter concerning electricity, from Mr. B. Franklin to Mons. Delibard, inclosed in a letter to Mr. Peter Collinson
1757 Franklin	An account of the effects of electricity in paralytic cases
1759 Wilson	A letter from Mr. Benjamin Wilson, to the Rev. Tho. Birch
1759 Symmer	New experiments and observations concerning electricity
1759 Beccaria	Experiments in electricity
1759 Wilson	Farther experiments in electricity
1759 Wilson	Experiments on the Tourmaline

(continued)

Table 4.5 (continued)

1759 Wilson	Farther experiments in electricity
1761 Canton	A letter from John Canton, to Benjamin Franklin, containing some remarks on Mr. Delaval's electrical experiments
1761 Nollet, Watson	An account of a treatise in French, presented to the Royal society, intituled, letters sur l'electricité
1761 Wilson	Observation upon some gems similar to the tourmalin
1761 Canton	A letter from John Canton to Benjamin Franklin, containing some remarks on Mr. Delaval's electrical experiments
1763 Wilson	A letter from Mr. B. Wilson to Mr. Aepinus
1764 Calandrini	Observations upon the effects of lightning, with an account of the apparatus proposed to prevent its mischiefs to buildings, more particularly to powder magazines; being answers to certain
1767 Lane	Description of an electrometer invented by Mr. Lane; with an account of some experiments made by him with it
1768 Priestley	An account of rings consisting of all the prismatic colours, made by electrical explosions on the surface of pieces of metal
1769 Priestley	Experiments on the lateral force of electrical explosions
1771 Priestley	An investigation of the lateral explosion, and of the electricity communicated to the electrical circuit, in a discharge
1771 Cavendish	An attempt to explain some of the principal phaenomena of electricity, by means of an elastic fluid
1772 Priestley	An account of a new electrometer, contrived by Mr. William Henly, and of several electrical experiments made by him
1773 Wilson	Observations upon lightning, and the method of securing buildings from its effects: In a letter to Sir Charles Frederick
1776 Cavendish	An account of some attempts to imitate the effects of the torpedo by electricity
1776 Cavallo	Extraordinary electricity of the atmosphere observed at Islington on the month of October, 1775
1777 Cavallo	An account of some new electrical experiments
1777 Cavallo	New electrical experiments and observations; with an improvement of Mr. Canton's electrometer
1780 Cavallo	An account of some new experiments in electricity, with the description and use of two new electrical instruments
1788 Cavallo	Of the methods of manifesting the presence, and ascertaining the quality, of small quantities of natural of artificial electricity
1788 Cavallo	Description of a new electrical instrument capable of collecting together a diffused or little condensed quantity of electricity
1788 Cavallo	Of the methods of manifesting the presence, and ascertaining the quality, of small quantities of natural of artificial electricity
1797 Pearson	Experiments and observations, made with the view of ascertaining the nature of the gaz produced by passing electric discharges through water
1800 Volta	On the electricity excited by the mere contact of conducting substances of different kinds

4.3.3.1 Electrification of Bodies

I will comment only Gray's two memoirs on electricity published in the Philosophical Transactions. The first of 1721, reported his experiences of electrification that led to an enlargement of the list of electrical materials, such as amber, the second ten years later, in 1731, that proved that electricity can be transmitted at distance. Gray published some other interesting memoirs in the Philosophical Transaction, three still in 1731 as a continuation of his experiments [110, 111, 113] and another in 1735 [114].

The memoir of 1721, *An account of some new electrical experiments*, of only four pages, classifiable with the categories of Peter Dear as a historical account, made almost no reference to theories concerning the nature of electricity, but only reported the results of the experiments, which proceed gradually, going from simple to more complex; result of the latter was suggested as a hypothesis by the former. Gray started from an observation that is presented as a random achievement. The feathers that had been attracted to an electrified glass tube, after the tube was removed behaved as an electrified material. With his words: "Having often observed in the Electrical Experiments made with a Glass Tube, and a Down Feather tied to the end of a small Stick, that after its Fibres had been drawn towards the Tube, when that has been withdrawn, most of them would be drawn to the Stick, as if it had been an Electrick Body, or as if there had been some Electricity communicated to the Stick or Feather" [109].[136]

These experiments provided an opportunity to verify if other materials than feathers could be electrified; in various ways by rubbing with hand, by heating before rubbing and also without rubbing. In the end Gray gave a list of substances he called electrical bodies; his paper ends with the words: "An Enumeration of the several Bodies mentioned herein, that we found to be Electrical: 1. Feathers, 2. Hair, 3. Silk, 4. Linnen, 5. Woollen, 6. Paper, 7. Leather, 8. Wood, 9. Parchment, 10. Ox-Guts [109].[137] These materials, Gray pointed out, apart from being electrifiable have another property which is common to glass; when they are rubbed in the dark a light follows the fingers when they are moved.

It must be said that Gray's memoir did not differ much from the contributions to the Philosophical Transactions of the early 1700s, in which the results of the most varied experiments were reported, some which today would be considered pure curiosities. The only timid reference that Gray made to the theories of electricity is when he talked about the electrification of paper: "finding them, after they are well heated before rubbing, to emit copiously their electric effluvia" [109],[138] a phenomenon far to be verified experimentally.

Gray was more expansive, in a long letter to Hans Sloane (then secretary of the Royal society), of 1707, referring the results of some experiments on electricity:

---

[136] pp. 104–105.

[137] p. 107.

[138] p. 106.

> Exp. 12th. or rather an addition to the second[139] if when the feather has left the Glass the hand or any other solid object be placed betwixt it and the Glass it will Return back to meet it and fix upon it provided the hand be nearer to it than any other object. I made this Experiment in order to confirm or overthrow my first Hypothesis concerning the Reason of this Phenomenon viz that Electricity Proceeded from an Emission and Reflection of its own Effluvia by external objects but this is contradicted by the now mentioned Experiment. I have therefore thought on an other Hypothesis, which at Present Seems to me somewhat more probable, that as all bodies Emitt soe they Receive part of the Effluvia of all other bodies that Inviron them and that the attraction is made according to the current of these Effluvia but then how rubing the Glass though it may cause a more copious and swift Eruption of the Effluvia yet that it should in like manner affect other distant bodies is hard to conceive. I am therefor far from thinking what I here offer is an Entire solution of the Phenomena this *I must leave to the consideration of the Learned* [emphasis added] [60].[140]

Gray first hypothesis ascribed the feather behavior to a mixture of flows, one direct from the tube, the other reflected from neighboring objects. The adherence of a hovering feather to a large object placed between it and the tube contradicts this hypothesis and consequently he suggested another one.

The final sentence in italic in the above quotation is a formula used many times by Gray (and Hauksbee); probably an indication of their feeling of cultural inferiority, but also a lack of interest in the formulation of not verified hypothesis.

4.3.3.2 The 'Transmission' of Electric Virtue

Of a very different tenor, perhaps not in the method, but in the results is his work published in 1731, *A letter to Cromwell Mortimer, M. D. Secr. R. S. containing several experiments concerning electricity*, which refers to experiments carried out mainly in 1729. The result of these experiments is summarized briefly by Gray at the beginning of his work:

> In the Year 1729 I communicated to Dr. Desaguliers, and some other Gentlemen, a Discovery I had then lately made, shewing that the Electrick Vertue of a Glass Tube may be conveyed to any other Bodies, so as to give them the same Property of attracting and repelling light Bodies, as the Tube does, when excited by rubbing; that this attractive Vertue might be carried to Bodies that were many Feet distant from the Tube" [112].[141]

This result is often presented by historians as the discovery of the transmission of electricity, neglecting to specify that Gray was studying electrostatic phenomena in which there is no electric current to transmit.

Also in this paper the experiments followed one another with small variations in a way that they seem completely natural, starting from a result that is presented as a casual observation, similar to what Gray had done in his work of 1721:

---

[139]Reference is to his second experiment referred to in the letter: "If when the feather is come to the Glass, it be held at about 6 or 8 in. Distance from the side of a wall edge of a Table Arme of a Chair or the like it will be drawn to it and thence to the Glass together without ceasing it flies to object at a greater Distance but then does not so often Return" [60], pp. 34–35.

[140]p. 36. Punctuation added.

[141]pp. 18–19.

> I then resolved to procure me a large Flint-Glass Tube, to see if I could make any farther Discovery with it, having called to Mind a Suspicion which some Years ago I had, that as the Tube communicated a Light to Bodies, when it was rubbed in the Dark, whether it might not at the same Time communicate an Electricity to them, though I never till now tried the Experiment, not imagining the Tube could have so great and wonderful an Influence, as to cause them to attract with so much Force, or that the attraction would be carried to such prodigious Distance, as will be found in the Sequel of this Discourse [112].[142]

The starting fortuitous event was the observation of the cork, which closed an electrified glass tube (see Fig. 4.16a), that attracted a feather as the tube itself did. In reality the experience described in the above quotation could not be completely fortuitous. The phenomenon described by Gray could not be noticed by a person who had no laboratory experience and who had not even the slightest idea about the acquisition of attractive power. In reality Gray had a basic theory, which conceived the propagation of the electric virtue as an effluvium coming out of the electrified body to reach the neighboring bodies. A theory certainly not new but that Gray had made his own since his first studies on electricity and that in the 1730s presented as if it were a new acquisition, certainly with a rhetorical device.

This idea of 'transmitting' the electric virtue is likely to underlie his idea of a medium to transmitting it from one body to another. From a strictly physical point of view this idea is not coherent; it is would have been more logical to consider the glass tube and the corks as a single body; the only peculiarity being the particular shape that this assembly of two bodies would have. This is the criticism made by Priestley to Gray: "This experiment shews that Mr. Grey had not properly considered the line of communication and the body electrified by it, as one and the same thing, in an electrical view, differing only in form, as they were both alike conductors of electricity [171].[143]

It is necessary to reflect, that although Gray presented his results as if he were writing a diary with a faithful account of what happened, also recording some failures, his exposition has a rhetorical character. Surely his experiment is not accidental as he wrote and perhaps it was not even been attempted by him first; it is a phenomenon that many experimenters have observed. Gray reflected critically on it and formulated some hypotheses about what could happen by making small variations of the experiment.

As already noticed, Gray's observation that the glass tube communicates his electric electricity to the cork that is able to attract a feather presented to it, was an opportunity for a series of experiments with the aim of verifying how far the electrical property of the tube could be transferred. But before these experiments Gray presented his 'electrostatic generator'. It was a very slender tube, 3 feet and 5 in. (about 104 cm) long with a diameter of 2 in. and 2/10 (about 5.6 cm), which is electrified by rubbing it with bare hands. The tube is closed at both ends by two corks to prevent dust from entering, which would compromise the experiments (Fig. 4.16a).

---

[142] pp. 19–20.

[143] p. 42.

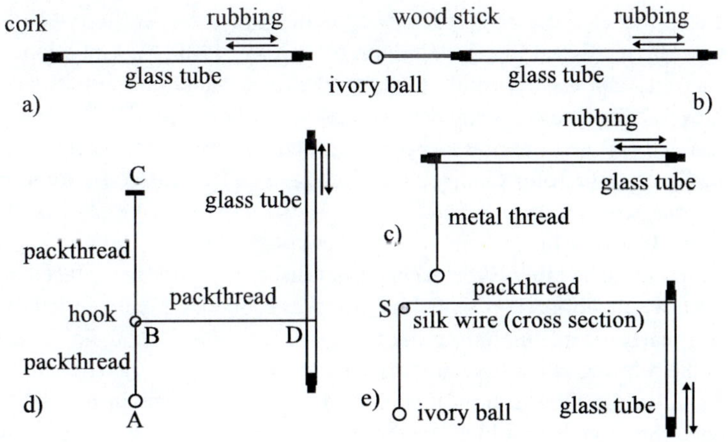

Fig. 4.16 Different arrangement of the tube and the receivers of electricity. Modified from [186], Appendix

Gray with his experiments had the aim of determining in a systematic way to which bodies the attractive virtue could be communicated. He also wanted to know how far he could pass on this property. After a preliminary test, he put a wooden stick 4 in. long (about 10 cm) into the hole made in an ivory sphere with a diameter of 1 in. 3/10 (3.3 cm). The other end of the stick was inserted into the cork, in turn stuck in the glass tube; notice that Gray's work is not accompanied by figures so it is impossible to accurately reconstruct the configuration of the apparatus he used. Figure 4.16b should fairly faithfully describe the configuration used in these early experiments. When he rubbed the tube, he observed that the ivory ball was attracting *the feather more vigorously than did the cork.* The length of the stick was then increased up to 24 in. (60 cm) and found that the attraction continued to be observed. He then replaced the wooden stick with iron and brass cables observing the same effects. He noticed that though the wire was closer to the cork its attracting power was not so strong as that of the ball [112]. Later, Gray increased the length of the cables up to 3 feet (90 cm), but at that point he observed many vibrations caused by the rubbing of the pipe that made the attractions hardly observable.

Then the ivory ball was hung to the tube by a metal string, iron or brass, according to what illustrated in Fig. 4.16c. When he rubbed the tube, the ivory ball attracted and repelled brass leaves under it, which Gray tended to replace to the feather. The same happened when he attached a ball of cork to the string and then an iron ball of 1 pound 1/4 (about 570 g). Following these procedures, he was able to communicate the electricity of the rubbed tube to different bodies connected to it with strings that were thus able to attract the brass leaves, for example a half-penny, a piece of block-tin, a piece of lead to fire-shovel, a copper tea-kettle or empty or full with water and so on, a chimney scoop, an empty or full of water copper teapot, a silver mug. He had finally managed to get metals to attract light bodies. No one had been able to achieve this effect in the 2000 years since the discovery of electricity "and so Gray succeeded at least in awaking their hidden electricity" [122].[144] Gray continued his research activity using ivory, flint-stone, sand-stone, load-stone, bricks, tiles, chalk and also several vegetable substances, as well green or dry [112].[145]

From May 1729 he continued his experiments in the country; some were made at Norton-Court with John Godfrey (nephew of John Flamsteed), some others at Otterden-Place with Granvil Wheler (1701–1770), a worthy member of the Royal society. With Godfrey he got positive results with a rod 24 ft (about 7,3 m) long, connected to the glass tube. Even at this great distance a cork ball placed at the end of the rod attracted a brass leaf when the tube was rubbed. Gray extended this length up to 32 ft (nearly 10 m), including the tube [112].[146] But once again the vibrations caused by the rubbing of the tube disturbed the experiment. He then decided to attach a string again to the tube to hold a sphere of cork or ivory, as in Fig. 4.16c. When Gray rubbed the tube, he could make the sphere attract the brass leaf even when the string used was 26 ft (8 m) long and had been suspended by him outside a balcony.

Later, he tried to increase the length of his device in the horizontal direction. First he made a loop at each end of two lines, and hanging one (BC of Fig. 4.16d) on a nail drove into a beam, the other line AD passed through the hook in B; the part BA of the line AD hanging downwards, supported the ivory ball at the end A; the other end D of AD was connected to the tube by a loop. Then the leaf-brass being laid under the ball, and the tube rubbed, yet not the least sign of attraction was perceived upon this. Gray concluded with the following explanation:

> When the Electrick Vertue came to the Loop that was suspended on the Beam, it went up the same to the Beam; so that none, or very little of it at least, came down to the Ball [112].[147]

It is possible that Gray suspected that something might not work. He had probably already had some experience of transmitting horizontally with a rope suspended from multiple strings that came down from the ceiling. The arrangement of Fig. 4.16d serves to highlight the simpler experimental situation to test the transmission of electric virtue: a horizontal line BD is suspended by a wire CB to the ceiling. Gray explained the failure of the transmission, by saying that the electric virtue was absorbed by the beam (which in modern language is seen as earth), coming close to the modern explanation.

In June 1729 Gray decided to continue the experiments with his friend Granville Wheler, with a small solid glass cane of about 11 in. (28 cm) long and 7/8 in. (2 cm) of diameter. Operating from a window, after rubbing the small glass cane with wires from 16 ft (4.9 m) to 34 ft (10.4 m) suspended from the cane, they got that the leaf-bras was attracted and repelled beyond what they expected.

According to Gray's narrative Wheler was desirous to try whether they could carry the electric virtue horizontally:

---

[144] p. 246.

[145] p. 22.

[146] p. 24.

[147] p. 25.

> I then told him of the Attempt I had made with that Design, but without Success, telling him the Method and Materials made use of, as mentioned above. He then proposed a Silk Line to support the Line, by which the Electrick Vertue was to pass. I told him it might do better upon the Account of its Smallness; so that there would be less vertue carried from the Line of Communication, with which, together with the apt Method Mr. Wheler contrived, and with the great Pains he took himself, and the Assistance of his Servants, we succeeded far beyond our Expectation [112].[148]

The first experiment of the horizontal transmission was made in the matted gallery July, 1719 in the morning. About four feet from the end of the gallery there was a cross line that was fixed by its ends to each side of the gallery by two nails; the middle part of the cross line was silk, the rest at each end packthread; then the line to which the ivory ball was hung, and by which the electric virtue was to be conveyed to it from the tube, being eighty feet and a half in length (25 m, about), was laid on the cross silk line, so as that the ball hung about nine feet (2.7 m) below it; then the other end of the line was by a loop suspended on the glass cane and the leaf-brass held under the ivory ball on a piece of white paper. When the tube was rubbed, the ivory ball attracted the leaf-brass and kept it suspended for some time [112].[149]

Figure 4.16e illustrates the disposition. The cross line of silk is represented by the circle S in grey (it is a cross section), the packthread is connected from one side to the tube, from the other side to a ball. Gray believed that the success of the experiment was due to the fact that the line was supported by a very thin thread, without giving weight to the fact that it was made of silk.

By using the silk support and making the packthreads made laps, the two friends managed to transmit the electric virtue up to 293 ft (90 m). Attempting a longer line caused the silk support to break. They then made attempts with supports of more resistant material, in particular brass wire always very thin to avoid, in the idea of Gray, the transmission of virtue to the gallery. With some surprises and though the tube was well rubbed, yet there was not the least motion or attraction given by the ivory ball, neither with the great tube which was used when the small solid tube was found ineffective.

This time Gray was forced to believe that it was not so much the subtlety of the support to prevent the dispersion of the electric virtue; it was rather the kind of material:

> By which we were now convinced, that the Success we had before, depended upon the Lines that supported the Line of Communication, being Silk, and not upon their being small, as before Trial I imagined it might be [112].[150]

The fact that there were materials that allowed the transmission of the electric virtue and other that did not was a very important discovery by Gray in the memoir of 1731. Though less striking than the transmission through long distances it involved the very nature of matter.

---

[148] p. 26.

[149] pp. 26–27.

[150] p. 29.

By using more than one silk support in order to reduce the effort they sustained, Gray and Wheler managed to transmit the electric virtue up to 765 ft (236 m), "and the attraction was not less perceivable" [112].[151]

A part silk Gray found that other substances could be used as support such as hair, glass and resin, while as receiver of the electric virtue he found soap bubbles in water, a map of the world, an umbrella. And also a person; an amazing experiment experiment that will be repeated in bourgeois salons. Here how this experiment, illustrated in Fig. 4.15, that become popular in the European Courts, is described by Gray:

> April 8, 1730, I made the following Experiment on a Boy between eight and nine Years of Age. His Weight, with his Cloaths on, was forty-seven Pounds ten Ounces. I suspended him in a horizontal Position, by two Hair-Lines, such as Cloaths are dried on: They were about thirteen Feet long, with Loops at each End. There was drove into the Beam of my Chamber, which was a Foot thick, a Pair of Hooks opposite to each other, and two Feet from these another Pair in the fame manner. Upon these Hooks the Lines were hung by their Loops, so as to be in the Manner of two Swings, the lower Parts hanging within about two Feet of the Floor of the Room: Then the Boy was laid on these Lines with his Face downwards, one of the Lines being put under his Breast, the other under his Thighs: Then the Leaf-Brass was laid on a Stand, which was a round Board of a Foot Diameter, with white Paper palled on it, supported on a Pedestal of a Foot in Hight, which I often made use of in other Experiments, though not till now mentioned Upon the Tube's being rubbed, and held near his Feet, without touching them, the Leaf-Brass was attracted by the Boy's Face with much Vigour, so as to rise to the Hight of eight, and sometimes ten Inches. I put a great many Pieces on the Board together, and almost all of them came up together at the same Time. Then the Boy was laid with his Face upwards, and the hind Part of his Head, which had short Hair on, attracted, but not at quite so great a Hight as his Face did. Then the Leaf-Brass was placed under his Feet, his Shoes and Stockings being on, and the Tube held near his Head, his Feet attracted, but not altogether at for great a Hight as his Plead: Then the Leaf-Brass was again laid under his Head, and the Tube held over it, but there was then no Attraction, nor was there any when the Leaf-Brass was laid under his Feet, and the Tube held over them [112].[152]

In June 1730 Gray and Wheler made an experiment showing that the attraction and repulsion was as strong, if not stronger, and that the effluvia might be carried to great lengths without touching the line with the glass tube [112].[153] This was precisely the confirmation of the idea that Gray had projected at the outset. After a first confirmation they made experiment to see how far the electric virtue could be carried forward in a line, without touching the same. They arrived to obtain an effect up to 886 ft (270 m).

Two more Gray's studies of effluvia in the paper of 1731 are worthy to be cited. The first investigation concerned whether solid objects receive a stronger communicated electricity than hollow objects. Two oak cubes, one solid and one hollow, were suspended and joined by a packthread; when the tube stood over the middle of the line connecting the two cubes, both attracted equally.

Gray expected that communicated electric virtue would be proportional to the quantity of matter in bodies; it was not the case (indeed the surface areas is a better

---

[151] p. 31.

[152] pp. 39-40

[153] p. 42.

candidate), but the negative result did not disconfirmed Gray's opinion about the theory of electric effluvia: "yet I am apt to think that the Electrick Effluvia pass through all the interior Parts of the solid Cube, though no Part but the Surface attracts; for from several Experiments it appears, that if any other Body touches that which attracts, its Attraction ceases till that Body be removed, and the other be again excited by the Tube"[112].[154]

The second investigation concerned the different power of attraction due to the shape of the support of the brass leafs to be attracted. With his words: In these experiments, besides the large stand above mentioned, I made use of two small ones. The tops of them were three inches diameter; they were supported by a column of about a foot in hight, their bases of about four inches and half. They were turned of lignum vitae (guaiacum); their tops and bases made to screw on for convenience of carriage. Upon the tops there were white papers. When the leaf-brass laid on any of these stands, I find it is attracted to a much greater hight than when laid on a table, and at least three times higher than when laid on the floor of a room [112].[155]

This is one of the first known description of the power of the points. That is, the electric force is stronger around sharp and pointed regions of conductors than around flat surfaces.

Others papers published on the Philosophical Transactions are summarized in [171].[156] To signal only some experiments to test the electric attraction in a void [111, 113].[157] Priestley, suggested that Gray probably could have avoided this experiment if he had known those carried out by Boyle: "In the first place, Mr. Grey made some experiments, which, probably unknown to him, had been made before by Mr. Boyle" [171].[158] Indeed, Gray cited Boyle's experiments with the air-pump [113][159] and probably knew also those related to electricity, thus he could have had his own reason to repeat the experiment.

4.3.4 The Leyden Jar

A typical modern design of the Leyden jar consists of a glass bottle with metal (gold) tin foils coating the inner and outer surfaces of the bottle. The foils stop short of the mouth of the jar. A metal electrode passing through the lid of the jar electrically connected by some means (usually a hanging chain) to the inner foil allows the charge to reach the inner foil; the outer foil is grounded. The inner and outer surfaces of the jar store opposite charges.

---

[154] p. 35.

[155] p. 42.

[156] pp. 40–42; 53–63.

[157] p. 423; 398.

[158] p. 40.

[159] p. 399.

Fig. 4.17 Leyden jar. **a** Redrawn from [172], p. 570; **b** Original drawing by Musschenbroek [160], p. 6

The original form of the device was just a glass bottle partially filled with water, with a metal wire passing through a cork closing it; the role of the outer plate was provided by the hand of the experimenter (Fig. 4.17).

A first version of the Leyden jar was due to the German ‘administrator and cleric’ Ewald Jürgen Georg von Kleist (1700–1748), and is known as *Kleist’s phial*, described by himself in a letter to Johann Gottlob Krüger (1715–1759), in the following way: “If a nail, a strong wire, etc., is introduced into a narrow-necked little medicine bottle and electrified, especially powerful effects follow. The glass must be very dry and warm. Everything works better if a little mercury or alcohol is placed inside. The flare appears on the little bottle as soon as it is removed from the machine, and I have been able to take over sixty paces around the room by the light of this little burning instrument" [133].[160]

Kleist was guided in his discovery by some false assumptions and got his result after various attempts by varying some parameters. From a modern point of view it can be said that Kleist arrived at his result by chance. This was, and still is, a quite common situation in modern physics, differently to what happened in the old natural philosophy where practically no discoveries occurred and theories served mainly to explain the why of known phenomena.

Kleist's first experiment occurred on October 1745 and its results was communicated to other people; but because he neglected to mention fundamental configurations, most probably because he did not know they were such, his correspondents failed to repeat the experience. In 1745–1746 Musschenbroek got an analogous apparatus and experience, which because of the place of the experience was called, and is still called, *Leyden jar*. To be precise the crucial part of the experience that led to the Leyden jar was due to Andrea Cunaeus, a layer of profession used to visiting Musschenbroek's laboratory.

Cunaeus wishing to electrify water, employed for this purpose a wide-mouthed flask, which he held in his hand, while a chain from an electric generator dipped in the water. When the experiment had been going on for some time, he decided to disconnect the water from the generator and for this purpose was about to lift out the chain; but, on touching the chain, he experienced a shock, which gave him the greatest consternation.

Cunaeus reported his discovery to Musschenbroek and his colleague Jean Nicolas Sébastien Allamand (1716–1787) who repeated the experiment, which succeeded. Two days later Musschenbroek himself repeated the experiment by replacing a globe for the jar. Musschenbroek, differently from Kleist, was most attentive in referring to his distressing experience and reported it to his appointed correspondent at the Académie des Sciences de Paris, René Réaumur (1683–1757):

> I would like to tell you about a new but terrible experiment, which I advise you never to try yourself, nor would I, who have experienced it and survived by the grace of God, do it again for all the kingdom of France. I was engaged in displaying the powers of electricity. An iron tube AB [see Fig. 4.18b] was suspended from blue-silk lines; a globe, rapidly spun and rubbed, was located near A, and communicated its electrical power to AB. From a point near the other end B a brass wire hung; in my right hand I held the globe D, partly filled with water, into which the wire dipped; with my left hand E I tried to draw the snapping sparks that jump from the iron tube to the finger; thereupon my right hand F was struck with such force that my whole body quivered just like someone hit by lightning [160].[161] (D.18)

The Leyden jar contravened all the principles on electricity then known, to the point of leaving Musschenbroek bewildered, as witnessed by the confession of ignorance expressed in his letter to Réaumur: "I understood nothing and could explain nothing about electricity". First convincing (sic) explanations, at least from the modern point of view, on the operation of the Leiden jar were provided only by Franklin and his collaborators.

---

[160] p. 176. English translation in [122].

[161] p. 6; critical transcription from [190], pp. 125–126. Translation in [122].

4.3.5 *A Comprehensive Theory of Electricity. Franz Ulrich Theodosius Aepinus*

While most English experimental philosophers pursued experimentation, leaving aside mathematics, in the Continent a comprehensive theory of electricity was developed in Germany by an experimental philosopher with good knowledge of mathematics: Franz Ulrich Theodosius Aepinus (1724–1802). He was born in Rostock, a city of Duchy of Mecklenburg, Germany. His father held the chair of theology and a his elder brother that of oratory at the University of Rostock. Aepinus studied medicine and mathematics at Jena, under the guidance of George Edward Hamberger (1697–1775), professor of botany, anatomy and surgery who should have an important role in Aepinus' education. Physics taught by Hamberger was all but Newtonian; inspired by Leibniz and Descartes with however a strong attention to experimentation. Aepinus took his M.A. in 1747 at Rostock with a dissertation on the paths of falling bodies.

Until 1755 Aepinus taught mathematics at Rostock, as a junior lecturer and published only on mathematical subjects: the properties of algebraic equations, the integration of partial differential equations, the concept of negative numbers. In 1751–1752 one of his auditors was Johan Carl Wilcke (1732–1796), who had come to Rostock for a clerical career but then oriented for physics and mathematics. A few years later Wilcke played a role in reorienting Aepinus career. In 1755 Aepinus became director of the observatory in Berlin and a member of the local academy of sciences. Here for two years was in close contact with Leonhard Euler. This would have removed his Leibnizian approach to natural philosophy and brought him closer to the Newtonian view.

Aepinus was neither especially interested nor experienced in astronomy and his closest published approach to the subject during his Berlin stay was a mathematical analysis of a micrometer. His main interest at the time was the study of the tourmaline, to which he was introduced by Wilcke, who had followed him to Berlin. Aepinus' first researches on the thermoelectric (modern term) properties of this stone were fundamental. He recognized the electrical nature of the attractive power of a warmed tourmaline and was particularly struck by the formal similarity between the tourmaline and the magnet in regard to polarity which inspired him to reconsider the possibility, then occasionally discussed, that electricity and magnetism were basically analogous; on tourmaline he wrote an important paper in 1756 [2]. In seeking an explanation for the strange behavior of this substance, Aepinus came to the anti-Franklinian idea of a Leyden jar without the glass (see below) and conceived an air condenser.

In 1756 Aepinus asked to be relieved of his positions in Berlin in order to accept the directorship of the observatory and the professorship of physics, vacant since the death of Richmann, at the imperial academy of St. Petersburg. Euler, with whom he boarded in Berlin, warmly recommended him for the job and interceded with Frederick II to procure his release, which occurred in the spring of 1757. The Petersburg academicians expected that Aepinus, as befitted Richmann's successor, would continue to work on electricity. They were not disappointed. Late in 1758 Aepinus completed the lengthy *Tentamen theoriae electricitatis et magnetismi*, herein after referred to as *Tentamen*.

In 1760 or 1761 Aepinus became instructor to the Corps of imperial cadets, a position that left him too little time to fulfill his scientific duties. The observatory was seldom used and the equipment in the physics laboratory deteriorated. These circumstances gave Mikhail Lomonosov (1711–1765) the opportunity for a furious attack on Aepinus. Despite such unfavorable conditions, Aepinus continued for a few years to produce papers on various mathematical and physical subjects. Among them perhaps the most interesting is a masterful discussion of the mercurial phosphorus and a critical examination of Mayer's theory of magnetism in the Novi commentarii of the Petersburg academy for 1766–1767. About that time Aepinus' scientific activity ceased almost entirely. He became preceptor to the crown prince, a member of the prestigious Order of St. Anne, an educational reformer, a diplomat, a courtier, and finally a privy councillor. In 1798, after forty years in Russia, he resigned his offices and retired to Dorpat where he died [5, 120].

4.3.5.1 Relationship Between Physics and Mathematics

As it can be seen from his biography, Aepinus received a traditional education in natural philosophy with a mechanicistic approach and in experimental philosophy. Furthermore, although mathematics was not his main interest, one cannot deny his competence, which is documented by many publications on mathematical subjects [86].

Aepinus interest in mathematics was indeed very strong; among the papers that he mentioned and most probably studied there were the *Géométrie* of Descartes and the *Arithmetica infinitorum* of John Wallis, together with l'Hôpital's *Analyse des infiniment petits*, papers by Varignon in the memoirs of the Académie des sciences de Paris, Guido Grandi's *De infinitis infinitorum*, Fontenelle's *Elemens de la geometrie de l'infini*, Newton's *Arithmetica universalis* (in the Leyden edition of 1732) and several minor works. Elsewhere in Aepinus' s early writings Euler's *Mechanica* was cited, as well as Daniel Bernoulli's *Hydrodynamica* and papers by Maupertuis and Clairaut on the solving of certain classes of differential equations.

Some of his ideas on the natural philosophy were developed in the introductory remarks of his *Tentamen*. Differently from Newton, Aepinus did not disdain to speak about hypotheses. He declared that from the agreement of a hypothesis about the causes of the registered phenomena one cannot be certain that he had reached the true cause. And although his theory about electricity (and magnetism) satisfy the majority of the phenomena: "I proceed more modestly than confidentially, and to put forward my proposition as probable rather than certain" [4, 5].[162]

[162] Introductory remarks, p. 5, p. 239.

Aepinus supposed that all electric and magnetic phenomena could be explained by means of some 'primitive' forces. He did not maintain necessary to inquire about their origins, declaring rhetorically that he left to further discussion of whom are happy to spend their time in speculations. He made the example of the approach by hypothesis of mathematicians who are satisfied when they succeeded to reduce their problems to the quadrature of the circle, even though it was still an open problem to be solved. The eminent Newton also, declared Aepinus, proceeded in this way. He demonstrated how the motion of the planets and satellites depend on centripetal forces, but he did not spend time to reach the roots of the forces [4].[163]

Differently from Newtonians, Aepinus declared that he believed as an indubitable axiom (pro axiomate enim indubitato) the proposition asserting that a being cannot act where it is not, and consequently he denied the possibility of an *actionis in distans*.

> If ever it is proved that the attraction or repulsion does not depend ultimately on some external pressure or impulse, then I judge that we are reduced to the point where we are forced to ascribe the execution and production of movements of that kind, not to corporeal forces but to spirits or beings with understanding of the things which act, and I cannot be induced to believe that this could happen in the world. So the whole of my opinion amounts to the fact that I consider the attractions and repulsions I have spoken about as phenomena whose causes so far lie hidden, but from which other phenomena depend and are derived. I think that considerable progress can be made in the analysis of the operations of nature by the scholar who reduces rather complicated phenomena to their proximate causes and primitive forces, even though the causes of those causes have not yet been detected. I trust that this declaration of my opinion will satisfy the most rigid censor [4].[164] (D.19)

Aepinus was thus brought to the introduction of forces apparently acting at distance not as a result of a his own conception of natural philosophy, but rather against them, because of the difficulty to explain mechanically many electrical phenomena, especially the behavior of the Leyden jar, for which Franklin's hypotheses on the absolute impermeability of glass to electric fluid, was not able to explain some experimental facts, for instance why the efficacy of the apparatus increased by decreasing the thickness of the glass of the jar.

It is worth reflecting that at the beginning of the theory of electricity—when the only phenomenon, or rather the main one, to be explained was the attraction by an electrified body of small pieces of any material—to explain this attraction by unexplained forces, would have been seen as a vicious circle, a return to the explanations of some pedantic representatives of schools. But at Aepinus' time the phenomena were many and complex and the use of a single principle, a force, even if unexplained, could be seen as a reasonable approach to an unitary treatment.

Aepinus reiterated his conceptions at the end of the *Tentamen*, in the last page of the discussion about the phenomena discovered by Richmann on the Leyden jar. Here he defended his approach largely based on mathematics.

[163] Introductory remarks, p. 7.

[164] Introductory remarks, pp. 7–8. Translation in [5].

> Some, who can barely tolerate such things in physics, will perhaps be displeased that I have mixed in so much mathematics. But I judge quite differently, and believe that this brief dissertation on Richmann's experiment contributes a great deal towards proving the truth of my principles of electricity. For an hypothesis grows greatly in plausibility if it can explain complicated phenomena, which can only be deduced from it by a long chain of reasoning [4].[165] (D.20)

Aepinus' theory received poor attention, at the beginning. It is often said that this was owing to the criticism, and almost unintelligible account, which Priestley gave of it in his influential summary of the electricity of the 18th century *History and present state of electricity* of 1767 [122],[166] where he declared the mathematical passages by Aepinus not worthy to be considered: "He that reads the first chapter, as well as many other parts of his elaborate treatise above mentioned, may save himself a good deal of time and trouble by considering, that the result of many of his reasonings and mathematical calculations cannot be depended upon; because he supposes the repulsion or elasticity of the electric fluid to be in proportion to its condensation [the quantity of electric fluid]; which is not true, unless the particles repel one another in the simple reciprocal ratio of their distances, as Sir Isaac Newton has demonstrated, in the second book of his *Principia*" [171].[167] Indeed the criticism of Priestley derived from a misunderstanding of Aepinus argumentations, as also a misunderstanding was the attribution to Newton what he said about the attraction of bodies [122].[168]

This is only a partial explanation of the poor attention on Aepinus' treatise, however. Independently of the bad publicity by Priestley, the *Tentamen* itself probably had too much mathematics, magnetism, Latin, and details for most electricians of the 18th century—Cavendish excluded. Moreover Euler, for example, thought the *Tentamen* relevant on experiments, but weak in theory. The acceptance of Aepinus' approach awaited the work of younger men, and for them its was fundamental.

Aepinus' approach was usual neither near the electricians, nor even among experimental philosophers. The only topics where the use of mathematics was accepted by all were those that today belongs more or less to mechanics (hydraulics, acoustics, pneumatics, dynamics, and so on). Other topics of what today belong to physics such as thermology, magnetism and electricity where considered not suitable to be mathematized by experimental philosophers; and even though they were presented in mathematical style, the proof of the statements, named usually theorem, were left to experiments and not to mathematics.

Benjamin Wilson, one of the leading English electricians, reported the views of his colleagues physicists when he wrote in patronizing terms to Aepinus that "the introducing of algebra in experimental philosophy, is very much laid aside with us,

[165] pp. 375–376. Translation in [122].

[166] p. 402.

[167] p. 426.

[168] p. 402.

as few people understand it; and those who do, rather cho[o]se to avoid that close kind of attention: tho' I make no doubt but I dar[e] say you had a very good reason for making use of that method" [5].[169]

A modern by reading Aepinus writings on electricity, including the *Tentamen*, is struck not by too much mathematics, to the contrary by too little of it. Indeed Aepinus used a rigor of a mathematician and all phenomena are interpreted by mathematical relationships. But he only (with a few exception) used simply algebraic passages: the expression of equation of balance among forces. No use is made of Calculus, because Aepinus did not analyze non-uniform distribution of electricity, and thus did not need infinitesimal element of ordinary matter of electric fluid do be accounted for; differently from what Coulomb will do some years later.

4.3.5.2 Tentamen Theoriae Electricitatis et Magnetismi

Aepinus' *Tentamen* is one of the most original and important book in the history of electricity, even though the contribution to the study of electricity was peripheral to that of magnetisms. It is the first reasoned, fruitful exposition of electrical phenomena based on forces whose origin is not questioned. The treatise appeared in 1759, rushed into print earlier than its author intended, by the pressing of Petersburg academicians, who were eager to show the world Aepinus' achievements; for this reason it contained many typos, which sometimes made it difficult to read.

As the book did not at first have much influence and had a small circulation, the Italian electricians did not read it at all or did not read it attentively until the late 1770s. Beccaria had not seen a copy as late as 1772; Volta appears to have developed this own views before he came across it. Few French physicists knew it before 1787. In Germany its knowledge was also scarce. In England several important electricians knew the *Tentamen*, or at least fondled it, during the 1760s, but apart from Cavendish, they did not understand it. Franklin complimented Aepinus several times for attempting to apply his principles of electricity to magnetism, but never recognized the transformation he operated [122].[170]

The edition of 1759 has a dedication to Count Kiril Razumovskiī, a preface and an introductory part. Four chapters follow, the first of them opens with the general principles he and Franklin assumed. The book ends with two appendices; the first discussed a phenomenon of the Leyden jar discovered by Richmann, the second certain paradoxical magnetic phenomena.

Aepinus rejected the current notion of electrical atmospheres. Three forces, according to him, create all the appearances of electricity: a repulsion among the particles of the electric fluid, an attraction among them and the corpuscles of common matter, and a repulsion among the corpuscles of the common matter. Although his exposition does not contain numerical values for the physical magnitudes he introduced, it is mathematical, with symbols used to indicate the excess or deficiency of

[169] p. 202.
[170] p. 401.

fluid and the associated forces. A part from the numerical values of the parameters, what is left indeterminate and makes not possible quantitative predictions, is the exact form of the law of the electric forces. He only assumed that forces decrease with distance and are proportional to matter and amount of electric fluid. The magnetic theory of the *Tentamen* operates on the same principles, except that the magnetic fluid can freely penetrate all substances but iron, in which it is so tightly held that it can neither increase nor decrease. A piece of iron is thus to the magnetic fluid what a perfect insulator would be to the electric. All magnetic phenomena depend on the displacement of the magnetic fluid within iron. Aepinus' analysis of magnetization is exactly analogous to his treatment of electrical induction (modern term, see below); it is adequate to all problems he considered except the formation of two magnets by the halving of one. Most notably it lead him to improve on Canton's and Michell's method of preparing artificial magnets and on the usual disposition of armatures.

4.3.5.3 The Principles of Electricity

At the beginning of Chap. 1 of the *Tentamen*, Aepinus listed what according to him are Franklin's principles of electricity:

- α There is a certain very subtle, full elastic fluid producing all electrical phenomena, called electric [electricum] for this, whose parts sensibly repel each other mutually even over rather sensible distances.
- β The particles of this fluid are attracted by the ordinary matter, from which all actual bodies are made.
- γ There is a great diversity in the way bodies act on the particles of the electric fluid; there are some bodies constituted so that the electric matter moves very easily in their pores because meets no resistance; other bodies, on the contrary, are such that electric matter admits a movement only with difficulty. Bodies of the former kind are called *non-electric per se*; those of the latter kind, *electric per se*.
- δ Electrical phenomena are of two kinds. Some originate from the transit of electric matter from one body which contains a greater amount of this fluid to another which holds less. These produce electric sparks and other phenomena concerned with the electric light; others occur without the actual transit of fluid, to which the attractions and repulsions which are exercised by electrified bodies must be referred to [4, 5].[171]

He avoided to add one more principle of Franklin, that according to which glass is fully impermeable to the passage of electric fluid. Indeed, Aepinus said he was establishing the general foundations of the theory of electricity and considered superfluous to mention the impermeability of glass since it cannot be considered as a peculiar property of glass only; on this point Aepinus will discuss lengthily in his treatise.

---

[171] pp. 9–10; p. 241.

4.3.5.4 The Natural State

A body which is not electrical is called in the natural state; in this state the amount of electric fluid is in the ‘right’ amount, the more the ordinary matter the more the electric fluid.

Aepinus using his principles and the equation of equilibrium was able to define the natural state and the right amount of electric fluid, in a clear way and to show that it is indeed natural. Let there be a body A containing a certain quantity of fluid, said Aepinus, whose parts are attracted by the ordinary matter of A, but which mutually repel each other. It is plainly obvious that a double action is exercised on a particle of this fluid such as B* clinging around the boundary of the body A. This particle B* will be attracted by the body A but will be also repelled by the fluid enclosed in its pores. If attraction and repulsion are equal, it is clear that there is no action on the particle B*. But if a supply of fluid is enclosed in the body A such that the repulsion overcomes the attraction, the particle B* will then have to yield to the former force; it must move quickly away and be separated from the body A. The same holds for the other particles of the fluid; it is thus clear that a continuous flow of fluid must take place from the body A [4, 5].[172] Because, continued Aepinus, when in excess the quantity of fluid enclosed in the body A is continuously diminished and as a result the repulsive force continuously decreased, the flow of fluid lasts until the repulsive force is reduced to equality with the attractive force (which meanwhile remains unchanged) and at this point all efflux must cease.

Today it its known that the dissipation (or acquisition) of electric fluid—actually electrons—from a body having an excess (or defect) of it is much more complex phenomenon. For instance in the case of air surrounding a charged conductor, when there is an excess of electricity in it, the air close to the surface of the conductor becomes ionized in positive and negative components; the positive components are attracted by the electrons in excess and neutralized; all is as there is a flux of free electrons but there is no in fact such a flux. In the case of vacuum the electrons can leave the conductor as γ rays, but only under particular conditions.

If finally it is assumed that the attraction exercised on the particle B* is greater than the repulsion, it is clearly established that the contrary ought to take place. For in this case particles of fluid will enter the body from every part until, because of the continuously increasing supply of fluid, the repulsion finally increases to the point where it becomes equal to the attraction, and then all further influx must cease.

It is clear from this that a fixed quantity of fluid can always be assigned so that there is neither efflux nor influx of fluid. But if a change takes place, fluid either escapes if its quantity has been increased, or if this has been diminished it increases until the equilibrium is restored. It seems that this fixed quantity can be fittingly called *natural*, since a body left to itself always returns (in the long time) of its own

[172] pp. 14–15; p. 244.

accord to the state where it contains neither lesser nor a greater quantity of fluid than is exactly sufficient to produce an equilibrium between the attracting and repelling forces [4].[173]

4.3.5.5 Ordinary Matter Repels

Let consider two bodies A and B, close to each other and constituted in natural state, and examine the action of A on B. It is evident that the electric fluid of A repels that of B, and thus the whole B, with a given force; let it be r. In turn the ordinary matter of A attracts the electric fluid of B, and thus the whole B, with a force a. Besides these forces there are other forces that acts on B. First the force A with which the electric fluid of A attracts the ordinary matter of B, second the force x that could act between the ordinary matter of A and the ordinary matter of B. Because the experience shows that B is at rest, the following balance hold: $a - r + x + A = 0$.

Then Aepinus introduced the equality $a - r = 0$, by asserting that "while a body is constituted in the natural state it in no way acts on particles of electric fluid outside of the body" [4].[174] This assertion according Aepinus is based on what he has proved in his Sections 7 and 8; where he had argued that on any element B* of electric fluid, in the boundary of a body A, the resultant of the repulsion r of all the electric fluid in A and the attraction a of all the ordinary matter of A, are such that the equilibrium $a - r = 0$ is verified. Here however he is arguing about a particular element B*—the whole body B indeed—which is outside A.

Aepinus was not bothered by this circumstance and went further. He assumed (1) that the electric fluid is evenly distributed in the body in the natural state, so that its quantity is proportional to its mass [5].[175] Another, unstated, assumption (2) is that the force between electric fluids is proportional to their quantities and, similarly, forces among ordinary masses and electric fluids and among ordinary masses themselves are proportional to their respective quantities.

On the basis of the assumptions reported above, by indicating with M, m and Q, q the masses and electric fluid in A and B respectively, Aepinus could write, for the first assumption, $M : m = Q : q$, for the second $a \propto Mq$; $A \propto mQ$, and thus $a : A = Mq : mQ$; which gives $A = a$ and consequently $x = -a$. That is a, r, A, x are all equal to each other, a part from the sign, and the particles of matter repel each other with a force equal to that with which the electric fluids repel.

Though Aepinus did not make explicit the dependence of the various forces on the distance, a part to say that them decrease with the distance, he made a further step in his analysis. Since, he said, experience shows that A and B remain at rest independently of their distance, this implies that the equality between the forces a, r, A, x is maintained, and thus they necessarily obey the same law with respect

173 Section 6, pp. 14–16.

174 Section 29, p. 37.

175 p. 247; see footnote 16.

to the distance [4].[176] There are however some ambiguities in the definition of the distance; mainly Aepinus intended the distance between two whole bodies and do not say as the distance is measured; for instance between the outer surfaces or the geometrical center. The only way distance has a clear meaning is when it concerns two points, for instance two particles of matter; but this is not the case in Aepinus.

At this point Aepinus made some consideration about his result, in particular about the finding that the ordinary matter repels, admitting that the fact should seem to his reader too hard to be accepted:

> I do not deny that when it first came to me right at the beginning of my thinking about Franklin's theory of electricity, I was somewhat horrified at it. But after I began to consider that it contain nothing contrary to the analogy of the operation of nature, I got used to it [4].[177] (D.21)

The reason for which Aepinus was horrified is that his findings contradicts the universal law of gravitation formulated by Newton, according to which the particles of matter attract in proportion of their masses. However, according to Aepinus, Newton's law holds for ordinary matter while his results only refer to electrified matter (sic). Moreover there is not contradictions at an ontological level as it could appear from the consideration that matter is endowed at the same time with two forces opposed in nature, one attractive the other repulsive, because the repulsive and attractive force are not considered as inherent to matter or essential to it [4].[178]

A modern reader, basing on the theory of matter of the early 20th century, cannot find strange that ordinary matter repel, on the condition that the ordinary matter is intended what remains after the electric fluid (the electron) is considered apart: that is protons and neutrons, and protons repel.

4.3.5.6 Forces Between Two Electrified Bodies

Assume two bodies A and B whose electric fluids in the natural state are respectively Q and q and that some fluid beyond the natural state, α and δ, are added in the order. The forces exercised by B on A (and *vice versa*) are:

1. Fluid in B attracts the ordinary matter of A by $\dfrac{(q+\delta)a}{q}$, where a is the force when B is in the natural state.
2. Fluid in B repels fluid in A by $\dfrac{(q+\delta)(Q+\alpha)}{Qq}r$, where r is the force when A and B are in their natural states.
3. Ordinary matter of B attracts fluid in A by $\dfrac{(Q+\alpha)}{Q}A$, where A is the force when A is in the natural state.
4. Ordinary matter of B repels the ordinary matter of A by R.

[176] Section 30, p. 38.

[177] Section 31, p. 39. Translation in [5].

[178] Section 31, pp. 39–40.

Thus the force with which B attracts A can be obtained by adding all the four contributions with the due sign. Considering that $A = a = R = r = a$ [4],[179] the relation is obtained where negative values means repulsion:

$$-\frac{\alpha\delta}{Qq}r$$

If $\alpha\delta > 0$ the two bodies repel, if, on the contrary, $\alpha\delta < 0$ the two bodies attract. When either $\alpha = 0$ or $\delta = 0$ the two bodies remain at rest. All goes as if instead of the lack of electric particles, there were particles of opposite electricity (as in the two fluids theory).

This relation is extremely interesting because explains experimental results that a naif application of Franklin's theory cannot justify. Franklin and the followers of one electric fluid theory, were not able to furnish a satisfactory explanation of the well known experimental results that two negatively charged body repel. Indeed there is no need that the lack of the electric repelling particle of the electric fluid could give raise to a repulsion. Some qualitative physical reasons were assumed, but none of them was considered as satisfactory by all. The previous relation obtained by a strict application of Franklin's rules and equilibrium equations, shows that the repulsion of negatively charged bodies depends on the balance of forces at play.

Besides this satisfactory explanation, the previous relation seems to indicate a flaw in Franklin's theory. The relation suggests the paradoxical result, contradicted by the common experience: When a not electrified body ($\alpha = 0$ or $\beta = 0$) is close to an electrified body it is not attracted by the electrified body. Aepinus insisted that this result is necessary when accepting Franklin hypotheses: "It is a most certain consequence of our theory that any body constituted in the natural state is neither attracted nor repelled by any body whether positively or negatively electrified" [4, 5].[180]

How is then possible to reconcile theory and experience? The only way left is to assume that a not electrified body when located near an electrified body becomes electrified in turn. Indeed the fluid of the electrified body A (with an excess α of electricity; see Fig. 4.18) repels the part of the non electrified body B close to A, *ab*, by displacing them to the other side of B, *bc*, and thus B is divided in two parts with equal and contrary excess of fluid β. If A is electrified positively β is positive, otherwise it is negative. The attraction of B by A is easily explained. The 'excess' of fluid α of A attracts the fluid $-\beta$ in B with a force r and repels the fluid β by a force r'. Because β is farther than $-\beta$ and thus $r' < r$, the total force between and B results in an attraction.

A modern recognizes in the explanation furnished by Aepinus the electrostatic induction, one of the most important phenomenon of the whole electrostatics; a phenomenon (if not the name) that in the second half of the 18th century time was

---

[179] Section 34, p. 46.

[180] Section 107, p. 114, p. 304.

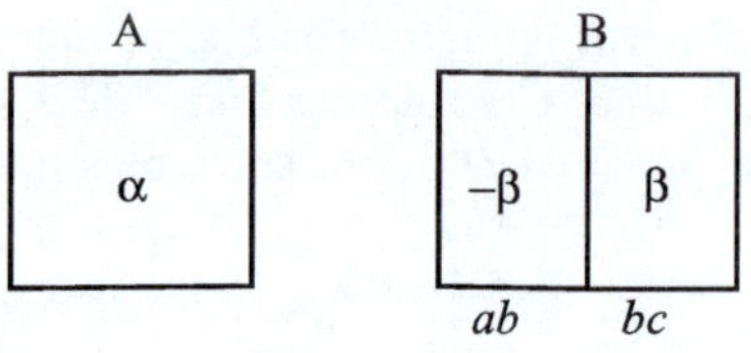

Fig. 4.18 The electrostatic induction

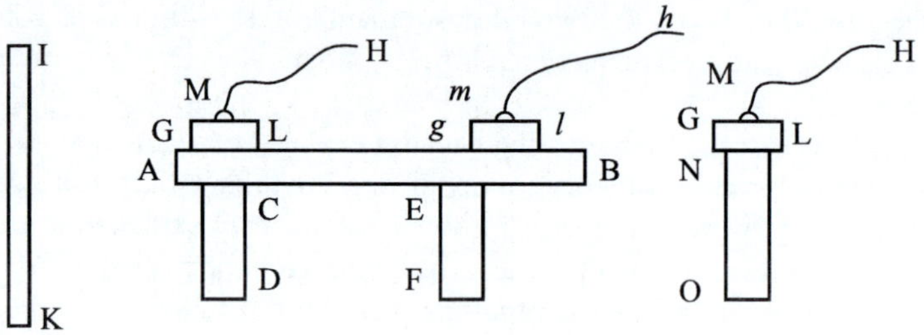

Fig. 4.19 Experimental verification of electrostatic induction. Redrawn from [4], Table I, Fig. 23

well known. To prove without any doubt that his explanation was correct, Aepinus proposed however an interesting experiment.

Place a metal rod AB, said Aepinus, about a foot long on glass supports CD, EF of Fig. 4.19, and on the end A place a metal piece GL about an inch and a half long fitted in the middle with a little hook M to which has been attached a well-dried silk thread HM. Then take the electrificatory glass cylinder IK, and, after it has been electrified by rubbing, move it to the end A of the rod, to a distance of about an inch and hold it there motionless. Lift the metal piece GL by means of the silk thread HM and place it on the glass support NO. If the body GL is examined with an electroscope it will be found to be electric and indeed negatively so. In the second experiment, let everything happen in the fashion described, and move the glass tube IK once more to the end A, with a body *gl* placed on the end B of the rod. If everything is done as in the preceding experiment, the body *gl* placed on the support NO will again be electric, but contrary to before, it will have a positive electricity [4].[181]

4.3.5.7 Interior and Exterior Electric Fluid in the Leyden Jar

Franklin theory of electricity, used in a consistent way by Aepinus by means of equilibrium equation, is shown to lead to results that contradict some of Franklin assumptions. For instance, according to the theory of Franklin, the interior and exterior face of a Leyden jar are equally electrified. Aepinus showed that it is not exactly the case. To prove the fact, instead of referring to a Leyden jar, he considered a sim-

[181] Section 124, pp. 127–128.

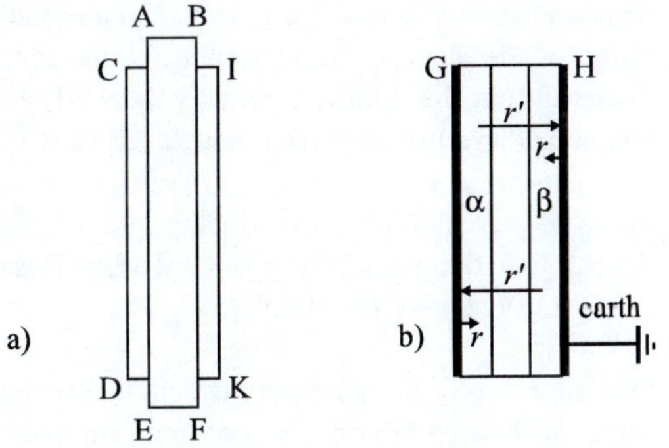

Fig. 4.20 A simplified version of Leyden jar; a simple condenser. **a** Original drawing redrawn from [4], Table I, Fig. 8; **b** A modern drawing

ple plane systems, which behaves similarly to the jar. Let consider, as in Fig. 4.20a a glass plate AB covered on both sides by two equal metal plates IK and CD; both the plates have the same normal amount Q of the electric fluid.

Let the plate IK be connected with the earth, while the plate CD electrified with the help of a chain that brings to it electricity. Let α be the excess of electricity communicated to the metal plate CD. The particles of the electric fluid that are on the exterior surface H of the plate IK (Fig. 4.20b) are repelled with a force given by $\alpha r'/Q$ by the electric fluid of the whole plate CD, and will be attracted by a force equal to $\beta r/Q$, where β is the electric fluid that has flowed from the plate IK toward the earth. So that the total flux, tending to expel these particles is given by:

$$\frac{\beta r - \alpha r'}{Q}$$

An equilibrium is reached when this force is equal to zero, that is for $\beta = \alpha r'/r$. Because the force r' is less than the force r, being the fluid α farther away from the particles in the layer H, it is $\alpha > \beta$; that is the two plates contain a different amount of electric fluid, and CD is more electrified than IK. After the equilibrium has been reached, the force acting on the particles of the layer G of the plate CD, given by:

$$\frac{\alpha r - \beta r'}{Q}$$

which for $\beta = \alpha r'/r$, that is at equilibrium, furnishes[182]:

$$\frac{\alpha(r^2 - r'^2)}{Qr}$$

This force increases continuously as α, that is the electricity of CD, increases, and eventually becomes so great that the air surrounding CD cannot any longer offer

[182] This equation has been corrected according the suggestion of [5], p. 269, footnotes 29, 30.

resistance to avoid the dissipation of it and the electric fluid escape and the accumulation of the electric fluid reached its maximum value [4].[183] To note that Franklin assumed that the equilibrium was reached when all the electric matter of the glass has been driven from its outer surface [122].[184]

4.3.5.8 Attraction When Both Bodies Possess Either Positive or Negative Electricity

One of Franklin's principle says that elementary electrical parts repels. It is the same for bodies which contains both an excess of electric fluid? Indeed in such a situation besides forces among electrical particles, always repulsive, there are also attractive forces due to the interaction between electrical and common matter, so the question is in principle left open. Aepinus probably knew from a work of the Eulers [87],[185] or from some experiments—though he did not refer about any of the two circumstances—that bodies with the same electricity can attract, and wished to prove that this result is in agreement with his theory

Consider bodies A and B of Fig. 4.21, both positively electrified whose natural quantities of electricity are $2q$ and $2Q$ and the excess of the electric fluid α and β respectively. When these bodies are put close to each other, the fluid in the parts NIOK and CDGH, which are the nearest, is repelled toward LNOM and GEFH of the amounts η and θ respectively.

The amount of exceeding fluid in the two bodies is resumed below:

Body	NIKO	NOML	GHDC	GEFH
Fluid	$e = \frac{\alpha}{2} - \eta$	$f = \frac{\alpha}{2} + \eta$	$a = \frac{\beta}{2} - \theta$	$b = \frac{\beta}{2} + \theta$

The forces that the four parts of the bodies A and B exchange are proportional to the amount of electric fluid e, f, a, b (see above) contained in them and inversely proportional to the normal quantities Q, q. They are given in the following:

Body	NIOK	NOML	NIOK	NOML
Body	GHDC	GHDC	GEFH	GEFH
Force	$-\frac{ea}{Qq}r$	$-\frac{fa}{Qq}\rho$	$-\frac{eb}{Qq}r'$	$-\frac{fb}{Qq}\rho'$

where r, r', ρ, ρ' are the forces the various parts of A and B would exchange, as indicated in Fig. 4.21b.

[183] Section 45, pp. 54–55.
[184] p. 338.
[185] p. 144.

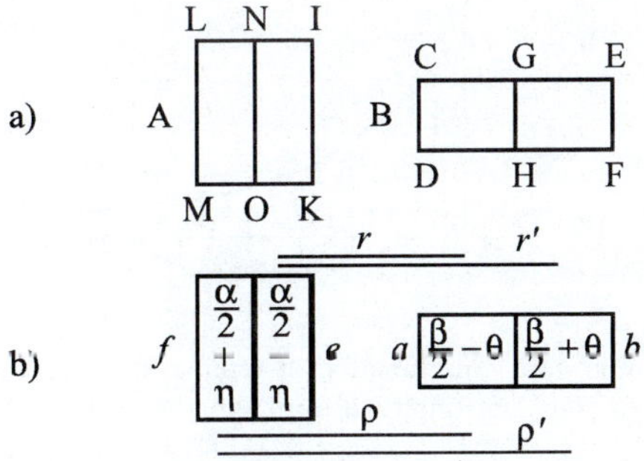

Fig. 4.21 Forces between two charged bodies. **a** Original drawing redrawn from [4], Table I, Fig. 22; **b** Arrangement of electricity and forces

By adding all the contributions, the force A and B exchange is:

$$-\frac{(ar + br')e + (a\rho + b\rho')f}{Qq}$$

It is evident that the values of e, f, a, b depend on the intensity of the repelling force among the particles of the electric fluid, which in turn depend on the distance among A and B and on the excess of the fluid α and β.

It is possible to show, on qualitative basis by considering a reasoning similar to that which proves the attraction between an electrical and a neutral body, see Sect. 4.3.5.6, that two bodies positively electrified sufficiently distant from each other repel, yet if they are put closer by an external force, they begin to attract. Indeed as the bodies are rather distant from one another, the repelling force on the fluid contained in the other is quite weak, so the quantities η and θ are small enough, so that a and e are both positive—of course f and b are always positive—and A and B repel. But when η and θ increase because the distance between them is decreased the two quantities a and e, or at least one of them, may become negative, and the force between the bodies changes its sign and they attract.

Aepinus underlined that his result seems paradoxical but it is a strict consequence of Franklin's view about electricity: "These phenomena would then contain enormous paradoxes were not their source and their friendly agreement with the fundamental laws of nature clear from our own theory. At the same time there is not a little probability in our theory, for when it is assumed, these paradoxical phenomena agree so aptly with the fundamental laws of nature; and I doubt very much whether this could as happily be the case with any of the other hypotheses developed to date and *alien to the analogy of the other operations of nature*" [4, 5].[186]

It is not difficult for Aepinus to reach a similar conclusion when the two bodies are negatively electric. More complicate is the analysis of the case when one of the body is negatively electric while the other is positively electric. In such a case it is impossible to reach any conclusion unless resource to experience is made "no satisfactory reply can be given to this question until we understand the function

[186] Section 133, pp. 135–136, 318.

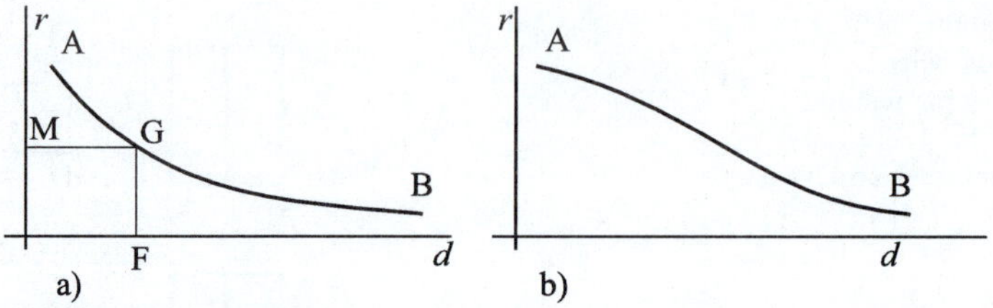

Fig. 4.22 The function of repulsive forces versus the distance among the two bodies A and B. Redrawn in simplified form from [4], **a** Table I, Fig. 24; **b** Table II, Fig. 25

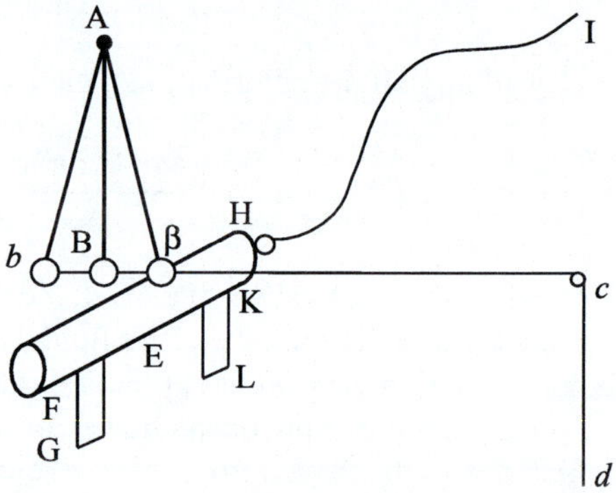

Fig. 4.23 Attraction between bodies with the same kind of electricity, redrawn from [4], Table II, Fig. 26

(functio) according to which electrical repulsion is exerted" [4].[187] All depends on the way the repelling force varies with distance. Aepinus considered the two possibilities shown in Fig. 4.22a, b. Only when the forces varies as in Fig. 4.23b is it possible that bodies oppositely electrified repel. But "if we consult experience, there can hardly be any doubt remaining that the often-mentioned graph of repulsions ought never to have a point of inflection" [4].[188]

Because some doubts would remain from these apparently paradoxical result; Aepinus suggested a simple experimental test, referred below in full.

Suspend a small globe of cork, B, about the size of a pea, from a silk thread AB as shown in Fig. 4.23. Below the pendulum AB, place a metal cylinder E about an inch in diameter on a glass supports FG, KL, and arrange the supports and the pendulum AB so that when the latter is vertical the globe B does not quite touch the cylinder E. To the globe B tie the silk thread *bcd* and pass it over the little hook *c*. Fix to the cylinder E an iron wire HI five or six feet long, properly supported on bodies electric *per se*. Electrify the globe B in the normal manner with the help of the glass tube and afterward, by moving the tube to the end I of the iron wire, make the cylinder E electric too with the same kind if electricity. It will then be observed that the globe B is repelled and the pendulum is raised to the position A*b*. Then pull the thread

[187] Section 138, pp. 139–140,.

[188] Section 142, p. 143.

bcd and force the globe B closer to the cylinder E, and after it has approached to a distance of 2, 3 or 4 lines, it will suddenly be observed that the repulsion has changed to an attraction, and the pendulum comes to the vertical position AB; if it is disturbed a little from this position by pulling the thread bed forward a half or whole line, it spontaneously returns again to its previous position. This experiment succeeds as well if a sulphur cylinder is substituted for glass [4, 5].[189]

Thus Aepinus can conclude: It is known among scholars dealing with electricity, the rules of differently electrified bodies attract one another and if possessing the same kind of electricity repel one another, then known as Dufay rules. The first rule, can be accepted; but the second, as it is usually promulgated (pronunciati), must be considered not completely true. Indeed if both bodies have the same electricity, they can at times repel one another, at other times have no action on one another, and at other times attract one another. "Thus one can generally admit the converse of the rule [bodies with the same kind of electricity repel] as true, that is, if two bodies repel one another, then the quantity of electric fluid in both bodies at the same time is either greater than or less than the natural. For it is easily established that only in this case can bodies mutually repel one another, since in other cases, as I have sufficiently shown already, they always attract each other and never repel" [4, 5].[190]

In general it can be said that from a theoretical point of view any possibility is open regarding the attraction or repulsion of two electrified bodies, they can attract, repel or remain at rest independently of their content of electricity. All depends on the balance of forces. Of course experience shows some particular occurrences, but all of them can be explained or predicted by the theory.

4.3.5.9 Air Condenser

Differently from Franklin, Aepinus found that the behavior of the Leyden jar could be reached even with matter different from glass, air in particular. The apparatus he considered was the one illustrated in Fig. 4.20 where the layer ABEF of glass has been replaced by air to form what is now called an air condenser. This system was presented already in 1756 at the end of Aepinus' memoir on the piezoelectric properties of tourmaline, *Mémoire concernant quelques nouvelles experience électriques remarquables* [2]. In his experiments, he suspended two surfaces covered with metal, so that they were parallel and the distance from one to the other in all their points was from one inch to 1 1/2 ($\sim$4 cm), without neither mediately nor immediately contact. Electric fluid (positive electricity) was driven from an electrified globe on one of these surfaces, the other was connected by means of a chain, to the floor so that the electrical matter, that was driven out by repulsion, flowed and the surface connected to the floor was able to acquire negative electricity. While these things were going on, said Aepinus, he felt a strong shock, quite similar to that which is commonly produced by means of glass. This experience would not succeed with small surfaces

[189] Section 144, p. 144; pp. 323–324.

[190] Section 147, pp. 146–147, 325.

and its effect comes far more sensitive when the surfaces used were large. The ones he used, each had 7 1/2 square feet (corresponding to a large square of ~80 cm side) and they were of wood covered with these sheets of tin that are applied to mirror glasses [2].[191]

By examining the matter more deeply, Aepinus said, that impermeability was involved in the essential property that makes bodies *electric per se*; it is by virtue of this property that in these bodies the movement of electric fluid is allowed only with difficulty through their pores. And it is quite strange that the celebrated Franklin, who admirably expounded the nature of bodies *electric per se*, did not perceive that the immediate consequence of the essential property of such bodies is that property which he named impermeability [4].[192] A property possessed also by air, even though to a less degree than glass.

4.3.6 The Italian School

While the discovery of the properties of the Leyden jar opened up new horizons for electrical research in France, Germany and England, in Italy, only after the circulation of the experiences and discoveries of Benjamin Franklin, systematic researches started on electrical phenomena. In truth, starting from the second half of the 18th century, Italian scholars had cultivated the study of natural and artificial electrical phenomena. In 1746, a small historical treatise was published anonymously: *Dell'elettricismo: ossia delle forze elettriche de' corpi* [180], which is commonly attributed to the physician Eusebio Sguario (fl. 1750), written from a Newtonian perspective. Sguario, separating the physical electricity from the medical one, opened the way for Beccaria and Galvani to go all the way. Some studies were made in Bologna, where a lightning rod was installed in the university tower, following the example of Franklin (1755), then removed because of population believing it attracted lightning and thus dangerous. Scipione Maffei (1675–1755) from Verona published the *Della formazione dei fulmini* [145], where he showed to know the most important experiences in Europe on electrical phenomena, including the experience of water electrification; he also described his own experiences with various electrical machines. These experiences, as well as those presumably conducted in Turin by Francesco Antonio Garro, can be framed in a predominantly educational and informative perspective.

However the news of the electric experiments conducted in Europe carried out in the laboratory fueled in many Italian physicians around 1745–1750, the opinion—sustained by experiments often carried out with not entirely correct methodologies—that the electric fluid could have beneficial effects on health and could therefore be used for therapeutic purposes. In Turin the medical pathologist Giovambattista Bianchi (1681–1761), in Venice Gianfrancesco Pivati (1689–1764) and in Bologna

[191] p. 120.

[192] Section 76, p. 83.

Giovanni Giuseppe Veratti (1707–1793), a member of the Accademia delle scienze dell'Istituto di Bologna, carried out a certain number of experiments and electrical applications on patients which attributed beneficial effects, resulting in healing according to their testimonies, gouty or sciatic rheumatic forms. The news of these electrical experiments and their beneficial effects on the human organism were made known to the public by letters or through the spread of printed works. They aroused the curiosity of Nollet who decided to come to Italy. This had a positive effect on the knowledge of electricity in Italy, as Nollet with his discussions certainly contributed to spreading new concepts.

Nollet in particular was impressed by Laura Bassi, wife of Veratti, who he met in Bologna. He dedicated to her one of his *Lettres sur l'electricité*. The letter to Madame Laura Bassi of the "Academie de l'Institut de Bologne" focused on some curious applications of electricity. According to Nollet, these applications could be further improved [185]. Laura Bassi has been almost totally ignored by historians of electricity, what is quite unjust considering that she presented no less than seven dissertations on electricity to the academia of Bologna, a number surpassed only by her husband's. She accepted the version of the single fluid theory proposed by Beccaria and in 1771 presented to the academia of Bologna the paper, unfortunately now lost, *Sopra l'elettricità vindice*, a strong point of Beccaria [54].

Studies on electricity should be signaled in the south Italy also. Some works were published in Naples. In 1747 the *Tentamena eletrice* of Georg Matthias Bose (1710–1761) was translated into Italian and a new edition of the *Dell'elettricismo* of Sguario was issued. Finally the work of Giovanni Windler *Tentamina de causa electricitatis quibus brevis historia de nonnullis cuctoribus qui hanc praecipue excoluerunt materiam, premissa est.* In 1748, one finds the work of Niccoló Bammacaro *Tentamen de vi electrica ejusque phaenomenis in quo aeris cum corporibus universi aequilibrium proponitur*. In 1750, the first edition of *Scienza della natura* by Giovanni Maria della Torre was published. This work was used as a manual in the teaching of experimental physics using the Newtonian approach. Although a treatise on physics, electricity was abundantly discussed. In 1761, Nollet's *Lettres sur l'électricité* was translated, most likely by Maria Angela Ardinghelli. After the 1760s no work on electricity was published in Naples [177].

However it was only with Beccaria, who had begun to experiment with his lightning rod possibly before Franklin, that electricity in Italy passed from living room, newspapers and gazettes subject, to a research topic within a few years. In the following I report the contribution of some Italian scholars who met a good international acknowledgment: Giambattista Beccaria, Gianfrancesco Cigna, Carlo Battista Barletti, Ruggero Giuseppe Boscovich. Before however a short mention to Frisi is due. There are no references instead in this book to the two greatest Italian electricity scholars of the 18th century, Alessandro Volta (1745–1827) and Luigi Galvani (1737–1798). The choice is motivated by the fact that the lines of research carried out by them fit better in the 19th century than in the 18th one.

Paolo Frisi, (1728–1784), Italian mathematician, astronomer, and physicist is best known for his work in hydraulics. Frisi was a member of the Barnabite religious order, a professor at the university of Pisa and the Scuole palatine at Milan and a member of most of the major scientific societies of his time. He was held in such esteem by his contemporaries that plans for nearly all the major hydraulic works constructed in northern Italy during his adult life were first showed to him for inspection.

In electricity his major work was the *De causa electricitatis dissertatio* which he presented at the Petersburg prize competition for 1755 of the academy of science of Petersburg [99]. The winner of the prize was Johann Albrecht Euler and Frisi classified as second. According to Pietro Verri (1728–1797), Frisi's friend and the editor of the famous newspaper *Il caffé*, the judges would have preferred Frisi but refused him the prize because he had written his name on his essay [191].[193] In any case Frisi's work was quite interesting. The reason that it has not found space here is because Frisi, as the Eulers, pursued an approach to electricity of Cartesian mould, that revealed sterile. Both Frisi and the Eulers applied mathematics to electricity, but it was more to show their ability in the field than to treat seriously electricity [122].[194] Another Frisi's contribution to electricity was the *Dei conduttori elettrici* published in 1781 [100], a short pamphlet devoted mainly to the problem of the defense against lightnings.

4.3.7 Giambattista Beccaria

Giambattista Beccaria (1716–1781), born in Mondovì, Piedmont, entered the order of Scolopi in Frascati as a novice and changed his baptismal name Francesco Ludovico into Giambattista. The concrete approach of the Scolopi schools had predisposed them to a rapid adoption of new scientific and philosophical trends, an aspect which was significant for the formation of Beccaria. His early knowledge concerned Galilean science and Leibniz's monadology, learned through Wolff. The subsequent reading of Antonio Genovesi and perhaps of Voltaire opened the mind of Beccaria to the knowledge of Locke and Newton. An itinerary not unlike that of other savants of the time.

After a teaching career started in 1737, eventually, in 1748, he was offered the chair of experimental physics at the university of Turin, previously held by Francesco Antonio Garro. The appointment also sanctioned the transition, at the university of Turin, from Cartesian physics to Newtonianism. An appointment which, moreover, did not occur without discontent and lacerations. Garro, supported by Giuseppe Roma—from Toulouse, belonging to the Order of Minims like Garro—tried in every way to hinder the newcomer. Exponents of a strongly Cartesian Turin tradition, in this devoted to the address given to their order by Mersenne, Garro and Roma on the one hand did not want to lose their place and, on the other hand, they did not intend to resign themselves to the Newtonian physics brought by Beccaria [164].

---

[193] pp. 23–24.

[194] p. 395.

Beccaria's appointment was due to Giuseppe Francesco Morozzo (1704–1767), university reformer in collaboration with the leaders of the pious schools. It was still Morozzo, who had heard from his envoys in Philadelphia about the recent electrical discoveries developed by Franklin and the experiences that proved them, to suggest Beccaria to strengthen his position from now on, distinguishing himself in this promising scientific branch. Beccaria got to work and in less than a year, in 1753, he produced his first important treatise *Dell'elettricismo artificiale e naturale* [23].

Beccaria had started his career as a professor teaching Galilean physics imbued with experimentalism and gathered around him a circle of young people, including Gianfrancesco Cigna (1734–1790), Joseph Louis Lagrange (1737–1813), Giuseppe Angelo Saluzzo (1734- 1810). He made Newtonian physics his own, but more in the methods than in the mathematical approach. There are numerous references to Newton in his texts. In particular, the idea of giving up the ultimate explanation of the origin of forces. "From the phenomena of nature to derive two, or three general principles of motion, and then explain, as the properties and actions of all bodily things follow from these manifest principles; this would be a great advancement of Philosophy; though the reason for these principles was not known yet" [23].[195] Beccaria also attributed to Newton the idea of an electric fluid [23].[196]

Although the research in electrical phenomena was the main affair of the intellectual life of Beccaria, he left, published and unpublished, many other writings on various topics: chemistry, meteorology, optics, astronomy, hydraulics, physiology. He was also entrusted with tasks of a practical nature, such as the revision of the system of weights and measures of the Sardinian states, the installation of lightning rods, the determination of a plant for the distribution of the waters of the Po. The most demanding of these works was the measure of the length of the meridian of Turin.

4.3.7.1 Treatises on Electricity

Of the many writings of Beccaria about electricity two treatises are particularly relevant, the *Dell'elettricismo artificiale e naturale* of 1753 and the *Elettricismo artificiale* of 1772.

The *Dell'elettricismo artificiale e naturale* was the first important Beccaria's work on electricity. It immediately was seen as a masterful work of synthesis, clarification and development. The starting point was the concept of a single fluid and of positive or negative electricity. Beccaria offered a rich experimental documentation, partly by himself and partly by others, exposing Franklin's theories in an organic and fairly rational way. The form of presentation used by Beccaria was the sparse and strictly logical one of mathematical texts, even if his language was not fluent. Apart from the logical form, the text did not substantially contain any technical application neither of geometry and algebra nor of Calculus; as was then customary for experimental

[195] p. 40.

[196] p. 126.

physics texts. This even though Beccaria was not completely lacking in mathematics as evidenced by his important work on geodesy, the *Gradus taurinensis* of 1774 [30].

The treatise, as suggested by the title, is divided into two books; the first addresses the problem of artificial electricity, the second the problem of natural electricity. The first book, divided into eight chapters and 464 paragraphs, illustrates the concepts and experiences concerning electric and non-electric bodies and the diffusion of electric steam (Chap. I); the motion of electrified bodies. Sparks and electric breeze (Chap. II); the speed with which electric steam spreads. Power of the points and electric atmosphere (Chap. Ill); the electricity of metals and glass (Chap. IV); air electricity (Chap. V); water electricity (Chap. VI); the electricity of plants and animals (Chap. VII); electrical phenomena and light and fire (Cap. VIII). These experiences can be conjectured as largely carried out by Beccaria starting from April 1752, or even earlier, that is from the moment when he became aware of the works of Benjamin Franklin published in London in the summer of 1751, or in Dalibard's French translation in the following year [173].[197] With clear reference to these experiences, Beccaria claimed however the originality and novelty of his experiments and in the presentation letter *To the readers* he wrote: "You will find, courteous readers, that in this work I did many times mention of the very famous electrician Beniamino Franklin and I would do it even more often, if his discoveries were not well known, and if I did not propose to write about electricity according to what experiences have shown me, but rather according to a broader understanding of phenomena, of which with a long and connected series of my own experiences I immediately ascertained myself" [23].[198]

The second book of the *Dell'elettricismo artificiale e naturale* has seven chapters. The first three chapters are dedicated to the description of the experiences with the *Franklin rod* and to the demonstration of the identities between atmospheric and artificial electricities, chapters IV, V and VI were instead an important contribution by Beccaria to the study of the effects of lightning in air and water. Chap. VII deal with the role of electricity on various natural phenomena, such as typhoons, earthquakes, waterspouts, Northern lights, cohesion, gravity, etc.

Galvani had studied Beccaria's text; it was also the first and main resource of the seventeen year old Alessandro Volta. In a Latin poem datable around 1764, Volta praised Beccaria, as one of the most recent discoverers of the electrical origin of lightning together with the American Franklin [10].[199] The Paduan Giuseppe Toaldo (1719–1797) was inspired by Beccaria for his treatise *Della maniera di difendere gli edificii dal fulmine* of 1772 (his pamphlet *Dei conduttori metallici a preservazione degli edifici dal fulmine* of 1774, contributed largely to remove the popular prejudices of the time against the use of the Franklinian rod). Beccaria also influenced, at least indirectly, Carlo Battista Barletti (1735–1800) who was in direct contact with Cigna, Beccaria's grandson and pupil and played an important role in the history of electricity in Italy, but not only.

[197] p. 252.

[198] To the readers, first rows.

[199] p. 14.

Franklin praised Beccaria as a master of method and saw in him the one who had systematically reduced the experiences and stances scattered in his own papers. He used Beccaria's text to counter Nollet's opposition to his theories. An important tribute was that granted by Joseph Priestley who did not hesitate, in his monumental *The history and present state of electricity, with original experiments*—released in the United Kingdom in 1767 but in Italy known above all in the French translation of 1771—to pay homage to him "Signior Beccaria, one of the most eminent of all the electricians abroad" [171].[200]

The *Elettricismo artificiale* of 1772, represents an overall recast of Beccaria's work on electricity. Franklin had it translated into English in 1776 as *A treatise upon artificial electricity* [31]. The book was formed by six chapters. Chapter I concerned the theory of artificial electricity, especially in *deferent* or conducting bodies,[201] deduced from the circulation of the electrical fire in the ordinary apparatus. Chapter II dealt with the theory of insulating bodies, with regard to the charging and discharging of them. Chapter III on pressing electricity (a term proper of Beccaria), or on electric atmospheres. Chapter IV on vivid electricity; otherwise on electric sparks. Chapter V on the electric tickling and wind; on the brush and the star, with interesting references to the medical uses, the subject of study a few years later of Tiberio Cavallo (1749–1809) in England. Chapter VI, the last, on electric motions and on the vindicating electricity, one of fundamental Beccaria's concept. In the subsequent sections the contents of Chap. III on the electric atmosphere and of Chap. VI on the vindicating electricity will be commented with some details.

Beccaria for his experiments, used an apposite electric machine. It consisted essentially by a rotating glass cylinder which was rubbed by the hand of a man. The complex was placed on a platform large enough to contain one or more men. From a theoretical point of view it was made by an insulator and two deferents. An insulator (the cylinder of glass), that is rubbed by a deferent (a man) and that goes to find the other deferent (a hollow tube in metal) with the rubbed part. Such an apparatus will be the more perfect as its size and exactness taken together, are the greater.

The machine, shown in Fig. 4.24, is described below using Beccaria words. In TS, between two poles an insulating body is to be fitted, which in the case of the figure is a cylinder of glass. The frame, made of little wood beams, and boards, ABCDMNOIK, is destined to support, first the insulating body TS, then the wheel R, and lastly it must be able to receive in M, one man who turns the wheel and another man who rubs the cylinder. The frame of wood and the man who turns the wheel, all communicate with the man who rubs. All together may be said to belong to the rubbing deferent body; their whole assemblage in the following is comprehended under the name *machine*. Lastly, the hollow tube of brass Y, a cannon, twelve feet[202] long (about 3.5 m), and a foot (about 30 cm) in breadth, which has one of its ends shaped like a hemisphere

[200]Vol. 1, p. 194.

[201]*Deferent* (deferente) is the term used by Beccaria for conductors. The English translation of his treatise replaced everywhere deferent with conductor.

[202]Beccaria in one occasion compared his measures, in particular his inch, with a physical magnitude, which allows to give it a value: "supposing the common height of mercury in the barometer

Fig. 4.24 Beccaria's electrostatic generator. Redrawn from [29], Fig. 1 of table I

and with the other terminating in a conic point, stands near the equator of the rubbed cylinder, is the other deferent body, in which the rubbed cylinder is to diffuse the electrical fire which it draws by the effect of the friction—the property of glass is to draw fire from the man that rubs, or from the machine, or through it, from the floor—This deferent body, of which the form and size may be varied at pleasure, is, indeed, usually called the prime conductor; however, in the following will be used the old name, *chain* [29, 31].[203]

4.3.7.2 The Electric Atmosphere

The concept of electric atmosphere was used in the first half of the 18th century by electricians, Franklin included, to explain many electrical phenomena. In the following it is presented the classical concept accepted by Franklin, and then the radical change introduced by Beccaria. According to Franklin, when an excess of electric fluid with respect to the normal one is added to a body, it does not enter inside, but will flow round the surface of the body and forms an *electrical atmosphere*. The form of this atmosphere is that of the body it surrounds. A shape which may be rendered visible in still air by raising a smoke from dry rosin.

in Turin to be twenty-seven inches and an half" [29], p. 165. Which gives for the inch the value of 2.76 cm, greater than the current English value.

[203]pp.10–11; p. 11.

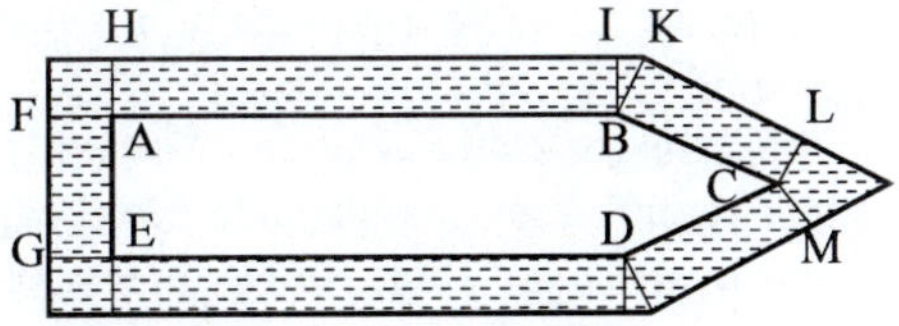

Fig. 4.25 Electric atmosphere around a pointed body. Redrawn from [94], Fig. 8 of the tables of figures

The form of the electrical atmosphere is that of the body it surrounds because it is attracted by all parts of the surface of the body, though it cannot enter the substance already saturated. Without this attraction it would not remain round the body, but dissipate in the air. The atmosphere of electrical particles surrounding an electrified sphere, is not more disposed to leave it or more easily drawn off from any one part of the sphere than from another, for symmetry reasons. But that is not the case with bodies of any other figure.

Thus let, said Franklin, for instance, a body shaped as A, B, C, D, E, of Fig. 4.25, be electrified, and consider every side as a base on which the particles rest and by which they are attracted. So the portion of atmosphere included in the portion of space H, A, B, I, has the line AB, for its basis. Now if one would draw off this atmosphere with any blunt smooth body and approach the middle of the side AB, he must come very near before the force of this smooth body exceeds the force or power with which AB holds its atmosphere. But the small portion of atmosphere between I, B, K has less of the surface to be attracted by (while at the same time there is a mutual repulsion between its particles), therefore here one can get it with more ease. And the easiest of all is between L, C, M, where the quantity of atmosphere is largest and the surface to attract and keep it back the least.

When one has drawn away some of these angular portions of the fluid, another succeeds in its place, from the nature of fluidity and the mutual repulsion of the electric fluid; so the atmosphere continues slowing off at such angle, like a stream, till no more is remaining. On these accounts we can conclude that electrified bodies discharge their atmospheres upon unelectrified bodies more easily and at a greater distance from their angles and points than from their smooth sides. Those points will also discharge into the air, when the body has too great an electrical atmosphere, without bringing any non-electric *per se* near, to receive what is thrown off. Indeed the air, though an electric *per se* yet has always more or less water and other non-electric *per se* matters mixed with it and these attract and receive what it is discharged [94].[204]

The concept of electric atmospheres offered a means for Franklin to explain various phenomena of electricity and the operation of lightning rods. The greatest difficulties probably arose in explaining the more traditional phenomena of attraction and repulsion. In the case of the repulsion of two positively charged bodies, when two bodies approach, their atmospheres interact and the bodies are rejected. The mechanism remains however vague. Are there acting forces at a distance? Or contact; but in this last case how can one explain the repulsion for sensible distances.

[204] pp. 56–58.

But, mainly, how to explain the repulsion between negatively charged bodies, where atmospheres do not exist at all.

Franklin generally accepted Newton's ideas, but he was not Newtonian enough to talk about forces at a distance. He accepted Newton's ideas, set out in the Queries of the *Opticks* for short-range forces, otherwise he could not speak of attraction between electric fluid and matter and repulsion between electric fluid and electric fluid. Franklin's ambiguity in the use of the concept of force is evident not only in the use of the idea of atmospheres, but also when he wanted to explain the distribution of the electric fluid within bodies. For example to explain the concept of normal quantity of electricity he started by a piece of common matter supposed entirely free from electrical matter. When a single electrical particle is brought nigh, it will be attracted and enter the body, and take place where the attraction is every way equal. If more particles enter, they take their places where the balance is realized between the attraction of the common matter and their own mutual repulsion [94].[205] But this explanation only makes sense if it is admitted that remote forces exist, otherwise a particle would remain where it is attracted by the particles of common matter.

Franklin introduced additional hypotheses about the properties of electric atmospheres to explain Canton's experiments of 1753, in which the discovery of electrostatic induction is recognized today. In particular:

1. Electric atmospheres, that flow round non-electric bodies, being brought near each other, do not readily mix and unite into one atmosphere, but remain separate, and repel each other.
2. An electric atmosphere not only repels another electric atmosphere, but will also repel the electric matter contained in the substance of a body approaching it; and without joining or mixing with it, force it to other parts of the body that contained it.
3. Bodies electrified negatively, or deprived of their natural quantity of electricity, repel each other (or at least appear to do so, by a mutual receding) as well as those electrified positively, or which have electric atmosphere [97].[206]

Beccaria radically changed Franklin's concept of atmosphere, transforming it from a static, or geometric, concept into a somehow dynamic concept. His more mature ideas on the point are expressed in Chap. III of the *Elettricismo artificiale*. At that time he probably had read Aepinus' *Tentamen*, as he mentioned it [29][207] and knew Richmann's experiments [29].[208]

Beccaria began to expose his ideas about the electric atmospheres by asserting that "*The electricity of a body A does not substantially diffuse into the ambient air*; that is to say, if a body A be electrified by excess, the fire added to it does not mix itself, even at a perceivable height, into the substance of the air around it, and if the body A be electrified by deficiency, the fire drawn from it has not been extracted from

[205] pp. 53–54.
[206] pp. 155–156.
[207] p. 175.
[208] p. 175.

the substance of the ambient air" [29, 31].[209] This is a position already expressed by Canton, at least since 1767, as reported by Priestley in his *The history and present state of electricity*: "It is now also Mr. Canton's opinion, that electric atmospheres are not made of effluvia from excited or electrified bodies, but that they are only an alteration of the state of the electric fluid contained in, or belonging to the air surrounding them, to a certain distance" [171].[210]

In essence, the vague idea of a cloud of external electric fluid, although adjacent, to the electrified body suggested by Franklin, is denied. Beccaria specified that this is a fact that he was the first to verify experimentally. The nature of electric atmosphere must therefore be explained in a different way, that is:

> The electricity of a body A, actuates the ambient air in such a manner, that by the means of the same it tends to introduce into the neighboring body B immersed in it, an electricity contrary to its own. And it is the air thus actuated, which constitutes what is commonly called the electric atmosphere [29].[211] (D.22)

Beccaria used the term *actuate*, which is not only his but belongs to the Italian literature of electricity, to indicate that something happens to the air surrounding an electrified body, but it is not clear what. Air is somehow electrified. Probably the idea that could make it clearer is to think of the air actuated as a force field. But not a field intended as a place of points where in fact forces were acting, but a Faraday-hypostatized force field. It is difficult to say that this was a concept clear to Beccaria; but it had a greater heuristic power than that of Franklin. Beccaria suggested that his reasoning was connected to a 1755 Franklin's essay on Canton's experience of 1753 [97].[212] and proposed a simple experiment to verify his thesis about the ontology of the electric atmospheres:

> I present a most fine flaxen thread to the conductor Y and if it be held at the distance of a foot, or more, from the surface of it, it will direct itself perpendicularly to it. I then present to this thread a rubbed stick of sealing-wax, and it flies from the conductor; I present to it a rubbed stick of glass, it runs again to it; which is the same as to say, the thread immersed in the atmosphere of the body or conductor A, which is electrified by excess, becomes itself electrified by deficiency. I repeat the experiment by the means, not of the chain, but of the [small] beams B or E (?) of the Machine; then the thread flies from the glass, and runs to the sealing-wax; whence it is likewise evident, that a body immersed in the atmosphere of another body electrified by deficiency, will itself become electrified by excess [29].[213] (D.23)

The conductor Y, the chain of his electrostatic machine, is charged by excess, or (+). So according to Beccaria's hypothesis the linen thread must be negatively electrified, that is (–). The experiment confirms the deduction since the linen thread is rejected by

[209] p. 173; p. 179.

[210] p. 244.

[211] p. 174. Translation into English in [31].

[212] p. 155.

[213] p.174. Translation into English in [31].

a sealing wax stick that has a resinous electricity (–), while it is attracted by the glass that has electricity (+). Vice versa when the linen thread immersed in the atmosphere of a negatively charged body it becomes (+).

After having introduced the concept of atmosphere and having carried out other experiments in confirmation of his hypothesis, Beccaria went on to consider the interactions between electric atmospheres of differently electrified bodies. The rules are:

1. When two bodies, impregnated with homologous electricities, meet, these electricities, by means of the atmosphere which they actuate, endeavor reciprocally to destroy each other.
2. When two bodies impregnated with contrary electricities, meet, these electricities, by means of their intervening atmospheres, reciprocally increase their intensities [29].[214]

To a modern readers these experimental (?) rules appears nothing but a trivial consequence of the electrostatic induction, and so they would have appeared to Aepinus.

Figure 4.26 shows a graphical representation of the electric atmosphere of spherical or cylindrical charged bodies (E, *e* mean excess, D, *d* defect and N neutrality). In Fig. 4.26-1 it is shown the atmosphere of a body E positively charged, whose shape is commented below:

> If a body E, be electrified by excess [Fig. 4.26-1], first a set of small lines springing against the ambient air from all points of the surface of the said body E (we shall see hereafter that both the excess and deficiency of electrified bodies only take place on their surface) may very well represent both the excessive fire of such body, and the increased tension exerted by the same fire against the natural fire of the ambient air; secondly, the ambient air may likewise be understood as being divided into successive exceedingly thin strata, and similar small lines may also, being understood to be applied to the convex surface of all those strata, express the direction and manner in which the excessive tension of the fire in excess on the surface of that body, propagates a corresponding tension or vibration in the fire inherent the ambient air. [29].[215] (D.24)

The small lines, named in the above quotation, could indicate for a modern reader, the lines of force of the electric field, which in the present case are represented by radial lines passing through the center of the circular body E. In Beccaria's words they represent the direction of the tension or vibration propagated into the air. The closed curves surrounding the body E are perpendicular to the lines of the force field. For a modern reader they represent equipotential surfaces. The thickness of the closed curves represent the length of the small lines, that is the intensity of the electric forces (on this point however Beccaria was not fully explicit).

In the case of Fig. 4.26-2, there is a body D electrified by defect. The graphical representation of the electric atmosphere is the same; its description is a little different however. But in this case too there is an electric atmosphere, which Franklin could

---

[214] p. 180.

[215] p. 181. Translation into English in [31].

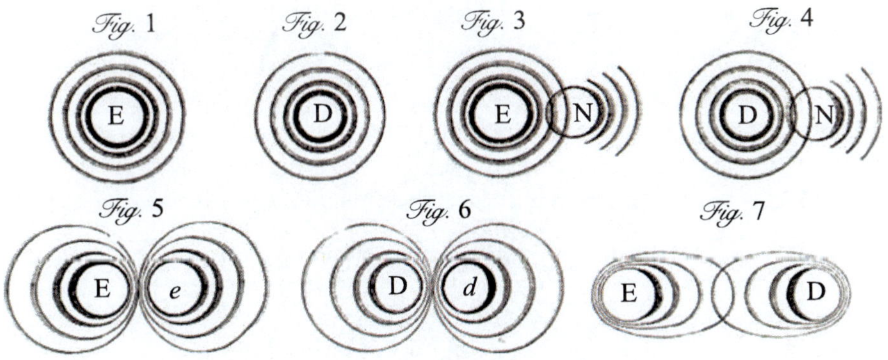

Fig. 4.26 Beccaria's electric atmospheres. Redrawn from [29], Figs. 1–7 of the table VIII

not admit because his atmosphere was identified by the excess electric fluid, which is not present here. One can think in such a situation only of a complex of small lines, which from all the points of the ambient air enter the body, and which mark the defect of fire, the defect of tension, which will exert the residual fire on the face of the body against the natural fire of the ambient air [29].[216]

More interesting is the case of Fig. 4.26-3, in which a non-electrified, or neutral body N, is presented to the atmosphere of an excess electrified body E (in modern term it represents a typical case of induction which involves a displacement of the charges of the body N towards the opposite part with respect to that exposed to the electrified body). In this case the small lines, which from the atmosphere of E following their direction enter the portion of the surface of N immersed in the atmosphere, will mark the natural fire, which will tend to leave that portion of the immersed face; and the small lines, which will arise from the portion of the not immersed face will mark the excess of fire fire, which is pushed away from the natural electric fluid of the immersed face [29].[217]

Figure 4.26-4 illustrates the case in which a neutral body N is immersed in the atmosphere of a body charged by defect D; a situation similar to that of Fig. 4.26-3. Figures 4.26-5, 6, 7 show the interactions of atmospheres of electrified bodies by excess and/or by defect.

Let consider in particular the case of Fig. 4.26-5, where E and *e* are two bodies electrified by excess. The small lines that come out of the face of the bodies protrude into the cavity of the successive layers of the adjacent air, marking the excess of tension that results from the excess of electric fire on the faces of the bodies and the excess of tension that arises from the fire of the air. The mutual opposition, to which the small lines proceeding from one of the body with respect to those proceeding from the other would indicate the force with which the two electricities will tend to destroy. These small lines in between the two bodies decrease as in the figure and

---

[216] p. 181.

[217] p. 181.

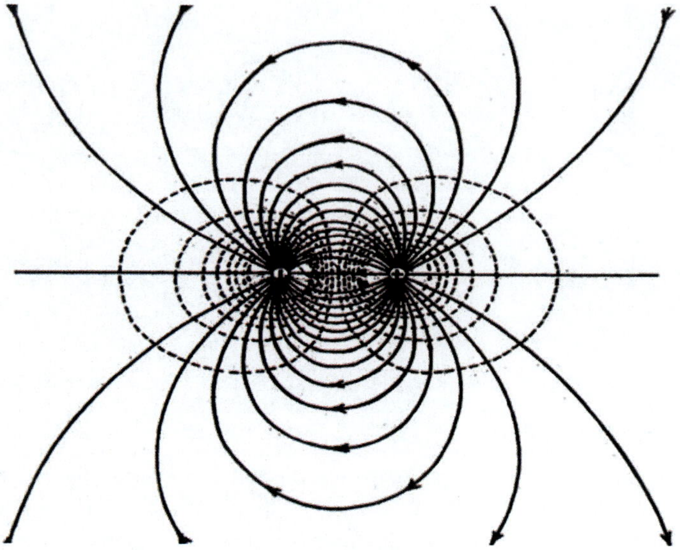

Fig. 4.27 Electric field of a dipole. Redrawn from [179], p. 6

the small lines on the external faces of the bodies and in the air corresponding to them grow, because precisely on them the excess of fire repelled from the inside will accumulate [29].[218]

Figure 4.27 shows potential (dashed lines) and force field for two equally and contrarily electrified bodies (a dipole) as shown in a modern textbook. The dashed lines should be compared with the closed curves of Fig. 4.26-7. There are similarities, but also difference. In particular the curves of Fig. 4.26-7 are rarefied in between the two electrified bodies and compressed externally; the contrary to what happens in Fig. 4.27.

A very interesting experiment related to the electric atmospheres, which was carried out using a device today known as *Beccaria well*, is worthy to be commented. The experiment shows both that the electric field (modern meaning) is zero inside a cavity of conductor and that the electricity is lodged on the surface only. This result predates that found by Coulomb in 1786 and by Cavendish in the 1770s. This fact, though well known is scarcely commented in the literature; for instance Heilbron in his very compressive treatise *Electricity in the 17th and 18th centuries* of 1979 did not cite Beccaria's well.

The electric well in object, shown in Fig. 4.28, is made by the vessel, or can, A of tin, fifteen inches (∼40 cm) high, and six and a half (∼17 cm) wide; to prevent its losing of electricity, its edge has been rounded by the means of an iron ring fixed to it. The can was insulated upon a small table T, raised upon a support of glass V,

[218]pp. 182–183.

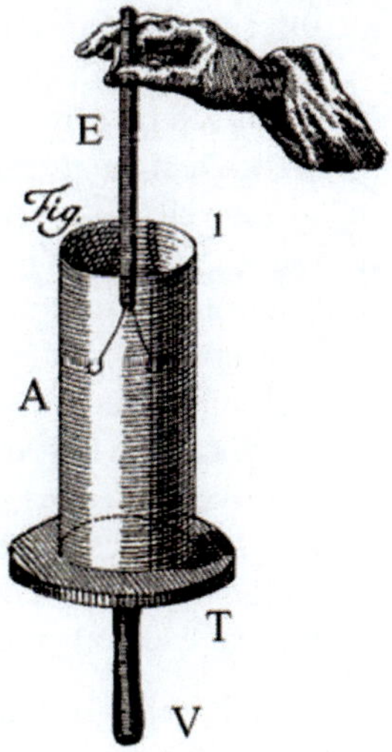

Fig. 4.28 Beccaria's well. Redrawn from [31], Fig. 1 of Table VI

and electrified by touching it with the hook (or with the coated bottom) of a charged Leyden jar, held by means of an insulated handle.

In this well two parts are distinguished; the first is the lower part, that is, that part of its cavity which reaches from the bottom to two third parts of the total altitude of it; the second is the upper part which, from thence, extends up to the edge of the well.

The electricity inside the well is measured by means of an electrometer, named *scrutator* (saggiatore) by Beccaria, annexed to a long stick of sealing-wax whose threads were exceedingly fine and only an inch and an half long; to them are fastened two bits of paper in order to render their motions within the cavity of the well sufficiently conspicuous. To perceive them the better, the bottom of the well was covered with a round plate of tin blackened cover. To note that because of this cover the lower part of the well behaves much like a closed cavity [29].[219]

The following results were detected:

I A man [a Beccaria's assistant] suspended the scrutator in the middle of the lower cavity of the well, in such a manner that it did not touch either the bottom or the sides; Beccaria then touched the well, at one time with the hook, and at another time with the outside of a Leyden jar; in both cases the threads remained unmoved.

II The person who held in his hand the threads of the scrutator now touched the bottom with them, and then the sides of the lower cavity, and the threads still remained unmoved.

III The same person suspended again the scrutator in the middle of the lower cavity, without touching either the sides or the bottom of the well. Beccaria then inserted into the well a small rod of brass, with a ball at its end, and presented it to the

[219] pp. 184–185.

threads of the scrutator, taking care not to touch either the edge or the sides of the well, and the threads flight to the ball. The electricity of the well was then destroyed by earthing it and the person who held the scrutator drew it out of the well. Then the threads manifested an electricity contrary to that communicated to the well. That is, if Beccaria had touched the well with the hook, the scrutator ran to the hook; if he touched the well with the outside of the bottle, the scrutator ran from the hook.

IV Beccaria annexed a short and very fine hair laterally to the lower cavity of the well; the scrutator was again suspended within it and the well electrified with the bottle. Seeing that both the threads of the scrutator and the hair remained unmoved, Beccaria put a brass rod (see previous test) into the well and presented it to the threads of the scrutator and when the latter ran to the rod the hair diverged a little from the well; if both the hair and the treads of the scrutator happened to be near each other, they immediately joined [29].[220]

Beccaria ended by noticing that Franklin had already proposed an experiment of a cork ball, which, hanging by a silk thread and lowered into a silver can till it touches the bottom of it, draws no electricity from it [97].[221] He also reported, perhaps with an excess of criticism, the results referred to by Priestley in his *History and present state of electricity* of 1767, saying that the author had made several attempts to analyze the experiment itself, but which had failed [29, 171].[222]

The idea of electric atmospheres allowed Beccaria to explain the attraction and repulsion of electrified bodies in a much more satisfactory way than Franklin could have done. The repulsion of two bodies both charged positively was usually explained by the repulsion of the excess of the electric fluid of the two bodies. This explanation seemed to work by admitting that the electric atmospheres of the two bodies come into contact, but it was not very convincing because a repulsion also occurred for sensible distances. But the repulsion of two bodies both charged negatively could not be explained by the mechanism of the traditional atmospheres, because in this case, at least according to Franklin there are no atmospheres: "Indeed Dr. Franklin himself ingenuously acknowledges, that he was a long time puzzled to account for bodies that were negatively electrified repelling one another" [171].[223]

And even to me, confessed Beccaria, "the same difficulty occurred at first; but it had not with me, and I think with great reason, the same weight it must have had with others, because I did not consider the divergences as being produced either by the excess, or the defect of the electric fire, absolutely considered; but indeed by the

[220] p. 185. Translation into English adapted from [31].

[221] pp. 325–326.

[222] p. 187; pp. 688–689.

[223] Vol. 1, p. 434.

inequality between the fire in the bodies, and that in the ambient air rather from the inequality between the amount of the electric fire of the bodies and the electric fire of the ambient air" [29].[224]

In truth Beccaria in his work of 1772 did not provide an explanation of the way the actuated air determines attraction or repulsion, limiting to the concise statement: "The fire inherent to the air tends to balance itself, but without mixing with the moveable fire in the deferent bodies [...] thus two bodies equally electric always will equally diverge, whether the excessive fire of the bodies surpasses the natural fire in the air or the natural fire in the air, surpasses the deficient fire in the bodies" [29].[225] For a detailed explanation of attraction-repulsion he referred to previous works, possibly to his letters of 1758 to Giacomo Bartolomeo Beccari [24], to the letter of 1757 to Benjamin Franklin published in the Philosophical Transactions for the year 1760 [25], where his idea of atmosphere was not yet defined, and to his memoir *De athmosphaera electrica* of 1769, still published in the Philosophical Transactions of 1771 [28].

Beccaria supposed that attraction (or repulsion) was due to an active role of air. That is, he sustained that attraction (or repulsion) could not occur in a vacuum, even tough this fact was contradicted by many experiences known to him; in particular from those conducted by Boyle. A modern reader, used to consider as natural electric forces in a vacuum, is puzzled by Beccaria's credence and is tempted to consider the fact as a defect in his argumentation. Indeed it is not the case because the experiences showing attraction and repulsion in a vacuum were not considered as definitive at the time because the experimental conditions made not possible to assert with certainty the existence of an absolute vacuum. And Beccaria was not the only one who supported this opinion [107].[226])

To prove empirically the need of air for the manifestation of electric forces Beccaria had already proposed an experiment, according to him definitive, reported in both his letters of 1757 to Franklin and of 1758 to Beccari. In the words of Beccaria: A slip of gilt paper of about eight inches long and four lines broad is rolled up, so as to form a little solid cylinder D, as shown in Fig. 4.29. It is suspended by a silk thread DG under an opening of an air pump IHK through the top of which the metal rod BC passes, having at its bottom the metal ball C. A metal rod LE is fixed, armed with a similar ball on the top, to the plate IK. Now the spheres C and E are in the same plane within the cylinder D and at equal distances from it. Then fixing the chain AB of the electrical machine to the rod BC and consequently the electrical fire being sent into the rod BC, it is observed: (1) that before the air is removed, the gilt paper cylinder D is agitated with the most violent vibrations between the two spheres E and C. (2) That while the air is removed the agitation is plainly diminished in proportion to the quantity of air removed. (3) When as much air as possible is removed the cylinder D hardly stirs. The air is then restored by degrees and the vibrations increase again in proportion to the restored air, and at length become as violent as before. Which

---

[224]p. 47.

[225]p. 48.

[226]Vol. 6., p. 486.

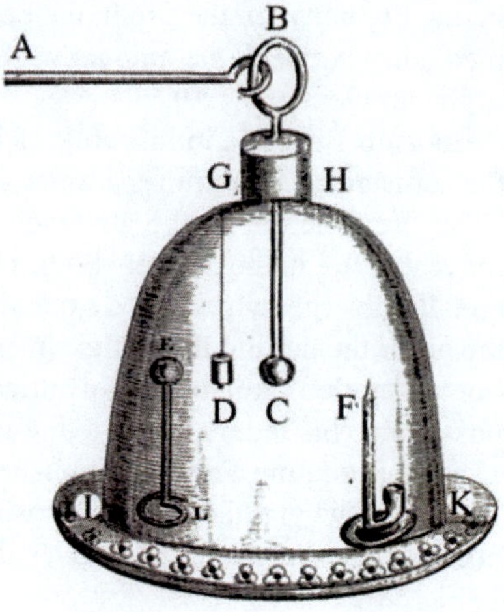

Fig. 4.29 Attraction and repulsion in vacuum and in air. Redrawn from [25]. Table XII

circumstances when contemplated in particular, and considered them also together, show that the quantity or greatness of electrical motions is owing to the air, either entire or in part [25].[227]

Beccaria can thus claim:

> To conclude, I attach so much value to this one experience, that for it alone I think that the oldest question of electricity [attraction and repulsion], considered as the most difficult, has largely dissolved. For if the electric movements diminish in the thin air and diminish in proportion to its thinning, it follows that these movements depend on some action of the electric vapor on the air [24][228] (D.25)

Then the intricate mechanicistic explanation follows of the attraction and the electrical repulsion of the bodies by means of the movements of the air, inspired to Cabeo, which do not take into account the idea of electric atmosphere, however. According to Beccaria, "the attractions of bodies, unequally electrified, are affected by the electric fire of the body in which it is more abundant, flowing out into the other body, through the intermediate air and even throwing off that. But that repulsions are caused by the proper fire of bodies expanding itself against the aerial fire, or by the aerial expanding itself more strongly against the proper fire; which expansion however, of the fire of one body overcoming another, seems to happen without the mutual mixture of one with the other" [25].[229]

---

[227] pp. 515–516. Translation into English adapted from James Parsons.

[228] p. 40.

[229] pp. 524–525.

4.3.7.3 Vindicating Electricity

In 1755 Jesuit missionaries in Peking sent a report of the strange results of an experiment closely related to Symmer's studies. A thin glass plate was electrified positively by friction and placed on the (glass) cover of a magnetic compass case. The needle of the compass immediately rose from its pivot and adhered to the underside of the cover for some time before returning suddenly to its normal position. If the glass plate was removed, the needle again rose and remained stuck to the lower face of the cover for a considerable time, but as soon as the glass plate was brought back, it felt away again. This could be repeated over and over again.

Aepinus described this experiment, with some others on electrical induction (modern term), in a paper he read at the St. Petersburg local academy in 1758, *Descriptio ac explicatio novorum quorundam experimentorum electricorum* [3]. Notwithstanding the relevance of the phenomenon observed, Aepinus made no mention of it in his *Tentamen*.

In the paper of 1758, published more or less while he was publishing his masterpiece, the *Tentamen*, however, Aepinus had given a complete explanation of the phenomenon. He stated that Peking experiments were certainly remarkable and it seemed quite paradoxical that electricity, became almost extinct and after some time, without further rubbing, could be resuscitated. Nonetheless, he thought, it could be shown that these effects followed from the Franklin's theory of electricity [3].[230]

Aepinus explanation, translated into a modern language and simplified, is as follows. The glass cover of the compass box (by polarization we would say) developed negative electricity above and positive below under the effect of the positively electrified glass plate (indeed this is a simplification of the phenomenon); the needle rose under the net force between the positive and negative electricities of the glass plane and cover. The positive electricity on the lower surface of the cover slowly leaks off through the needle, that receive positive electricity, and drops. On removal of the positively electric glass plate the needle reascends because of the attraction of the net negative electricity remaining on the cover. It falls again by replacing the glass plate because the lower surface of the cover, becoming again positively electric, repels the positively electric needle. This simple explanation did not satisfy everyone however. Beccaria, in particular, attempted an alternative account in terms of a peculiar notion that he called *vindicating electricity* (*elettricità vindice*) that is electricity the vindicator; an explication that however remained at the phenomenological level.

Beccaria presented is idea of vindicating electricity in the short treatise *Experimenta, atque observationes, quibus electricitas vindex late constituitur, atque explicatur* of 1769 and in more extended form in the *Elettricismo artificiale* of 1772 [27, 29]. In the following it is reported as he introduced the vindicating electricity in the *Elettricismo artificiale*. Two insulating bodies, or an insulating and a deferent body contrarily electrified, after joining together remain strongly united; in which they differ from two deferent bodies, which having in their mutual contact equalized their respective quantities of fire, repel each other in proportion to the fire that remains

---

[230] p. 278.

in both. Moreover, after an insulating body has joined with another insulating body, or with a deferent body, they apparently loose their contrary electricities (so far as these were equal). As soon as one of the bodies is taken from the other, it recovers its former electricity. "It is this property which could be expressed by the words, *vindicating electricity*" [29].[231]

Beccaria considered the vindicating electricity a principle of electrostatics, to be added to Franklin's principles. In a pamphlet written in Latin in form of letter to Franklin in 1767, he had written: "I thought they are wrong, who think your theory is invalidated by Symmer's experiment; that all those experiments [...] demand a new principle, which may be joined with yours and which especially agrees with yours" [26].[232]

Indeed Beccaria considered his principle much more acceptable than to admit two different types of electricity as suggested by Symmer. Here a severe criticism against the two fluid theory:

> This is the character of a philosophy illusive, and too lazy to inquire into the causes of things, thus to imagine fluids that must have such motions as cannot take place in bodies. Bodies differently electrified mutually attract each other: Why? Because they are animated by two fluids that mutually attract each other. Bodies similarly electrified mutually repel each other: Why? Because they are animated by two fluids, the particles of each of which repel each other, and that in a most similar manner. But that these are by no means the principles and ways of nature, is manifest from the exact unity of the effects of those supposed fluids, which unity is absolutely repugnant to the supposed diversity of their natures; from the impossibility of their separation, on which separation, nevertheless, every one of their manifestations must depend; lastly, from the manner after which they should unite again; which would be entirely opposed to all phenomena [29].[233] (D.26)

Most modern reader would not agree with Beccaria. Indeed they do not consider that of vindicating electricity a very simple idea and not even correct at a phenomenological level. This was also the opinion of Alessandro Volta, who did not share Beccaria's conception [193].

Very interesting is the complex experiment reported first in the work *Experimenta, atque observationes, quibus electricitas vindex late constituitur, atque explicatur* of 1769, which I will summarize below with the explanation, more perspicuous, given by Cavallo [53].

With reference to Fig. 4.30a, b, let AB*ab*, represents a plate of glass, coated on both sides with the two metallic coatings, CD*cd* which are not stuck to the plate, but only laid upon it (this is what is known as *Franklin square*). From the upper coating CD (Fig. 4.30b), three silk threads proceed, which are united at their top H, by which the coating may be removed from the plate in an insulated manner and may be presented to an electrometer in order to measure its electricity. FG is a glass stand, which insulates and supports the plate. Let the plate AB*ab* be charged by means of an electrical machine, so that its surface AB may acquire one kind of electricity, (which

[231] p. 401.

[232] p. 1.

[233] p. 48. Translation in to English [31].

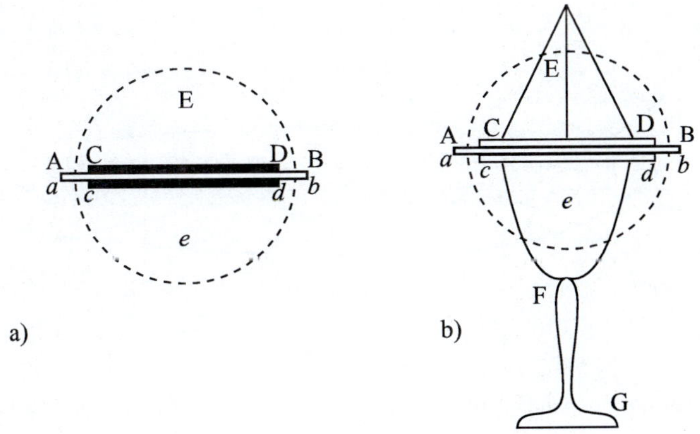

Fig. 4.30 Experiment on vindicating electricity. Redrawn from: **a** [27], Fig. 1 of the table of figures; **b** [53], vol. 2, Fig. 6 of table IV

may be called K) and the opposite surface *ab* may acquire the contrary electricity, (which may be called L).

Then, if the coating CD is removed from the plate and presented to an electrometer, it will be found in the posses of the electricity K, that is of the same kind with that which was communicated to the surface AB of the glass plate; from whence it is deduced, that the surface AB has imparted some of its electricity to the coating. Now, this disposition of the charged plate to give part of its electricity to the coating, is what Beccaria denominated the *negative vindicating electricity*.

If the coating be again and again alternately laid upon the plate and removed, its electricity K will be found to decrease gradually, till after a number of times (which is greater or less, according as the edges of the plate insulate more or less exactly), the coating will not appear at all electrified. This state is called the limit of the two contrary electricities; for if now the above-mentioned operation of coating and un-coating the plate be continued, the coating will be found possessed of the contrary electricity, that is the electricity L. This electricity, L, of the coating is weak on its first appearance; but it gradually grows stronger and stronger till a certain degree; then insensibly decreases, and continues decreasing until the glass plate has entirely lost every sign of electricity. By this change of electricity in the coating, it is deduced, that the surface AB of the glass plate changes property; and whereas at first it was disposed to part with its electricity, now, (that is beyond the limit of the two contrary electricities) it seems to vindicate its own property, that is, to take from the coating some electricity of the same kind with that of which it was charged: hence this disposition was called by Beccaria the *positive vindicating electricity* [25, 53][234]

Figure 4.31 shows graphically what referred to above. FO and F*o* respectively represent the electricity of the upper and lower surfaces of the glass plate as furnished

[234]pp. 196–199; 1–3.

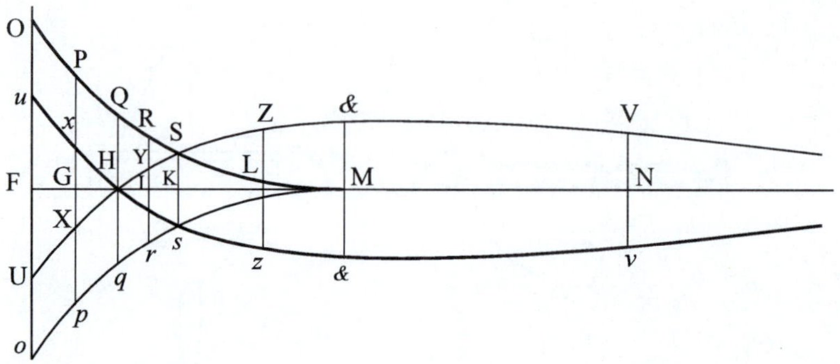

Fig. 4.31 Experiment on vindicating electricity. Redrawn from [27], Fig. 2 of table of figures

Fig. 4.32 Volta's electrophore [46]

by an electric machine; in the following reference is made only to the upper surface, whose curves are represented with a thicker lines, a specular situation holds for the lower part represented with lighter lines. Fu is the electricity measured on the coating CD when it is removed; more generally the curve u, x, H, y, s, z, &, v represents the electricity measured in the various times on the coating when it is again and again alternately laid upon the plate and removed. The curve OQLMN represents the electricity measured when the coating is on the glass plate [25].[235] The curve HYZ&V represents the electricity measured on the upper surface of the glass plate.[II]

From M on, A behaves like the shield of an electrophore; that is when the coating is off the glass plate and the cover are electrified, when the coating is on the whole

[235] pp. 1–3.

does not manifest electrification. The electrophore, shown in Fig. 4.32,[236] whose invention was announced by Volta to Priestley in 1775 [193], consisted of a metal dish containing the dielectric (insulator) cake B, a wooden shield A covered with tin foil rounded to remove all corners and points, and an insulating handle. One can charge the electrophore readily by rubbing the dielectric while grounding the plate, a procedure that became standard. The functioning of the electrophore is easily explained with the theory of induction. When the shield is superimposed to the insulator, for instance positively electrified, the face of the shield in contact with the dielectric becomes charged negatively, the upper face positively, by induction. By grounding the upper face, for instance by touching it with one hand, the shield is fully charged with negative electricity. At this point the shield is detached from the dielectric with the insulating handle. This gives a certain amount of negative electricity that one can use at will. Once this electricity is finished, the operation can be repeated. The process can last for a long time, also depending on the properties of the matter used for the insulator B. If there were no dispersions, the shield could furnish negative electricity for an infinite number of times.

4.3.8 Carlo Battista Barletti

Carlo Battista Barletti (1735–1800) only sixteen wore the dress of the Scolopi, after a school career where more than mathematics he learned grammar and rhetoric. He first taught experimental physics in Chiavari then in Albenga and Savona, finally in Milan from where in 1772 he was called to the chair of experimental physics of the university of Pavia. Here he worked alongside men such as the biologist Lazzaro Spallanzani (1729–1799) and the mathematician Gregorio Fontana (1735–1803). He owed this call to his valuable works on electricity dedicated to Carlo Firmian (1716–1782) governor of the Austrian Lombardy. In 1778 because his poor health, the chair of physics was split into general and experimental; Barletti was entrusted with that of general physics, while that of experimental physics was assigned to a young Alessandro Volta, linked to him by cordial friendship testified by close and important correspondence.

In 1796 after the French army occupied Pavia, Barletti, with other professors from the Pavia university including Gregorio Fontana, Francesco Antonio Alpruni, Lorenzo Mascheroni, joined the Cisalpine Republic, established by the French government. Elected for the municipality of Pavia in June 1797, two months later he left this position to assume that of commissioner at the central administration of the Ticino department, without much success in reality. In 1799 the Austro-Russian army reoccupied Pavia. Barletti was arrested on charges of collaboration. During one of the interrogations that he underwent in the following months, he presented a long self-defense in which, among occasional certificates of loyalty to the Austrian

[236] p. 619.

regime, there was the confession of that inexperience and that moralism, which were the causes of his political failure as commissioner. He died while still being detained for the inquest on his political activity.

Barletti became interested in electricity since the early 1770s; he did so in the way experimental physics was then practiced by most; that is without any substantial use of mathematics and any attempt to formulate mathematical laws. In 1771, after an intense laboratory work, he published *Nuove sperienze elettriche secondo la teoria del Sig. Beniamino Franklin e le produzioni del P. Beccaria* (herein after *Nuove sperientie*). In 1773–1774 he collaborated with the Encyclopédie d'Yverdon [134]. In 1776 his *Dubbj e pensieri sopra la teoria degli elettrici fenomeni* (herein after *Dubbj e pensieri*) was issued followed in 1780 by a pamphlet on lightening and medical use of electricity entitled *Analisi d'un nuovo fenomeno del fulmine ed osservazioni sopra gli usi medici della elettricità* [17].

From 1782 to 1794 he published important works in the Memorie di matematica e fisica della società italiana, which consolidated his anti-Franklin position to which he eventually arrived [18–21]. With the publication of a treatise on physics, which had the title *Fisica particolare, e generale* Barletti proposed the ambitious aim of gathering the various chapters of physics in an organic context, tracing them back to Galileo's method and Newton's principles. The dedications of the published volumes all bear the date 1785; instead, the year of printing of the individual volumes is missing [52, 135].

Barletti, son of his time, claimed the right to be interested also in literature and mundane things. In a letter of 1782 to Mario Lorgna, founder of the nascent Italian society of sciences, later called Accademia dei XL, of which Barletti will be part, after talking about scientific books, Barletti continued praying Lorgna to send him the pastoral drama *La fida ninfa* by Scipione Maffei, *L'Alceo* by Antonio Ongaro and *La Filli di Sciro* by Guidubaldo Bonarelli. Then, fearing objections from the interlocutor, he added: "You will laugh because a physicist gives a mathematician such gallant commissions and you will believe that they are for some elegant physiologist. On the other hand, I tell you frankly that they are for myself [...]. In this century of humanity, who can not be gallant?" [136].[237]

4.3.8.1 From Franklin to Symmer

From 1771 to 1776, in the years from *Nuove sperienze elettriche* to *Dubbi e pensieri*, Barletti changed his opinion on electrical phenomena, moving from Franklin to Symmer position. In the *Physica specimina* of 1772, he still stated: "with a general agreement, Franklin in electric matter should be considered as Newton in the heaven system" [15].[238] And a little later, in the same work, he repeated that the double electricity—vitreous or positive, and resinous or negative—had to be explained according to Franklin's hypothesis, as caused respectively by excess and defect of

[237] p. 53.

[238] p. 6.

the electric fluid. Eventually in the *Dubbj e pensieri*of 1776 the one fluid theory was openly repudiated. The conversion was considered sensational by the world of scholars of electrical matters and led Barletti to clash with Beccaria, whom he considered his teacher and to find himself at the center of controversies that increased his fame anyway.

Nuove Sperientie

The treatise *Nuove sperientie* of 1771 opens with the presentation of the electric generating machine used in the various experiments, shown in Fig. 4.33. The generator is made of a glass disk fixed to the wooden axis EE, which is made to rotate by the handle A. The disk is electrified by rubbing thanks to two leather cushions LL. Barletti added that an even more effective result could be obtained by rubbing the disc with the fingers of the hand. The electricity generated is collected by the chain NN, a brass tube, loaded by means of the two PP brass balls that are close to the disk, without touching it. Barletti did not indicate any measurements, but from the drawing, assumed to be in scale, comparing it with the hand, the glass disk seems to have a diameter of about one meter.

Barletti gave a definition of the electric atmosphere, which although similar to that of Beccaria, was devoid of any ontological connotation, "It is said *perceivable atmosphere* of any electric body that distance, in which the signs are observable around it, that is the action on the bodies immersed in it" [14].[239] Differently from Beccaria, that saw his atmosphere represented by actuated air, Barletti referred implicitly to an action at a distance, in the footprint of Aepinus, even though he was not cited in the text.

This definition of an electric atmosphere allowed him to formulate a law (of induction) formally similar to that formulated by Beccaria:

> An electric body has the power to change the natural amount of the electric fire in bodies immersed in its atmosphere and it induce in it electricity of opposite nature [14].[240] (D.27)

Assertion he proved with simple experiments.

Barletti's book attracted Priestley's attention, who recommended it to Franklin. Franklin in turn said he was interested in checking the experiments referred to there and in the letter of reply to Priestley of 4 May 1772 he declared: "I intend soon to repeat Barletti's experiments, being provided with the requisites and shall let you know the result" [98].[241]

Dubbi e pensieri

Barletti's most important work, however, is normally considered the short treatise *Dubbi e pensieri* of 1776 even though not very known abroad. The work is divided into two letters, the former addressed to the physicist Felice Fontana (1730–1805), brother of the mathematician Gregorio Fontana dated 11 February 1776; the latter to

---

[239] p. 55.

[240] p. 58.

[241] Vol. 6, p. 343.

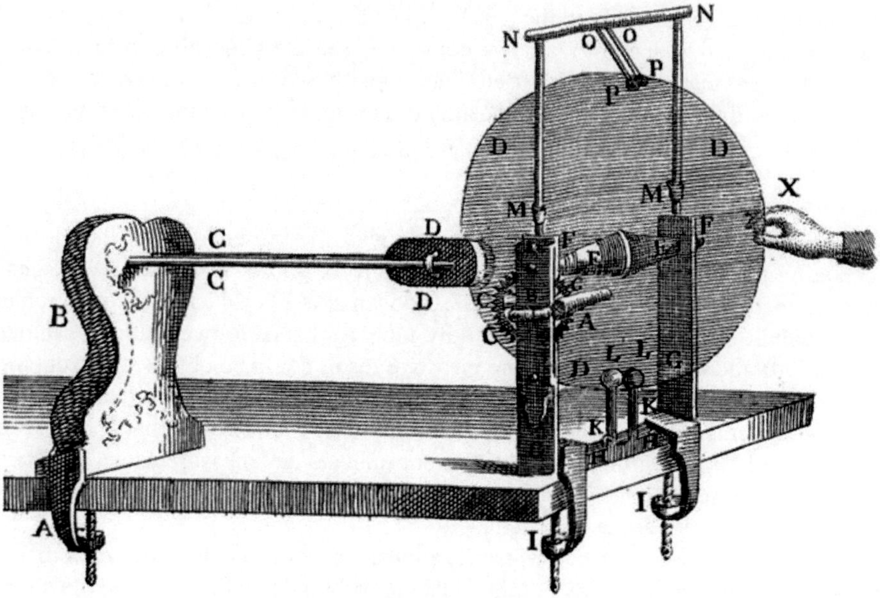

Fig. 4.33 The electric generator used by Barletti, [14], Fig. 1 of the table of figures

Alessandro Volta dated March 24, 1776. In both the letters Barletti gave large space and resonance to Aepinus' work (the *Tentamen*), very original for the application of analytical calculation to electrical laws and filled with original experiences and declared it necessary to consider again and with less prevention the main experiences that led him to the Franklin's theory [16].[242] For historical reason the reference to Aepinus in the letter to Volta is relevant, because the latter had not known Aepinus until then.

Below it is reported how Barletti explained his change of ideas on the nature of electricity: It is therefore necessary in this development, and reciprocal action of electricity on the opposite faces of the Leyden jar, to recognize two very different fluids, one of which is the positive, the other the negative electricity form. Nor it seems strange this way of conceiving the charge of the electric jar, to be contrary to Franklin's theory; since it is not an effect of instability to debate, nor an incentive to try things new, but necessary consequence of truth and evidence, "to which I willingly sacrifice any my opinion; I speak of that evidence, which in your experiences shows the existence of two streams of electricity in the discharge of the Leyden jar, and renders evident the real and positive efficiency of that kind of electricity, which is believed to be negative by Franklin's supporters" [16].[243]

[242]p. 27.
[243]p. 27.

On the basis of his reflections Barletti formulated the following points:

1. Exciting, or developing electricity, is nothing but than breaking down the union of the two fluids that constitute the fixed state in the different bodies.
2. Each body requires a certain quantity of those fluids, either united and fixed, or disunited and dissolved. When the two fluids are disunited, there is electricity manifested by the known electrical signs. However, they naturally tend to come together, and in reason that they return to their original union, the electricity balances itself and all signs of it vanish.
3. The nature of the fluid already developed, directed by means of the conductors to an insulating armed face, determines the nature of the dominant electricity, which is excited by inflow in the insulating layer. The fluid, which first develops and exits from a rubbed insulating face, determines its species.
4. According to the different state, or way of union of the two fluids in the bodies, which make, or receive rubbing, or other equivalent action, one or the other of the two fluids develops and then different signs result.
5. If through a conductor they have free continuous access, the two opposite electricities will balance at any moment. And if such a conductor joins two fluids already developed and divided, they will reach their natural union with effects corresponding to their forces and masses; and this is what the discharge of a Leyden jar does, that is the electric shock.
6. The mutual attraction of the two fluids makes the idea of their effort more clear than the idea of expansive force, that is elasticity. For this reason, the effort of a developed fluid would always be exercised equally in every part [16].[244]

The need of the existence of two fluids, as well as by experiments, is also suggested by metaphysical reasons. In fact, Barletti wondered: "Who can conceive of a nothing in the negative face, as he can of the excess of fluid in the positive? [...] But why am I wasting time and words in such trifles, while the impotence of nothing is well known, and the need for true and real opposing and equal forces, so that all their efforts are reduced, or kept in balance?" [16].[245]

After having exposed his conceptions on the nature of electricity Barletti went on to interesting comments on the behavior of bodies electric *per se* (that is dielectrics), with a particular reference to Volta's electrophore. To explain the complex phenomena exhibited by electrified dielectrics he made reference to the distinction due to Gianfrancesco Cigna, between Franklinian and Symmerian electricity, see Sect. 4.3.8.2. Meaning by the former the part of electricity free and mobile that can be equilibrated easily as in conductors; by the latter the other electricity much less mobile and difficult to equilibrate, found in electrified dielectrics.

To test the complex way the Symmerian electricity behaves, Barletti recalled an experiment by Aepinus, without a clear specification wether or not he repeated it himself. Let AB be the glass tube of Fig. 4.34a, fixed horizontally on the edge of a table by the end B, and pushed out by the remaining length of a few feet. With

[244]pp. 28–30.

[245]p. 45.

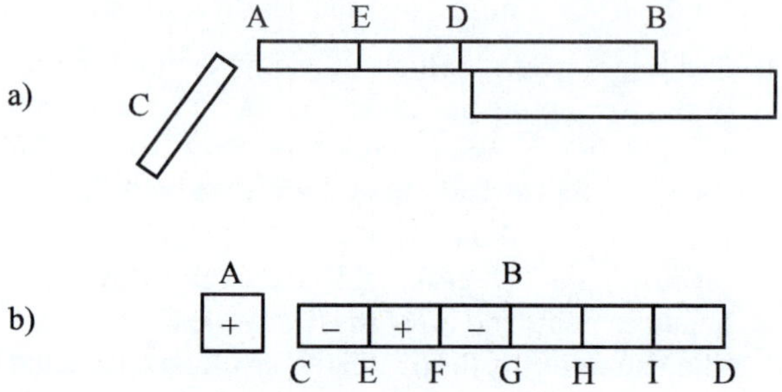

Fig. 4.34 Alternating electricity. **a** Redrawn from [4], Fig. 42 of table II. **b** Redrawn from [16], Fig. 4 of the table of figures; also [4], Fig. 43 of table II

the electrified tube C one repeatedly touches the end of the tube A. After this, by exploring the action of electricity along the entire length of the tube AB, one can recognize positive electricity from A to E for four or five inches, which is followed by about two inches of negative electricity from E to D; and after DB the tube resumes positive electricity, although somewhat weaker, but still quite notable. This experiment, repeated several times, constantly presented the same phenomena, even when instead of the glass tube AB, or C, a solid sulfur cylinder was used [16].[246]

The explanation of this interesting behavior of the body AB, had been given by Aepinus in his *Tentamen* with a thought experiment. With reference to Fig. 4.34b, suppose that the body B, is electric *per se* and of considerable length and that a positively electric body A is moved close to it. The electric fluid of B, because of the repulsion of A, recedes from the end C toward the interior in such a way that it leaves a part CE negatively electric while the fluid propelled from CE crosses into EF. This fluid should stop here for a moderate period of time at least, since because of the difficulty with which it moves through the pores of the body electric *per se* it cannot flow swiftly through the remaining parts of the body. And for this reason, once EF has become positively electric, it acts like a positively electric body and repels the electric fluid contained in FG. Applying the same reasoning to the part GH, it is clear that an alternation of positive and negative electricity follows.[247] Of course a similar phenomenon occurs if the body A is negatively electric.

Notice that Barletti introduced his experience without any reference to the care needed to obtain the illustrated result, unlike Aepinus who pointed out that the experiment at first had provided results contrary to the theory and he had to force the experience. This leaves some doubt as to whether Barletti actually carried out the experiment. One more reason for this suspect is the value of the measures of the electrified portion of tube Barletti referred to, very close to those referred to by Aepinus. According to Aepinus, when he moved the electrified body C to the end A and took it away he did not discover in the tube AB the state he had hoped to induce in it. So he changed the method of conducting the experiment somewhat. After everything had been done as just described, he not only moved the electrified body C to the

[246]p. 61.

[247]Section 200, pp. 192–193.

end A of the tube, but he also rubbed this end several times with C. After the test had been carried out he found that the end AE was positively electric for a length of one or five inches; beyond this, a part EF, about two inches long, was sensibly negatively electric, and beyond this the tube was again positively electric: weakly so, yet significantly enough. He repeated this test and had the same success, and similar things were observed when using a solid cylinder of fused sulphur instead of either the tube C or the tube AB [4].[248]

Barletti ended the exposition of Aepinus' experiment by praising him:

> Although the illustrious Author speaks very quickly on the results of these experiences of his, it seems to me nevertheless, that they derive quite naturally from the electric theory, and thus bring more light into it, than the whole volume. I compare similar experiences to those traits of a master in drawing, where the cross-section of a large building is presented, which explains better at a glance both the parts and the whole, which would not be understood by turning it several times separately [16].[249] (D.28)

Further Writings

After the publication of the *Dubbj e pensieri*, Barletti continued his studies on electricity. In his treatise of 1780 on lightening, Barletti paused to examine the effects caused by a lightning, which had hit the wind vane of the church of San Siro and San Sepolcro in Cremona. By examining the margins of the holes caused by lightning he believed he could deduce that opposite electricities had simultaneously hit the wind vane from opposite sides; from this he drew confirmation of Symmer's hypothesis. But his most important contribution was left to the Memorie di matematica e fisica della società italiana already referred to. This contribution, even though very interesting, had little circulation in Europe, both for the difficulty of language, Italian, and the poor spreading of the journal where they were published.

The memoirs of 1788 and 1794 [20, 21], were devoted to the criticism of Franklin's theory on the Leyden jar, and are partially of speculative nature. The memoirs of 1782 and 1784, both having the same title *Introduzione a nuovi principj della teoria elettrica dedotti dall'analisi de' fenomeni delle elettriche punte*, and divided in part I and part II, referred instead very numerous and accurate experiments on the power of points [18, 19]. In total the two memoirs count just less than two hundred pages; a small treatise indeed.

The first part of the memoir published in 1782 is divided into two sections; the first shows a series of experiments, the second entitled *Introduction to the theory of points in the electrical phenomena*, refers theorems and corollaries. They are demonstrated not by a deductive method starting from the first principles of electricity, as Aepinus had done, but, in the tradition of experimental philosophy, they are a generalization of the experiments reported in the first section and are verified with further experiments.

The experiments of the first section of the memoir tend to measure the effects of electricity discharged or absorbed by points against a Franklin square. The measurement was indeed carried out according to a scarcely reliable criterion—on this point Barletti was very vague—on the basis of the observation of the electrical signs

[248] Section 202, pp. 194–196.

[249] pp. 62–63.

perceived as discharges and brightness. The apparatus used to produce electricity is not described in the memoir; it is only said that it has a glass disk. Most likely it is the same electric machine used in the *Nuove sperienze elettriche* of 1771. The electricity produced by the machine is collected by a cannon that can extract it from the leather pads that rub the glass disk, and in this case Barletti speaks of resinous electricity, or from the glass disk itself, and in this case it is a question of vitreous electricity.

The point was made by a brass wire with a diameter of one line (nearly 2 mm) and three inches and nine lines long (less than 10 cm), terminated by a very sharp extremity. The Franklin square was made up of a glass sheet of Germany of about 17 × 12 in. (nearly 40 × 30 cm), covered with a tin leaf from the two opposite sides. One of the faces of the square was opposite the point, the other was grounded. The experiment consisted in measuring the effect of the point when it was loaded with a fixed and constant number of turns of the glass disk of the electric machine while the distance between the point and the glass plate changed. In the first experiment—named by Barletti *point of resinous electricity*—carried out with resinous electricity, one started from close to the glass plate and then moved away, in the second experiment—named point opposed to the resinous electricity—one started from a distance and then got closer. The experiments were repeated considering vitreous electricity. For the sake of space I do not report the details of the experiments; I just say that they gave different results—even a lot—depending on whether one started first from near or from far and for the two types of electricity.

The second part of the memoir published in 1784 refers to more complex experiments, carried out with a more powerful electric machine, endowed with a greater glass disk, so that one turn of it gave the same effect of fifteen turns of the previous machine. In the memoir were considered points opposed to points and surface opposed to surfaces. The memoir ended with a severe criticism of Franklin's theories, while Franklin himself is treated with much respect. The criticism should also be seen as an autocriticism to his previous ideas about the nature of electricity.

> Franklin's ideas became popular, and he became the founder of a school and the leader of a systematic sect, and it was enough for all his conjectures, and even his most daring suspicions were transformed into dogmas by the bigoted Franklinians and proposed and defended with a language of fanatical persuasion [...] But in the midst of many definitions, and many statements, the electric science did not acquire nothing but ample paraphrases of the first facts indicated by Franklin, pronounced with as much confidence and persuasion, as it was the perplexity and the uncertainty of their author.
>
> [...]
>
> It is neither necessary, nor appropriate to associate to the error the names of the authors, who contributed to propagate it; especially when their names don't affect the reality of things; and all the better when, when they had the naivety to confess to having erred in defining the starting point; is thus hoped, that with more mature discussion of facts they could recognize their blunder [19].[250] (D.29)

[250]pp. 120–121.

4.3.8.2 Gianfrancesco Cigna

Gianfrancesco (Giovanni Francesco) Cigna (1734–1790), professor of anatomy at the university of Turin, was Beccaria's relative, student and assistant in the early fifties, along with Lagrange and Saluzzo. He combined medical teaching and medical practice with a good success. To his researches in medicine, physiology and chemistry, he combined all his life researches in electricity, which although of poor health, he continued to pursue with tireless activity. He was extremely thorough and rarely happy with what he got. His works are not very numerous, but in his drawers at the death a remarkable number of finished but unpublished works was found.

In 1757 Cigna, Lagrange and Saluzzo, who attended seminars of Beccaria, had a dispute with the master about the possibility that combustion and animal life lasted for a long time in a confined and closed space; it seemed to them that this was possible and they drew from it a refutation of the phlogiston theory, supported by Beccaria. The thesis of the three young people was erroneous; Cigna himself will later write a special memoir to deny it, the *De causa extinctionis* of 1767 [62], but the disagreement with the master, energetic temperament and not very tolerant of criticism continued. The conflict between Beccaria and Cigna did not end after the publication of the denial. Only in 1778 Beccaria, seriously ill, asked Cigna for a medical consultation and the two were pacified. In any case the three young people, Cigna, Lagrange and Saluzzo, continued the experimentation, establishing a private society which would later become the Accademia reale delle science of Turin, to which Beccaria never wanted to join. For some more biographic information on Cigna see [13, 189].

Even though only occasionally Cigna wrote on electricity, he was noticed abroad, especially by Nollet, who saw him as an allied against Beccaria, and by Priestley who read and appreciated his presentation of the two fluid theory. This international recognition was surely due to the great ability of Cigna as an experimenter, but it also depended on the fact that electricity was then a new and fertile field and the first who cultivated it, even though not a specialist in the field, could receive a good harvest.

Cigna had already included electrical experiments, following Beccaria's approach before the contrast, in his doctoral theses of 1757 that he intended to submit for the aggregation to the medical college of Turin. After a break he continued his studies on electricity and magnetism and in 1759 published the *De motibus electricis* with comparative analysis of the theses of the unique nature of the electric fluid, the most common at the time, and of the one recently proposed by Symmer, which explained the totality of electric phenomena from the contrasting action of two opposite fluids. The conclusion was that both theories could adequately explain the experimental appearances, so that there was no way, with the knowledge currently available, to carry out an *experimentum crucis* that conclusively could decide on their validity. To reach such an apparently agnostic result, Cigna compared the evidences adduced by each of the two theories, by classifying them; especially that of Symmer so that Priestley, having read Cigna's memoir, considered it a presentation and a development more important than the original formulation: "They were diversified, and pursued much farther by Mr. Cigna, of Turin, who has also explained them upon the principles of Dr. Franklin theory; though he was of opinion, that no experiments that had yet been made were decisive in favour of either of the two hypotheses" [171].[251]

In the following, I will illustrate some experiences carried out by Cigna in his work *De novis quibusdam experimentis electricis* [63], widely commented by Priestley [171].[252] One set of experiments reproduced in a varied form Symmer's. In one of the experiments, Cigna insulated a plate of lead and bringing an electrified silk ribbon near it, observed that it was attracted very feebly. Bringing his finger to the lead, a spark issued out of it, upon which it attracted the ribbon rigorously; after this lead and silk both together showed no signs of electricity. Upon the separation of the ribbon, they again both appeared to be electrified and a spark was perceived between the plate of lead and the finger [63, 171].[253]

In another experiment, two plates of glass laying upon a smooth grounded conductor were rubbed in the same manner as the ribbons had been rubbed, they likewise acquired electricity and adhered firmly to one another and to the conductor. When they were together, they showed no signs of electricity. When they were separated from the conductor and kept together, they showed, on both sides, a vitreous electricity and the conductor, if it had been insulated, contracted a resinous electricity. The two plates of glass when separated from each other, got the two electricities; the upper of the vitreous and stronger, and the lower of the resinous and weaker [63, 171].[254]

Priestley gave relevance to a new method invented by Cigna of charging a Leyden jar. He insulated a smooth plate of lead, and while he brought an electrified body, as a stocking, to it, he took a spark with the wire of a phial from the opposite side; and removing the stocking, he took another spark with his finger, or any conductor communicating with the ground. Bringing the stocking nearer the plate a second time, he took a second spark, with the wire of the jar, as before; and, removing it again, he took another, in the same manner, with his finger. This operation he continued, till the jar was charged; which, in favorable weather, may be done with very little diminution of the electricity of the stocking [171].[255]

Cigna offered further examples, however, and one deserves attention because it suggested the principle of the electrophore. Approach a charged ribbon to a lead plate; draw a spark from the lead, endowing it with an equal dose of electricity of the opposite sign; lay the ribbon on the plate and all signs vanish. Remove the ribbon, a small spark passes, ribbon and plate regain their electricities; the plate's residue may be drawn off entirely by grounding, but the ribbon's cannot. Indeed, it can still charge the metal many times by induction (modern interpretation); with Cigna Words "It is possible to find an easy method of multiplying the electricity without rubbing [sicque modum inveniemus facilem electricitatem absque frictu multiplicandi]" [63].[256]

---

[251] pp. 247–248.

[252] pp. 259–266.

[253] p. 43; p. 263.

[254] p. 52; p. 263.

[255] pp. 270–271.

[256] p. 51.

When Cigna read the letter of 1775 in which Volta described the electrophore to Priestley a device that reproduced results very close to those illustrated above, he considered himself a victim of plagiarism, and complained publicly about it. The question was debated and resolved favorably to him by Beccaria and the physicist Antonio Maria Vassalli Eandi (1761–1825): "This is the electrophore expressed by our doctor Cigna in the ribbon and in the insulated lead sheet [...]. The perpetuity that Mr. D. Alessandro [Volta] has attributed to his electrophorus is only a greater durability of the electricity impressed in the resin" [188].[257]

Cigna made a considerable difference between the electric fluid which gives the shock, and that on which some other phenomena of coated glass depend. He laid two plates of glass, well dried, one upon the other, as one piece; the lower of them being coated on the outside; and, when they were insulated, he alternately rubbed the uppermost plate with one hand, and took a spark from the coating of the lower with the other hand, till they were charged; when the coating, and both the plates adhered firmly together. Giving a coating to the other side, and making a communication between that and the other coating, the usual explosion was made. But the plates, though thus discharged, still cohered; and though, while they were in this state, they showed no other sign of electricity; yet, when they were separated, they were each of them found to be possessed of an electricity opposite to that of the other. If the two plates were separated before they were discharged, and the coating of each were touched, a spark came from each; and when put together, they would cohere as before, but were incapacitated for giving a shock [63, 171].[258] He, therefore, compared the electricity which gives the shock to the electricity of a metal plate; which is lost with taking one spark and the electricity by which the two plates of glass cohere. The one (named Franklinian) is dispersed at once, but the other (named Symmerian) slowly; the one existing, as he supposes, in the conductors, or upon the surfaces of the electrics, and the other in the substance itself [63, 171].[259]

4.3.8.3 Ruggero Giuseppe Boscovich

Ruggero Giuseppe Boscovich (1711–1787) was a polymath; astronomer, mathematician and physicist, Jesuit, philosopher, diplomat and poet born in the Venetian Republic of Ragusa. His national belonging is a source of discussion. He was a Croatian native speaker; father originally from Herzegovina and mother of Italian ancestry. He however was educated in Italy and wrote nearly all his papers in Italian (and Latin of course). From this point of view there is any meaning in listing him among Italian scholars.

Boscovich was a typical 18th century example of a cosmopolitan intellectual: though he studied, lived and worked mainly in Italy, he spent ten years in France, and stayed in various capitals, including Warsaw, London, Vienna and Constantino-

[257] p. 132.
[258] p. 55; p. 272.
[259] pp. 55–56, p. 272.

ple. He dealt mainly with mathematics and physics, but he was good in other fields also, poetry included. In his study of the shape of the Earth he used the idea of minimizing the sum of the absolute values of the deviations. He introduced a procedure for calculating the orbit of a planet based on observations of its position and also gave a procedure to determine the equator of a planet. Also important are his contributions to mechanics and statics [51]. Nearly all Boscovich works are published in the Boscovich National Edition directed by Edoardo Proverbio [175].

One of his most famous and lasting work was the *Philosophiae naturalis theoria redacta ad unicam legem virium in natura existentium* (simply the *Theoria* in the following) of 1758, which was reissued both in Latin and English in 1763. Boscovich's treatise can be compared, for what concerns his approach to the philosophy of nature, to the texts of mathematical physics of Euler and d'Alembert. In these, as in the *Theoria*, there is basically no reference to experiments devised to verify the proposed theories, as is instead typical of the experimental philosophy of the period. Empiric knowledge naturally enters, but essentially as background knowledge.

The work was not immediately an international success, but it was read in Italy and influenced numerous scholars including Alessandro Volta [165].[260] It was however much appreciated later on and possibly influenced the conception of matter at the turn of the 20th century [184]. The *Theoria* dealt only marginally with electricity, a topic that was not central to Boscovich but nevertheless attracted his attention. Boscovich knew about the work of Beccaria and Franklin. Current studies have found no evidence that Boscovich directly or indirectly knew the works of Aepinus, the *Tentamen* in particular, which was published more or less at the same time as the *Theoria*. Nevertheless, I will argue that it is to Aepinus that his approach is closer.

In the *Theoria* Boscovich carried out an operation of consistency on the concept of force introduced by Newton, in the *Principia* and in the Queries of the *Opticks*. In the *Principia*, as commented in Chap. 1, Newton seemed to see force from a phenomenological point of view; it exists but its causes are unknown and perhaps not even interesting. The most important example of force was gravity. In the *Opticks* Newton thought that at the microscopic level forces of various kinds could intervene different from gravity, which could be both repulsive and attractive. Newton seemed to attribute essentiality to these forces at some time; that is, the forces had an ontological value and not just a phenomenological status, as Newton himself had tried to credit in the *Principia*. Boscovich simplified Newtonian physics by introducing a single type of force, whose cause was of no interest, showing that with it one could obtain the expressions of all the forces of nature: gravity, capillarity, magnetism, elasticity, plasticity, etc.

Boscovich acted as a mathematician who takes metaphysics into his hands in a situation in which a canonical philosopher does not have sufficient tools to allow him an elaboration. That is, a metaphysics that serves to give reason to actual experiences. In some way he performed an operation similar to that carried out by Euler for the notions of space and force. And he did so perhaps in a more articulated and convincing way. For what force is concerned Boscovich took a reverse path with

[260] pp. 87–88.

respect to Euler. For Euler force was indeed a primitive concept, but its existence, its cause, was justified by the impenetrability of bodies. For Boscovich it was the impenetrability of bodies which depended on forces, given as existing in themselves, which prevent the interpenetration of the bodies because as they approach to each other, an increasingly large repulsive force intervenes.

Boscovich's idea of force and matter is well described in the summary he gave at the beginning of his *Theoria*. According to him matter consists of points that are perfectly simple, indivisible, of no extent, separated from one another. Each of these points p has a property of inertia and in addition a mutual active force depending on the distance between p and another point q in such a way that, if the distance is given, both the magnitude and the direction of the force are given. If the distance is diminished indefinitely, the force is repulsive, whilst if the distance is increased, the repulsive force will be diminished, vanish, be changed to an attractive force that first of all increases, then decreases, vanishes, is again turned into a repulsive force, and so on many times over like a sinusoid; until at greater distances it finally becomes an attractive force that decreases approximately in the inverse ratio of the squares of the distances [38].[261] Figure 4.35 shows a possible trend of the force-propensity (acceleration?) between two points.

The change in sign of force was most probably suggested to Boscovich by Newton's Query 31 of the *Opticks*, well known to him: "And as in algebra, when affirmative quantities vanish and cease, there negative ones begin; so in mechanics, where attraction ceases, there a repulsive virtue ought to succeed" [161].[262] But while in Newton one could see the possibility that attractive and repulsive forces stem from different causes, in Boscovich, who is concerned only with effects, this problem did not exist, and one cannot see but a unique function of forces varying with distance.

About the nature of force Boscovich was officially agnostic. Despite he gave greater importance to the idea of force than Newton did, he avoided hypostatization and considered force as simply a name to be given to the fact that there are approaches and departures between bodies even in the absence of visible causes. He started his presentation of force by asserting the Newtonian principle of inertia for which an insulated point admits an inherent propensity to remain in the state of rest, or of uniform motion in a straight line, in which it is initially set. But if there is also another point anywhere, there is an inherent propensity to compound the preceding motion with the motion which is determined by the interaction of the two points.

Indeed any two points of matter are subject to a determination to approach one another at some distances and in an equal degree recede from one another at other distances. This determination is called *force* by Boscovich, which may be repulsive or attractive if the propensity is to increase or decrease their distances respectively. Force does not denote the mode of action, but the propensity itself, whatever its origin, of which the magnitude changes as the distances change. this is in accordance with a certain definite law, which can be represented mathematically [38].[263]

---

[261] p. 17.

[262] p. 370.

[263] p. 39.

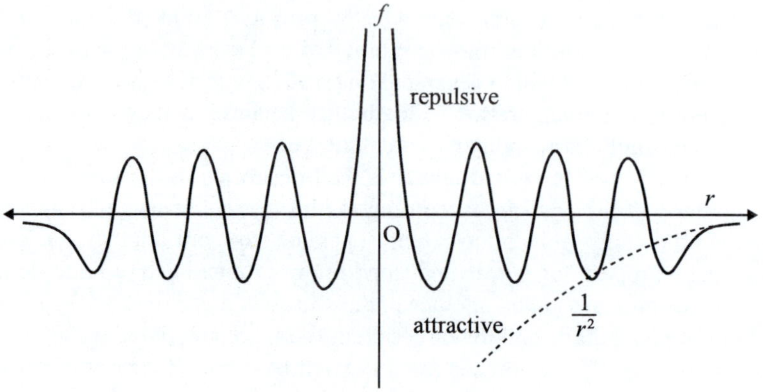

Fig. 4.35 A possible function of force-propensity f versus distance r for two points. Adapted from [38], p. 41

Boscovich spent pages and pages to show that his concepts of force and matter were able to explain every phenomenon and were fully satisfactory from a metaphysical point of view. And he tried to show that his idea of an immaterial point was not so strange as it might appear at first sight; indeed it was an essential condition for a complete coherence of his system.

What will attract the attention of physicists towards the *Theoria* was not so much his idea of a point without extension, which remained hard to accept and which could appear interesting only for a philosopher or a metaphysician, a concept in which one usually sees the influence of Leibniz. It was rather the unifying power contained in it. A single function that describes the trend of force as a function of distance, although complex and strange, was able to explain all the phenomena of nature. And the immaterial point was often replaced by a very small point without creating any embarrassment.

In the following I will not comment on the degree to which Boscovich succeeded in achieving his unifying purpose, for which reference can be made to [115]; rather I will briefly explain his conception of matter, which then serves to relate to his ideas on the nature of electricity. Only a metaphysical argumentation on the nature of force is reported below to show Boscovich's way of reasoning.

> The objection is frequently brought forward against mutual forces that they are some sort of mysterious qualities or that they necessitate action at a distance. This is answered by the idea of forces outlined in Art. 8, & 9. In addition, I will make just one remark, namely, that it is quite evident that these forces exist, that an idea of them can be easily formed, that their existence is demonstrated by direct reasoning, & that the manifold results that arise from them are a matter of continual ocular observation. Moreover these forces are of the following nature. The idea of a propensity to approach or of a propensity to recede is easily formed. For everybody knows what approach means, and what recession is; everybody knows what it means to be indifferent, & what having a propensity means; & thus the idea of a propensity to approach, or to recede, is perfectly distinctly obtained. Direct arguments, that prove the existence of this kind of propensity, have been given above. Lastly also, the various motions

> that arise from forces of this kind, such as when one body collides with another body, when one part of a solid is seized & another part follows it, when the particles of gases, & of springs, repel one another, when heavy bodies descend, these motions, I say, are of everyday occurrence before our eyes. It is evident also, at least in a general way, that the form of the curve represents forces of this kind. In all of these there is nothing mysterious; on the contrary they all tend to make the law of forces of this kind perfectly plain [38].[264]

According to Boscovich, all the non extended points the matter is made of, are identical; that is they do not differ from each other. Boscovich believed that the points cannot touch, and one point interacts with another by the attractive and repulsive forces according to the law exemplified in Fig. 4.35. Boscovich's elementary points are different from the atoms of the Greek philosophers and also from the contemporary concept of the chemical atoms, simply because traditional atoms have a size. It so could be wrong to call Boscovich's elementary points as *Boscovich atoms*, as sometimes it occurs. As a matter of fact Boscovich never called his points atoms. He used the term atom in his *Theoria* however, but it was to mean particles composed of elementary points, which remain together owing to the mutual forces [38].[265]

Two or more particles may be situated, in respect of distance and position so to constitute a particle of first order. Two or more of these particles may constitute a particle of the second order, and so on. And though all the elementary particles of matter are equal to each other, from their combination and the joint action of the elementary forces, there must result an inexhaustible variety of mutual actions fully capable to explain all the phenomena of the material world.

There could be a huge difference amongst different groups of points which form the different kinds of particles of which bodies are formed. First difference comes from the number of points that make up the particle, then the different disposition of points. Indeed by the assembly of many equal points, it can result particles that act on each other with forces that may have any trend, that is attractive repulsive and even changing from one kind to another by varying the distance, and in the limit particles may be as inert [38].[266] The forces of the particles may be such that one side is able to attract a second particle, while the other side repels. Indeed, there may be any number of places in the surface of even a spherical particle, which attract another particle placed at the same distance from the centre, whilst others repel and still others have no action at all. For, at these places there may be a greater or less number of points than in other places, and these may be situated at different distances from the centre and from one another [38].[267]

In a few sections Boscovich dealt with electricity, by naming Franklin and Beccaria: "Further, it is clear that from these principles there can be derived an explanation of all the chief phenomena in electricity; the theory of these, discovered by Franklin in America with truly marvelous sagacity, has been greatly embellished & confirmed, & even further developed at Turin by Fr. Beccaria, a most learned man, in his excel-

[264] p. 95.
[265] pp. 310–313.
[266] pp. 298–301.
[267] p. 303.

lent work on this subject, published some years ago" [38].[268] Boscovich summarized Franklin's theory without presenting any substantial criticism, also apparently accepting the idea of electric atmospheres, what it was not necessary for his theory. It seems as Boscovich renounced to consider long range forces, necessary to explain in a simple way the phenomenon of induction and all experiments connected with it, as Aepinus did.

A concept that Boscovich regained from his idea of fire is that of relative saturation. That is there is a natural amount of electric fluid that can be contained in a body; if it has more or less of this amount the body shows electric signs.

> Thus, of two of these bodies, of which the saturation corresponding to their natures is not the same, one will be electric by excess, & the other by defect, with respect to one another. If these bodies approach one another to within that distance, for which the particles surrounding the bodies, & adhering to them like atmospheres, can act upon one another; then, from the body that is electric by excess this fluid will immediately flow towards the one that is electric by defect, until equality is reached. During this flow, the substances which respectively yield & receive the fluid will simultaneously approach one another, if they are light enough, or if they are freely suspended; & if the motion of the concentrated matter is vigorous, there will be explosions, & sparks, & even lightning, thunder, & thunderbolts [38].[269]

Boscovich also commented the experiment of the Leyden jar and Franklin's square and remembered Franklin's principle, for which when bodies that are *naturally electric* (in modern term dielectric) have a very small thickness, such as a thin glass plate, a much greater amount of the fluid can be collected on one of the surfaces and at the same time from the other surface exactly opposite to it there can be withdrawn an equal amount of the fluid.

Below Boscovich explanation, nearly verbatim follows: When the plate is thin, there can be a repulsion, exerted by the particles of the fluid situated on one of the surfaces, acting on particles situated near the other surface. Still, it may be that this is not sufficient to overcome the attraction by which the particles (of common matter) adhere to those that are next to them. But, if this is assisted on the one side by the attraction of a conducting body moving towards it, the repulsive force can overcome the attraction. Now, when this is the case, part of the fluid will flow off from the surface and enter the new body that has been brought close to it. And since part of the repulsive force ceases owing to the removal of this part of the fluid, there will adhere to the nearer surface a greater amount of the electric fluid brought to it by the water or the gold; until, however, communication being restored by means of a series of bodies that are merely electric by communication, the flow of the fluid from one surface to the other will be unhindered. Moreover, this explanation is confirmed by the fact that, if the experiment is tried with a plate that is too thick, it will not succeed [38].[270]

---

[268] p. 361.

[269] p. 361.

[270] p. 363.

Boscovich ended his consideration by commenting: "Whether these things are indeed as stated cannot be determined, unless it can be shown at the same time that it is impossible for them to be otherwise [38].[271]

Boscovich continued to be interested in electricity even after the publication of the *Theoria*. He discussed the subject with Nollet during his stay in Paris, between 1759 and 1760. He also discussed about it with English electricians in his long stay in London, where he arrived in 1760 from Paris. In London he met Benjamin Wilson (1721–1788) and Edward Delaval (1729–1814). At the end of 1760 he also related to John Michell (1724?-1793), who showed him a series of experiments carried out with artificial magnets [174].[272] But mainly he met Benjamin Franklin, who already knew Boscovich by name, having more or less direct knowledge of *Sopra il turbine che la notte tra gli xi e xii giugno del 1749 danneggió una gran parte di Roma* published by Boscovich in 1749 [37]. The first meeting with Franklin took place in the 1760s. The two met again, much later, in Paris in the 1770s [174].[273]

After the meetings of London with the British electricians, Boscovich continued in his interests on electrical phenomena. In 1764, he interacted with Giambattista Beccaria for the installation of a protection system for the spire of the Milan cathedral from the action of lightning [50]. In 1767 he wrote a letter to Giovan Stefano Conti (1756–1768) in which he reported on the latest experiences on electricity at his knowledge [39].[274] In the letter he commented on the results of Beccaria in particular the account of "an experiment [of] the Jesuits of Peking [...]. They electrified a glass placed it on a compass, that is on the glass of it: the magnetized needle raised with his tip, and attached itself to the glass, but after a while it felt back, repulsed with greater force than the one, which gravity only exercises, and detaching [...] there are also the phenomena of Vimmer's [Symmer's] socks, and even more varied [...] but truly the good P. Beccaria does not have the gift of clarity"

Then Boscovich concluded: "You need a tome to follow the theory, and give behind a thousand varied cases, and a thousand effects [...]. The book [*Elettricismo artificiale*] will come, but I don't know when [...]. This subject is now more than a science; but in all the experiments I have seen or heard, the same general laws hold true" [39].[275]

4.3.9 Charles Augustin Coulomb

Charles Augustin Coulomb (1736–1806), one of the major figures in the history of physics and engineering, during his youth attended lectures at the Collège Mazarin and the Collège de France in Paris. Here he received a good classical grounding

---

[271] p. 363.

[272] pp. 12–15.

[273] p. 16.

[274] vol V/1, pp. 332–334.

[275] p. 332–333.

in language, literature, philosophy and the best available teaching in mathematics, astronomy, chemistry and botany. Then Coulomb moved to Montpellier where joined the Société des sciences of Montpellier as an adjoint member in 1757 and read several scientific papers. At this stage his interests were mainly in mathematics and astronomy as testified by several papers on these topics he read at the Societé.

He came back to Paris in the autumn of 1758, seeking the tutoring necessary to enter the École du genie at Mézières, at the time the foremost technical school of Europe, and found it in Charles Étienne Louis Camus (1699–1768). After some months of study he passed the entrance examination. At about this time he formed lasting friendships with Jean Charles de Borda (1733–1799) and Charles Bossut (1730–1814), his teacher of mathematics; he also attended Nollet's physics course. This comprised general notions of matter and gravity, mechanics, optics, astronomy, electricity and magnetism. Coulomb graduated in 1761 with the rank of *lieutenant en premier* in the Corps du génie and shortly after, in 1764 he was sent to Martinique where he remained until 1772.

In 1774 Coulomb began to work on a memoir on magnetic compasses that subsequently shared the first prize in the Paris academy's competition for 1777 [68]. After 1779, being sent to the shipyards at Rochefort as a military engineer, Coulomb engaged in a lengthy series of experiments on friction. In 1781 the result of these researches won the double first prize at the Académie des Sciences de Paris and gained Coulomb election to the academy as *associé géomètre* a position that changed in that of *associé mécanicien* in 1784. The year 1781 marked a decisive break in Coulomb's life and career. Permanently stationed in Paris, a member of the Académie, he had solved his problems for a living. Since then he was able to devote the major portion of his time to researches in experimental physics. Coulomb the engineer became the physicist and public servant. He read twenty-five scientific memoirs at the Académie des sciences de Paris (and at its successor, the Institute de France) from 1781 to 1806. In addition to his physics research Coulomb participated in 310 committee reports to the Académie concerning machines, instruments, canals, engineering and civic projects.

Coulomb devoted much time to public services. His last employment was as inspector general of public instruction from 1802 until his death in 1806, in which office he played a significant role in supervising the establishment of the French system of lycées. Two decades of field duty in the Corps du génie accustomed Coulomb to a modest style of life. The probate description of his personal belongings accords with this. He was not a men of letters; his library contained 307 books, 238 of which were technical volumes issued by the Académie des sciences de Paris.

Coulomb's major interest was long engineering and mechanics; his researches in the field included fundamental memoirs on rupture of beams and masonry piers, soil mechanics, friction theory, and ergonomics. In these he can be considered one of the greatest engineer of the 18th century Europe. His most important memoir on engineering was also his first, *Essai sur une application des règles de maximis et minimis à quelques problèmes de statique, relatifs à l'architecture* of 1773 [67]. One of the purpose of the memoir, was "to determine, as far as a combination of mathematics and physics will permit [to understand], the influence of friction and of cohesion in some problems of statics" [67].[276] In the memoir of 1773 there is almost an embarrassment of riches, for Coulomb proceeded to discuss the theory of rupture of masonry piers, the design of vaulted arches and the theory of earth pressure, where he developed a generalized sliding wedge theory of soil mechanics that remains in use today in basic engineering practice of soil mechanics [123]. Still in this memoir Coulomb tackled the problem of the resistance of inflexed beams [48].[277] The fact that Coulomb dealt with the problem of friction is most probably due to the interest that friction has in applications. Friction had always been ignored by mathematicians and (experimental) physicists and regarded as a disturbance, to be overlooked in ideal conditions. An important exception is represented by Guillaume Amontons (1663–1705) with his *De la resistance causée dans les Machines, tant par les frottemens des parties qui les composent, que par roideur des cordes qu'on y employe, & la maniere de calculer l'un & l'autre* of 1699 [9]. For Coulomb friction became a fundamental phenomenon to be studied when practical problems were to be solved.

Another Coulomb's celebrated study was the *Théorie des machines simples*, with which he won the Gran Prix from the Académie des sciences in 1781 [79]. He investigated both static and dynamic friction of sliding surfaces. Another subject of much interest to Coulomb was the question of efficiency human work, and in this field (now named ergonomics) he made one of the most significant contributions before the studies of the American Frederick Winslow Taylor (1856–1915), a century later. His works, started in the 1770s, are summarized in the *Résultats de plusieure expériences destinées à déterminer la quantité d'action que les hommes peuvent fournir par leur travail journalier, suivant les différent manières dont ils employent leurs forces* of 1799 [78].

However this portion of Coulomb's work suffered of a relative neglect and with the exception of his friction studies, most of his mechanics memoirs were little known until the early 19th century. A reason was that in formulating methods of approach to fundamental problems in structural mechanic he introduced advanced mathematics, in particular the variational calculus, rather than to give numerical solutions to specific problems as preferred by the 18th century engineers and physicists. It required that group of Polytechniciens, teachers and students, in the early nineteenth century to appreciate the importance of his work in the context of the new engineering mechanics.

His work in physics was integrally tied to his work in mechanics. In 1784 he presented at the Académie des sciences de Paris the *Recherches theoriques et expérimentales sur la force de torsion et sur l'élasticité des fils de métal* [69]. This is now classified as a work in structural mechanics but in turn it provided him with a means to investigate quantitatively electricity and magnetism.

Coulomb's major contributions on electricity and magnetism, a part from the prize-winning memoir *Recherches sur la meilleure maniere de fabriquer les aiguilles aimantées* on magnetic compasses of 1777 [68], can be be found in the famous series

[276]p. 343.

[277]pp. 80–82.

of seven electricity and magnetism memoirs read at the academy of Paris from 1785 to 1789 [70–75, 77]—collected also in a single volume [76]—(here the dates are those appearing in the front cover of the Mémoires of the Académie, printed in the period 1788–1793) and several magnetism memoirs prepared after the French Revolution.[278]

It has been noticed that Coulomb made his first mention to electricity in his work on the magnetic compass by suspecting a possible electrical disturbance in the motion of the magnetic needle. And rather than to explore the electricity, his researches on the subject would be motivated also to eliminate electric disturbance in his compass [84].[279] Personally I cannot imagine Coulomb dedicated only to applications and measurements.

4.3.9.1 Coulomb's Role in the History of Electricity: Physics, Metaphysics and Mathematics

The significance of Coulomb's contribution to electricity and the esteem in which he was held, is amply testified by the adoption of the Coulomb as the unit of electric charge and the universal reference to the law of force in electrostatics as Coulomb's law.

Indeed, almost every modern book of classical electrostatics follows a physical mathematical approach based on a few principles drawn from the real world. It is postulated: (1) the existence of electrical particles of two different types; the former mobile and negatively charged, the electrons, the latter fixed and positively charged, the protons; (2) they attract or repel each other by obeying the law of Coulomb. All the properties of electricity, at least in conducting bodies, induction, capacitance, charge density, etc. are justified theoretically. Recurse to experiments is needed only to determine the numerical values of some characteristic constants. It is thus not strange that Cavendish's and Aepinus who gave fundamental contributions to electricity before Coulomb are, today, almost forgotten.

But, as it is so common in history, the lapse of time has idealized what Coulomb actually did compared with what Coulomb's name signifies. The actual historical development is hidden in the axiomatic formulation of modern electrostatics. Coulomb studied a very limited class of phenomena, especially the distribution of charges on the surface of conducting bodies of simple form, put into contact with each other and possibly separated. Even if he could manage very small charges, a small sphere could equally have a potential of many volts, while Volta, for instance, was studying minute potential differential. And he, and his pairs, derived his inspiration more from Franklin, Canton, Aepinus and Cavendish than from Coulomb.

In any way Coulomb's memoirs on electricity, notwithstanding their limits, had a great impact on his contemporaries, much more than those of Cavendish for instance who presented results not less relevant than Coulomb's. Cavendish's published papers

---

[278] Most of biography information comes from [106].

[279] p. 8.

were written in a highly condensed, austere style, not very easy to be read by experimental physicists. Moreover they were published on a journal, the Philosophical Transactions, whose readers, a least for electricity, were more interested in experiments than in theory. On the contrary, Coulomb published on the Mémoires de l'Académie des Sciences de Paris, a journal whose readers appreciated much a mathematical physical approach. Moreover in Coulomb's writings, there was a different rhetoric. Formal arguments were interspersed among practical details of apparatus, result tables, brief speculative remarks and comments, which made his memoirs much more pleasant and simple to read than those of Cavendish. The problems of electricity were contained in a milieu of much more familiar mechanical ideas. Indeed one might say that Coulomb brought electricity into the familiar terrain of mechanics, rather than introducing mechanics, when necessary, into the domain of electricity [84].[280] Coulomb's memoirs were generously sprinkled with mathematical calculations and arguments, but nowhere did he seriously attempt the abstraction that characterized Cavendish's argument. He was thus able to convince many of his contemporaries of the feasibility of a mathematical approach to electricity.

Coulomb had no difficulty to treat with actions at a distance, then also accepted in France, Descartes's home. He conceived electricity as due to corporeal fluids and did not stress whether they were of a continuous nature or made up of particles; in both cases volumes can be considered small enough to represent elementary particles of electricity. The particles exchange among each other forces at a distance, of attraction or repulsion, whose cause was not indicated. Unlike Aepinus, who considered actions among finite portions of electric fluid, Coulomb dealt with infinitesimal elements in order to apply Calculus. What allowed him to go beyond Aepinus was however the fact that he was able to specify the form of the function of the electric force. Only in this way he could develop effectively his mathematical approach. His was a great step which would not be allowed by experiments alone, but strongly suggested by the application of the new mathematics for the force of gravity, for instance.

According to Coulomb electric forces are exerted through a space whose nature is not specified. In particular, in no case did Coulomb refer to an empty space and in no case was its Euclidean nature questioned. He did not operate at a cosmological level, like Newton, and his electric bodies are naturally immersed in air. No reference is given to the interaction among the electrical forces and the particles that make up air. Of course Coulomb cannot be blamed for not having understood the complex role that air plays in the transmission of electricity. At the time there was no widespread agreement on the role of air. Cavendish conceived, for example, dry air as a non-conductor, but due to the inevitable presence of humidity, the air could become a conductor.

Coulomb made extensive use of mathematics in his electrical works; the simplest situation occurred in the first two memories of electricity, aimed to specify the form of the function of force between two elementary charges. A more complex situation occurred when he came to deal with the density of electricity on the surface of simple conducting bodies, spherical, in most cases, and cylindrical. He, like Cavendish,

[280] p. 27.

had a good knowledge of mathematics of the time, in particular he knew well the latest developments of differential and integral calculus. But he had to face complex problems, too difficult to deal with even the refined tools he mastered; problems that will be addressed later by Poisson basing on Laplace's studies on gravitational problems. As an engineer, however, Coulomb intended to propose an albeit approximate solution, not so much of calculations but of physical modeling. His models were brilliant, sometimes even ingenious, but at the same time they could be coarse and inconsistent.

Coulomb's law in modern notation has the expression:

$$f = \epsilon \frac{q_1 q_2}{r^2}$$

with ϵ a constant of proportionality depending on the units of measurement and on the medium surrounding the two electric elementary charges (positive or negative), q_1 and q_2, placed at a distance r. Although the analogy with Newton's universal gravitation equation is evident Coulomb never referred to it explicitly. Actually Coulomb did never formally write an equation similar to that reported above; he rather used the language of proportions. Moreover he did not comment the possibility of medium different from air, accounted for by the constant ϵ.

Coulomb's law was proved into two steps; before the dependency on r was verified experimentally, then the dependence on q_1 and q_2 was someway argued.

Coulomb demonstrated the dependence on the inverse of the square distance, taking it as a hypothesis and then testing it experimentally. Cavendish had provided a very elegant and sufficiently rigorous proof in some of his 1770s manuscripts, which however remained unpublished (did Coulomb know them?). His proof was however indirect (see Fig. 4.13) and the forces, mechanical magnitudes, were evaluated only by measuring non-mechanical magnitudes, such as electric charges. The proof of Coulomb was instead of the direct type through the measure of a force by means of the torsion balance; a decidedly more intuitive approach for his contemporary physicists.

There were historical, metaphysical and empirical reasons that pushed Coulomb toward the hypothesis of the dependence on the inverse of the square of the distance, the same that had motivated Newton in his law of universal gravitation. Indeed, before Coulomb, many other scholars besides Cavendish had suggested the inverse square law: Michell (1750), Lambert (1758), for instance. In [105] ten instances are documented, some of which were known to Coulomb; but probably a careful investigation should find much more instances. Priestley, by quoting a Franklin's experiment which showed that inside an electrified cup there is not electric activity, commented: "May we not infer from this experiment, that the attraction of electricity is subject to the same laws as that of gravitation; therefore according to the squares of the distances. Since it is easily demonstrated, that were the earth in the form of a shell, a body in the inside of it would not be attracted to one side more than another" [171].[281]

The problem of precursors is a classic in the history of science; on this matter it is noteworthy to cite I Bernard Cohen who refers to Franklin :

> Often, once a theory has proved successful, jealous contemporaries (like later historians) can show how each part of the theory had already been stated; yet it is to be noted that such prior statements achieve at this later date a significance they never had before. many of the elements of Franklin's theory and a number of the new facts he and his Philadelphia friends discovered could be found in the literature of the 1740s—but found more easily after Franklin's work than before it [65].[282]

Coulomb's attribution is justified by the fact that he chose the law of inverse square among many other possibilities that at the time were considered equally plausible and he was the only one who made a definitive and motivated choice. Coulomb chose the inverse square law and 'proved' it with very accurate measurements and after him the law of the inverse square was substantially undisputed in the applications, even though some doubts about its exactnesses remained. Indeed nearly a hundred years later when Maxwell was editing Cavendish's unpublished papers he found that the direct experimental evidence for this law was no better than it was one hundred years earlier. Maxwell himself subjected this law to new experimental test; but it was Cavendish's not Coulomb's methods that he used.

The dependence on q_1 and q_2 could be proved in principle experimentally by using the torsion balance, but Coulomb did not do so and this part of the law was left unproved. He declared that the dependence was obvious (see below). Indeed this proportionality was given for granted by Aepinus as well, but most probably the inspiration came from Newton gravitational law. Not for nothing Coulomb referred to q_1 and q_2 as the electric masses.

4.3.9.2 Memoirs on Electricity

What a modern reader first notices in the memoirs of Coulomb on electricity is the almost complete absence of references to the theories of his time on the matter and the almost total absence of quotations, with a few exceptions. He admired, and cited, Aepinus for his effort to reduce electricity and magnetism to mathematical analysis and Musschenbroek for his ability as an experimenter. He cited only once Franklin [68],[283] even though he met him and operated with him. He was also nearly completely silent about causal explanations, even though for the schools he attended he might have had a good knowledge of natural philosophy and perhaps also of metaphysics. But in his writings on electricity, he avoided entering the field of philosophy, as instead did the 'mathematician' d'Alembert, of only one generation older than him, long a colleague of his at the Académie, for general physics and mechanics.

[281] p. 689.

[282] pp. 298-299.

[283] p. 258.

There could be many reasons that may have suggested his approach to Coulomb. He may not have been thoroughly familiar with the theories of magnetism and electricity to risk exposing them. Or, more probably, it may have been that he had decided that these theories were of little interest and in any case the subject of extensive discussions and not worth repeating, dedicating the limited space offered by the magazine to present new results. Of course the debate of 'savants' on the nature of electricity influenced Coulomb's ideas. But he saw in the approach advocated by Aepinus the only way to proceed. He accepted the hypothesis of an electric fluid, or rather of two, but instead of going to hypothesize causal mechanisms he preferred to refer to proximate causes, that is to attractive and repulsive forces.

Even he was never explicit about the theories he accepted about electricity, his position is clear and the language he adopted was different from that of Aepinus who married the hypothesis of a single fluid. Coulomb spoke of two types of electricity, one positive and the other negative, but never in the sense of excess or defect of the electric fluid as Franklin and Aepinus did. He spoke of electric density (*densité électrique*), using the term as electricity per unit of volume or surface; but sometimes simply as an amount of electricity. He also spoke of electric masses (*masses életriques*) to refer to the total amount of electricity [71];[284] but he never used the term *electric charge*, used instead by Franklin.

The First Two Memoirs on Electricity. Experimental Proof of Coulomb's Law
Coulomb's first two memoirs on electricity officially had the aim to evaluate experimentally the mathematical expression of the forces between two electrically charged spheres versus the distance between their centers. A very particular problem at first glance and perhaps uninteresting. In reality, the spheres, when they are small, lend themselves very naturally to represent elementary electric masses. So measuring the forces between two small spheres is equivalent to measuring the force between two elementary electric masses. Coulomb declared his true purpose only at the end of the second memoir, but without giving it particular emphasis.

Before even introducing experiments on electrified spheres, Coulomb presented his apparatus to measure forces: the torsion balance, one of the most interesting device used by him, exposed in the final form in his paper *Recherches théoriques & expérimentales sur la force de torsion, & sur l'élasticité des fils de métal* of 1784 [69].

Coulomb was not the first to propose the use of torsion elasticity of wires to measure torques. John Michell's is considered as a possible precursor [132],[285] but there are not convincing proof of the fact and thus Coulomb at the moment should be considered as the inventor of an effectively working torsion balance [105].[286]

Coulomb's torsion balance shown in Fig. 4.36, is constructed on a glass cylinder ABCD, having both diameter and height of 12 in. (32 cm).[287] In the upper part of

[284] p. 579.

[285] p. 191.

[286] pp. 162–165.

[287] The measurements are expressed in French inches, equal to about 2.7 cm.

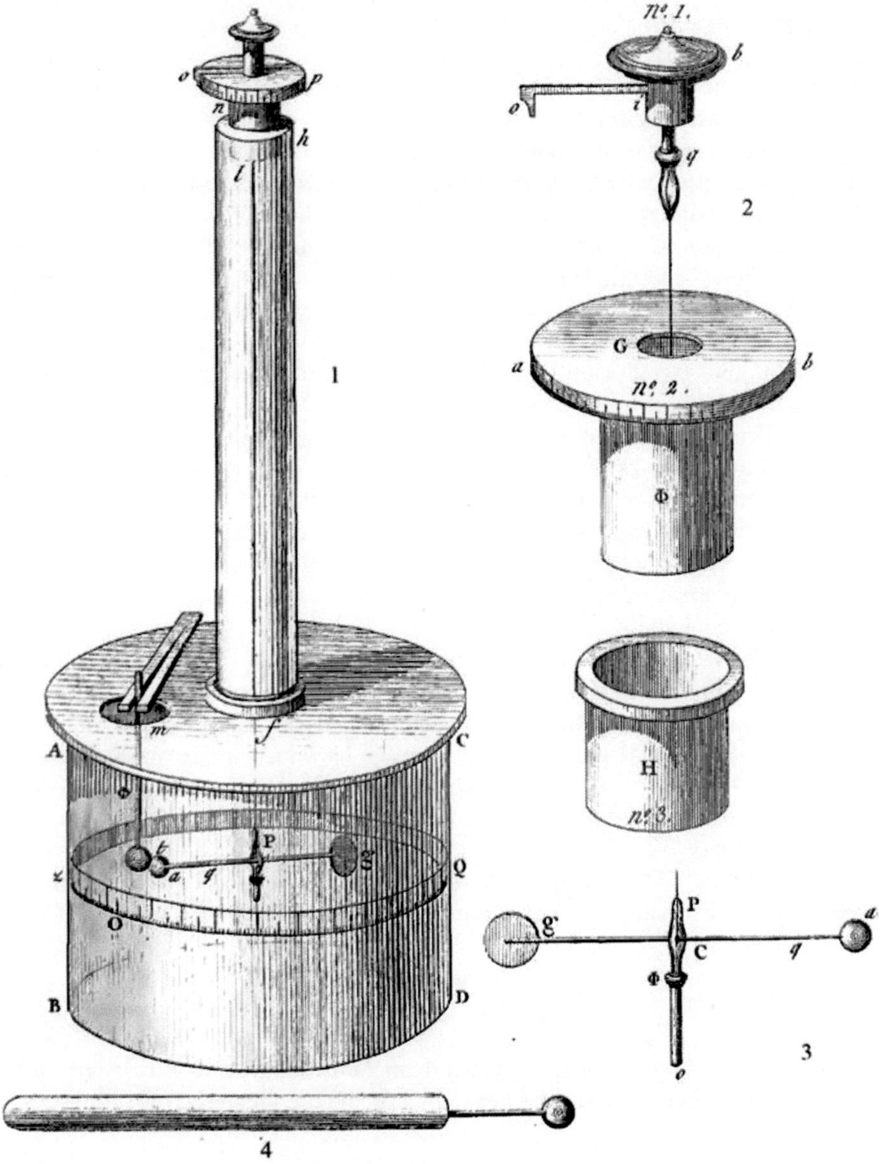

Fig. 4.36 Coulomb torsion balance. From [70], Fig. 1 of Table XIII

the cylinder there is a glass plate with two equal holes m and f having a diameter of 20 *lignes* (4.5 cm).[288] At the central hole f a vertical tube 24 in. (65 cm) long is cemented. In the top of this tube it is placed the torsion micrometer op. It contains several parts, including a circumferential scale divided into degrees, a knob with scale pointer oi, and a pincer q to hold the torsion wire. The matter used for the torsion wire used was either silver, copper, or silk. At the lower extremity of the wire, it is attached a thin metal pincer P, one ligne (2.2 mm) in diameter, which weighs enough to keep the torsion wire taut. The pincer held a thin, horizontally suspended needle of eight inches (21.5 cm). A tiny, gilded, elder wood pith-ball a, about two lignes (4.5 mm) in diameter, was fixed at one end of the straw and a small, vertical paper plane g at the other. The paper plane served both as a counterweight and to damp out oscillations. A graduated scale ZOQ, marked in degrees, was attached around the outside of the large glass cylinder.

The balance was then employed in the following way. The pointer oi of the micrometer at the top of the cylinder raising from f is turned until the horizontal needle ag containing the pith-ball a, is lined up with the zero on the graduated scale ZOQ, meantime the angle measured by the micrometer reads zero. Then a thin, insulated rod mt with a second, identical pith-ball t, mounted at its end was inserted through the hole m so that it touches the pith-ball a. The balance is now ready for electrical experiments.

The torsion wire, used in the first memoir of 1785, was of silver 28 in. long (about 75 cm) such that its mass was of 1/16 of grain (corresponding to nearly 0.02 mm of diameter, about as thin as a human hair) and with a lever arm of 4 in. of the needle, a force of only 1/40800 grains[289] (1.28×10^{-8} Newton) sufficed to turn the balance through three degrees, which is according to Coulomb the maximum precision allowed by his balance, due to environmental disturbance [70].[290] A precision that even today is not easy to pair.

The sensitivity of the balance referred to above was evaluated by Coulomb not experimentally but rather calculated using the formula found in the memoir of 1784:

$$\frac{\mu B \phi^4}{l}$$

for the torque of the wire, where ϕ is the diameter of the wire,[291] l its length, B the angle of rotation and μ a constant depending on the used material [69].[292]

[288] A *ligne* is about 1/12 of an inch, or 2.2 mm.

[289] One French grain is about $5 \times 10^{-4} N$, the weight of a mass of 50 mg.

[290] p. 574.

[291] Coulomb used the symbol D, which is here avoided to not confuse with the product of electric masses introduced later.

[292] p. 247.

In the first of his memoirs Coulomb's declared objective was to prove the following proposition:

> The repulsive force between two small spheres electrified with the same type of electricity is inversely proportional to the square of the distance between the centers of the two spheres [70].[293] (D.30)

The proposition is assumed as an hypothesis to be verified experimentally and not as a result of induction from a series of experiments. Indeed only the result of three experiments, or tests, are reported, too few for the validity of induction.

In the first test, Coulomb charged an insulated pin by a Leyden jar or an electrostatic machine. He next inserted the pin through he hole *m* to touch to the two pith-balls *a* and *t*. These, having received the same amount and kind of electricity, separate by a certain distance *d*. Coulomb noted the separation in degrees on the lower scale ZOQ. In the second and third tests the micrometer was rotated so that the distance between the two pith-balls *a* and *t* was forcedly reduced. The results of the three tests are reported in the Table 4.6 [70],[294] with the angles defined in Fig. 4.37a. This figure illustrates the position of the needle and the two pith-balls *a* and *t*. Straight line $\chi\chi'$ represents the needle in the virtual initial position (that is the position which *a* would have for the rotation of the micrometer only), $\phi\phi'$ it in the final position; α is the angle measured in the micrometer, *d* the angle measured in the scale ZOQ—or equivalently the distance between the two pit balls—and *x* the total angle of torsion of the wire, calculated as $d + \alpha$.

The last row of the table, which is not due to Coulomb but mine, reports the product of the total torsion angle, proportional to the torque of the wire (and thus to the force of repulsion) and the square of the distance between the two pith-balls; if Coulomb's hypothesis, for which the electric force is proportional to the inverse square of distance, is true this product should be the same for all the three texts.[295] Indeed, this is substantially the case: "It results indeed from these three tests that the repulsive force the two balls electrified with the same kind of electricity follow the inverse ratio of the square of their distances" [70].[296]

Two factors must be accounted for in measuring large-angle separation of the pith-balls. The first is that the force between the balls goes not as the arc *d* in degrees but as the chord (dashed segment *ta* in Fig. 4.37a). The second that the arm of the lever, the segment upon which the pith-ball *a* acts is not equal to *Ca* (see Fig. 4.36, right bottom) but actually this value multiplied by $\cos d$. For a pith-ball separation not very much greater than 30°, however, as in the present case, Coulomb noted that these two factors tended to cancel and that the forces can be calculated simply from the torsion separation in degrees.

---

[293] p. 572. Translation in Gillmor 1971.

[294] pp. 572–573.

[295] If the electric force is proportional to the inverse square of distance, that is $f \propto \frac{1}{d^2}$, the product fd^2 is constant.

[296] p. 174.

Table 4.6 Angles of torsion in the test of repelling electrified bodies, [70], pp. 575–576

Micrometer setting (α)	0	126	567
Separation of pith-balls (*d*)	36	18	8.5
Total torsion angle ($x = d + \alpha$)	36	144	575.5
$x \times d^2$	46656	46656	41580

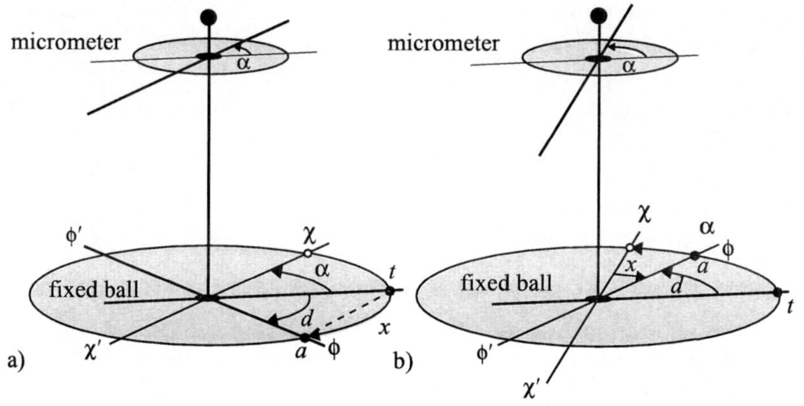

Fig. 4.37 Positions of the pith-balls and angles of torsion

Coulomb did not comment the discrepancy of the third test; a part from error of measurements and the leakage of the electricity, one reason could be due to the large angle of torsion, over a full tour of 360°, and in this case the linear relation between torque and torsion angle is no longer strictly valid.

Notice that Coulomb believed he had tried his proposition using only a few measures, which among other things only confirmed it in an approximate way. A modern reader, even a not shrewd one, has the doubt that Coulomb reported only the measures that best fitted the result and had discarded others. It must be taken into account, however, that the first memoir on electricity is only one of a series of memoirs in which the theme of the experimental determination of the inverse square law occurs more or less directly. Without these further memoirs, perhaps Coulomb's proof could not be considered nothing but an uncertain statement.

It must be signaled that in the recent past it has been doubted that Coulomb have reported the actual results he found in the experiment, because attempt to reproduce them failed. As often occur in the history of science, as in the case of the experiment of the inclined plane of Galileo for instance, later researches has shown instead that Coulomb's results could possibly be reproduced and that most probably they were those found by Coulomb [157].[297] This suggests that experimental investigations, carried out by specialists in the Paris academy of sciences, were carried out under high standards of accuracy.

[297]p. 547.

In the second memoir Coulomb faced the more complex case of bodies with the opposite kind of elasticity and thus attracting. This time it was very difficult to use the same apparatus employed to measure the repulsion because the resistance to torsion of balance varies (approximately) linearly with the torsion angle while the force of attraction with the inverse of the square of distance (and thus of the angle of separation) of the two pith-balls. This can determine situations of unstable equilibrium and it can happen that the pith-balls attract each other until they come into contact, thwarting the experience. Coulomb explained the problem with the use of the laws of mechanics. With reference to Fig. 4.37b the following equilibrium equation can be written for rotation:

$$nx = \frac{D}{(\alpha - x)^2}$$

in which the first member represents the resisting force of the torsion wire (notice the force and not the torque), being x the total angle of rotation and n a constant of proportionality. The second member of the equality concerns the electric active force, where D is what Coulomb called the product of the electric masses, that is, with Franklin's nomenclature, the product of the charges of the two pit balls [71][298] and $(\alpha - x)$ represents the distance d between the two pith-balls.

The previous equation can be rewritten as $D = nx(\alpha - x)^2$, which is a function always positive in the interval $[0 - \alpha)$; its values at the extremity of the interval are zero, and thus D has a positive maximum. Coulomb found this maximum at $x = 1/3\alpha$, which gives $D_{max} = 4/27n\alpha^3$. If the value of the electric masses and the initial position α are such that $D > D_{max}$ no equilibrium is possible and the two pith-balls get into touch. But even for $D < D_{max}$, when $x > 1/3\alpha$ the equilibrium becomes unstable and the two pith-balls merge [71].[299]

Coulomb attempted with some caution to use the same procedure of the first memory, but soon realized that, because of the instability of the equilibrium, the operation was too delicate and thus he moved from a static to a dynamic measurement approach, returning to the method of time oscillations he had used in his 1777 on the magnetic compass.

The dynamic balance, illustrated in Fig. 4.38, consists in a large globe Gr, made of cardboard covered with a tin stain sheet, centered in G. A gilded paper disk l is attached to a shellac-thread needle lg suspended by means of a tin silk tread sc, g is a counterweight to maintain horizontal the needle. After charging the large globe the small disk l is grounded so that an electric density of sign different from that of the globe is inducted in it. Then the suspended needle is set into motion and the oscillations are measured for different distances rl between the gilded paper disk and the globe. Coulomb made some assumptions:

[298] p. 579. Coulomb used the symbol D instead of M; it has be changed here not to confuse it with the diameter of the torsion wire.

[299] pp. 579–580.

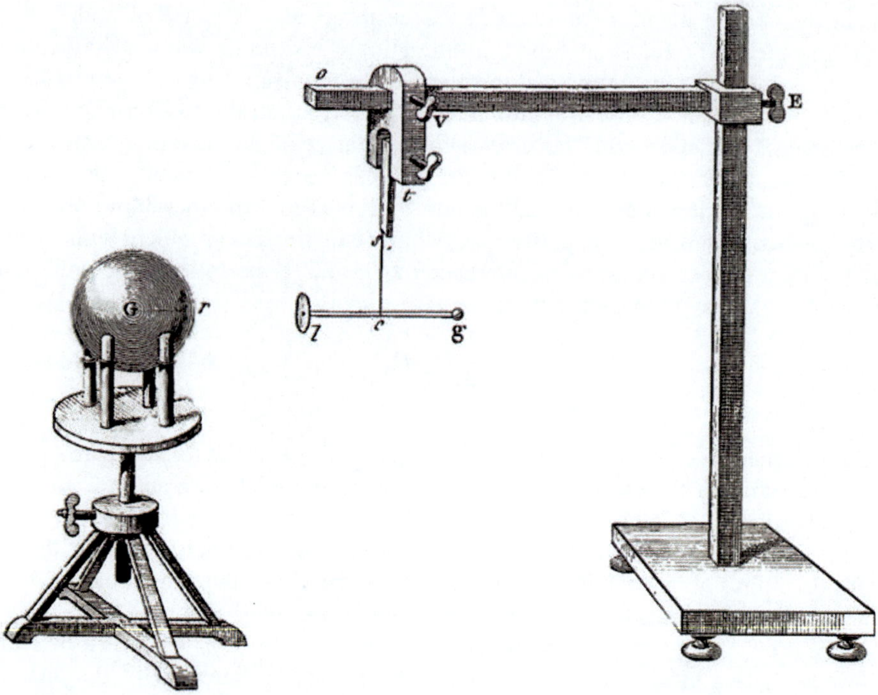

Fig. 4.38 Timing oscillating set. Partially redrawn from [71], Fig. 1 of Table XIV

1. The electricity is uniformly distributed on the surface of the globe Gr (on this point he will argue in the fourth memoir).
2. The torque of wire sc is small with respect to the force exchanged between the two electrified bodies, thus its resistance to torsion plays no role.
3. Given the quite large distance Gl, the electric force on l has a constant value and direction.

Because of the first assumption, and from Newton's results in the *Principia*,[300] to which however there is no explicit reference, Coulomb could assert that "when all the points of a spherical surfaces, act with an attractive or repulsive force, inversely proportional to the square of the distances, the force on a point located at a whichever distance from the sphere, is the same as when all the spherical surface is concentrated in its center" [71].[301] This means that all goes as if on the paper l acted a force due to the electricity of the sphere located at the distance Gl. And because of the second and third assumptions all goes as the needle gl were a normal gravitational pendulum,

[300]Book 1, section XII, Proposition 71, Theorem 31. Actually, because of the induction between the globe Gr and the small disk l, the electricity is not exactly uniformly distributed on Gr and Theorem 31 does not hold exactly [64], vol. 1, p. 43.

[301]p. 583.

Table 4.7 Period of oscillation versus distance [71], p. 584

Distance Gl (inches)	9	18	24
Time for 15 oscillations	20	41	60
Predicted results	20	40	54

whose period of oscillation T is inversely proportional to the square root of the force ϕ, playing the role of gravity acceleration, or $T \propto \sqrt{\phi^{-1}}$.

In substance if the experimental results would show a relation of proportionality between T and d, they proved the law of the inverse square [71];[302] indeed if $T \propto d$ and $T \propto \sqrt{\phi^{-1}}$, it immediately follows $\sqrt{\phi^{-1}} \propto d$, or equivalently $\phi \propto d^{-2}$. That is force varies as the inverse of the square of the distance.

Table 4.7 reports the experimental results (only three as in the case of the first memoir), which substantially support the assumption.

Coulomb noticed that if one could take into account the lost of the electricity occurred between the first and the last experience (four minutes) the time of the latest oscillation would be a little less than the reported value and the agreement with the theory would increase.

Although Coulomb 'proved' directly by experiment that the electric force f laws vary inversely as the square of the distance, he never explicitly demonstrated that they are also proportional to the product of the respective electric masses. That is, using a modern notation, Coulomb had only demonstrated that $f \propto 1/r^2$, but not that $f \propto m_1 m_2/r^2$, where m_1 and m_2 are the electric masses.

At the end of the second memoir, he formulated the proposition that is today known as Coulomb's law:

> The electric action, whether repulsive or attractive of two electrified globes and consequently of two electrified molecules, is in the ratio compounded of the densities of the electric fluid of the two electrified molecules and inversely as the square of the distances [71].[303] (D.31)

Without any further comment.

Only in the part of the second memoir, related to magnetism, he was a little bit more explicit, by declaring that the proportionality with respect to the electric masses is evident in itself: "The magnetic fluid acts by attraction or repulsion according to the ratio compounded directly of the density of the fluid and inversely of the square of the distances between its molecules. The first part of this proposition does not need to be proved" [71].[304]

Notice the reference, in the statement (D.31) above quoted by Coulomb of his law, both to globes and molecules. All Coulomb's experiments in the first two memoirs concerned attraction or repulsion between (small) globes, or spheres; the conclusion

[302] p. 584.

[303] p. 611. Translation in [105].

[304] p. 593.

at the end is a law that is assumed to be valid also for small (elementary) electric masses.

Actually Coulomb statement appears to be contradictory to a scrutiny. Indeed if Coulomb's law were valid for molecules it could not be exactly valid for globes also. This because when two globes are put close to each other their electric molecules interact, producing a polarization. Simple calculations shows that in the case of globes Coulomb's law needs a corrective factor, less than unity, which is the lower the greater the ratio between radius of the globes and their distance [84].[305]

The validity of Coulomb's law for elementary particles of electricity and not only for two spheres transforms the problem from one-dimensional—force between two spheres in the direction of the two centers—into a multidimensional one; in fact, multiple forces can act on an elementary particle and in different directions from neighboring particles. The parallelogram rule for the composition of forces can be adopted; in this way it is implicitly assumed that the electricity value of the individual particles does not change due to the interaction with the others. This assumption is certainly acceptable when conceiving the phenomenon of electricity associated with a material fluid, whose electrical characteristics are invariable (law of conservation) and the electric forces deriving from this fluid depend only on its quantity.

Still in his second memoir [71][306] and in the fifth [74],[307] to be commented in a more extended way below, Coulomb measured the force on a charged globe (body) for a given distance from an electrified body C, then touched the globe (body) with an identical uncharged globe (body). By symmetry, he could take the two globes to divide their electricity equally. After separation, he could measure the force between one of the globes and the electrified body C and found it has halved. But Coulomb was not explicit on the point. If he did he would have suggested a way to give a measure to the electric masses, assuming an arbitrary unity of measure (that is a given sphere charged always with the same procedure).

Third, Fourth and Fifth Memoirs: Mathematics and Experiments

In his third memoir, Coulomb examined losses due to leakage of electric fluid, a phenomenon to which he gave great relevance. He used Calculus, starting from the assumption that electricity loss is proportional to the electric mass, in the elementary time dt. Which gives an exponential trend. Coulomb saw leakage as taking place by direct contact on a molecular level, through electricity-sharing either with adjacent air molecules or across the small *idio-électrique* (that is dielectric) supports holding the electrified body. The first effect was for him usually the most relevant. Experiments with electricity leakage and conceptions of material behavior led Coulomb to the theory that in nature there is probably no perfect *idio-électrique*; that is, all bodies have a limit above which they cannot resist the passage of electricity. Incongruences of Coulomb's theory of leakage can be read in [22].[308]

---

[305] p. 40.

[306] p. 586.

[307] pp. 443–444.

[308] p. 228.

The fourth memory has a more theoretical trend; its purposes are declared at the beginning:

> Here it is proved two main properties of the elastic fluid:
> The first: that this fluid does not spread over any body by a chemical affinity or by an elective attraction, but that it distributes itself, between different bodies put in contact, solely by its repulsive action.
> The second: that in conducting bodies, the fluid which has reached a *state of stability* [emphasis added][309] is spread over the surface of the body and does not penetrate into the interior. [73].[310] (D.32)

In the following only the second properties is discussed. Its proof is carried out both with empirical and theoretical approach. This property at Coulomb's time was an acquired results; however not a really convincing proof was presented.

Coulomb needed a more sensible torsion balance to perform his experiments and was able to obtained one with the sensitivity of 1/60000 of grain for a rotation of an angle of 90^0, that is more than forty times more sensitive than the balance used in the first memoir. The balance was used as an electrometer with the gilded paper attached to the needle suspended to the torsion wire electrified permanently. Then the body electrified to test, having the same kind of electricity as the gilded paper, is presented to the electrometer.

The electricity density was measured in the different points of the electrified body by means of a small gilded paper disk (known as Coulomb *proof plane*) with a diameter of 1 line 1/2 ($\sim$3 mm) and 1/18 of lignes ($\sim$0.1 mm) of thickness attached to the end of a shellac thread of 1 ligne (2.2 mm) of diameter (see Fig. 4.36, object n. 4). Coulomb assumed that the gilded paper put into contact with an electrified body picks up a density of electricity equal to that of the conductor where it is touched, because of its smallness [73].[311]

In the experiment described below, Coulomb employed a conductor made of a solid wooden cylinder 4 in. (11 cm) in diameter pierced with several shallow holes 4 lignes (1 cm) in diameter and in depth. After charging the wooden cylinder, Coulomb touched the proof-plane to points on the surface and then noted the resultant deflection of the electrometer by the proof-plane. If he applied the proof plane to the bottom of one of the holes in the wooden body and then presented it to the electrometer he found no signs of electricity. That is, presented to the outside of the body, the proof-plane indicated electricity there, but on the inside it indicated no electricity.

Coulomb will return in the sixth memoir to the experimental proof that electricity in a conductor is only lodged on the surface. Insulate a conducting body and electrified it, said Coulomb, and then form an envelope cut into two parts, which leaves a little play between it and the body. Whether this envelope has or not the same figure as the body, no matter the success of the experiment. With the body placed on an insulator,

[309] By making explicit that the distribution of electricity was in a stable state, Coulomb avoided the then almost impossible to solve problem of determining the laws of spatial distribution of electricity during the process of charging.

[310] p. 67. Translation in [105].

[311] p. 74.

and enclosed between the two parts of the envelope, he connected the envelope to the body by iron sticks. Removing the two envelopes, he found, by means of the electrometer, that all the electricity of body went to the envelopes and that the body, either had preserve nothing, or keep only an imperceptible part [74].[312] One recognizes here, albeit in a different form, the famous experience of Cavendish left in the drawer [22, 58].[313] Did Coulomb see it?

The experimental result about the distribution of electricity in a conductor is followed by an analytical proof, presented as a theorem:

> THEOREM. Whenever a fluid, enclosed in a body where it can move freely, acts by repulsion in all of its elementary parts [...] and the repulsive action of the fluid elements which produce its elasticity is greater than the inverse of the cube, as, for example, we have found for electricity, which is as the inverse square of the distances, then the action of the masses of electric fluid placed at a finite distance from one of the elements of this fluid being not infinitely small relative to the elementary action of the points in contact, all the fluid must move to the surface of the body and there must not remain any at all in its interior [73].[314] (D.33)

Coulomb proof, quite simple, consists in a reduction to the absurd. The hypothesis to be contrasted is: There is some electric fluid inside the body *and* all the electric particles are at rest. With reference to Fig. 4.39, let consider a small element of surface dae of an electrified body and a point a on it. From a trace the normal ab to the surface and through the point b pass a normal plane dbe. Then consider a small element dce symmetric to dae with respect to de. Now, said Coulomb, as "the law of continuity holds", it is necessary, as dae is made infinitely small, that the fluid density at point c either be equal to, or differ only infinitesimally from, the density at point a. Consider now the point b coinciding with the barycenter of $dcea$; the forces that the two symmetric parts dae and dce exert on it are equal and contrary. Because b is in equilibrium, and the sum of all the forces acting on it must be zero this means that the force of the fluid in the rest of the volume (AFBecd) on b is zero. But as the electric forces among the electric particles vary proportionally to the inverse square of the distance, this is not possible—unless the spatial density of the electricity everywhere inside and on the surface of the body is zero which is not the case by the first part of the hypothesis—a result that Coulomb had proved in a note of the second memoir [71].[315] So the forces of the rest of volume (AFBecd) acting on b are both zero and non zero, which is absurd. Thus admitting equilibrium there cannot be electric particle inside the body.

Coulomb proof is however partially ambiguous. Apart from the validity of the demonstration referred to in the second memoir, about whose validity I do not pronounce, regarding the value of forces of the rest of volume (AFBecd) on b, the

[312] pp. 620–621.

[313] pp.105–107; p. 232.

[314] pp. 75-76. Translation in [105].

[315] pp. 587–588. Coulomb proof is strictly connected to that reported by Cavendish in his paper of 1771 [55], pp. 586–587. This suggests that even though Coulomb did not cite him he knew Cavendish's writings.

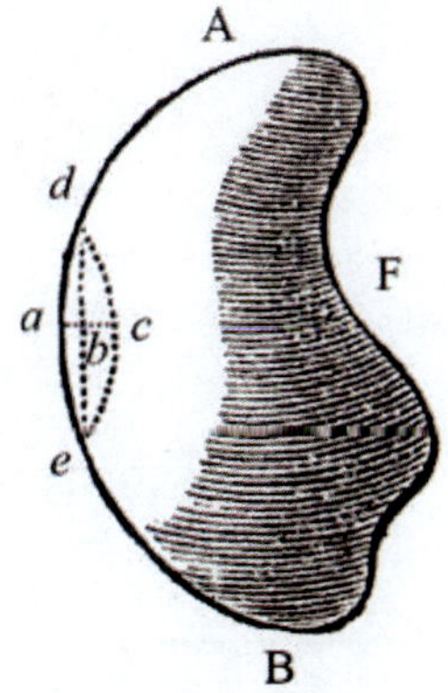

Fig. 4.39 Equilibrium inside a conductor. Redrawn from [73], p. 76

proof of Coulomb is valid only if the particles have all the same nature. But because Coulomb, as it will be clear in the sixth memoir, postulated the presence of particles with opposite electricity the proof does not concern a true body. Coulomb neglected the fact and assumed the absence of any form of electricity inside a body [49].

The fifth and sixth (1790) memoirs on electricity and magnetism were devoted to the experimental investigation of the distribution of electricity among conducting bodies of differing sizes and shapes, during and after contact. Following the measurement of electricity distribution, Coulomb attempted to develop analytical justification for his results using various approximate formulations.

New, much larger balances were constructed for this purpose, simpler to use and that allowed to measure electricity on bodies (spheres), relatively large.

Figure 4.40 shows one of the torsion balance used. Here a represents the body whose electricity is to be measured while b is the paper disk previously electrified with the same kind electricity of the body a. Thus the balance works as an electrometer.

The purpose of the fifth memoir was to determine in what ratios the mean value of electricity is shared between two bodies of similar shape but unequal size and also to determine the density of the electric fluid on their surface. Coulomb employed two experimental methods for this investigation. To measure the overall electricity ratios, or better the ratios of mean densities, he used the torsion balance similar to that of Fig. 4.40; to measure the density of electricity at each point on a body, he employed the proof plane introduced in his fourth memoir and his most sensitive balance still of the fourth memoir.

Coulomb conducted his first set of experiments on spheres with radii ranging from 1/12 (2.2 mm) to 12 in. (32 cm). He placed two spheres having different diameter in touch one with other, charged and then separated them. After separation he measured the (mean) value of electricity on each sphere. Finally, after many experiments, he compiled a table of ratios of mean densities of electricity on the separated spheres.

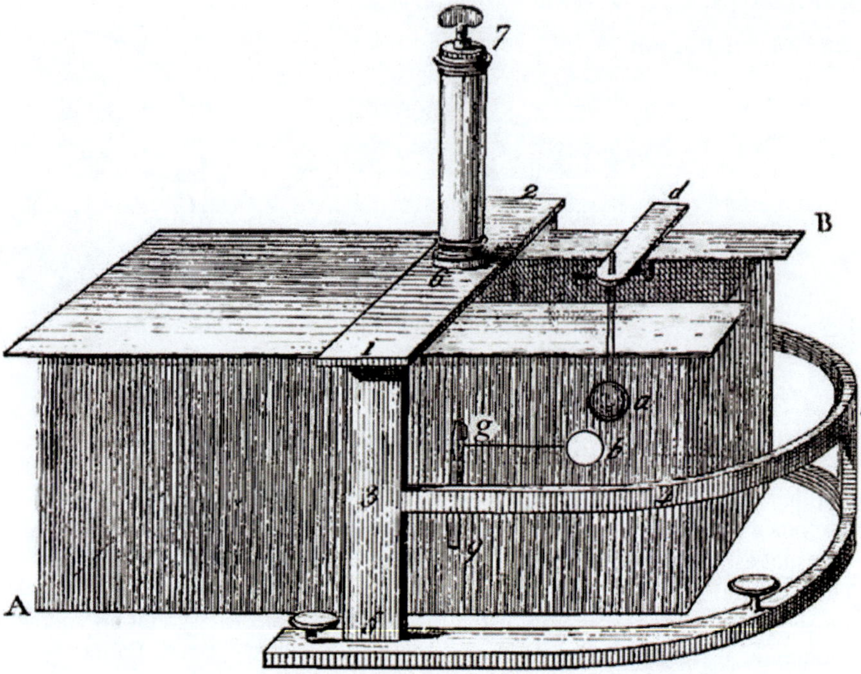

Fig. 4.40 Torsion balance for large electrified bodies. Redrawn from [74], Fig. 1 of table VIII

Coulomb defined the density of electricity by the equation (his symbols):

$$D = \frac{Q}{S}$$

with Q the quantity of electricity and S the surface of the sphere.

This is an explicit definition of a term, density, he had used ambiguously in the preceding memoirs. The definition is meaningful even though no value can be assigned to Q. Aepinus had made the same in his calculations with electric forces. The quantity of electricity was indeed a well defined concept in the fluid theories; in principle it is given simply by the volume of the electric fluid. The fact that it was difficult to measure was considered a minor problem, for many cases it was enough to evaluate the ratio of two quantities of electricity.

Coulomb could indeed evaluate the ratio of the densities of two spheres characterized by electricity and surface respectively Q, Q', S, S' without knowing the absolute values of these quantities, with the relation [74][316]:

$$\frac{D'}{D} = \frac{Q'}{Q}\frac{S}{S'}$$

[316] p. 430.

He proved that when a very small sphere touches a very large one, or better when the ratio of the radii is infinity—the values actually used by him were for the smaller sphere a radius 1/12 in. (2.2 mm) and 4 in. (11 cm) for the larger—its density δ after the contact is double than that of large body D, that is $\delta/D = 2$. In such a case because the amount of electricity left in the smaller body after the contact may be very small, the leakage can nullify reliable measurements so that much attention should be paid in performing the test [74].[317]

In a second series of experiments Coulomb examined the electricity distribution on bodies during contact. Using pairs of globes with diameters in the ratios 1 : 1, 1 : 2, 1 : 4, Coulomb placed them into contact, charged them and then measured the electricity density at various points on their surfaces by using his proof plane. The first experience of the series concerns two equal spheres with 8 in. in diameter, as shown in Fig. 4.41a. After being put into contact and electrified, the density of electricity is measured at different points. It is found in particular that in the vicinity of the contact zone, within 20°, there is no electricity. It grows gradually up to 90°. By measuring electricity at 180° there are values similar to those recorded at 90°.

For two unequal spheres, the bigger of radius R and the smaller of radius r, Fig. 4.41b, the density of electricity was found to increase on the smaller sphere as the ratio of the sphere radii was increased. In the limit case of an infinite ratio of R/r, the ratio between the density δ at a point 180° on the small sphere and the mean density D of the large sphere becomes four; that is $\delta/D = 4$ [74].[318] In this case the smaller sphere acted only as a minor perturbation of the field pattern on the larger one, while the larger sphere had a major effect on the distribution over the smaller.

Besides experimental results on the distribution of electricity on the surface of conductors, Coulomb presented a theoretical analysis also. This reveals his great ingenuity but even his limit, as will be discussed below.

Before presenting a model for his theoretical analysis, Coulomb illustrated a new proof that in conductors electricity is distributed on the surface only. And proposed one more experiment that had a twofold purpose. On the one hand it should confirm that electricity is logged on the surface, on the other hand it should suggest that the force between two electrified body is proportional to the amount of electric fluid. Place, said Coulomb, on a very dry day, a solid body (a) in the big torsion balance, and read the position of the needle, be it θ_0, If you touch this body, after having electrified it, by another body (b) that has the exact same figure, by bringing back one of the two bodies near the electrometer it is found that the needle rotate of about $1/2\theta_0$ as (b) has absorbed a half of the electricity of (a), which should have been located at the surface because the contact occurred in a very short time and only through the surface. As a further comment, Coulomb added, that the electric fluid would not have, as the theory indicates, an infinitely thin thickness; but as there is not a perfectly smooth surface and the air is not fully impenetrable to electricity, the electric fluid is diffused around the body in a layer of some thickness, thickness

[317] pp. 432–436.

[318] p. 457.

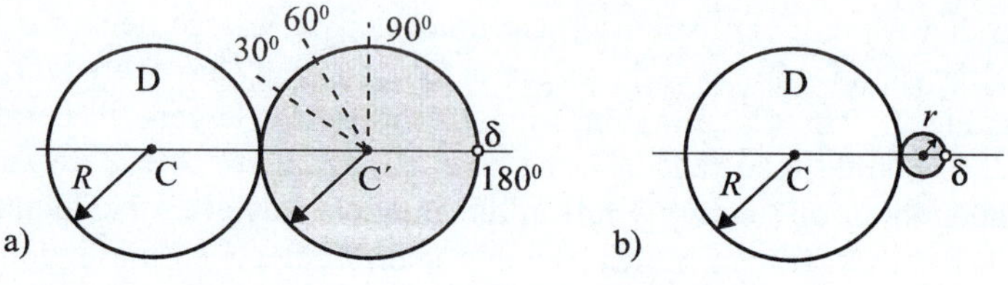

Fig. 4.41 Distribution of electricity on the surface of two sphere in contact

which varies with the density of the electric fluid and the state of the air, but which in general is so small, especially in the very dry days, to be possibly neglected [74].[319]

Coulomb started his theoretical analysis by studying the distribution of electricity in the three spheres shown in Fig. 4.42, which are put into touch and then electrified. Let the common electric density of the two bigger equal spheres be D, and δ that of the smaller sphere.

To make easy the analysis, Coulomb assumed that the density of electricity is uniformly distributed on the surface of the spheres, even though the experimental tests have clearly shown that this is not the case because of the mutual interaction of the spheres. As the purpose is however that to consider only the mean densities, the assumption can be considered not very strange. Notice that for his reasoning Coulomb made use of a plane representation of the three dimensional spheres (Fig. 4.43). This as a matter of fact is correct because the three spheres are symmetrical with respect to the axis $\mathrm{C}x\mathrm{C}'$ connecting their centers, and the density of electricity is the same on the circles cut on the spheres by planes orthogonal to $\mathrm{C}x\mathrm{C}'$.

Coulomb imposed the equilibrium of the particle of electricity located at point b, at the boundary of the sphere A and the sphere φ.[III] Imposing that the summation of all the forces action on b vanishes gives [74]:[320]

$$D\left(1-\frac{2R^2}{(R+2r)^2}\right)=\delta$$

Coulomb commented his result as follows: "Although this first formula is not based on a rigorous theory, but only slightly approximated, it is good to see how far it is from the truth, comparing it with experience" [74].[321]

A few line below Coulomb however acknowledged that the agreement with experience is not very good for $r \ll R$, when the value of δ furnished by the previous relation becomes negative. Indeed experiment showed at most negligible value of δ, but never negative. Coulomb attributed the error of his relation to the interaction of the electricities of the two larger spheres, not taken into account, so that near the small sphere their density of electricity is much smaller. He suggested a refined analysis, based always on a very crude model. The larger spheres are divided into two parts,

---

[319] p. 443–444.

[320] p. 445.

[321] p. 445.

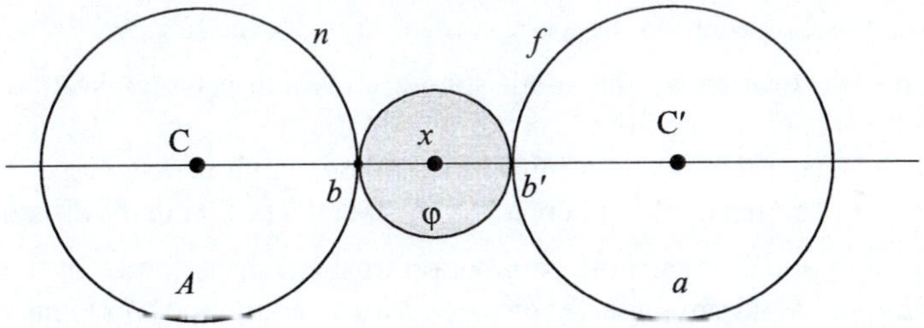

Fig. 4.42 Distribution of the mean density of electricity on three electrified spheres. Redrawn from [74], Fig. 6 of table IX

Fig. 4.43 Distribution of density of electricity on the surface of two electrified spheres. Redrawn from [74], Fig. 9 of table IX

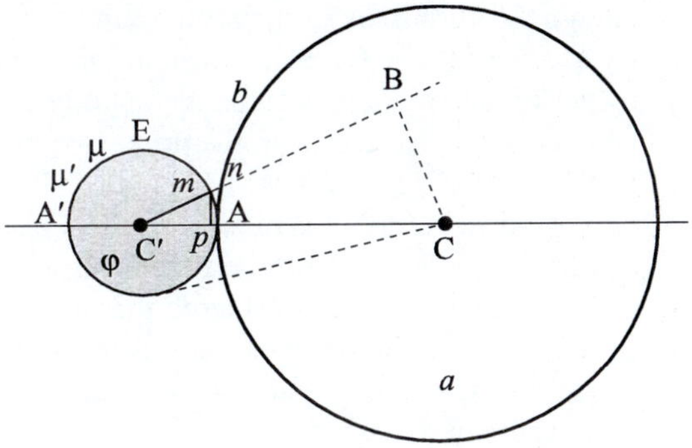

one of which, the one closest to the small sphere, has zero electrical density. This time the calculations shows that δ is always positive [74].[322]

A more intriguing analysis, whose ingenuity exceeds any imagination, Coulomb devoted to find the distribution of the electricity on the surface of two spheres put in to touch and then electrified. Coulomb reasoning is not strict and may be wrong, but it is clever and fascinating. His is not the reasoning of a professional mathematician, though his mathematics is updated and quite good; but his physical models though ingenious are not free from errors. This is typical of modern theoretical physicists and engineers.

Figure 4.43 shows the two spheres of which the distribution of electricity is looked for. To obtain his purpose Coulomb imposed the equilibrium at the point m located on the surface of the smaller sphere. As in the previous analysis it is assumed that the electricity is uniformly diffused on the surface of the two spheres with density respectively D (the larger sphere a) and D' (the smaller φ). A further infinitesimal sphere is imagined located around m having density δ and radius dr. To justify the introduction of this infinitesimal sphere Coulomb said that he surface of conductors can be imagined as covered by infinitesimal spherical electric masses.

[322]pp. 446–447.

According to Coulomb on the electric mass at m act the forces:

1. The force $f_{A'}$ exerted by the small sphere φ, which acts as mass point whose electricity is concentrated in the center C and thus at distance r from m.
2. The force f_m exerted by the infinitesimal sphere acting at distance dr.
3. The force f_A exerted by the sphere a, acting at distance Cm in the direction Cm.[IV]

By imposing equal to zero the summation of these three forces in the direction C′mB, Coulomb could obtain the expression of δ as a function of D and D'.

Some criticisms on Coulomb approach are needed, in increasing order of criticality:

1. An uniform distribution of electricity on the two spheres is assumed to be even though the purpose is to study unevenness.
2. By considering the force due to an infinitesimal electrified sphere, Coulomb as a matter of fact admitted a force that a body exerted on itself as an external force; which is not consistent with the principles of statics.
3. If the equilibrium is imposed in another direction, for instance in the direction orthogonal to C′mB a different result is obtained for δ; or in other words no equilibrium is possible for the particle m with the electrical forces alone. Indeed other forces should be considered that prevent the particle m to leave the small sphere φ. Today these forces are looked for in the atoms, at Coulomb day, and by himself, a role could be attributed to the pressure of air; but here Coulomb did not mention it.

In the case of two equal spheres the analytical results show that for an angle $f < 23°$, δ is negative; experience gives instead a zero or a very weak positive value. For points located at 90° or 180° the agreement with the theory is found quite satisfactory. This confirms as a wrong model can sometimes furnish satisfactory results.

Coulomb analysis was repeated in a much more refined way and with the use of different analytical tools, by Poisson in his memoirs on electricity published in the Mémoires de la Classe des Sciences Mathématiques et Physiques de l'Institute Impériale de France for the year 1811 [169, 170].[323] In the first of them, *Mémoire sur la distribution de l'électricité à la surface des corps conducteurs*, by substantially confirming Coulomb's experiments [169].[324] Poisson's memoirs opened a research topic not yet closed in the field of mathematical physics .

In the sixth memoir Coulomb continued the study of the surface distribution of electricity and extended the investigation to groups of conducting spheres, cylinders, planes and variations of these groupings. He found for instance that the density of electricity on large numbers (24) of small spheres (2 in., 5 cm) placed in contact along a line with a large sphere (8 in., 22 cm), as shown in Fig. 4.44 (Fig. 3) decreases with the increase of the distance from the large sphere, following an exponential decay, as shown in Fig. 4.44 (Fig. 4).

---

[323] The year 1811 is the date attributed to the publication of the Institute; in [126], the true dates of presentation are referred to.

[324] pp. 60, 66, 80.

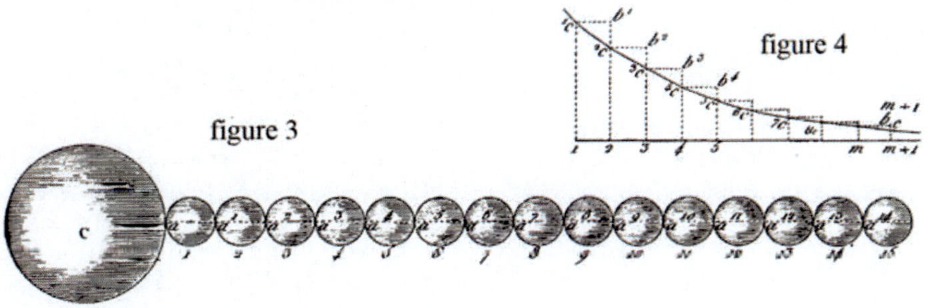

Fig. 4.44 Distribution of density on a row of small spheres. Redrawn from [75], Figs. 3 and 4 of table XXX

Experiments were flanked with theoretical analysis giving satisfactory results. In the following however I only will discuss the proposition:

> When, for example, a globe of eight inches of diameter is touched with a small plane [...], it takes at each of these surfaces an electric density equal to that of the surface of the globe. That is, this small plane [...] is charged with a quantity of electricity twice that of the portion of surface of the globe which it touched [75].[325] (D.34)

This proposition, according to Coulomb, is corroborated both by theory and experiments. An experimental result is the following. Consider an isolated 8 in. (22 cm) diameter sphere of a conducting material placed on an insulating support whose amount of electricity is such to rotate the needle of electrometer by 144°. Put into contact the sphere with an insulated conducting disk 16 in. in diameter (the thickness is not specified). As a result, the sphere loses a certain amount of electricity and when it returns to the electrometer it produces a rotation of the needle of 47°; this means that the disk absorbed from the sphere a quantity of electricity measured by $144 - 47 = 97$, more or less double than that remained in the sphere. Keeping in mind that the surface of the sphere $S_S = 4\pi R_S^2$ ($R_S = 8$) is one half of the total surface (two faces) of the disk $S_D = 2\pi R_D^2$ ($R_D = 16$), it happens that the electricity is distributed between the sphere and the plane in proportion to their total surfaces, or that the electricity density is the same in both conductors. It must be said that though Coulomb declared he found similar results for disks of different sizes he did not report results in details. He stated however that the proposition holds true for a globe 8 in. in diameter and a small plane insulated of 6 lines in diameter.

Anyway Coulomb concluded that, if the sphere is much larger than the disk, as is the case with the proof plane, the electricity lost by the sphere is negligible and therefore the density assumed by the proof plane is the same as the initial density of the sphere. In other words, the amount of electricity absorbed by the test plane is double that contained in a portion of an area sphere equal to that of the test plane, because the proof plane has two faces [75].[326]

---

[325] p. 675.

[326] pp. 674–675.

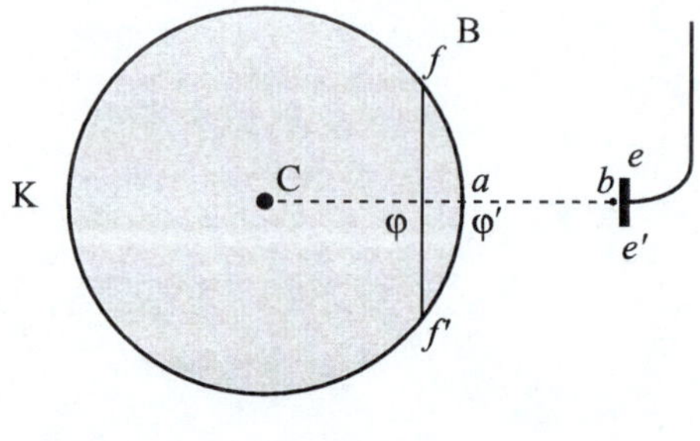

Fig. 4.45 Density of electricity taken by a small disks. Redrawn from [75], Fig. 8 of table XXXI

The theoretical proof is carried out with reference to Fig. 4.45.

A small plane ee' is located at a small distance ab from a sphere of center C and radius R, such, said Coulomb, that the layer of air cannot impede to the electric fluid of the sphere to pass to the small disk. The problem is to find the density of electricity $\underline{\delta}$ on ee', measured as usual with respect to the area of one face only, given the density of electricity D of the sphere. The result is obtained by imposing the equilibrium of the point b, located at an infinitesimal distance from the plane ee'. The forces to be considered are that of the sphere and that of the plane. The first force is given by the expression $f_1 \propto 2DR^2/(R+ab)^2$,[327] which because ab is very small reduces to $2D$. For the force of the plane Coulomb furnished the expression:

$$f_2 \propto \underline{\delta}\left(1 - \frac{a}{(R'R' + aa)^{\frac{1}{2}}}\right)$$

where R' is the radius of the disk ee' and a the distance between b and the disk ee'. Coulomb assumed this expression as known. Indeed he had already presented it in his fifth memoir, even though without any proof [68][328]; however the result is correct, or better in accord with modern electrostatics. Because a is infinitesimal and thus negligible, it is $f_2 \simeq \underline{\delta}$. For the equilibrium between f_1 and f_2 its is: $\underline{\delta} = 2D$, which sometimes is referred to as Coulomb theorem, "in perfect accord to the experimental results" [75].[329]

The proof, even though ingenious, also in this case leaves something to be desired in terms of rigor. For instance Coulomb considered the possibility that the electric fluid could overcome a small gap of air (with a spark?); but at the same time he assumed that the situation was equivalent to that of a direct contact between the small disk and the sphere. So even though the calculations are correct, some problems can be raised for the physical modeling.

The result presented by Coulomb, corroborated, according to him, by both analytical and experimental proofs, creates bewilderment in a modern reader, both because apparently strange and in an evident contradiction to what Coulomb had affirmed in the fourth memoir, were the proof plane is introduced for the first time: "I made

[327] The origin of factor 2 in the expression of f_1 is commented upon in the endnote III.

[328] p. 447.

[329] pp. 675–676.

the proof plane (le petit plane) of gilded paper touch the surface of the cylinder, as this plane has only 1/18 of line of thickness; it becomes a part of the surface of this cylinder and consequently assumes a quantity of electric fluid equal to that contained by a part of the surface equal to that of the proof plane" [73].[330] Coulomb did not comment the contradiction but Maxwell did; he after noticing that, "some objections have been raised to Coulomb's use of the proof plane" [64],[331] devoted a full section of his *A treatise on electricity and magnetism* of 1783 [64] to the study of the value of the density assumed by the proof plane when put into contact with an electrified body. He used tools of mathematical physics, as made by Poisson in his memoirs of 1811, of scalar nature as the electric potential which allowed a much more refined analysis, and recognized that the density assumed by the proof plane depends on its shape. In the case of very very tiny circular proof plane the density—evaluated with one face only—is equal to that of the point of the surface of the body to be measured. For a circular proof plane with a ratio of about 10 between the radius and the thickness, like that used by Coulomb, the density is more than four times that of the electrified body—thus twice of the value suggested by Coulomb [64].[332] It must be said however that the result about density distribution presented by Coulomb are ratios of density; so what counts is that the density of electricity absorbed by the proof plane is proportional to the density of the charged body to be tested; and this is surely valid using always the same proof plane.

In the following pages, art. 45, Coulomb extended his result to the case of a conductor having a whatsoever shape [75][333]:

> The result just found, by experience and theory, for a small plane put in contact with a globe, is general for all bodies ending in a whichever convex surface. Actually, whatever the shape of the body is, experience learns that a small plane put into contact with these surfaces, always takes, at the moment it is removed from contact, a quantity of electricity twice that of the affected area portion. Experience gives the same double ratio again, by presenting a very small plane to a large electrified plane [75].[334] (D.35)

Nature of Electricity

Coulomb was stingy of theories about electricity and magnetism. In his papers published on the memoirs de l'Académie des sciences de Paris, he dealt the subject in three occasions only. At the end of his sixth memoir where he confessed his preference for the two fluids theory, even though its validity cannot be decided on experimental basis, in the introduction of the memory of 1777 on magnetism, where he criticized the mechanicistic Neocartesian explanations and in the seventh memoir on electricity and magnetism, entitled *Du magnétisme*, of 1789, where he suggested

---

[330] p. 74.

[331] Vol. 1, p. 277.

[332] Vol. 1, p. 281. The expression for the ratio between the density of the proof plane and that of the electrified body is given by $1 + 8\frac{z}{r}\ln\frac{8\pi r}{z}$, with z the thickness and r the radius.

[333] pp. 677–678.

[334] p. 676.

some changes in Aepinus' s theory [77].[335] Below a summary of Coulomb ideas regarding the nature of electricity only.

Whatever may be the cause of the electricity, said Coulomb, all the phenomena are explained, in the sense that calculations are in accordance with experiments, supposing two electric fluids (the parts of the same kind of fluid repelling each other in inverse proportion of the square of the distances and the parts of the different kind of fluid attracting in the same inverse proportion of the square of the distances). According to this supposition, the two fluids in conducting bodies, always tend to move until that there is equilibrium, that is to say, until the attractive and repulsive forces are mutually compensating each other. This is the situation in which all bodies are found in their natural state. But if, by any operation, a given quantity of one of the two electric fluids is made to pass into an insulated conducting body, it will electrify, that is to say, it will repel electrical parts of the same nature and will attract electrical parts of a different nature because of the superabundant fluid with which it is charged. If the electrified conducting body is brought into contact with another insulated conducting body, it will share with it the electric fluid which is superabundant; but if it is communicated to a non-isolated body, it will in a moment lose all its electricity, since it will divide it, with the globe of the earth, whose dimensions relative to it are infinite.

Coulomb referred to Aepinus, who had supposed that there was only one electric fluid, the particles of which repel each other and were attracted by the molecules of the body with the same force as they are repelled. But to explain the state of the bodies in their natural situation, as well as the repulsion and attraction, it is necessary to suppose that the molecules of the bodies repel each other with the same force as they attract the electric particles. It is easy to feel that the supposition of Aepinus gives, by means of calculation, the same results as that of the two fluids.

There are for Coulomb two reasons in favor of the two fluid theory. The first has a metaphysical nature: "however that of the two fluids which has already been proposed by several physicists, because it seems to me contradictory to admit at the same time inside bodies, an attractive force in inverse ratio to the square of distances as gravity, and a repulsive force in the same inverse ratio of the square of distances; a force that would necessarily be infinitely large, relative to the attractive action from which gravity results" [75].[336] The second was drawn from recent experiments. The supposition of the two fluids is indeed in conformity with all the modern discoveries of chemists and physicists, who have made us acquainted with different gases, the mixture of which in certain proportions suddenly and entirely destroys their electricity; an effect which cannot take place, for Coulomb, without something equivalent to a repulsion between the parts of the same gas, which constitutes their elastic state, and to an attraction between the parts of the different gases, which makes them lose their elasticity all at once. "As these two explanations [one or two electric fluids] have only a greater or lesser degree of probability, I warn, in order to put the theory that follows, away from any systematic dispute, that in the supposition of the two

[335]pp. 488–491.

[336]p. 672.

electric fluids, I do not have other intention than to present with the least possible element the results of calculation and experience, and not to indicate the true causes of electricity" [75].[337]

4.4 Quotations

D.1 Tous les systemes modernes sur la Lumiere peuvent être reduits à deux. Car, ou les mouvemens du corps lumineux sont transmis jusqu'à l'oeil, seulement parce qu'ils se communiquent à la matiere qui est entre le corps lumineux & nous, de même que les fremissemens d'un corps sonore ne parviennent jusqu'au tympan de l'oreille, que parce qu'ils ont excité dans l'air un semblable mouvement; ou l'agitation du corps lumineux produit en lui une émission, & un écoulement de corpuscules, qui viennent frapper l'organe de vûë de la même maniere à peu prés, que les parties invisibles qui se détachent d'une fleur, viennent frapper nôtre odorat. Il n'y a pas de milieu dans cette alternative, ou le corps lumineux renvoye vers nous des particules de sa substance, ou il n'en renvoye pas; il faut necessairement que la Lumiere se repande de l'une de ces deux manieres, ou avoir recours aux qualitez occulte.

D.2 Je crois donc que la matière lumineuse consiste en un *soulfre*, très subtil & très agité, & que ce n'est autre chose que le *principe actif* des Chimistes, ainsi nommé, parce qu'il agit seul, & qu'il fait agir les autres. Car des cinq principes de Chimie, il n'y a que le *soulfre*, qui ait cette proprieté: parmi les autres quatre, on en, compte un purement *passif*, qui est la *Terre*, laquelle ne fait que servir de receptacle ou de guaîne aux autres, & trois *moyens*, qui sont le *Sel*, l'*Eau* & le *Mercure*, lesquels deviennent capables d'agir lors qu'ils sont joints au *soulfre*.

D.3 Que chaque couleur des corps ne consiste dans la figure & dans l'arrangement particulier des parties qui le composent, qu'entant qu'ils sont par là plus propres à rompre & à absorber dans leurs pores la Lumiere d'une certaine couleur, & à réfléchir celle d'une autre couleur. Ainsi le Carmin, par exemple, est fort rouge, parce qu'il ne réfléchit que la Lumiere rouge, & que toutes les autres especes de Lumiere se rompent & se perdent dans ses pores sans se réfléchir.

D.4 Coroll. 39. Donc les différentes vîtesses, les différentes degrés de réfrangibilité, & les différentes couleurs de la lumière, ne sont en elle, & hors de nous, qu'une seule & même propriété, ou n'expriment que la gradation des effets dés à une même cause.

D.5 Mais si le corps *M* est tel, que la matière subtile réfléchie ait ses vibrations moins promptes dans certains degrez; que je ne crois pas qu'on puisse determiner exactement; on aura quelqu'une des couleurs qu'on appelle *primitives*; le jaune, le rouge, le bleu, si toutes les parties du corps *M* diminuent également

[337] p. 673.

les vibrations que cause Ia flamme dans Ia matière subtile. Et l'on verra toutes les autres couleurs qui se font par le mélange des primitives, selon que les parties du corps *M* diminueront inégalement Ia promptitude des vibrations de la lumiere.

D.6 Mais si le corps *M* est tel que la matière subtile réfléchie excite dans l'oeil des vibrations plus ou moins promptes dans certains degrés, que je ne crois pas qu'on puisse determiner exactement, on aura quelqu'une des couleurs simples homogènes ou primitives, comme le rouge, le jaune, le bleu, etc., et l'on aura les antres couleurs composées, et même la blancheur qui est la plus composée de toutes, selon les divers melanges des rayons dont les vibrations auront diverses promptitudes. Je dis que la blancheur est la plus composée de toutes parce qu'elle est composée de l'assemblage des vibrations différentes en promptitude, que produit dans la matière subtile chaque partie differente de la flamme. Comme tout est plein et infiniment comprimé, chaque rayon conserve dans toute sa longueur la meme promptitude de vibration qu'a la petite partie de la flamme qui le produit. Et parce que les parties de la flamme ont un mouvement varié, les rayons des couleurs ont nécessairement des vibrations et font des refractions différentes. Mais il faudrait voir sur cela les experiences qu'on trouvera dans l'excellent ouvrage de M. Newton.

D.7 Je me figure présentement, que tout cet amas de petits tourbillons qui remplit les vastes espaces du Monde, est parsemé de corpuscules très subtils, durs ou solides, laissant entre eux des intervalles, si vous voulés, mille fois plus longs que le diametre d'un de ces corpuscules, je n'en determine pas Ia longueur, il suffit que je conçoive très-clairement que chaque ligne droite tirée d'un point à l'autre, enfilera une infinite de ces petits corpuscules, dont je puis supposer les intervalles à peu-près égaux, puisque ses corpuscules font uniformément disperses parmi les petits tourbilons, quoique les corpuscules eux-mêmes puissent être de differente grandeur.

D.8 Un corps mis dans un centre d'équilibre forcé, s'il en est déplacé par quelque cause que ce soit, jusqu'à un petit intervalle dans la direction des deux ressortes ou forces motrices opposées, il retournera sur ses pas, & fera des vibrations en temps égaux en forme d'oscillations tautochrones.

D.9 Deinde etiam vidimus pulsum, postquam semel est formatus, in directum promoveri, siquidem medium suerit uniforme; unde simul rectitudo radiorum lucis intelligitur. Ipse autem pulsum promotio oritur ab agitatione particularum medii elastici, ubi pulsus versatur, quae cum ubique secundum determinatam directionem vergat, pulsui secundum eandem plagam motum inducit.

D.10 Cela s'entend des couleurs pures & hautes, telles que l'arc en Ciel & le Prisme nous les présentent. Les autres couleurs mêlées ou basses ne different entr'elles que comme les tons de diverses octaves. Ainsi au cas qu'un rayon rouge fasse 10000 vibrations dans une seconde, des rayons qui font 5000, ou 2500, ou 1250, ou 625 vibrations dans le meme tems, produiront aussi une couleur rouge, mais moins haute que Ia premiere. Par consequent il y aura plusieurs couleurs differences de chaque nom, comme on a dans un Clavecin plusieurs tons qu'on exprime par la même lettre.

D.11 Nunc quo pacto conversa ratione idem eveniat, explicemus. Immota iuxta alteram mensae extremitatem tabella, post eam charta constituatur quae utriusque lucernae radios excipiat: tum lucernae ipsae transferantur, dispari tamen a charta intervallo, sic ut geminae lucernae intercapedo sit dupla eius qua simplex distat lucerna: palam igitur chartam intuenti geminae lucernae lumen clarius apparere quam simplicis; utque paria fiant ea quae in charta spectantur lumina, oportere geminam lucernam longius abduci, quia per duplex spatium minus quam duplex factum fuit luminis decrementum.

D.12 Il n'y a en effet qu'à prendre un flambeau, & l'éloigner ou l'approcher jusqu'à ce qu'il éclaire successivement de la même maniere que les deux lumières que l'on veut comparer; & si on prend les quarrez des deux differens éloignemens ces quarrez mis dans un ordre renversé, exprimeront le rapport des lumieres.

D.13 Utique inter hypotheses quas mathematicas vocabo, & eas, quae physicae sunt, maxima adest differentia. Physicae ita plerumque assumuntur, ut qua in re a vero aberrent haud constet, unde fit ut suo quaeque ordine iterum reiiciantur, prout earum a vero aberratio successu tantum temporis detegitur. In mathematicis fere semper non modo constat, quanam in parte a vero recedant, verum & plurimis casibus in antecessum definire licet aberrationis momentum.

D.14 Commune id esse videtur scientiae humanae fatum, ut ea sint ab intellectu remotiora, quae vel maxime obversantur sensibus. Huius certe nobis effati praeclare sistit exemplum luminis theoria. Plures enim eaeque gravissimae in pervestiganda ipsius vi atque natura occurrunt difficultates vix ac ne vix superandae, ut mirum sit, in eadem re, quae fons est ipse claritatis, tanta adhuc cognitionem nostram circumfusam esse caligine, tantasque in ipsa luce remanere tenebras.

D.15 In omnibus mundi corporibus duae propositae sunt causae, sive principia ex quibus ipsa corpora producta sunt, materia & forma; electrica motione a materia magnetica vero a forma praecipua invalescunt, longeque inter se differunt, dissimilesq; evadunt; cum altera nobilitata plurimus virtutibus sit, & praepotens; altera obscura, & minoris potentiae, & carceribus quasi quibusdam plerunq; conclusa

D.16 Igitur ex frictione non foedante, effluvium non immutatum ab ardore, sed quod suum est, vnitionem facit cohaerentiam, apprehensionem, & ad fontem confluentiam, [...] Effluvia vires extendunt, quae propria sunt, & peculiaria, & sua, diversa a communi aere, ab humore genita, motu calorifico ab attritu & attenuatione excitata, tanquae materiales radii quae rerinet & attollunt paleas, festucas, & ramenta, donec extinguuntur, aut evanescunt; quae tum rursus soluta (corpuscula) a terra ipsa allecta, ad terram delabuntur.

D.17 Nimirum vorticem quendam agitatae materiae invisibilis circa floccum adhaerescere, qui vortex ortus fit et transditus a vortice circa sphaerulam excitato postquam attritu panni incaluerit. Hic flocculi vortex impedit ac prohibet ipsum admotae sphaerulae adjungi, quia et ipsa suum habet verticem. At semel admoto ad floccum digito aliave re, disturbatur ejus vortex atque abigitur, unde tunc non difficulter ad sphaerulam accedit.

D.18 Lubet tecum communicare Experimentum aliquod novum, sed terribile, quod velim ne ipse capias, nec oblato licet mihi toto Galliae Regno, repetarem impetum expertus, et Dei propitio favore servatus. Occupabar in detegendis Electricitatis viribus: Tubus ferreus AB ex filis servicis coeruleis suspendebatur, Globus vitreus celeriter circumactus fricatusque, ponebatur prope alteram extremitatum A, et suam vim electricam cum tubo hoc ferreo AB communicabat: Prope alteram extremitatem B positum erat filum Oeneum CD, libere pendulum; manu dextra F capiebam globum vitreum D aliquo usque impletum aqua in quam propondebat filum; manu sinistra E conabarelicere Scintillas crepitantes quae ex tubo ferreo avolant indiguum; mox manua dextra F afficitur tanta vi ut contremiscat totum corpus, non aliter quam si quio a fulmine percuritur.

D.19 Sique unquam probetur attractionem quandam aut repulsionem, a pressione externa aut impulsu absolute non pendere, tum eo reductos nos iudico, ut adstruere cogamur motus eiusmodi, non a viribus corporeis, sed a spiritibus sine entibus, eorum quae agunt intelligentibus, peragi atque produci, quod quidem, quod in mundo, locum habeat, ut credam, induci non possum. Tota itaque mea sententia eo redit, ut attractiones atque repulsiones de quibus locutus sum pro phaenomenis habeam, quorum hactenus causae latent, a quibus vero alia ph[a]enomena pendent ac derivantur. Existimo autem, non contemnendum facere in analysi operationum naturae progressum qui phaenomena complicatiora, ad causas sua proximas viresque primitivas reducit, quamvis causae causarum istarum nondum sint detectae. Hanc, quam proposui, sententiae meae declarationem, et rigidissimo censori satisfacturam, confido.

D.20 Displicebit autem forte quibusdam, quod ad disquisitiones mathematicas aliquantum prolixiores delapsus hic sim, sunt enim, qui aegre talia in physicis tolerant. Ast ego quidem aliter iudico ac brevem hanc dissertationem, experimenti *Richmanniani* phaenomena ad causas suas revocantem, ad probandam veritatem principiorum theoriae elettricae a me evolutorum, plurimum valere existimo. Magna enim hypothesi physicae accedere verosimilitudo credendas est, si deprehendatur phaenomenis complicationibus, iis nempe, quae non nisi per longiorem ratiociniorum seriem ex ipsa deduci possunt, apte cohaerere.

D.21 Sub initium, cum cogitare de electricitatis theoria Frankliniana inceperam, me ad ipsam quodammodo exhorruisse. Postquam autem considerare cepi, nihil in ipsa contineri, quod analogiae operationum naturae contrarium esset, facile me ad ipsam assuefeci.

D.22 La elettricità d'un corpo A attua sì fattamente l'aria ambiente, che per mezzo di essa mira ad indurre la elettricità contraria nel corpo B immerso in essa. Ed è l'aria così attuata, che costituisce ció, che comunemente si chiama atmosfera elettrica.

D.23 Presento un pelo di lino al conduttore Y (Tav. I. Figure 1.); alla distanza di, uno, o più piedi esso si dirige normalmente alla faccia del conduttore; gli presento a lato un bastone di ceralacca stropicciata, se ne discosta; gli presento una canna di vetro stropicciata, vola ad essa; cioè il pelo immerso nell' atmosfera del corpo elettrizzato per eccesso diviene elettrico per difetto. Replico lo sperimento rispetto a' travicelli B, od E della macchina pneumatica; e il pelo fugge dal vetro, vola alia ceralacca; onde il corpo immerso nell' atmosfera del corpo elettrizzato per difetto diviene elettrizzato per eccesso.

D.24 Se un corpo E sia elettrizzato per eccesso primamente un complesso di lineette, che sporgano contro dell' aria ambiente da tutti i punti della faccia di esso corpo E (vedremo nell' articolo seguente, che appunto l'eccesso, o difetto di fuoco non s' induce che nella superficie de' corpi) potranno segnare e il fuoco eccessivo di esso corpo, e l'eccesso di tensione, che eserciterà il fuoco stesso contro il fuoco naturale dell'aria ambiente; in secondo luogo poi si potrà intendere divisa in successivi picciolissimi strati l' aria ambiente, e *simili lineette* che si applichino alle convessità di tutti essi strati, potranno segnare Ia direzione, e il modo, con che la eccessiva tensione del fuoco ridondante sulla faccia del corpo propaga una tensione, o vibrazione corrispondente nel fuoco inerente nell' ambiente aria.

D.25 In somma io tanto di valore attribuisco a quest'unica esperienza, che per essa sola penso sciogliersi in gran parte la più antica quistione dell'Elettricità, e che sino ad ora è stata riputata la più difficile. Imperocché se i movimenti elettrici scemano nell'aria diratada e scemano a proporzione di esso diradamento, ne segue che tali movimenti dipendano da alcun'azione del vapor elettrico sull'aria.

D.26 Questo è il genio della filosofia pigra, e lusinghiera: fingere quantunque fluidi, che abbiano in se i movimenti, che non s'intendono ne' corpi. I corpi diversamente elettrici si attraggono: perché? Perché sono animati da' due fluidi diversi, che ti attraggono. Gli elettrici similmente si respingono: perché? Perché sono animati dall'uno, o dall'altro de' due fluidi, le parti di ciascuno de' quali si respingono similissimamente. Ma che quello non sta il genio della natura, anche in questo caso si scorge e dalla esatta unità di tutti gli effetti, che dovrebbero essi produrre inconciliabile con la natura loro diversa, e dalla impossibilità della segregazione loro, donde per altro dipende ogni loro effetto, e dalla maniera, con che si riunirebbero, che discorderebbe affatto da' fenomeni.

D.27 Un corpo elettrico ha la forza di mutare la naturale dose di fuoco ne' corpi immersi nella sua atmosfera e introdurre in essa contraria elettricità.

D.28 Quantunque l'illustre Autore parli con molta rapidità sui risultati di queste sue sperienze, mi sembra ció non ostante, che indi ne derivino assai naturalmente, e portino cosi nell'elettrica Teoria maggior luce, ed espressione, che non gl'interi volumi. Io paragono simili sperienze a que' tratti da maestro nel disegno, ove si presenta lo spaccato d' un grande edifizio preso in tale profilo, che ne spiega meglio in un colpo d'occhio e le divisioni, e l' insieme, che non si comprenderebbe con girarlo più volte a parte a parte .

D.29 Presero voga le Frankljniane idee, e diventó Franklin Capo scuola, e duce di sistematica Setta, e tanto bastó perchè tutte le sue congetture, e per fino i suoi più azzardati sospetti venissero dai Franklinjanj Settarj trasformati in dogmi, e proposti, e difesi con linguaggio di fanatica persuasione [...] In mezzo peró a tante definizioni, e tanti stabilimenti. non acquistó l'elettrica scienza, che ampie parafrasi de' primi fatti indicati da Franklin, pronunciate

con altrettanta confidenza, e persuasione, quanta fu la perplessità, e l'incertezza del loro Autore.
[...]
Non è punto necessario, nè opportuno di associare all' errore i nomi degli Autori, che hanno contribuito a propagarlo; massimamente quando i loro nomi non influiscono nella realtà delle cose; e tanto meglio quando, come ebbero essi l'ingenuità di confessare di aver errato nel definir la luce di partenza così è sperabile, che con più matura discussione de' fatti potessero egualmente riconoscere il loro abbaglio.

D.30 La force répulsive de deux petits globes électrisés de la même nature d'électricité, est en raison inverse du carré de la distance du centre des deux globes.

D.31 Que l'action, soit répulsive, soit attractive de deux globes électrisés, & par conséquent de deux molécules électriques, est en raison composée des densités du fluide électrique des deux molécules électrisées, & inverse du carré des distances.

D.32 Où l'on démontre deux principales propriétés du Fluide électrique:
La première, que ce fluide ne se repande dans aucun corps par une affinité chimique ou par une attraction élective, mais qu'il se partage entre differens corps mis en contact uniquement par son action répulsive;
La seconde, que dans les corps conducteurs le fluide parvenu à l'état de stabilité, est répandu sur la surface du corps, & ne pénètre pas dans l'intérieur.

D.33 Toutes les fois qu'au fluide renfermé dans un corps ou il peut se mouvoir librement, agit par répulsion dans toutes ses parties élémentaires [...] est plus grande' que l'inverse du: cube, telle, par exemple, que nous l'avons trouvé pour l'électricité en raison inverse du carre des distances pour lors, l'action des masses du fluide électrique placée à une distance finie d'un des élémens de ce fluide, n'étant pas infiniment petite relativement à l'action élémentaire des points en contact, tout le fluide doit se porter à la surface du corps, & il ne doit point en rester dans son intérieur.

D.34 Lors'que l'on touche, par exemple, le globe de 8 pouces de diamètre, avec un petit plan [...] il prend à chacune de ces surfaces une densité électrique égale à celle delà surface du globe, c'est-à-dire, que ce petit plan [...] se charge d'une quantité d'électricité double de celle de la portion de surface du globe qu'il a touchée.

D.35 Le résultat que nous venons de trouver par l'expérience &; par la théorie, pour un petit plan mis en contact avec un globe, est général pour tous les corps terminés par une surface courbe, convexe d'une figure quelconque. Quelle que soit en effect la figure du corps, l'expérience apprend qu'un petit plan mis en contact avec ces surfaces, prend toujours, au moment qu'on le retire du contact, une quantité d'électricité double de celle de la portion de surface touchée. L'expérience donne encore ce même rapport double, en faisant toucher un plan très petit à un grand plan électrifié.

Notes

[I]For a modern reader the linearity of forces is simple enough to accept, since elastic forces for small oscillations can always be approximated by a linear function.

Bernoulli has less confidence in the theory of functions and used a fairly elaborate reasoning. With reference to Fig. 4.3, consider a particle subject to two springs arranged on one side and on the other on a straight line to which the particle belongs. Curves AF and CA represent the forces exerted on the particle according to its position on the line MN. When the particle is in P the two forces are in equilibrium, being equal and opposite with a value PB. If the particle passes in G the two forces will be GL and GE respectively and there will be an unbalanced force equal to EL which will tend to return the particle to P, causing oscillations. Bernoulli says that when the oscillation excursion is small the surface BEL can be approximated with a triangle; this implies the linear link between the restoring force El and the displacement PG [35], p. 18.

[II]Here a modern explanation of the phenomenon described by Beccaria; for the sake of simplicity in the explanation the role of polarization of the glass plate is only hinted on. Let the original free charge on the surface A of the glass plate be Q, that on the coating CD, q; assume Q and q both positive. FO represents $(Q + q)$. Remove CD: the effective charge near A drops by q, or Fu, and CD's potential increases enough to leak a fraction of its charge, yq say, into the air. Replace CD: the net charge at A becomes PG $= Q + (1 - y)q$. Remove CD once more, decreasing the effective charge near A by XG $= q(1 - y)$; the coating, having again lost to the air, returns with approximately $q(1 - y)^2$; and (after touching cd) the net charge at A is QH $= Q + (1 - y)^2 q$. At QH the charge which CD brings away has become insensible. This explain the curve OQLMN.

To explain the curve u, x, H, y, s, z, &, v and the positive vindicating electricity, one has to assume that CD (and cd) is touched (alternately) when it is assembled, and left when it is stripped; this is not clearly explained neither by Beccaria nor by Cavallo. Grounding the coating CD, by touching it, removes the last of its original charge q and endows it with a negative one (by induction) proportional to QH. Remove CD: A appears to revindicate positively (IY)—notice that from H on the light lines HYSZ& being symmetrical to u, x, H, y, f, z, &, v also represents the positive electricity vindicated by A, see below—with the withdrawal of the counterbalancing negative coat. Touch the coatings alternately: CD obtains a larger induced charge than before (say LZ) because of the depolarization, which lags behind the removal of the coercing charge; at M, where the maximum revindication occurs at denudation (M&), the depolarization is complete. Thereafter A shows no force when covered and gradually loses its power of positive vindication as the free charge on its surface dissipates [122], p. 409, footnote 15.

[III]Point b is subject to the following forces, positive sign is from left to right:

1. The force exerted by the sphere A, which acts as mass point whose electricity is concentrated in the center C, and thus at distance R from b:

$$f_A \propto \frac{DR^2}{R^2} = D$$

.

2. The force exerted by the small sphere φ acting at distance r:

$$f_\varphi \propto -\frac{\delta r^2}{r^2} = -\delta$$

.

3. The force exerted by the sphere a, acting at distance $R + 2r$:

$$f_a \propto -2\frac{DR^2}{(R + 2r^2)}$$

where it is taken into account that the force f_A acting on a point of the surface is one half of the force f_a acting on a point far from it [22, 75], p. 621; pp. 232–233.

[IV] These forces have the expressions:

$$f_{A'} \propto \frac{D'r^2}{r^2} = D'.$$

$$f_m \propto -\frac{\delta dr^2}{dr^2} = -\delta.$$

$$f_A \propto -2D\frac{R^2}{\overline{Cm}^2} \times \frac{\overline{mB}}{\overline{\overline{Cm}}}$$

In the expression of f_A, the last multiplying term is due to evaluate the component of the force applied by A on m—acting in the direction Cm—in the direction C′mB.

References

1. Achinstein P (ed) (1991) Particles and waves. Oxford University Press, Oxford
2. Aepinus FUT (1756) Mémoire concernant quelques nouvelles experience électriques remarquables. Histoire de l'Académie Royale des Sciences et Belles Lettres de, Berlin, pp 105–121
3. Aepinus FUT (1758–1759) Descriptio ac explicatio novorum quorundam experimentorum electricorum. Novi Commentarii Academiae Scientiarum Imperialis Petropolitanae 7:277–302
4. Aepinus FUT (1759) Tentamen theoriae electricitatis et magnetismi. Academiae Scientiarum, Saint Petersburg

5. Aepinus FUT (1979) Aepinus's essay on the theory of electricity and magnetism. Introduction by Home RW, translated into English by Connor PJ. Princeton University Press, Princeton
6. Aguilon F (1613) Opticorum libri sex philosophis iuxta ac mathematicis utilis. Moreti, Antwerp
7. Algarotti F (1737) Il newtonianismo per le dame. Ovvero dialoghi sopra la luce e i colori. Unknown, Napoli
8. Allchin D (1994) James Hutton and phlogiston. Ann Sci 5(6):615–635
9. Amontons G (1699) De la resistance causée dans les machines, tant par les frottemens des parties qui les composent, que par roideur des cordes qu'on y employe, & la maniere de calculer l'un & l'autre. Mémoires de l'Académie Royale des Sciences de Paris, pp 206–222
10. Arecco D (1747) Da Newton a Franklin. Giambattista Beccaria e le relazioni scientifiche fra Italia e America nel sec. XVIII. Con una scelta di documenti. Accademia Urbense, Ovada
11. Aristotle, (2018) Physica. The internet Classical Archive, Translated into English by Hardie RP, Gaye RK
12. Badcock A (1962) Physical optics at the Royal society 1660–1800. British J Hist Sci 1(2):99–116
13. Baldini U (1981) Giovanni Francesco Cigna. In: Various (ed) Dizionario biografico degli italiani, vol 25, Istituto della Enciclopedia Italiana, Roma
14. Barletti C (1771) Nuove sperienze elettriche secondo la teoria del Sig. Beniamino Franklin e le produzioni del P. Beccaria. Galeazzi, Milan
15. Barletti C (1772) Physica specimina. Joseph Galeatium, Milan
16. Barletti C (1776) Dubbi e pensieri sopra la teoria degli elettrici fenomeni. Galeazzi, Milan
17. Barletti C (1780) Analisi d'un nuovo fenomeno del fulmine ed osservazioni sopra gli usi medici della elettricità. Bianchi, Pavia
18. Barletti C (1782) Introduzione a nuovi principj della teoria elettrica dedotti dall'analisi de' fenomeni delle elettriche punte. Parte prima. Memorie di Matematica e Fisica della Società Italiana 1:1–54
19. Barletti C (1784) Introduzione a nuovi principj della teoria elettrica dedotti dall'analisi de' fenomeni delle elettriche punte. Parte seconda. Memorie di Matematica e Fisica della Società Italiana 2:1–122
20. Barletti C (1788) Della supposta eguaglianza di contraria elettricità nelle opposte facce del vetro, o di uno strato resistente per ispiegare la scarica, o scossa della boccia di Leyden. Memorie di Matematica e Fisica della Società Italiana 4:304–309
21. Barletti C (1794) Della legge d'immutabile capacità, e necessaria contrarietà di eccesso, e difetto di elettricità negli opposti lati del vetro, e di altro strato resistente supposta da Franklin per la spiegazione della carica, e della scarica elettrica nella boccia leidense. Memorie di Matematica e Fisica della Società Italiana 7:444–461
22. Bauer E (1949) L'électromagnétisme. Hier et aujourd'hui, Michel, Paris
23. Beccaria GB (1753) Dell'elettricismo artificiale e naturale libri due. Campana, Turin
24. Beccaria GB (1758) Dell'elettricismo. Lettere di Giambattista Beccaria dirette al chiarissimo Sig. Giacomo Bartolomeo Beccari. Tipografia di Colle Ameno, Colle Ameno
25. Beccaria GB (1759–1760) Experiments in electricity: in a letter from father Beccaria, professor of experimental philosophy at Turin, to Benjamin Franklin. Philos Trans R Soc Lond 51:514–526
26. Beccaria GB (1767) De electricitate vindice Joannis Baptistae Beccariae ex scholis piis ad Beniaminum Franklinium virum de re electrica, & meteorologica optime meritum. Epistola. Fontana, Turin
27. Beccaria GB (1769) Experimenta, atque observationes, quibus electricitas vindex late constituitur, atque explicatur. Stamperia Reale, Turin
28. Beccaria GB (1771) De athmosphaera electrica. Philos Trans R Soc Lond 60:277–301
29. Beccaria GB (1772) Elettricismo artificiale. Stamperia Reale, Turin
30. Beccaria GB (1774) Gradus taurinensis. Ex Typografia Regia, Turin
31. Beccaria GB (1776) A treatise upon artificial electricity. Nourse, London
32. Beer A (1738) Grundiss des Photometrishen Calcüles. Vieweg, Brownschweig

33. Bernoulli J (1701) Disquisitio catoptrico-dioptrica exhibens reflexionis & refractionis naturam, nova & genuina ratione ex aequilibrii fundamento deductam & stabilitam. Acta Eruditorum, pp 19–26
34. Bernoulli J (1732) Meditationes de chordis vibrantibus, cum pondusculis aequali intervallo a se invicem dissitis, ubi nimirum ex principio virium vivarum quaeritur numerus vibrationum chordae pro una oscillatione penduli datae longitudinis D. Commentarii Academiae Scientiarum Imperialis Petropolitanae 3:13–28
35. Bernoulli J II (1736) Recherches physiques et géométriques sur la question: Comment se fait la propagation de la lumière. In: Recueil des pièces qui ont remporté les prix de l'Académie royale des sciences, vol 3. Jombert, Paris, pp 1732–1741
36. Birch T (1756–1757) History of the Royal society of London (4 vols). Millar, London
37. Boscovich RG (1749) Sopra il turbine che la notte tra gli XI, e XII giugno del 1749 danneggió una gran parte di Roma. Pagliarini N and M, Rome
38. Boscovich RG (1922) Theoria philosophiae naturalis redacta ad unicam legem virium in natura existentium. Latin-English edition, Open Court, London
39. Boscovich RG (2008) Carteggio con Giovan Stefano Conti (2vols). In: Proverbio E (ed) Edizione nazionale delle opere e della corrispondenza di Ruggiero Giuseppe Boscovich. Digital publication, Corrispondenza
40. Bouguer P (1729) Essai d'optique sur la gradation de la lumiere. Jombert, Paris
41. Bouguer P (1760) Traité d'optique sur la gradation de la lumiere. Guerin & Delatour, Paris
42. Cabeo N (1629) Philosophia magnetica, in qua magnetis natura penitus explicatur, et omnium quae hoc lapide cernuntur, causae propriae afferuntur: nova etiam praxis construitur, quae propriam poli elevationem, cum suo meridiano, ubique demonstrat, multa quoque dicuntur de electricis, & aliis actractionibus, & eorum causis. Kinckium, Cologne
43. Cabeo N (1646) In quatuor libros meteorologicorum Aristotelis commentaria (2 vols). Corbelletti, Rome
44. Cannon J, Dostrovsky S (1981) The evolution of dynamics: vibration theory from 1687 to 1742. Springer, Dordrecht
45. Canton J (1753) Electrical experiments, with an attempt to account for their several phaenomena, together with observations on thunder clouds. Philos Trans R Soc Lond 48:350–358
46. Cantoni G (1878) Elementi di fisica, 4th edn. Vallardi, Milan
47. Cantor G (1983) Optics after Newton. In: Theories of light in Britain and Ireland, 1704–1840. Manchester University Press, Manchester
48. Capecchi D (2003) Storia della scienza delle costruzioni. Progedit, Bari
49. Capecchi D (2019) Some incongruences in Coulomb's memoirs on electricity 1785–1788. Forthcoming. In: Rossi P (ed) XXXIX Congresso SISFA, Pisa
50. Capecchi D, Tocci C (2012) Il sentimento di Ruggiero Boscovich sulla Gran Guglia del Duomo di Milano. Palladio 49:115–128
51. Capecchi D, Tocci C (2016) Three technical reports of R.G. Boscovich on the statics of domes. In: Various (ed) Further studies in the history of construction. Proceedings of the third annual conference of the construction history society. Construction History Society, Cambridge, pp 251–262
52. Cappelletti V (1964) Carlo Barletti. In: Various (ed) Dizionario biografico degli italiani, vol 6, Istituto della Enciclopedia Italiana, Roma
53. Cavallo T (1786) Complete treatise on electricity, 3rd ed, vol 2. Dilly, London
54. Cavazza M (2009) Laura Bassi and Giuseppe Veratti: an electric couple during the enlightenment. Contrib Sci 5(1):115–128
55. Cavendish H (1771) An attempt to explain some of the principal phaenomena of electricity, by means of an elastic fluid. Philos Trans R Soc Lond 61:584–677
56. Cavendish H (1776) Attempts to imitate the effects of the torpedo. Philos Trans R Soc Lond 66:196–225
57. Cavendish H (1798) Experiments to determine the density of the earth. Philos Trans R Soc Lond 88:469–526

58. Cavendish H (1879) Experimental determination of the law of electric force. In: Clerck-Maxwell J (ed) The electrical researches of the honourable Henry Cavendish, Cass. London, pp 105–113
59. Cavendish H (1879) Experiments on the charges of bodies. In: Clerck-Maxwell J (ed) The electrical researches of the honourable Henry Cavendish, Cass. London, pp 114–143
60. Chipman RA (1954) Unpublished letter of Stephen Gray on electrical experiments, 1707–1708. Isis 45(1):33–40
61. Chipman RA (1958) The manuscript letters of Stephen Gray, F.R.S. (1666/7–1736). Isis 49(1):414–433
62. Cigna G (1760–1761) De causa extintionis flammae et animalium in aere interclusorum. Miscellanea Philosophico-Mathematica Societatis Privatae Taurinensis 2:168–203
63. Cigna G (1762–1765) De novis quibusdam experimentis electricis. Miscellanea Philosophico-Mathematica Societatis Privatae Taurinensis 3:31–72
64. Clerk-Maxwell J (1873) A treatise on electricity and magnetism, vol 2. Clarendon, Oxford
65. Cohen IB (1954) Neglected sources for the life of Stephen Gray (1666 or 1667–1736). Isis 45(1):41–50
66. Cohen IB (1956) Franklin and Newton. Harvard University Press, Cambridge
67. Coulomb CA (1776) Essai sur une application des règles de maximis et minimis à quelques problèmes de statique relatif à l'architecture. Mémoires de mathématique & de physique, présentés à l'Académie Royale des Sciences par divers savans 7:343–382
68. Coulomb CA (1780) Recherches sur la meilleure maniere de fabriquer les aiguilles aimantees. Mémoires de mathématique et de physique présentés à l'Académie Royale des Sciences, par divers savans 9:167–264
69. Coulomb CA (1784) Recherches théoriques & expérimentales sur la force de torsion, & sur l'élasticité des fils de métal, &c. Mémoires de l'Académie Royale des Sciences de Paris, pp 229–269
70. Coulomb CA (1785) Premier mémoire sur l'électricité et le magnétisme. Construction et usage d'une balance électrique, &c. Mémoires de l'Académie Royale des Sciences de Paris, pp 569–577
71. Coulomb CA (1785) Second mémoire sur l'électricité et le magnétisme. Où l'on determine suivant quelles lois le fluide magnétique ainsi que le fluide électrique agissent, &c. Mémoires de l'Académie Royale des Sciences de Paris, pp 578–611
72. Coulomb CA (1785) Troisième mémoire sur l'électricité et le magnétisme. De la quantité d'électricite qu'un corps isolé perd dans un temps donné, &c. Mémoires de l'Académie Royale des Sciences de Paris, pp 612–638
73. Coulomb CA (1786) Quatrième mémoire sur l'electricité et le magnétisme. Oú l'on démontre deux principales propriétés du fluide électrique, &c. Mémoires de l'Académie Royale des Sciences de Paris, pp 67–77
74. Coulomb CA (1787) Cinquième mémoire sur l'électricité et le magnétisme. Sur la manière dont le fluide électrique se partage entre deux corps conducteurs mis en contact, &c. Mémoires de l'Académie Royale des Sciences de Paris, pp 421–467
75. Coulomb CA (1788) Sexième mémoire sur l'électricité et le magnétisme. Suite des recherches sur la distribution du fluide électrique entre plusieurs corps conducteurs, &c. Mémoires de l'Académie Royale des Sciences de Paris, pp 617–705
76. Coulomb CA (1789) Mémoires sur l'électricté et le magnétisme. Bachelier, Paris
77. Coulomb CA (1789b) Septième mémoire sur l'électricité et le magnétisme. Du magnétisme. In: Coulomb (ed) Extraits des mémoires de l'Académie Royale des Sciences de Paris, publiés dans le années 1785 à 1789, avec planches et tableaux, Bachelier, Paris, pp 455–505
78. Coulomb CA (1799) Résultats de plusieure expériences destinées à déterminer la quantité d'action que les hommes peuvent fournir par leur travail journalier, suivant les différent manières dont ils employent leurs forces. Mémoires de l'Institute national des sciences et arts-Sciences mathématiques et physiques 2:380–428
79. Coulomb CA (1821) Théorie des machines simples. Bachelier, Paris

80. Darrigol O (ed) (2012) A history of optics from Greek antiquity to the nineteenth century. Oxford University Press, Oxford
81. Dello Preite M (1979) L'immagine scientifica del mondo di Johann Heinrich Lambert. Dedalo, Bari
82. Desaguliers JT (1742) A dissertation concerning electricity. Innys & Longman, London
83. Descartes R (1724) Les principes de la philosophie. Le Gras, Paris
84. Devons S (1975) Coulomb electrical measurements, ERIC number ED182150, https://eric.ed.gov/?id=ED182150
85. Devons S (1975) Henry Cavendish. The law of force of electricity, https://stwww1.weizmann.ac.il/wp-content/uploads/2016/08/Henry-Cavendish----the-law-of-Force-of-Electricity.pdf
86. Dhombre J, Pensivy M (1988) Esprit de rigueur et présentation mathématique au XVIIIème siècle: le cas d'une démonstration d'Aepinus. Histoira Mathematica 15(1):9–31
87. Euler AJ (1757) Recherches sur la cause physique de l'elettricité. Mémoires de l'Académie Royale des Sciences et Belles Lettres de Berlin 13:125–159
88. Euler L (1739) Tentamen novae theoriae musicae ex certissismis harmoniae principiis dilucide expositae. Academiae Scientiarum, Saint Petersburg
89. Euler L (1746) Nova theoria lucis et colorum. Opuscula Varii Argumenti 1:169–244
90. Euler L (1746) Recherches physiques sur la nature des moindres parties de la matiere. Opuscula Varii Argumenti 1:287–300
91. Euler L (1746) Sur la lumiere & le couleurs. Histoire de l'Académie des Sciences et Belles Lettres de Berlin pp 17–24
92. Euler L (1754) Essai d'une explication physique des couleurs engendrees sur des surfaces extremement minces. Mémoires de l'Académie des Sciences et Belles Lettres de Berlin 8:262–282
93. Euler L (1770–1774) Lettres à une princesse d'Allemagne sur divers sujets de physique & de philosophie (3 vols). Steidel & Compagne, Mietau, Leipzig
94. Franklin B (1751) Experiments and observations on electricity, made at Philadelphia in America. Cave, London
95. Franklin B (1752) A letter from Mr. Franklin to Mr. Peter Collinson, F. R. S. concerning the effects of lightning. Philos Trans R Soc Lond 48:289–291
96. Franklin B (1758) Des Herrn Benjamin Franklins Esq. Briefe von der Elektricität. Translated into German by Wilcke JC. Kiesewetter, Leipzig
97. Franklin B (1769) Experiments and observations on electricity, made at Philadelphia in America. Henry, London
98. Franklin B (1836–1840) The works of Benjamin Franklin: containing several political and historical tracts not included in any former edition and many letters official and private, not hitherto published: with notes and a life of the author, Sparks, J (ed), vol 10. Wittemore and Niles and Hall, Boston
99. Frisi P (1757) De causa electricitatis. In: Euler A, Frisi P, Laurentii B (eds) Dissertationes selectae Jo. Alberti Euleri et, Paulli Frisi et Laurentii Béraud, quae ad imperialem scientiarum petropolitanam Academiam an. 1755 missae sunt, cum electricitatis caussa & theoria, praemio proposito, quaereretur, Junctinum, Saint Petersburg, pp 41–131
100. Frisi P (1781) Dei conduttori elettrici. In: Frisi P (ed) Opuscoli filosofici, Galeazzi, Milan, pp 27–48
101. Galilei G (1890–1909) Le opere di Galileo Galilei (National edition), Barbera FA (ed), vol 20, Florence
102. Geikie A (1905) The founders of geology. Macmillan, London
103. Gilbert W (1600) De magnete, magneticisque corporibus, et de magno magnete tellure; physiologia nova. Short, London
104. Gilbert W (1893) On the loadstone and magnetic bodies and on the great magnet the earth. Translated into English by Mottelay PF, Quaritch, London
105. Gillmor CS (1971) Coulomb and the evolution of physics and engineering in eighteenth-century France. Princeton University Press, Princeton

106. Gillmor CS (1970) Coulomb, Charles. In: Gillispie CC (ed) Complete dictionary of scientific biography. Scribner, New York
107. Gleig G, Macfarquhar C (1788–1797) Encyclopaedia Britannica, or, a dictionary of arts, sciences, and miscellaneous literature (18 vols + 2), edited by Gleig and Macfarquhar
108. Gray JJ, Tilling L (1978) Johann Heinrich Lambert, mathematician and scientist, 1728–1777. Historia Mathematica 5:13–41
109. Gray S (1721) An account of some new electrical experiments. Philos Trans R Soc Lond 31(366):104–107
110. Gray S (1731) A letter concerning the electricity of water, from Mr. Stephen Gray to Cromwell Mortimer, M. D. Secr. R. S. Philos Trans R Soc Lond 37(422):227–230
111. Gray S (1731) A letter from Mr. Stephen Gray to Dr. Mortimer, Secr. R. S. containing a farther account of his experiments concerning electricity. Philos Trans R Soc Lond 37(423):285–291
112. Gray S (1731) A letter to Cromwell Mortimer, M. D. Secr. R. S. containing several experiments concerning electricity. Philos Trans R Soc Lond 37(417):18–44
113. Gray S (1731) Two letters from Mr. Stephen Gray, F. R. S. to C. Mortimer, M. D. Secr. R. S. containing farther accounts of his experiments concerning electricity. Philos Trans R Soc Lond 37(426):397–407
114. Gray S (1735) Experiments and observations upon the light that is produced by communicating electrical attraction to animal or inanimate bodies, together with some of its most surprising effects; communicated in a letter from Mr. Stephen Gray, F. R. S. to Cromwell Mortimer, M. D. R. S. Secr. Philos Trans R Soc Lond 39(436):16–24
115. Guzzardi L (2019) Points, distances, determinations: Ruggiero Boscovich's theory of natural philosophy. Springer, Dordrecht
116. Hakfoort C (1995) Optics in the age of Euler. Cambridge University Press, Cambridge
117. Hankins T (1967) The influence of Malebranche on the science of mechanics during the eighteenth century. J Hist Ideas 28(2):193–210
118. Hawes JL (1968) Newton's revival of the aether hypothesis and the explanation of gravitational attraction. Notes Records R Soc Lond 23(3):200–212
119. Hawes JL (1971) Newton's two electricities. Ann Sci 27(1):95–103
120. Heilbron J (1970) Aepinus, Franz Ulrich Theodosius. In: Gillispie CC (ed) Complete dictionary of scientific biography. Scribner, New York
121. Heilbron JL (1976) Robert Symmer and the two electricities. Isis 76(1):7–20
122. Heilbron JL (1979) Electricity in the 17th and 18th centuries. University of California Press, Berkley
123. Heymann J (1972) Coulomb's memoir on statics: an essay in the history of civil engineering. Cambridge University Press, Cambridge
124. Home R (1988) Leonhard Euler's 'anti-Newtonian' theory of light. Ann Sci 45(5):521–533
125. Home RW (1982) Newton on electricity and the aether. In: Bechler Z (ed) Contemporary Newtonian researches. Reidel, Dordrecht, pp 191–214
126. Home RW (1983) Poisson's memoirs on electricity: academic politics and a new style in physics. British J Hist Sci 16(3):239–259
127. Hujer K (1952) Father Procopius Divis? - The European Franklin. Isis 43(4):351–357
128. Hutton J (1794) Dissertation upon the philosophy of light, heat, and fire. Cadell and Davis, Edinburgh
129. Hutton J (2019) Mactutor history of mathematics archive. http://www-history.mcs.st-and.ac.uk/index.html
130. Huygens C (1888–1950) Oeuvres complètes de Christiaan Huygens (22 vols). Nijhoff, The Hague
131. James FA (1984) The physical interpretation of the wave theory of light. British J Hist Sci 17(1):47–60
132. Jammer M (1957) Concepts of force: A study in the foundation of dynamics. Harvard University Press, Cambridge
133. Krüger JG (1746) Geschichte der Erde in den allerältesten Zeiten. Lüderwald, Halle
134. Laguzzi A (1992) Carlo Barletti e le "Encyclopédies". Studi Storici 33(4):833–862

135. Laguzzi A (1994) Per una biografia di P. Carlo Barletti, fisico del '700 e patriota repubblicano. Memorie dell'Accademia Urbense 12
136. Laguzzi A (2012) Carlo Barletti. Epistolario, accademia Urbense. https://www.archiviostorico.net/libripdf/Barletti_Epistolario.pdf
137. Lambert JH (1760) Photometria, sive, De mensura et gradibus luminis, colorum et umbrae. Detleffsen, Augsburg
138. Lambert JH (1770) Mémoire sur la part de la photométrique de l'art due peintre. Histoire de l'Académie Royale des Sciences et Belles Lettres de Berlin, pp 80–108
139. Lambert JH (1779) Pyrometrie oder vom Maasse des Feuers und der Wärme. Haud & Spener, Berlin
140. Lambert JH (1784) Systême du monde. Duchesne & Durand, Berlin
141. Lambert JH (2001) Photometry, or, on the measure and gradations of light, colors and shade, Translated into English by DiLaura DL. Illuminating Engineering Society of North America, New York
142. Le Compasseur G (1742) Traité d'optique. Ou l'on donne la théorie de la lumiere dans le système Newtonien, avec de nouvelles solutions des principaux problèmes de dioptrique & de catoptrique. Jombert, Paris
143. Lechalas G (1884) L'oeuvre scientifique de Malebranche. Revue Philosophique de la France et de l'Étranger 18:293–312
144. Mach E (1926) The principles of physical optics. An historical and philosophical treatment. Translated into English by Anderson JS and Young AFA. Methuen & Co, London
145. Maffei S (1747) Della formazione de' fulmini trattato del sig. marchese Scipione Maffei raccolto da varie sue lettere, in alcune delle quali si tratta anche degli'insetti rigenerantisi, e de' pesci di mare su i monti, e più a lungo dell'elettricità. Tumemani, Verona
146. Maignan E (1648) Perspectiva horaria sive de horographia gnomonica tum theoretica, tum practica libri quatuor. Rubeus, Rome
147. de Mairan D (1717) Dissertation sur la cause de la lumiere des phosphores et des noctiluques. Brun, Bordeaux
148. de Mairan D (1722) Recherches physico-mathématiques sur la réflexion des corps. Mémoires de l'Académie Royale des Sciences de Paris, pp 6–51
149. de Mairan D (1723) Suite des recherches physico-mathématiques sur la réflexion des corps. Mémoires de l'Académie Royale des Sciences de Paris, pp 343–386
150. de Mairan D (1737) Sur la propagation du son dans les differens tons qui le modifient. Mémoires de l'Académie Royale des Sciences de Paris, pp 1–60
151. de Mairan D (1738) Troisiéme partie des recherches physico-mathématiques sur la réflexion des corps. Mémoires de l'Académie Royale des Sciences de Paris, pp 1–64
152. de Mairan D (1740) Quatrieme partie des recherches physico-mathématiques sur la réflexion des corps. Mémoires de l'Académie Royale des Sciences de Paris 1:1–58
153. Malebrache N (1879) De la recherche de la vérité. In: Bouillier MF (ed) 2 vols. Garnier, Paris
154. Malebranche N (1699) Reflexion sur la lumiere et le couleurs, et la generation du feu. Mémoires de l'Académie Royale des Sciences de Paris, pp 22–35
155. Malebranche N (1958-1967) Correspondance. Actes et documents, 1690–1715. In: Robinet A, als (eds) Oeuvres complètes de Malebranche (20 vols), vol 19, Vrin, Paris
156. Martin C (2013) Causation in Descartes' Les météores and late Renaissance Aristotelian meteorology. In: Garber D, Roux S (eds) The mechanization of natural philosophy. Springer, Dordrecht, pp 217–236
157. Martinez AA (2006) Replication of Coulomb's torsion balance experiment. Arch Hist Exact Sci 60(6):517–563
158. McCormmach R (2004) Speculative truth: Henry Cavendish, natural philosophy, and the rise of modern theoretical science. Oxford University Press, Oxford
159. Mersenne M (1636) Harmonie universelle. Cramoisy, Paris
160. van Musschenbroek P (1746) Observations de Monsieur Musschenbroek lues par Monsieur de Reamur. Mémoires de l'Académie Royale des Sciences de Paris Procès-verbaux 65:4–6

161. Newton I (1730) Opticks: or, a treatise of the reflections, refractions, inflections and colours. Innys, London
162. Nollet JA (1746) Essai sur l'électricité des corps. Guerin, Paris
163. Nollet JA (1754-1765) Leçons de physique expérimentale (6 vols). Arkstée et Merkus, Amsterdam & Leipzig
164. Pace A (1970) Giambattista Beccaria. In: Various (ed) Dizionario biografico degli italiani, vol 7, Istituto della Enciclopedia Italiana, Roma
165. Pancaldi G (2003) Volta. Science and culture in the age of enlightenment. Princeton University Press, Princeton
166. Pastorino C (2020) Alchemy and the electric spirit in Isaac Newton's general scholium. In: Ducheyne, Mandelbrote S, Snobelen S (eds) Newton's general scholium after 300 years, Forthcoming
167. Pedersen KM (2017) The velocity of light and the colour changes of Jupiter's satellite. Res Publ Sci Stud 40:1–41
168. Playfair J (1802) Illustrations of the Huttonian theory of earth. Creech et als, Edinburgh
169. Poisson SD (1811) Mémoire sur la distribution de l'électricité à la surface des corps conducteurs. Mémoires de la Classe des Sciences Mathématiques et Physiques de l'Institute Impériale de France 12:Part. I, 1–92
170. Poisson SD (1811) Second mémoire sur la distribution de l'électricité à la surface des corps conducteurs. Mémoires de la Classe des Sciences Mathématiques et physiques de l'Institute Impériale de France 12:Part II, 163–274
171. Priestley J (1775) The history and present state of electricity, with original experiments, Bathurst et al (ed) 4th edn, corrected and enlarged, London
172. Privat-Deschanel A (1876) Elementary treatise on natural philosophy. Appleton, New York
173. Proverbio E (2000) Sulle ricerche elettriche di Giovanbattista Beccaria e sui suoi rapporti con Ruggiero Giuseppe Boscovich nelle applicazioni dell'elettricismo naturale e artificiale. In: Congresso XX (ed) Schettino E. SISFA, Naples, pp 231–280
174. Proverbio E (2003) Gli interessi scientifici di Ruggiero G. Boscovich per i fenomeni elettrici e i suoi incontri con Benjamin Franklin ed altri elettricisti inglesi e francesi. Quaderni di Storia della Fisica 11:3–48
175. Proverbio E (2006–2019) Boscovich National Edition. Edizione nazionale delle opere e della corrispondenza. http://www.edizionenazionaleboscovich.it/2018
176. Rohault J (1723) System of natural philosophy. Illustrated with Dr. Samuel Clarke's notes, vol 2. Knapton, London
177. Schettino E (2000) Franklinists in Naples in the second half of the 18th century. In: Congresso XX (ed) Schettino E. SISFA, Naples, pp 347–352
178. Scriba C (1973) Lambert, Johann Heinrich. In: Gillispie CC (ed) Complete dictionary of scientific biography. Scribner, New York
179. Sette D (1967) Lezioni di Fisica. Elettromagnetismo, Veschi, Rome
180. Sguario E (1747) Dell'elettricismo: ossia delle forze elettriche de' corpi. Di Simone, Naples, first edition Venice, 1746
181. Shapiro AE (1973) Kinematic optics. 'A study of the wave theory of light in the seventeenth century'. Arch Hist Exact Sci 11(2/3):134–266
182. Sheynin OB (1971) J.H. Lambert's work on probability. Arch Exact Sci 7(3):244–256
183. Steck M (1970) Bibliographia Lambertiana. Gerstenberg, Hildesheim
184. Stoiljkovich D (2014) Contribution of Boscovich's theory to modern comprehension of the structure of matter. In: Anderton R, Stoiljkovich D (eds) Roger Boscovich-The founder of modern science, Lulu, Raleigh, pp 4.1–4.11
185. Talas S (2012) Physics in the eighteenth century: new lectures, entertainment and wonder. In: Cifarelli L, Simili R (eds) Laura Bassi. Emblema e primato nella scienza del Settecento, Editrice Compositori, Bologna, pp 177–188
186. Torres Assis AK (2017) I fondamenti sperimentali e storici dell'elettricità. Bollettino trimestrale dell'Associazione per l'Insegnamento della Fisica 26(2):3–223

187. Truesdell CA (1960) The rational mechanics of flexible or elastic bodies. In: Leonhardi Euleri Opera omnia (1911-, in progress), 2, vol 1, part 2, Teubneri GB; [then] Fussli Turici; [then] Birkhäuser, Basel
188. Vassalli-Eandi AM (1783) Memorie istoriche intorno alla vita ed agli studi del padre Giambatista Beccaria delle scuole pie, professore di fisica sperimentale nella R. Università di Torino. Stamperia Reale, Turin
189. Vassalli-Eandi AM (1821) Memorie istoriche intorno alla vita ed agli studi di Gianfrancesco Cigna. Memorie della Reale Accademia delle Scienze di Torino 26:XIII–XXXVI
190. Veil H (2019) Elektrisches Feuer 1746: Ein bizarrer Streit um Wissenschaft und Öffentlichkeit in Leipzig. Humanities Online, Frankfurt am Main
191. Verri P (1787) Memorie appartenenti alla vita ed agli studj del signor don Paolo Frisi. Marelli, Milan
192. Volk O (1980) Johann Heinrich Lambert and the determination of orbits for planets and comets. Celest Mech 21:237–250
193. Volta A (1775) Articolo di una lettera del signor don Alessandro Volta al signor dottore Giuseppe Priestley, opere di Alessandro Volta, vol 3. https://echo.mpiwg-berlin.mpg.de/content/electricity
194. Watson W (1746) A sequel to the experiments and observations tending to illustrate the nature and properties of electricity. Davis, London
195. Whittaker ET (1910) A history of the theories of aether and electricity, From the ages of Descartes to the close of the nineteenth century. Longmans, Green Co, New York

Chapter 5
The Emergence of the Science of Engineering

Abstract Modern technology historians identify the birth of a new figure in the 18th century, the scientific engineer. His goal was the rationalization of design and implementation of processes. For this purpose, he used hypotheses and experimentations, as in the (mathematical) physical sciences. The need for such a new figure derived from the tumultuous developments of science (physics and chemistry in particular) of the 18th century, consequence and cause of the economic development. With its dizzying growth in the 18th century, science revealed the possibility of applications to areas never thought of before. However, scientists were dealing with general problems. Their solutions did not provide for an immediate application. Thus, there was the basic need for an intermediate operator between the scientist and the final user. More precisely, there was a need for a sufficiently large body of qualified engineers. After general considerations on the relationship between mathematics, natural philosophy, new physics and technology, the chapter goes on to look for the reasons that led to this process.

5.1 Science, Technology and Engineering

Science and *technology* are two terms that in the current language indicate different activities, in the sense that if one asked some people if a problem pertains to science or otherwise to technology they would give the same answer in most cases. However, there may be some dispersion and sometimes the answer would be: I do not know.

If instead one asked what is the difference between science and technology he would receive an uncertain answer. There are indeed several standard suggestions, provided by the experts; philosophers of science and technology, scientists and technologists and even politicians. Suggestions that in some way have been metabolized by not educated public and have become commonplaces. A simple answer is that science is more theoretical, technology more practical. That is scientists are studying much more and making very complex experiments in a clean laboratory, perhaps by putting on white coats, while technologists study less and make more applicative experiments, perhaps even using blue overalls. In these judgments there are two motivations. One of epistemological nature, the other of sociological nature. On the

D. Capecchi, *Epistemology and Natural Philosophy in the 18th Century*,
History of Mechanism and Machine Science 39,
https://doi.org/10.1007/978-3-030-52852-2_5

one hand one sees in science a more noble activity because engaged in disinterested knowledge of nature, of the truth in itself. On the other hand this activity does not require manual labor. And thus, according to the mentality rooted in the West since the ancient world, that of scientists being closer to the activity of the rulers, that typically do not carry out manual works, is noblest.

The use of a term in historical accounts cannot be the free choice of a single person, for example of a historian to introduce a category, but it must be based in some way on common usage taking into account its history. Almost always there are nuances in the accepted use and the task of any scholar is to specify the use he himself intends to favor. In the case of *technology*, from the Greek τέχνη, there are also linguistic difficulties arising from problems of translation, in particular related to the difference between technique and technology, that in some languages are given by the same words [52]. Here this difference is not stressed and technology is the term used to cover possible difference between technique and technology. There are in any case some different acceptations of the term; the main of which are resumed below in very general way [46, 51][1]:

1. Technology as a set of artifacts or systems of artifacts.
2. Technology as a form of knowledge (for the design, production, maintenance and use of technological artifacts and systems).
3. Technology as a range of activities (designing, producing, maintaining and using artifacts).
4. Technology as an expression of the will of its makers, designers and producers (volition).

In the following the meaning 2 is generally considered.

Technology, in this meaning and according a very diffuse idea, is basically synonymous with engineering. There are however broader meanings of technology and engineering, now widespread even among non-experts. On the one hand, engineering, when defined as the activity of the engineer, is only a part of technology, together with medicine, agriculture, domestic economy, communication, etc. On the other hand engineering can be defined as the whole process of making, whose actors are not only engineers or technologists but also contractors and politicians, and thus engineering contains technology. Below a possible distinction between science, technology and engineering:

> Viewed from their achievements, scientific activity, technological activity, and engineering activity yield three different kinds of achievements. Scientific activity results in academic achievements, or new scientific knowledge, such as new theories, new scientific principles, new scientific concepts, or academic dissertations. Technological activity results in technological achievements, or new technological knowledge, such as a patent for an invention, "know-how", or Engineering activity results in engineering achievements, or material products and material facilities that mean material wealth, such as a power station building, a new railway, a new car, or a new computer. Scientific knowledge belongs to all mankind, while

[1] p. 4.

> technological knowledge, such as patents belongs to their inventors or a certain company. As to engineering achievements its nature is not knowledge, but material wealth [10].[2]

In the following I will consider engineering in the more restrict way, that is as the activity of engineers and as such a subset of technology.

The figure of the engineer is nuanced as always happens when one wants to search in the past profiles, quite well defined today but not in the past. The English term engineer is of the 14th century, with an etymology that is not completely clear; it would derive from the late Latin *ingenium*, or devices.[3] Whereas the terms scientists, physicists were coined by academicians, *engineer* originated in every day usage. In the following I will mainly refer to the role of the engineer of the 18th century, when though the term was common it did not have the same meaning of today. According to the Encyclopédie:

> The job of an engineer requires a lot of study, talent, ability and genius. The basic sciences of this state are arithmetic, geometry, mechanics and hydraulics.
> An engineer must have some skill in drawing. Physics is necessary to judge of the nature of the materials used in the buildings, of the waters, and the different qualities of the air of the places which one wishes to fortify.
> It is very useful for him to have general and particular knowledge of the civil architecture, for the construction of military buildings, as barracks, magazines, arsenals, hospitals, quarters of the staff, &c. whose engineers are usually charged [26][4] (E.1).

The engineer is thus distinct from craftsman not only because a designer and not just a performer but also because he has a basic education that includes elements of mathematics and may be of natural philosophy. Among engineers, a not too much numerous category if according to the Encyclopédie in France at the time there were three hundred of them, there were characters of different education varying from those who possessed a good education in natural philosophy and mathematics and those who were little more than craftsmen.

5.1.1 Technology Versus Applied Science

Strict definitions of science and technology, in the sense of expressing one or two characteristics that constitute necessary and sufficient conditions to characterize them, are hard to come by. Essentialist (intensive) definitions of science and technology can be proved questionable, simply because definitions of science and technology are not shared. An alternative view to an essentialist definition, could be the adoption of a nominalistic-empirical strategy and see as science and technology were considered

[2]pp. 35–36.

[3]In the classical Latin, *ingenium* means, talent and genius. This is a possible origin for the Italian *ingegnere*. For example, Leonardo da Vinci referred to himself as an (*ingegnero*). In the Middle Ages, c. 1292, the term was used to indicate devices or machines, from this the English engine. In the classical Latin device is translated as *machina*.

[4]Article *Ingénieur*

as such by people. But also this strategy may fail, first of all because it presupposed at least a weak, or working, form of (essentialist) definition. Concluding both the two strategies should be considered [60].[5]

A common definition, that can be considered either as essentialist or nominalist, is that technology is applied science. This view is today generally not yet defended by historians and philosophers of technology; but it is still the view shared by most people and probably even by the majority of technologists. A criticism of this view mostly depends and the meaning of the term *applied science*. Leaving aside a precise definition, that will attempted below, here it is suggested that the equation applied science = technology means that that technology is a quite straightforward application of results obtained by science:

> The terms 'technology' and 'applied science' will be taken here as synonymous [...]. The method and the theories of science can be applied either to increasing our knowledge of the external and the internal reality or to enhancing our welfare and power. If the goal is purely cognitive, pure science is obtained; if primarily practical, applied science. Thus, whereas cytology is a branch of pure science, cancer research is one of applied research [14].[6]

However the equation applied science = technology is at least misleading and all considered false. This can be argued in two conflicting views, both of them are defended below as in an exercise of rhetoric. On the one hand it can be argued that technology is not exhausted by science; on the another hand it can be argued that the differences between science and technology are not very great, neither at a methodological nor at a epistemological level.

5.1.1.1 Difference Between Science and Technology

As regards the first type of argument, that is that technology is not exhausted by science, history shows that technological development has been largely independent of the development of a formalized science, carried out by a small group of people largely dedicated to a contemplative activity and endowed with a good education both in natural philosophy and in mathematics. There are entire societies, such as the Roman and Chinese empires, which had a very advanced technology without, at least apparently, having a substantial formal theoretical elaboration behind it. In the 18th century, the introduction of hydraulic and thermal machines are usually mentioned as examples of science-free technological developments; in the 20th century the invention of the transistor is also cited as an example. Though it could be argued that most of these situations belong to the past, when science was not yet developed, no one can reasonably deny that similar cases could happen again.

Science, as it is commonly understood (for the past the situation was more nuanced), tends to develop general theories or theories that are always valid, but that cannot be connected in a simple way to the concrete world. The only point of

[5] p. 68.

[6] p. 329.

contact on the part of the theories with reality is the agreement with the experimental results. Technology is instead rooted in the real world.

To understand the interplay between science and technology some considerations connected to structural engineering are reported below. Structural engineering is largely based, or at least it was such until the middle of the 20th century, on the theory of elasticity. This is a scientific theory, in particular a physical mathematical one, which draws its inspiration from an 'experimental' law known as Hooke's law, according to which, at the macroscopic level, the displacements of the points of all bodies are proportional to the applied forces, the more approximately the more the displacements are small. The theory of elasticity in its traditional form, developed by French mathematicians, including Augustin Cauchy and Siméon Denis Poisson, is strongly idealized.

Material bodies are assimilated to mathematical continua, without entering into the merit of their actual microscopic corpuscular structure. On these continua are defined generalized displacements and forces, known as stresses and strains, introduced with the extensive use of Calculus. It is assumed that these quantities, stress and strain, which are in themselves not measurable experimentally, are linked by a generalized Hooke's law: stresses are linearly proportional to strains. We thus arrive at a highly axiomatized theory, which today is referred to as the *mathematical theory of elasticity*, which is not based on experimental data, as in principle it is true for classical mechanics and optics, but it is rather a hypothetical deductive theory validated by the experimental confirmation. Notice that the theory of elasticity was inspired not only and perhaps not even mainly by technological needs, but also by theoretical studies such as those related to the propagation of light (and sound). According to the wave theories, dominant in the 19th century, light propagates through an elastic medium. And the theory of elasticity was useful for the study of this propagation. Augustin-Jean Fresnel (1788–1827) and George Green (1793–1841) were mathematicians and physicists that moved in this direction.

The mathematical theory of elasticity leads to differential equations connecting the variables of displacement of the points of a given body to the external forces applied to it. These equations could be solved for relatively simple bodies and distributions of forces only and therefore were not useful in the practical calculation of structures (today, thanks to computers it is possible to solve these equations, albeit numerically, in every situation).

In order to apply the theory of elasticity to structural calculations, less general and less rigorous theories began to be developed, which were initially addressed at the examination of the fundamental elements of structures, in particular those in steel in vogue in the construction of industrial buildings and bridges of the 19th century: beams. The scholars involved in these studies were either mathematicians or engineers skilled in mathematics, such as those formed at the École polytechnique. Among them were Claude Louis Navier (1785–1836) and Adhémar Jean Claude Barré Saint Venant (1797–1886) [18]. The beam appeared as a particularly interesting structural element, because it was relatively simple and the foundation of almost all steel constructions. The theory of beam was carried out satisfactorily by Saint Venant. His results, however, were too complex to be used by a practitioner, even a good one.

So many simplifications were introduced by engineers and mathematicians. However, there was a critical problem to be solved: how could the results obtained for the single beam be applied to the case of frames, that is assemblies of beams? From the middle of the 19th century, relatively simple approaches were developed by scholars commonly classified as engineers. Among them played a fundamental role Christian Otto Mohr (1835–1918) and Alberto Castigliano (1847–1884), who completed the work of their predecessors. Mohr generalized the principle of virtual works for the calculation of elastic structures; Castigliano introduced an energy method that required minimizing a function of a limited number of discrete variables. The two approaches led to the same solution: an algebraic system of linear equations [18]. The cycle at this point was essentially closed. A very general theory, the theory of elasticity, developed by mathematicians and physicists, elegant but useless in itself, was transformed into less and less general theories, sometimes justified only by the fact that they worked, developed by skilled engineers.

Although each structure poses problems peculiar to its requirements and environment, it shares salient features with other structures similar to it. Thus civil engineering structures are susceptible to generalization and mathematical representation and structural analysis was the first engineering science to mature. In contrast, machines serve countless complex functions. Their forms and operations, diverse and specialized to their functions, are less prone to generalization and theorization. General principles of heat or fluid flows that underlie various types of engines were discovered later and only with significant input from engineering practices. Mechanical engineering has tended to be more empirically oriented, especially in its early days when the relevant sciences were embryonic. Natural science was of course relevant, but its contribution was more in the use of systematic reasoning and controlled experimentation than in specific theories [6].[7]

The engineer, as most technologists, has often to solve a problem more or less defined in a given time; that is, he must quickly find a satisfactory solution at all costs. To carry out his work he draws on his knowledge that can derive from science commonly said or from the elaborations provided by engineer-researchers who dealt with problems similar to his. However, it may happen that such knowledge is not enough for him; indeed it is usually so. In this situation he can decide to study the problem and propose new theories, if he has enough time and culture. Or, and this is the most common situation, he can resort to experience not interpreted by any theory. This experience can be codified in specific norms, today very articulated, arranged by engineering companies. If the engineer does not find indications, he must recur to his personal experience and common sense (which often coincide). In the case of structural engineering, the choice is simple. To design a structure that he did not understand very well, the engineer can decide to make it 'very robust', even if much more expensive.

A scientist is almost never found in the situation of the engineer. Usually he does not have a time limit or to face a well defined problem. Furthermore he must understand the profound reasons of a phenomenon and not only suggests gross relations

[7]p. 34.

that offer predictions satisfactory in many cases, but not always. I said 'almost never found', because in modern organizations the scientist also often has time constraints, on pain of cutting funding for his research, and this can put him in a situation similar to that of the engineer, leading to formulate not well corroborated theories.

In the recent past has been proposed a new point of view which sees a strict link between science and technology, reproducing somehow the equation applied science = technology: the category of *finalized science*, carried forward in the 1970s by some German philosophers and scientists [11, 67]. Finalization is intended as a process through which goals, external to science, become the guidelines of the development of the scientific theory itself. Finalized science is seen as a precise characterization of applied science:

> The term 'applied science' gives the misleading impression that goal oriented science simply involves the application of an existing science, rather than the creation of a new theoretical development. This in turn feeds the misconception that pure science is superior to applied science [39].

Agricultural chemistry is a typical example used to illustrate the concept of finalized science. It illustrates a number of interconnections between social needs, cognitive patterns in science and strategies for the institutionalization of science which are relevant to the development of science and technology. In particular, reflection on agricultural chemistry allows considerations of the interaction between, the existence of a social problem, the perception of this problem, the limited ability of science to offer a solution to the problem at a particular stage of development, the development of experimental techniques and models, the institutionalization of a variant in science [39].[8]

The theory of finalized science assumes natural sciences as paradigmatic and states that these sciences go through three successive phases. First, the exploratory phase, which corresponds to Kuhn's pre-paradigmatic phase [40]. In this phase, a well structured theory is not (yet) available and the research methods are mainly empirical either quantitative or taxonomic rather than theoretical and explanatory. Next, the paradigmatic phase guided by a general theory that structures the field of phenomena and directs the way in which it should be investigated. This corresponds to Kuhn's normal science, the goal is the validation of central theoretical ideas. These second-stage developments can continuously evolve into fully developed or *closed theories*. In general two things can be said of a closed theory:

1. Scientists repeatedly express the conviction that all the essential work in a certain field has been done.
2. Theories whose internal potential for development has been exhausted retain their paradigmatic significance in pragmatic contexts. In particular, they remain the basis for the development of technologies, where they are regarded as suitable instruments for dealing with particular classes of empirical questions. As far as solving such pragmatic problems is concerned, the improvements represented by

[8] p. 46.

their successors are irrelevant. The orbits of space vehicles are still calculated using Newtonian, not Einsteinian, mechanics for instance [11].[9]

Older theories achieved the status of closed theories. An example of closed theory is furnished by Clifford Ambrose Truesdell for classical mechanics:

> The word 'classical' has two senses in scientific writing; (1) acknowledged as being of the first rank or authority, and (2) known, elementary, and exhausted ("trivial" in the root meaning of that word). In the twentieth century mechanics based upon the principles and concepts used up to 1900 acquired the adjective "classical" in its second and pejorative sense, largely because of the rise of quantum mechanics and relativity. "Fundamental" in physics came to mean "concerning extremely high velocities, extremely small sizes, or both". Physicists gave less and less attention to classical mechanics because they thought nothing more could be learned from it and nothing new discovered about it, although of course they continued to use it in the design of the experimental apparatus with which they claimed to controvert it. At about the same time "applied" in mathematics came to refer not to the object studied but to the originality and logical standards of the student, again in a pejorative sense. Engineers still had to be taught classical mechanics, because in terms of it they could understand the machines with which they worked and could devise new machines for new purposes. Research in mechanics came to be slanted towards the needs of engineers and to be carried out largely by university teachers who regarded mathematics as a scullery-maid, not a goddess or even a mistress. Leading exponents of applied mechanics were Ludwig Prandtl (1875–1953) and Geoffrey Ingham Taylor (born 1886) [82].[10]

According to the theory of finalization the close theories are more or less complete. However, they can further develop in a third phase, in which they are oriented towards external objectives and interests through the development of "special theories" in order to realize certain application technologies. It is at this point that science is finalized. Contrary to Kuhn, at this stage the "practical merit" and the "approval outside the group of specialists" are primary values, and yet achieving this merit requires the development of truly new theoretical knowledge [60].[11]

But also equating finalized science and technology has its own difficulties. Mainly even the knowledge gained in the third stage of science is only a part of the knowledge required for design, production and maintenance of technological artifacts or systems.

5.1.1.2 Similarity Between Science and Technology

To counter the equation applied science = technology, one can also show that science and technology are not so different and that pure science is not detached from the activities considered typical in technology and vice versa that technology, at least in its post 18th century version, does not lack the theoretical elaborations proper of science. Some philosophers, probably motivated more by a sociological ideology than by a profound knowledge of science and technology are sustainers of the equation

[9]pp. 131–132.

[10]Part II, pp. 127–128.

[11]p. 75.

science is technology; it is for instance the case of Martin Heidegger and Jürgen Habermass [60].[12]

One aspect of science that has a technological feeling is experimentation. A scientific theory refers to quite schematic situations; when it has to be tested it is usually necessary to introduce some assumptions which allow the predictions of the theory for the experiment be carried out. That is, as a matter of fact, in science one is not confronted with two types of activity (theoretical and experimental), but with three: the elaboration of the theory, its development to make possible empirical tests and the design and performance of experiments [60].[13]

In science the experimental activity requires usually instrumental action. A characteristic feature of experimental science is that access to its objects of study is mediated through instruments and/or other equipments or devices. In an experiment, it is searched a correlation among some readings (usually numbers) of instrumentation. Important necessary conditions for the success of the experiment is its stability and reproducibility which implies a control of the experimental set and its environment that should be maintained throughout. Classically three kinds of interactions are distinguished. The *required interactions*, which make the experiment to behave according to its design; the *forbidden interactions*, which might disturb the intended experiment and the *neutral interactions*, which neither enable nor disturb the experiment. Like experiments, making technology to work is necessary that its process be stable and reproducible and the control of the relevant interactions with the environment constitute a necessary condition for this goal [60].[14]

According to Srdjan Lelas, the activity of observing and experimenting is essentially technological in nature and constitutes an essential ontological element of science. A theory could not be treated as a simple contemplative activity of experimental outcomes. A theory should be rather considered as a condensed set of instructions of how to build an experimental apparatus, or, better, how to guide the production of experimental artifacts [43].[15] A different view is that of Bruno Latour; he sees the scientists as builders of facts and the technologists as builder of artifacts. The problem of the builder of 'facts' is the same a that of the builder of 'artifacts': how to convince others, how to control their behavior, how to gather sufficient resources in one place, how to have the claim or the object spread out in time and space [42].[16]

In modern times, in many fields of research, it is still more challenging to distinguish between science and technology. Science has becoming *big science* and has acquired the format of an industrial organization, which is typical of technology. Moreover, nano-sciences, molecular biology, the new frontier of natural sciences, need complex machineries whose use and construction typically imply technological aspects. And scientists and technologists, engineers in particular, generally differ very little in the choice between knowledge and utility. That between science and

---

[12] p. 83.

[13] p. 72.

[14] p. 81.

[15] p. 442.

[16] p. 131.

technology is a fuzzy distinction with important overlap, which is larger in complex projects [6].[17]

5.1.1.3 Epistemology of Technology

Strictly connected with theme of the difference between science and technology, actually a particular point of view of it, is the debate that has been going on for some time about whether technology has its own epistemology, if instead it shares this with science or if the question is meaningless because technology does not represent a form of knowledge. The answer is far from being shared. The possibility that technology has its own epistemology distinct from that of science is supported by various authors in modalities that are well summarized in [35]. The following possibilities are discussed:

1. Is the thesis that science points to truth and technology to utility true?
2. What is the cognitive value of the theories that express approximate models of reality. Do they have only an instrumental character? What relationship is there between instrumentalism in technology and science?
3. Is the form of knowledge of technology tacit, that is, part of the knowledge produced by technological practice impossible to make fully explicit in declarative statements, but only acquired through personal experience?
4. Finally is it true that technology differs from science because the former has a descriptive character, the latter a prescriptive one?

In [51] it is suggested a different point of view, by distinguishing between *engineering philosophy* of technology and *humanities philosophy* of technology. The former addresses the problem in the way presented in the foregoing list. The latter sees primacy of humanities over technology and bears on interpreting the meaning of technology with respect to man. A modern starting point of this view can be routed in Jean Jaques Rousseau's *Discourse sur les sciences et les arts* of 1750. Subsequently Romanticism saw modern technology as somehow obscuring essential elements of life. The same view can be found in some modern philosophers of the 20th century. The humanities point of view has implications on the epistemology of technology, but it is not considered here; the interested reader can find suggestions in [51]. Below only the first two lines of the previous list will be briefly discussed

The very idea that science aims at truth while technology aims at use, has its roots in Greek philosophy; for instance in the distinction made by Aristotle between episteme and techne, a distinction which has long influenced the epistemology of science and technology in the western world. Usually episteme is translated as theoretical (pure) knowledge, whereas techne as art or (experience-based) practical knowledge. Below as Aristotle defined episteme and techne, in the order, in his *Ethica Nichomachea*:

> Therefore the object of scientific knowledge is of necessity. Therefore it is eternal; for things that are of necessity in the unqualified sense are all eternal; and things that are eternal

[17] p. 16.

> are ungenerated and imperishable. Again, every science is thought to be capable of being taught, and its object of being learned. And all teaching starts from what is already known, as we maintain in the Analytics also; for it proceeds sometimes through induction and sometimes by syllogism. Now induction is the starting-point which knowledge even of the universal presupposes, while syllogism proceeds from universals. There are therefore starting-points from which syllogism proceeds, which are not reached by syllogism; it is therefore by induction that they are acquired. Scientific knowledge is, then, a state of capacity to demonstrate, and has the other limiting characteristics which we specify in the Analytics, for it is when a man believes in a certain way and the starting-points are known to him that he has scientific knowledge, since if they are not better known to him than the conclusion, he will have his knowledge only incidentally. Let this, then, be taken as our account of scientific knowledge [4].[18]

> Art [techne] is essentially a reasoned state of capacity to make, and there is neither any art that is not such a state nor any such state that is not an art, art is identical with a state of capacity to make, involving a true course of reasoning [...]
>
> All art is concerned with coming into being, i.e. with contriving and considering how something may come into being which is capable of either being or not being, and whose origin is in the maker and not in the thing made; for art is concerned neither with things that are, or come into being, by necessity, nor with things that do so in accordance with nature (since these have their origin in themselves). Making and acting being different, art must be a matter of making, not of acting. [...] Art, then, as has been is a state concerned with making, involving a true course of reasoning, and lack of art on the contrary is a state concerned with making, involving a false course of reasoning; both are concerned with the variable [4].[19]

In these quotations Aristotle considered episteme and techne as two different forms of knowledge with distinct objects as their aims. Episteme refers to eternal objects, ungenerated and imperishable, that is objects that do not undergo changes. Techne to objects that have their origin in the maker and are ephemeral, generated, perishable, that is objects that undergo changes.

In substance in the *Ethica Nichomachea* Aristotle saw the distinction between science (episteme) and technology (techne) in the ontology of the objects the two forms of knowledges aim. In other writings Aristotle suggested however a different view. The difference between episteme and techne is seen in the way the cause of a thing is known. In the *Analytica posteriora*, Aristotle said that one possesses episteme when he knows the cause with which the thing is, and that thus cannot be otherwise [3].[20] The example he gave for the case one has episteme is geometry. In the *Metaphysica*, Aristotle undermined the possibility of episteme in the strict sense of the *Analytica posteriora*: "The minute accuracy of mathematics is not to be demanded in all cases, but only in the case of things which have no matter. Hence method is not that of natural science; for presumably the whole of nature has matter. Hence we must inquire first what nature is: for thus we shall also see what natural science treats of (and whether it belongs to one science or to more to investigate the causes and the principles of things)" [5].[21] Instead of grasping what is eternal

[18] 1139b, Book 6, part 3.

[19] 1140a, Book 6, part 4.

[20] 71b, Book1, part 2.

[21] 995a, Book 2, part 3.

and necessary, knowledge can only grasp what happens for the most part, that is the regularity of nature, to which there can be exceptions [12].[22]

Aristotle exemplified in his writings the difference between techne and episteme when he explained the kind of knowledge in medicine. According to Aristotle this form of knowledge is represented both by episteme, when it studies health, and techne, when it produces health. Here it is clear that Aristotle is not using his classification based on ontology. While he is contrasting techne and episteme, meanwhile he saw similarity among them. A person who possesses technique is superior to whom has only experience (empiria), because he uses universal judgement.

> So the person with techne is like the person with episteme; both can make a universal judgment and both know the cause, which is not a necessary, universal truth but knowledge of the regularities of nature, to which there are exceptions. In other words, scientific and technological knowledge start to overlap when concerning objects that are considered as stable and regular but not so much as necessary and universal [12].[23]

About the assumption that the epistemology of technology has an instrumental character (line 2 of the previous list), it can be said that even some scientists and philosophers of science, a minority in truth, see scientific theories as purely instrumental and do not attribute them any cognitive value of 'reality'. But even though one embraces a realistic view of scientific theories, he must confess that they always present an instrumental aspect. Indeed the relationship between a scientific theory and the external world has either a cognitive or an instrumental nature. The first case occurs when using the data provided by the experience to validate the theory; the second when the theory is used to predict a phenomenon. And from this last point of view the difference between science and technology seems to be very little.

But a different answer can also be given, showing that even the instrumental aspect of technology is not so fundamental. It is true that the technologist, the engineer for instance, has a primary interest in the use of theories and often rather than trying to understand phenomena he is content to provide a their description, often a mathematical relationship. This relationship may be approximate only, even crudely; in such a case adequate safety coefficients are assumed. For example, if an engineer has to design a cable to support a weight, experiments are performed on various kinds of cable that could be used for the purpose, by estimating the average value P^* of the weights that lead them to failure. For safety reasons, given the dispersion of experimental results of the breaking tests, and the uncertainties on the value of the weight to be supported, it is assumed that the cable should be subjected not to the weight P^* but to a lower value P^*/ν, where ν is the safety coefficient, greater then 1, which can sometimes even reach 2 or 3. But the engineer-researcher will generally try to refine his knowledge of the phenomenon to be studied, because the only way to be able to produce an accurate and economic project is to study in depth the phenomena in order to understand them as the scientist does, so he can use a unitary safety coefficient ν.

---

[22] p. 53.

[23] p. 54.

Concluding, it can be said that the main difference between science and technology is that the former considers more relevant the cognitive moment, the latter the predictive one.

5.1.2 A Historical Perspective

5.1.2.1 Technology as Magic

Technology as a form of knowledge for the production of objects useful for the mankind is as old of the mankind. Ancient technology was driven by two engines. One represented by the craftsmen, the mason masters etc, who carried on their work based on knowledge handed down from father to son. Knowledge that even when transmitted in written form or by means of drawings, did not have an organic character. Another engine was represented by educated men, that could be either mathematicians (mostly), natural philosophers, physicians, magicians and alchemists.

The attention of historians of technology has focused so far on mathematicians and natural philosophers, and, even though to a lesser extent, on physicians. But an important role was also played by magicians (and alchemists) at least since the Renaissance, a period in which the activities of those who today are classified as engineers and those as magicians overlap.

During the Renaissance the only form of magic allowed by Church was *natural magic*, understood as a practice that put into action the knowledge about what was then considered as nature but avoided any interventions of demons. Natural magicians could get effects that were natural but that nature left alone would not have produced or would have produced with extreme difficulty. Pietro Pomponazzi (1462–1525) drastically limited the meaning of the term natural magic: natural magic is simply the activity that results in the study of the occult (in the sense of unknown) virtues of physical entities. Experience allows one to record the action a body exercises on another body and to classify individual bodies according to their properties. It is not possible to explain the "why" of the occult virtues, because they work according to "a subtle quality that we do not know", but we know that alterations follow persitent laws of nature.

Natural magic, although generally devoted to arouse the wonder of people, had a very little esoteric core, known as artificial or mathematical magic [32],[24] in which there was no reference to hidden causes and in this case the magicians operated like engineers with the application of known rules of technology, both of medieval and Hellenistic tradition. Using optical, hydraulic and mechanical techniques rather than celestial influences, they created contrivances that rivaled and outdid the power of nature herself. Cornelius Heinric Agrippa (1486?-1535) in his *De incertitudine et vanitate scientiarum atque artium declamatio invectiva* printed in 1527 and translated

[24] p. 11.

into English in 1676 [1] as *The vanity of arts and science*, described the role of mathematical magic as follows:

> Of Mathematical Magick.
> There are besides these, many other imitators of Nature, wise inquirers into hidden things, who without the help of natural Virtues and Efficacies, confidently undertake, onely by Mathematical learning, and the help of Celestial influences, to produce many miraculous Works, as walking and speaking Bodies; which notwithstanding are not the real Animal: such was the wooden Dove of *Archytas*, which flew; the Statues of *Mercury*, that talk'd; and the Brazen Head made by *Albertus Magnus*, which is said to have spoken. In these things *Boetius* excell'd, a man of a large Ingenuity, and manifold Learning; to whom *Cassiodorus* writing upon this Subject, Thou, saith he, hast propounded to thy self to do great things, and to know the most difficult: by thy ingenious skill Metals are heard to roar, Brazen *Diomed* sounds a Trumpet, a Brazen Serpent hisses, Birds are counterfeited, and they that are incapable of a voice of their own, yet are heard to make a sweet noise: We relate but small things of thee, that hast so great a power to imitate Heaven. Of these delusory Sciences may be said that which we read in *Plato*'s tenth Book of Laws: Art is given to Mortals, which enables them to produce certain posterior and succeeding Inventions, neither pertaking of Truth or Divinity, but certain Imitations somewhat akin thereto: Wherein Magicians have adventured to proceed so far, by the help of that ancient [1].[25]

For their part, the engineers, in addition to carrying out public utility works, civil and military, also had to take care of entertainments, every time a ruler entered a city subjected to him, married a child and so on. For instance, to inspire awe in the spectators Filippo Brunelleschi (1377–1446) made angels and the same Christ fly on a mechanical mandorla, and many other engineers equipped their inventions with similar automata, and from this point of view they acted as magicians.

Giovanni Battista Della Porta (1535–1615) among the supporters of magic was perhaps the one for which magic and technology were mostly mingled. Della Porta wrote a book on magic, *Magia naturalis*, in his early twenties; the text was a huge publishing success in Italy and abroad and saw a second expanded edition in 1589, after more than 30 years. It was written into Italian and published in 1611 under the title *Magia naturale* [22]. If a modern reader looks at the *Magia naturale* skipping the preliminary pages in which the essence of magic is presented in a traditional way, he will soon wonder in what sense the text of Della Porta could lead to the term magic in the title. Of course there are chapters devoted to astrology and others to alchemy, topics that today are regarded with suspicion and associated with superstition. But the rest of the chapters have a technological character that gives the solution of more or less interesting practical problems.

In Book XVII (20 books in total) of *Magia naturale*, optical topics are considered. Dalla Porta opened the book declaring it of mathematical nature and asserting that optics is the most important among the mixed mathematics, emphasizing its practical utility: "But what can be imagined more ingenious, than to mathematical demonstrations, imagined by the soul, to which most certain experiences follow?" [22].[26]

[25] p. 113.

[26] p. 472.

What differentiates the treatment of Della Porta from those of traditional optics is his focus on strange phenomena that can appear amazing but actually follow the rules of geometrical optics; this is typical of magic treatises. Here are some phenomena that Della Porta explained more or less correctly:

That the face of who looks at appears divided in the middle.
That the face of who looks at appears like that of a donkey, of dog or pig.
Make that farther letter can be read in the walls.
How to make a mirror that represents only the image you want [22].[27]

Magic lost part of his fashion gradually. One of the last book on the subject by a relevant scholar was *Mathematicall magick, or, the wonders that may be performed by mechanical geometry in two books*, see Fig. 5.1, first published in 1648 by John Wilkins (1614–1672), one of the founder of the Royal society of London [83, 87]. In the foreword to the reader, Wilkins explained the reason of the term magic in the title. It aimed to allude to "vulgar opinion, which doth commonly attribute all such strange operations unto the power of Magick" [87].[28]

Wilkins' book, which had a prevailing didactic purpose, does not lack of interesting philosophical suggestions. It is divided into two books, the former concerning the mechanical powers and entitled *Archimedes*, the latter concerning mechanical motions and entitled *Daedalus*. *Archimedes* deals with the six classical machines in the order, the balance, the lever, the wheel, the pulley, the wedge and the screw; it looks like to books of the previous century, in particular to the *Mechanicorum liber* by Guidobaldo dal Monte (1545–1607). Only in the final chapters the term magic finds its reason, but not in the sense of occult but in that of wonder.

In Chap. 12 Wilkins discussed about the possibility to raise the earth by a lever, on the footsteps of the famous assertion attributed to Archimedes. For the weight of the earth he assumed the figure of 2.4×10^{24} pounds, "as Stevinus had calculated" [87].[29] By supposing to raise the earth with a weight of 100 pounds, more or less the weight of a man, from the law of the lever it results a ratio, between the displacement of the earth and that of the man pressing the lever, of 2.4×10^{22}. This means that for what large may be the displacement imposed by the man the motion of the earth is not perceptible. In the end in these situations only mathematics can show that the effect is real: "Therefore though such extreme slowness may seem altogether impossible to sense and common apprehension, yet this can be no sufficient argument against the reality of it" [87].[30]

In the second part of the book the sense of wonder is much more pronounced. This part is devoted to automata, that for him had not the modern meaning (that is robots or androids), but simply indicated machines useful for men activity. Some space is given to the human flight and the wonder it aroused, one of the symbol for

[27] pp. 473–476.
[28] To the reader.
[29] p. 82.
[30] p. 116.

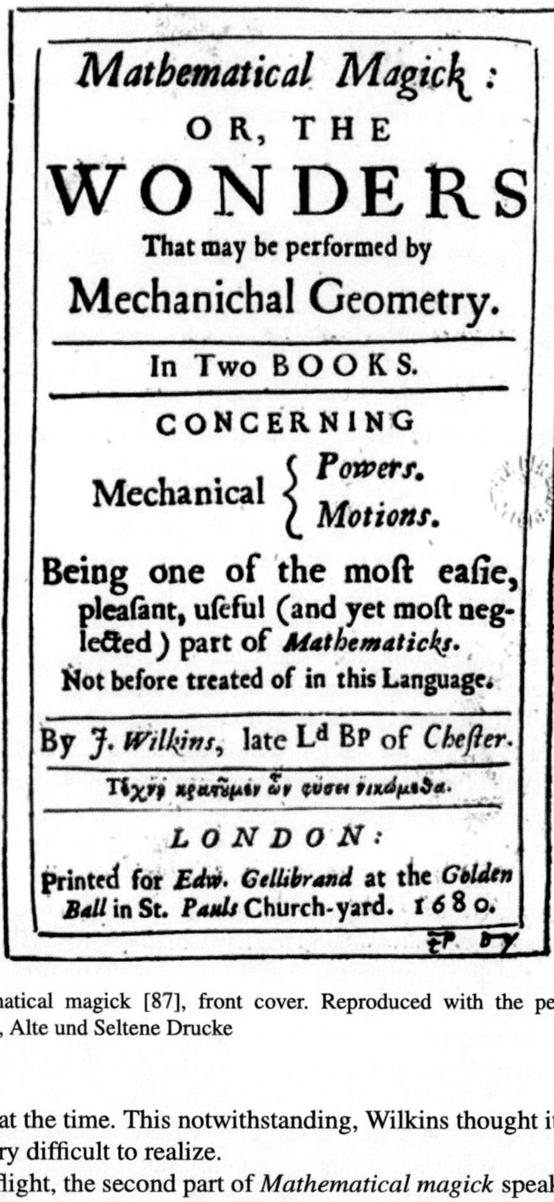

Mathematical Magick :

OR, THE

WONDERS

That may be performed by

Mechanichal Geometry.

In Two BOOKS.

CONCERNING

Mechanical { *Powers.* / *Motions.* }

Being one of the moſt eaſie, pleaſant, uſeful (and yet moſt neglected) part of *Mathematicks*.

Not before treated of in this Language.

By *J. Wilkins*, late Ld Bp of *Cheſter*.

Τέχνῃ κρατοῦμεν ὧν φύσει νικώμεθα.

LONDON:

Printed for *Edw. Gellibrand* at the *Golden Ball* in St. *Pauls* Church-yard. 1680.

Fig. 5.1 Mathematical magick [87], front cover. Reproduced with the permission of ETH-Bibliothek Zürich, Alte und Seltene Drucke

the impossible at the time. This notwithstanding, Wilkins thought it really possible, even though very difficult to realize.

Apart from flight, the second part of *Mathematical magick* speaks of many wonderful items: the sailing chariot, the submarine navigation, the perpetual motion; in any case Wilkins' automata are located at the borderline of the possible and impossible to arouse a sense of wonder but not of impossibility. Still, Wilkins' machines are not just fictional devices, functioning only on the page of a book. The possibility to actually realize these machines is considered important. In certain passages, Wilkins inveighed against speculation and against the limited nature of notional contrivances [83].[31]

In the 18th century magic was no longer in fashion among the learned men. It remained diffuse among the common populace, leaving however its technological contents and maintaining only its exoteric and mysterious aspect. Here as the Encyclopédie described magic:

> But we are gradually recovering from this former attitude and one can say that the awareness of this so-called natural magic is, even in the eyes of the multitude, continually retreating. Under the light of science we are, happily, continuously discovering the secrets and systems of nature, supported by many sound experiences which show humanity of what it is capable itself and without magic . Thus, we see the compass, the telescope, the microscope and, in our own time, polyps and electricity. In chemistry and physics, the most beautiful and useful discoveries will immortalize our era and if Europe were to fall back into the barbarism from which it has finally emerged, we will seem like magicians to our barbarous successors [26].[32] (E.2)

5.1.2.2 Medicine Between Science and Technology

Medicine is considered below mostly for its epistemological aspects in the early modern era, when the problem of finding its epistemological status was part of the more general problem of Humanism to revaluate techne over episteme and to be free from the Aristotelian epistemology. Partly because the humanists favored the active role (though not manual) over the contemplative; partly because they believed that science in the Aristotelian sense was not possible and thus supported a probabilistic approach, in which rhetoric took the place of dialectic and syllogism. In any case the difference between science and technology, or art as was called at the time, was maintained and was more or less the same as considered by Aristotle.

One of the first scholar who posed the problem was Francesco Petrarca (1304–1374) who denounced the unseemly besetting state of medicine that was seen more as harmful than useful by many. Petrarca came in about epistemological debate that remained very lively at least until the 17th century (and still continue today): is medicine a techne (technology, art) or an episteme (science)? or is it something else again? The answer of the scholars of the Humanism and Renaissance was variegated; some dealt with medicine as science, some others as techne and still others neither of them.

Art, in its widest meaning was any set of coherent rules suited to direct human activity. In the limit just a list of precepts, but also—and this was the view of the humanists as as that of Aristotle himself—some form of theoretical elaboration. Since I century AD one distinguished between manual or mechanical arts and liberal arts. The liberal arts were nine: grammar, rhetoric, logic, arithmetic, geometry, music, architecture and medicine. The list shows that it was not respected the division of Aristotle who would have considered some of them as sciences rather than arts. Later,

[31] p. 475.

[32] Article: *Magie*. Translation by Steve Harris

in the V century AD, the liberal arts were reduced to seven, eliminating from the previous architecture and medicine, considered not worthy to be practiced by a free man because they were, as any technology, too attached to the manual.

It must be said that the discussions on the epistemological status of medicine was partly a problem of names. There was no doubt that the physician had knowledge in various fields, natural philosophy, anatomy, mathematics, alchemy etc, which he often helped to develop; and in this sense the physician was certainly a philosopher, mathematician, and so on. And if one called medicine everything the physician did, it was clear that medicine was superimposed to science. But if one considers medicine in a more restricted sense its relation with science is less clear. When the physician was beside a sick person, he had in any case to suggest a therapy. Rarely could he follow the axiomatic approaches of the disciplines he had studied and helped to develop, he had rather to make choices that were not always verbally codified. That is he acted as a technologist.

Medicine, broad and restricted meaning, of the 16th century saw a strong overlap with the Aristotelian natural philosophy; and if the physician could agree to consider the activities they practiced, medicine, as technology, they also saw themselves as learned men. This was one of the reasons that pushed physicians to treat surgery as a separate medical activities; this was in fact an activity mostly manual that would inevitably have led the medicine out of the most prestigious 'theoretical activities', such as natural philosophy.

Niccolò Leoniceno (1428–1524), professor of medicine at Ferrara and humanist, was among the first to pursue the attempt to separate medicine from philosophy of nature and to classify it as an art. Leoniceno supported his argument by referring to the epistemology of Galen which for him was different from Aristotle's [17].[33]

Connected to the problem of classification of medicine as a science or art there was that of division between theory and practice. Regardless of the separation of the professions, the contrast between practical and theoretical medicine had also an epistemological character, investing the mutual role of experience and theory, presenting again the ancient diatribe dogmatic/empirical. The physicians of the humanist tradition gave the major contribution to bring the problem to the conceptions of Greek physicians like Hippocrates of Kos (c 460–c 370 BC) and Galen (II century AD), in whom this division did not exist. This is for example the position held by the physician Giovanni Mainardi (1462–1536) who, though not a direct disciple of Leoniceno, followed him in considering medicine as an art. Professor at the University of Ferrara, he quoted Galen in his *In primum Artis parvae Galeni librum commentaria* of 1536: "In this case the authority of Galen pushes me, who wherever states that all medicine is either productive or repairing and never he makes any division into theoretical and practical" [45].[34]

Giovanni Battista Da Monte (1498–1551), professor of medicine first in Ferrara and then in Padua, considered medicine as a science, albeit of practical nature and thus distinct from natural philosophy. Medicine is subalternate to natural philosophy

---

[33]p.132.

[34]p. 122.

as it considers the human body—that is the object of study by natural philosophy—exclusively only from what health is concerned; like mixed mathematics which are subalternate to mathematics. For him medicine could not be an art because it had no its own methodology. According to Da Monte, all the principles of medicine should be directed to the end, that is to health. In fact the best physicians, as Hippocrates of Kos and Galen, considered all things in medicine as they were physicians and not philosophers. In medicine therefore one has to consider everything with respect to the health, one seeks to preserve, if present, or to recover if it has been lost [45].[35]

Da Monte addressed his criticism toward modern empirical physicians. For him they had only countless recipes to sell but not a real method. The empirical physicians was deaf and blind; they lent their care based on their previous observations and did not know the general rules. They healed only diseases they already knew [45].[36]

A counter current approach in the Renaissance was that of the physician Leonardo Fioravanti (1518–1588), an outsider to the academic environment, who had a decidedly empirical approach and insistently recalled to address to practices, with the idea that nature is the teacher of all things. He went so far as to say that even animals had their own medicine: "This medicine is common in all the people of the world and a part belongs to the irrational animals, a part to the peasants, another part to the women and another part to the rational physicians who posses it by means of the theory. It is the weakest of the all others and we can never use it, if we did not test it with the experience, which is what is proper of the peasants" [27].[37] Experience, is thus considered so much the fundamental element of any medical knowledge that the three quarters of it is owned by any people, while the fourth part of it, it is that possessed by physicians

To explain the difference between empirical and theoretical approaches, Fioravanti appealed to the example of the continue fevers. As science requires the knowledge of the cause of infirmity, the physicians, to be faithful to their status, are required to indicate which is precisely the cause of such fevers, locating them in the "decay of blood, which is being corrupted in the veins". Fioravanti did not deny at all that conclusion, indeed he inclined to accept it, since it seemed absolutely 'likely'. But if it appeared such to him, that was not because of some syllogism constructed starting from first principles of physics, but rather because the experience showed that when blood is taken out from the veins of feverish people, it is altered and corrupted [45].[38]

Fioravanti did not give up a theoretical horizon of reference, which served to reprocess the data of experience and to derive operational rules or hypotheses to be tested with further experience. That is, he was a supporter of techne and not only *empiria*. In the case mentioned above of continue fevers, for example, after stating that one had to rely on experience rather than science, he added that, despite everything, he could still believe in science, which was not very different from the experience in this case. Because true medicine is nothing else, that the theory of experience,

[35] p. 139.

[36] p. 142, note 17.

[37] p. 8v.

[38] p. 164.

as well one can see from those who are experts in this profession. Medicine never had principles from other things than from experience, because before the virtues of herbs, stones and animals were discovered one never found anyone who had written some theory [45].[39]

The empirical approach proposed by Fioravanti was not shared by most; it forced physicians to test the actual efficacy of their results, that is to make experiment, or to use an expression of the time, *periculum facere* who one could come up with to justify a certain treatment and its possible failure. This approach, however, posed serious ethical problems. Not for nothing the medical schools highlighted the other meaning of the term periculum, that is probable damage, according to the interpretation of the Hippocratic derivation of the *experimentum periculosum*, considered by Galen as a warning to the physicians because they did not abuse of their power venturing innovative interventions on their patients. Indeed, those on which he was operating were human beings and not, as in the case of the carpenter or locksmith, pieces of wood or metal, that once damaged could be repaired or otherwise thrown away without special torment: "*vita brevis, ars longa*" [29].[40]

5.1.2.3 The Emergence of the Scientific Engineer

The relationship of science with engineering has a more complex history than that of science with medicine. Until the second half of the 20th century, there was no real debate about the epistemology of engineering among scientists, technicians and philosophers. There was, especially since the end of the 17th century, a debate on the social role of science, on whether it should be oriented to pure knowledge or to the satisfaction of the needs of the human race. A debate must be said that interested more philosophers than physicists, mathematicians, chemists, etc., who had always been concerned with finding applications for their studies. Among the philosophers promoting the social utility of science there was certainly Francis Bacon, whose influence was strong in England, flanked by the Puritan ideology that saw science as a means of improving the welfare of men. This implied, in some way, more than the involvement of scientists in practical activities, a greater interest of technicians toward science.

The evolution of economy and society of the modern era was strongly influenced by three fundamental discoveries: circum-navigability of the earth, gunpowder, printing. The ability to circumnavigate the globe was perhaps the most important one, leading to a boost of the economy of nations. It also entailed the development of navigation techniques with the use of the compass, the representations of geographic maps, the improvement of astronomy for navigation looking at the stars, the crafting of ships, which no doubt provided a stimulus to the improvement of many applied sciences [70]. The spread of modern artillery based on the propellant effect of gunpowder was important, especially for the history of mechanics. This was true espe-

[39]p. 169.

[40]Vol. 17/2, p. 353.

cially since the XVI century, when artillery had become extremely effective. The development of artillery had as a natural consequence the development of defensive techniques. This gave birth to bastioned fortresses, first appeared in Italy, then a battleground for national and foreign armies. Perhaps even more than artillery, fortress design led to the development of methods of construction and a better understanding of the strength of materials and thus to the affirmation of a new class of technicians.

Alongside the external economic and social thrusts, associated with these developments, there was an internal thrust due to a cultural awakening, favored among other things by the schools of the abacus, which led the most curious and intelligent craftsmen to see their profession differently. The activity of craftsmen were regulated by rules imposed by the corporations of arts. The increasing complexity of military and civil enterprises urged the rulers to assume specialized qualified technicians with skills in mathematics. The same held for the public administrations.

Starting from the 15th century stable administrative structures were created to control engineering activities. Besides the traditional division of work, between master and apprentice, other levels born. This led to the birth of a new category of scholars. Though it is difficult to precise the characteristic of this figure, the word engineer is the most recurrent, associated to a specification: artist-engineer, scientist-engineer, architect-engineer, administrator-scientist-engineer.

The emergence of the figure of the engineer, seen as a technician in some way educated in sciences, is a characteristic feature of the 15th century and the first half of the 16th [31]; in a period in which the reduced creativity (real or apparent) of 'pure' scientists was counterbalanced by the great creativity of 'applied' scientists. A short list of engineers from Italy, the nation where the technological development was relevant, is sufficient to give an idea of the dimension of the phenomenon: Mariano di Jacopo, better known as Taccola, (Siena, 1381–1458), Leon Battista Alberti (Genoa, 1404–1472), Francesco di Giorgio Martini (Siena, 1439–1501), Leonardo da Vinci (Vinci, 1452–1519), Vannuccio Biringuccio (Siena, 1480–1539), Francesco de' Marchi (Bologna, 1504–1576), Giovanni Battista Bellucci (San Marino, 1506–1554), Daniele Barbaro (Venice, 1513–1570).

The first step in the new era for the development of engineering was probably the introduction of a design phase of the artifact, clearly distinct from the execution phase, by means of drawings, which after the invention of the perspective became more and more understandable. The role played by what is now considered the science of the time (natural philosophy and mixed mathematics together with magic and alchemy) is more difficult to evaluate.

In the 16th and 17th centuries, a period little explored by the historians of engineering, with the development of sciences, or rather of mixed mathematics, the great engineering works were entrusted to figures skilled in mathematics and philosophy of nature. Often these scholars were called as consultants to assist the technicians of the local structures. And there was a coexistence of technicians who had taken possession of theoretical tools and theorists who also dealt with practical problems.

Correspondence of some leading exponents of early modern science, shows how often they were required to intervene as experts in the discussion of engineering problems. A particularly significant example is that of Simon Stevin (1548–1620), active in various fields of engineering. He worked in the service of the prince Maurice of Orange-Nassau and was part of various commissions for the realization of engineering projects. Throughout his life Stevin took care of the construction of mills and other hydraulic works, obtaining numerous patents for the invention of mechanical devices. His attempts at theoretical analysis of complex engineering problems resulted in the compilation of short treatises dedicated to topics such as, for example, the prevention of wear in the design of gear wheels. In Stevin's unpublished writings there are also detailed attempts to calculate the efficiency of windmills used for water drainage.

Galileo, the prototype of 'scientist', gave advices to the Arsenale of Venice, he built lenses and military compasses, he taught the art of fortifications. He operated as a consultant for the water regulation of the river Bisenzio for the Grand Duke of Tuscany. Torricelli also, normally considered a pure mathematician, had a technologic counterpart as a builder of lens and other scientific instruments. Not even Newton disdained manual works. It must be said, however, that for most technicians the influence of science, even the most developed at the time, mechanics, was moderate and therefore in some way science and technology moved separately, although some important interactions occurred.

Things changed in the 18th century due to the economic pressures that in the second half of the century, especially in England, led to the industrial revolution. In this century, mechanics, including hydraulics, acoustics, fluid dynamics, was very developed and electricity thermodynamic and modern chemistry were growing. Specialization in the sciences was certainly greater than in the previous centuries, but the figure of the pure theoretician was still rare. Most scholars also devoted themselves to aspects classifiable today as technological, and in many cases it was difficult to say which of the two roles, scientists and engineer, was prominent.

Toward the end of the 18th century the demand for qualified technicians became so high that sporadic interest in the application aspects by physicists and mathematicians was no longer sufficient. In France the government, to satisfy the increasing requests of military and civil projects, founded the École royale des ponts et chaussées in 1747, the École royale du génie de Mézières in 1748 and in 1783 the École des mines. In England the new bourgeois classes encouraged the establishment of associations devoted to practical applications of the new results of science. Through masonic lodges, coffee house lectures, dissenting academies, mechanic institute and many private societies, scientific ideas, in particular Newton's ones, were disseminated [20].[41]

In 1771 there was the first known formal meeting of professional engineers in England, when seven of the leading engineers of the time agreed to establish a *Society of civil engineers*. The leading light of the new society was John Smeaton commonly referred to as the father of civil engineering in modern day; the other founding members were Thomas Yeoman, Robert Mylne, Joseph Nickalls, John Grundy, John Thompson and James King. In the first year they were joined by John Golborne, William Black, Robert Whitworth and Hugh Henshall and these eleven

---

[41] p. 126.

were known as the original members. The society existed as a dining club which facilitated intellectual discussions and communication of ideas and knowledge of different disciplines within civil engineering. While this was seen as a step forward, the society maintained an informal status, held a limited number of technical meetings and comprised a restricted membership policy. The Latin motto "Omnia in numero, pondere et mensura" was added to the summons card in 1793.

There were three classes of membership: (i) First class; those who were actually employed in designing and forming works of different kinds, in the various departments of engineering; in substance engineers properly said in modern terms. (ii) Second class; men of science and gentlemen of fame and fortune (honorary members). (iii) Third class; various artists, whose professions and employments, were necessary and useful thereto as well as connected with civil engineering (honorary members) [37].

Below as the foundation of Society of civil engineers is described in the preface to the *Reports of the late John Smeaton*.

> In all the polished nations of Europe, this was, and is, a profession of itself, and by itself. Academies, or some parts of such institutions, were appropriated to the study of it, and of all the preparatory science and accomplishments necessary to form an able artist, whose profession comprehends the variety of objects on which he is employed; and of which the present work is an example and a proof.
>
> In this country, however, the formation of such artists has been left to chance; and persons leaned towards the public call of employment, in this way, as their natural tum of mind took a bias. There was no public establishment, except common schools, for the rudimental knowledge necessary to all arts, naval, military, mechanical, and others.
>
> Civil Engineers are a self-created set of men, whose profession owes its origin, not to power or influence, but to the best of all protection, the encouragement of a great and powerful nation; a nation become so from the industry and steadiness of its manufacturing workmen, and their superior knowledge in practical chemistry, mechanics, natural philosophy, and other useful accomplishments [...].
>
> The same period. gave rise, also, to an association of some gentlemen, employed as above mentioned. They often met accidentally, prior to that union, in the Houses of Parliament, and in courts of justice, each maintaining the propriety of his own designs, without knowing much of each other. It was, however, proposed by one gentleman to Mr. Smeaton, that such a state of the Profession, then crude and in its infancy, was improper: and that it would be well, if some sort of occasional meeting, in a friendly way, was to be held; where they might shake hands together, and be personally known to one another; that thus, the sharp edges of their minds might be rubbed off, as it were, by a closer communication of ideas, nowise naturally hostile; might promote the true end of the public business upon which they should happen to meet in the course of their employment; without jostling one another with rudeness too common in the unworthy part of the advocates of the law, whose interest it might be to push them on perhaps too far, in discussing points in contest [78].[42]

In France, after the Revolution, a state institution was born with the declared objective of modernizing the figure of the engineer, radically modifying the didactic and cultural approach of the pre-existing engineering schools, such as the École des ponts et chaussees. An École centrale des travaux publics (renamed later *École polytechnique*) was then created in 1794, which will become a model for all the schools of

[42]Vol. 1, pp. 5–6.

engineering on the Continent, which officially sanctioned the birth of what is now called scientific engineer.

A fundamental role on the birth of this school was played by Gaspar Monge, know today as the father of descriptive geometry and ranked in a too simplistic way as a mathematician. Monge took as a model the teaching he carried out at the École royale du génie de Mézières with the aim of remodeling the engineering profession. This profession was badly defined at the time; one could only say that the engineer conceived an artwork and controlled its execution, while the artist or craftsman executed it. Engineer training traditionally took place in the field, often handed down from father to son. In the École royale des ponts et chaussées the organization of teaching remained similar to that of the Ancien regime. The mathematics lessons were given by the most skilled students and the practice was acquired by the engineers who practiced their profession. In the Mézières school, the theoretical teaching was instead of a much higher level.

The École centrale des travaux publics, according to Monge's idea had to be completely different and break radically with the old schools of engineering. From his experience as a professor at Mézières, Monge had gained great distrust toward the *corps savants* of the army and thus decided that teachers should be professors. According to Monge, the school should not only form engineers but also spread the passion for the exact sciences throughout France. It had to be a school of savants that had to take into account also the practical applications of acquired notions. The teaching was organized around magisterial courses, in which mathematics played the fundamental role, held by specialists of various sectors, mathematicians, physicists, chemists and engineers; between them it is enough to quote: Berthollet, Lagrange, Laplace, Prony. Monge's idea clashed with the tumultuous political life of the time, with him having to abandon the project because his situation had become critical and called for arrest. The school survived but Monge's project was downsized. It ceased to be the encyclopedic school that he had conceived and simply became a formidable preliminary school to those of the bodies of engineers. This change was sanctioned by the law of September 1st 1795, which also decreed the change of name from École centrale des travaux publics to École polytecnique [7].

Far from being a mere place of learning, the École polytechnique was a recruitment and training area where a social type, a culture of the service of the state with its codes, its practices, was built which made the School a true state institution. This approach originated the process of construction and legitimization of a new social elite imposing its power at the heart of the administrative system which was thus unveiled. Originally created by the mountain government to train engineers, an intermediate form of the 'artist' and the 'scientist' the School was rapidly changing, in a place of formation of the bureaucratic elite. By subordinating practical knowledge to theoretical knowledge (one of the reasons for the traditional criticism of 'technocrats'), the aim was to build the progressive empowerment of the social figure of the ancient state engineer [8].

5.2 Scientists or Technologists?

In the preceding pages, science and technology has been opposed, without being able to identify a clear difference between the two forms of knowledge; the debate is still open in current literature. Things become even more complicated if instead of opposing science and technology, the actors operating in the two disciplines are opposed, namely scientists and technologists, understood as human beings. In this opposition there is often a fundamental misunderstanding in current discussions. Let consider the engineer for instance; today the engineer is generally conceived as a professional, like the physician and the lawyer for example. The need of modern society requires a high number of engineers dictated by economic development. Instead, the scientist is conceived as a member of a restricted elite. As far as training is concerned, it can be said that most of those who work as engineers have a degree, or it is supposed to have, in engineering. Those who carry out the activity of scientist have instead a degree in 'scientific' disciplines, such as physics, chemistry, biology for example and maybe have a research doctorate. Most graduate at the faculty of engineering become engineers, while only a small number of graduates in scientific disciplines become scientists. Many of them work in industry, public administration, teaching, or even today, they are unemployed.

When scientists are opposed to engineers, homogeneous categories should be considered; thus one should not contrast the graduate in engineering with the graduate in physics, for instance, but rather the researchers who deal with engineering problems, even if sometimes they carry out professional activity, with researchers dealing with physics, which generally only do this. Both of them have often a PhD.

The above is true for today's world; in the past at least until the end of the 18th century, the distinction between scientist and engineer was even more ambiguous; not for nothing the term *scientist* did not exist yet and *engineer* had a rather vague meaning. Still in the 18th century we often find ourselves faced with characters not easy to classify. For instance Navier and Coulomb in France, Smeaton and Watt in England and Plana and Poleni in Italy, can be classified either as engineers or scientists. The fact that it was the same person who exercised the two activities makes it difficult to think that these are very different activities.

The preparation of engineers was different from country to country. In England, as already mentioned, one moved with private associations. Engineers had a professionally oriented preparation, in which the study of mathematics and physics was important but not very profound. In France, instead, the state was interested in the training of engineers, who had a strong background in mathematics and other scientific disciplines.

In the following, for reasons of space, I will dwell only on the English situation. Here, the industrial revolution began around the middle of the 18th century and the request of civil engineers—the military ones were not prepared—capable of designing the increasingly demanding number of civil and industrial structures, including bridges and new steam machines, as well as of managing many other aspects connected with the industry becomes very compelling.

Many skilled English engineers operated in England after the 1750s; I will refer here to Smeaton and Watt, not so much because they are the most famous, but because in them one can observe at the higher degree the interaction between the scientist and the engineer, which however happens differently than in previous centuries. Now was the engineer who led the scientist way around and not *vice versa*, as usually happened before.

5.2.1 A Civil Engineer: John Smeaton

John Smeaton (1724–1792), one of the foremost British engineers of the 18th century, also gained a reputation as a man of science. At his beginning he was encouraged to follow a legal career and after a sound elementary education, with element of mathematics also, at Leeds Grammar school he was sent to London for further employment and training in the courts. An early inclination toward mechanical arts soon prevailed, however, and he became a maker of scientific instruments, a pursuit that allowed ample scope for both his scientific interests and his mechanical ingenuity.

Early in the 1750s Smeaton began experiments that constituted his chief contribution to science; and during this period he also was busied with several technical innovations, including a novel pyrometer with which to study the expansive characteristics of various materials [71]. By the end of the decade it had become evident that engineering works were more profitable than making scientific instruments, thus Smeaton established himself as a consultant in these fields. During the last thirty-five years of his life he was responsible for many engineering projects, including bridges, steam engine facilities, power stations run by wind or water, mill structures and machinery, and river and harbor improvements. He was a charter member of the first professional engineering society, the already named Society of civil engineers (not to be confused with the later Institution of Civil Engineers), which, founded in 1771, after his death became known as the Smeatonian Society.

Though he did not receive a regular education in natural philosophy and mathematics, Smeaton attended regularly meetings at the Royal society. Under Charles Cavendish's proposal he was elected in 1753 a fellow of the Royal society and not only for his skill in mathematics, mechanics and natural philosophy, but also for his excellence as an instrument maker.

In his research on waterwheels Smeaton reopened the vexed question of the relative efficiency of undershot and overshot wheels. Through experiments on a model wheel he showed that, contrary to common opinion, overshot wheels are twice as efficient as undershot. Beyond this empirical generalization Smeaton displayed his scientific bent by speculating on the cause of the greater loss of energy in the undershot wheel by concluding that it was consumed in turbulence. He published his paper *An experimental enquiry concerning the natural powers of water and wind to turn mills, and other machines, depending on a circular motion* on the Philosophical Transactions of the Royal Society in 1759 [72]. Smeaton's paper was so well received to award the author with the Copley Medal, the highest honor of the Royal society. Following this initial success in research on applied mechanics, Smeaton's interests drifted toward physics and he devoted two experimental investigations to the vis viva dispute and the laws of collision (see below) in 1776 and 1782.

Smeaton considered himself as an independent engineer and began to call himself (among the first) a *civil engineer*. The first official mention of the term civil engineer appeared in a 1763 London directory addressed to Smeaton and Thomas Yeoman (1709?-1781). As a pioneer, Smeaton was self-taught. For his design work he usually prepared sketches that were elaborated by his draughtsman [38]. Smeaton's career provides an early example the interaction of engineering and science. His technical interests influenced the direction of his scientific research and *vice versa*; for instance he used the results of his research in his own waterwheel designs, consistently favoring breast wheels and overshot wheels and almost never using the undershot system. There is reason to believe that Smeaton's work led other designers to forsake the long-preferred undershot wheel.

Smeaton performed extensive tests on the Newcomen engine, optimizing its design and significantly increasing its efficiency and was unquestionably the greatest of Newcomen's successors [79]. These studies, however, though very interesting were soon overshadowed by James Watt's invention of the separate condenser, and for this little explored by historians. A minor contribution to observational astronomy, which shows some skills in mathematics, rounded out Smeaton's scientific work [73, 74].

5.2.1.1 Experiments on Mills

With his memoir *An experimental enquiry concerning the natural powers of water and wind to turn mills, and other machines, depending on a circular motion* of 1759, Smeaton opened up a substantially new perspective for experimentation. Until then, at least in most cases, the experiments had the role of verifying, or discovering, directly or indirectly, laws or principles of a general nature, that is valid in every situation. The experiments could be qualitative or quantitative. Their motivation was generally dictated by the curiosity of the individual scholar, even if boosts from technology had their value. Smeaton carried out experiments on artifacts, on machines, with the aim of discovering or verifying laws of a particular nature, valid only for the machine with which he was experimenting or machines similar to it. Experimentation was necessary because the use of general laws was too complex for the computation capability of the times to be used and also because some general laws, for example those of fluid dynamics, were not yet well known; at least in most scientific and technical environments. There was of course the scholar's curiosity to discover new phenomena; but there was mainly the need of the engineer to improve, and to do it in useful times, the design of his machines which at the time was based partly on empirical rules and partly on clearly unreliable theoretical results.

Without any knowledge of the theory of models, which will be developed only in the 20th century, Smeaton still realized that the situation for real machines can be very different from that of models and a thing that does very well in a model, could not answer satisfactory in large. And indeed, though the utmost circumspection is used, real working of machines cannot be fully ascertained but only by making trials with them, when made of their proper size. It is for this reason, that Smeaton deferred offering his results obtained in 1752–1753, by using models, till he had an opportunity of putting the deductions made therefrom in real practice, in a variety of cases, and for various purposes [72].[43]

The execution of experiments on models was not new at the time, for example Desaguliers made intensive use of them in his physics lessons, but they were very idealized models, only useful to illustrate the validity of some physical laws. Machine builders and engineers of the past had also resorted to scale models; but no one before Smeaton had carried out a rigorous and systematic experimentation, certainly influenced by the new conceptions of experimental philosophy, and few had the manual skill to make models with the accuracy like that of Smeaton for his past as an instrument maker.

Smeaton's memoir is divided into two parts, one dedicated to water wheels and the other to windmills. Of the first part I have already reported on another occasion, [16][44] and therefore I will limit myself to a brief summary while I will dedicate a few more words to the second part. Although he gave very few bibliographical references—only Euler is cited for the first part and Parent (misprinted as Parint) and Maclaurin for the second—Smeaton showed to know well the state of the art about mills, even if it is unlikely he knew the theoretical works of hydrodynamics by Euler, the Bernoullis and d'Alembert.

Water Wheels

In the first part of the memoir, the one dedicated to water wheels, the reference is certainly to the work of Parent of 1704 [58]. And from Parent Smeaton took up the idea of measuring the efficiency of machines by the ratio of useful and available power. Smeaton however differently from Parent defined the power not basing on forces and speeds, but on more directly measurable quantities, as explained below.

Smeaton defined the *original power* of the water as the product between the quantity of water released in a given time (from a river for instance) and the height that water comes down from. The *effect* of the machine is the sum of the weight raised by the action of this water and the weight necessary to overcome the friction, multiplied by the height the weight will be raised to in a given time; the effect as well the original power are proportional to power (modern meaning), because the work in an assigned fixed interval of time; the efficiency is the ratio between effect and original power. Smeaton found that, for the undershot wheels, the maximum may be more than 30% [72],[45] nearly the double than that provided by Parent (15%) [58].[46]

[43]p. 101.

[44]pp. 423–428.

[45]p. 115.

[46]p. 333.

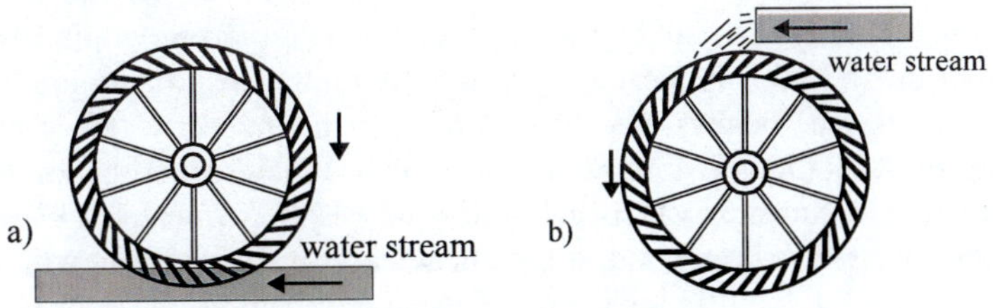

Fig. 5.2 Examples of undershot wheel (**a**) and overshot wheel (**b**)

However, the most interesting part of Smeaton's memoir concerns the determination of overshot wheels performance (Fig. 5.2 shows two typical cases of under and overshot wheels). He found that using the same wheel with plane blades, the efficiency was double that of the undershot wheels and confirmed the results obtained by Antoine Deparcieux (1703–1768) before him, that the efficiency of the wheel increased by slowing its speed [15].

He explained the phenomenon by considering the particles of water as non-elastic soft bodies that deform and because of this deformation some mechanical power is lost: "The effect therefore if overshot wheels, under the same circumstances of quantity and fall, is at a medium double to that of the undershot: and, as a consequence thereof that non-elastic bodies, when acting by their impulse or collision; communicate only a part of their original power; the other part being spent in changing their figure in consequence of the stroke" [72].[47] Smeaton returned on the problem of the dissipation of mechanical power as consequence of the change of shape in a paper of 1782, that will be commented later. The mechanism of dissipation proposed by Smeaton though fascinating, was forgotten after Joule's researches that attributed the lost of mechanical power in the water to the production of heat because of the turbulent motion.

Thanks to Smeaton, the overshot wheels reached a high efficiency and contrasted the success of new-born steam machines.

> However much Mr. Smeaton's valuable observations may have been disregarded by authors, they have not been lost to practical men [...]. [As a result of his experiments] he determined to apply the water, in all cases, so that it should act more by its weight, and less by its impulse; and the advantage gained by that improved construction was found to be fully equal to his expectation. It was afterwards so generally adopted and improved upon by himself and by other engineers in this country, that although undershot water-wheels were, about fifty years ago, the most prevalent, they are now rarely to be met with; and wherever economy of power is an object, no new ones are made [63].[48]

Wind Mills

The second part of Smeaton's memoir of 1759, dedicated to wind mills, is generally considered as less interesting than the first. The judgment, however, derives from the fact that the first part is the one that most influenced the technology of the time. In

[47] p.130.

[48] p. 291.

reality, from a strictly theoretical point of view, the memoir on windmills is richer and more original. For example, there are precise references to Antoine Parent and Colin Maclaurin (see below) who had written on the subject a few years earlier. The presence of citations is important both because they reveal the knowledge of recent literature by Smeaton, and because they suggest that Smeaton's knowledge of mathematics must have been good, sufficient at least to read works in which modern Calculus was used. Even if in his memoir Smeaton did not use it.

This part of the memory opens with the presentation of the model/machine used. According to Smeaton a wind can be obtained in two ways; either to make the air move toward the machine, or to make the machine to move against the air. But moving the air against the machine, in a sufficient volume, with steadiness and the required speed, was not easily put in practice the. And carrying the machine forward in a right line against the air, would require a larger room than one could conveniently meet with. What is most practicable, therefore, was, to carry the axis, whereon the sails were to be fixed, progressively round in the circumference of a large circle, as shown in Fig. 5.3. Smeaton confessed that the idea of this model was not his and named (Samuel?) Rose and Ellicott as predecessors. However he perfected the model, for instance by introducing the pendulum XV to regulate the rhythm with witch the hand in Z has to pull the rope ZH and consequently maintain a constant rotation speed. The rotation of the rotor support arm (FG) was accomplished by the hand (Z) at the left, pulling the cord that turned the barrel on the shaft (DE). Speed was adjusted so that the support arm made one tum in the time the pendulum (VX) made two vibrations. Thus even though the whirling-arm apparatus was not a new idea, in his customary manner Smeaton improved on others' work in its construction.

Smeaton's model rotor had a sail-tip radius of 53 cm, a sail length of 46 cm, and a sail breadth of 14 cm. The maximum 'wind' speed developed appears to be about 2.7 m/s; hence, the Reynolds number—an index to measure turbulence—for these tests was very low, about 25 000. Less then in the field.[69].[49] This may have affected his conclusions quantitatively if not qualitatively.

Experiments Smeaton carried out regarded the evaluation of what he called the *effect* measured by the product of the weight p raised by the wheel multiplied by its turns in a given interval of time (for an assigned model the number of turns is proportional to the vertical displacement of the weight p, but more simple to measure). The tests can be divided into two groups. In the first the magnitudes to be varied were the configuration of the sails, the value of the weight to raise and the duration of the experiment; in the second group the speed of wind was varied.

Results regarding the first group of tests are reported in the table shown in Fig. 5.4. From column 9 of the table (here 'product' is for effect) it can be seen that the optimal angle of the plane of the blades with respect to the plane perpendicular to the axis of rotation is not 35° as calculated by Parent [59],[50] but much lower, precisely 15° (compare the value of row 1 with that of row 3). Rows 5–19 concern different shapes of the sails. Here too it is evident how the effects obtained with sails designed

[49] p. 29.

[50] p. 530. Here Parent referred to [57], pp. 300–302.

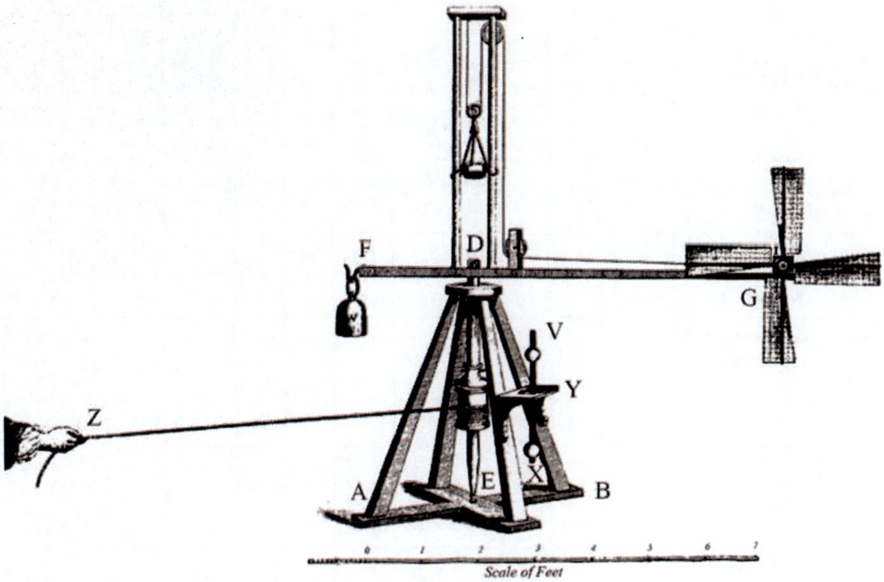

Fig. 5.3 Model of the wind mill. Redrawn from [72], after p. 174

with empirical criteria are superior to those designed with mechanical theories; for example those suggested by a Maclaurin's theorem, which furnished a formula for the angle with which the plane of the sail must vary along the radius of the blade to take into account the different speeds of its points [44].[51]

Very interesting are also the results Smeaton obtained when the wind speed changes. They are expressed through general propositions which he referred to as *maxims* also known as *Smeaton's law* [89]. In particular:

> Maxim 3d. The effects of the same sails at a maximum are nearly, but somewhat less than, as the cubes of the velocity of the wind.
> Maxim 8. The effect of sails of similar figure and position, are as the square of the radius [72].[52]

In the modern literature on wind mills, besides Smeaton's law, there is reference to a Smeaton's coefficient, that is the constant k, that relates the force f on a plane of surface S acted by a orthogonal wind moving with speed v ($F = kSv^2$. Actually Smeaton did not calculate this coefficient, but it could be extracted from his data. The value obtained is much higher than the value currently used; this notwithstanding k is still named Smeaton's coefficient.

Smeaton ended his considerations on wind mills with the following conclusions. In trying all the experiments, the different density of air, which is found at different times, will cause a difference in the load, proportional to the difference of specific

[51] p. 176.

[52] pp. 155, 160.

The kind of ſails made uſe of.	N°.	Angle at the extremities.	Greateſt angle.	Turns of the ſails unloaded.	Turns of ditto at the maxim^m^.	Load at the maximum.	Greateſt load.	Product.	Quantity of ſurface.	Ratio of greateſt velocity to the velocity at a maximum.	Ratio of greateſt load to the load at maximum.	Ratio of ſurface to the product.
		°	°			*lb.*	*lb.*		*Sq. In*			
Plain ſails at an angle of 55°.	1	35	35	66	42	7,56	12,59	318	404	10:7	10:6	10:7,9
Plain ſails weather'd according to the common practice.	2	12	12		70	6,3	7,56	441	404		10:8,3	10:10,1
	3	15	15	105	69	6,72	8,12	464	404	10:6,6	10:8,3	10:10,15
	4	18	18	96	66	7,0	9.81	462	404	10:7,	10:7,1	10:10,15
Weathered according to *Maclaurin's theorem*.	5	9	26½		66	7,0		462	404			10:11,4
	6	12	29½		70½	7,35		518	404			10:12,8
	7	15	32½		63½	8,3		527	404			10:13,
Sails weathered in the *Dutch* manner, tried in various poſitions.	8	0	15	120	93	4,75	5,31	442	404	10:7,7	10:8,9	10:11,
	9	3	18	120	79	7,0	8,12	553	404	10:6,6	10:8,6	10:13,7
	10	5	20		78	7,5	8,12	585	404		10:9,2	10:14,5
	11	7½	22½	113	77	8,3	9,81	639	404	10:6,8	10:8,5	10:15,8
	12	10	25	108	73	8,69	10,37	634	404	10:6,8	10:8,4	10:15,7
	13	12	27	100	66	8,41	10,94	580	404	10:6,6	10:7,7	10:14,4
Sails weathered in the *Dutch* manner, but *enlarged* towards the extremities.	14	7½	22½	123	75	10,65	12,59	799	505	10:6,1	10:8,5	10:15,8
	15	10	25	117	74	11,08	13,69	820	505	10:6,3	10:8,1	10:16,2
	16	12	27	114	66	12,09	14,23	799	505	10:5,8	10:8,4	10:15,8
	17	15	30	96	63	12,09	14,78	762	505	10:6,6	10:8,2	10:15,1
8 ſails being *ſectors* of *ellipſes* in their beſt poſitions.	18	12	22	105	64½	16,42	27,87	1059	854	10:6,1	10:5,9	10:12,4
	19	12	22	99	64½	18,06		1165	1146	10:5,9		10:10,1
	1.	2.	3.	4.	5.	6.	7.	8.	9.	10.	11.	12.

Fig. 5.4 Results of the experiment for a given wind speed. Table III. [72], p. 144

gravity of air, though its velocity remains the same. Notice that a variation of specific gravity may arise not only from a variation of the weight of the whole column of atmosphere, but also by the difference of temperature of the air concerned in the experiment, and possibly of other causes; yet the irregularities that might arise from a difference of specific gravity were thought to be too small to be perceivable. Moreover, as all the experiments were tried in the summer season, in the day-time, and under cover it may be supposed that the principal source of error would arise from the different weight of the column of the atmosphere at different times, which seldom varies above 1/15 part of the whole: "yet as all the principal conclusions are drawn from the medium of a considerable number, many whereof were made at different times, it is presumed that they will nearly agree with the truth, and be altogether sufficient for regulating the practical construction of those kind of machines, for which use they were principally intended" [72].[53]

[53] p. 174.

Technology of mills and science
If one had to summarize the influence of science on the technology of water-wheels and wind mills of the XVIII century, not only in Smeaton but in general, he would be tempted to say that it was modest, almost negligible as claimed by some historians [33, 63]. The impression also derives from the perception that the water wheels and wind mills are simple machines, which have existed since ancient times.

But if one reflects on the improvements introduced in the understanding of their operation, useful in the decision-making phase if and where to place one and in the efficiency of their way of operating, he cannot but admit that they are not so simple and could not be the result of a common craftsman.

Indeed there was a fruitful interaction between science and technology and, though the application to the wheels of the results coming from rational mechanics based on a high formalization had a limited impact, on the contrary, less formalized theoretical considerations, such as those of Parent and de Borda (1733–1799), had a decisive role, despite their high degree of idealization. Considering Smeaton's peers as foreign to science, as was done by historians asserting the low influence of science on technology, is certainly debatable and cannot be shared by all, and indeed it was not [55]. Moreover if one also conceded that specific theoretical achievements of mechanics and fluid dynamics were scarcely relevant, he meantime had to recognize that the methodology of experimental science was fundamental in the conduct of experiments on mills.

By reading his memoirs, it is clear that Smeaton's experimental work was not that of an intelligent craftsman with good dexterity, who performed experiments more or less casually. Smeaton had a good knowledge of mechanics and also a remarkable culture of experimental philosophy. For example, in identifying the magnitudes to measure the performance of the water wheels he realized that the product of the weight raised by the height to which it is raised had a precise mechanical meaning, which he called power, and is not very different from the modern concept of mechanical energy. Even the idea of estimating a measure of friction presupposes a basic knowledge of mechanics; at least a modern concept of force. Smeaton knew that water falling from a height h acquires a speed v, such that $v^2 \propto h$. His embryonic concept of mechanical energy allowed him to affirm that the mechanical efficiency of the undershoot wheels is lower than that of the overshot wheels because a part of the mechanical power is lost in the undershot wheels during the impact with the wheel blades. Moreover there was an underlying epistemology that guided him. The laws of physics are invariant for him and even if the causes of the various phenomena are not known, their effects can be measured, resulting in mathematical laws, valid for all cases similar to the one studied. Only this faith can justify a repetition of the same experience, possibly by slightly changing the parameters involved.

The consequences of Smeaton's experiments on hydraulic wheels were important and made it possible to make some improvements to the rules of design. Smeaton gave summary suggestions on the number of blades to use and their shape. However, the acquisition of the fact that overshot wheels are much more efficient than undershot if appropriate provisions are adopted, was fundamental for the technology of the time.

Smeaton's work on wind mills, however imprecise in measurement, suggested basic insights into the process of wind energy conversion, and much of it is still valid today. In spite of all Smeaton's good work in analyzing performance, the builders of the times who continued with their established designs and practices, seem to have taken little notice of it. Of course, some of the labor-saving improvements were made, but otherwise, there were only minor changes. The steam engine put a brake on the use of wind power, although the established design of the windmill did hang on for a long time, continuing to be built in the early twentieth century.

In the same period, similarly to Smeaton but independently of him, the mathematician Johann Albrecht Euler (1734–1800), a son of Leonhard, used father's theory of fluid dynamics for waterwheels and arrived at conclusions similar to those of Smeaton about the superiority of the overshot wheels. Johann Euler's work won a scientific prize but was largely ignored by engineers because it was opaque to them and offered no more practical advice than Smeaton's empirical results. However, his mathematical theory was helpful to suggested a more general type of hydraulic machine and formed a step in the subsequent development of water turbines of the 19th century [6, 16].[54]

In the XIX century the interaction between technology of mills and science was more profound and clear. The development of hydraulic and wind machines was brought in the frame of applied mechanics, where theory, the rational mechanics, and practice, experiments in the laboratory and in the field, were carried out by the same people, the modern engineers, determining a great improvement in the efficiency of all kind of machines.

These engineers were deeply involved in mathematics and physics to consider themselves more as scientists than as practitioners; for instance they addressed their memoirs to the Académie des science or the Royal society instead of to technological journals. They made great recourse to experiments, but not so much to verify the general mechanical theories behind their designs. The experiments had rather two main scopes. On the one hand to highlight some minor defects of the machines to be corrected after a theoretical review of the problem; on the other hand to evaluate numerical values of some correcting coefficients which allowed to pass from theoretical to practical formulas. This was due not to errors in theory but to simplified assumptions. For example very often the conservation of living forces—or work—was assumed and friction was not modeled distinctly; its effect was taken into account when performing experiments under various operating conditions and arranging tables of correcting coefficients.

5.2.1.2 Theoretical Works Versus Engineering

In 1776 and 1782 Smeaton wrote two interesting memoirs, both more or less directly related on the way of measuring the force, that is either as mv or mv^2, where naturally m is the mass of a body moving with speed v. He recognized that his conclusions

[54]p. 36, p. 428..

were in opposition to those favored by the disciples of Newton, but diplomatically, and correctly from a modern point of view, specified that both mv and mv^2 were useful when properly interpreted. He explicitly said that his two memoirs took inspiration form his experimental work on mills, which furnished results in contrast with many theoretical analysis, that of Parent for the water wheels for instance.

Mechanical Power and Living Force

In his paper *An experimental examination of the quantity and proportion of mechanic power necessary to be employed in giving different degrees of velocity to heavy bodies from a state of rest* of 1776 [75], Smeaton faced directly the problem of the correct measure of force, intended as capacity of acting, that is if it should be given by mv (referred to as old opinion) or by mv^2 (new opinion). Smeaton cited some authors that wrote about mills: Parent, Maclaurin, Belidor, Desaguliers by remembering that for the latter "the dispute was a dispute about the meaning of words"; in any case the meaning of words has to be explained [24].[55]

Smeaton's memoir was published in a prestigious journal, the Philosophical Transactions, and thus considered possessing a certain degree of originality at the time. This originality does not appear very evident to a modern reader, who is tempted to consider the results found by Smeaton experimentally as an obvious consequences of the theorem of the conservation of living forces; a theorem already widely known and applied—but not proved—for example by Johann Bernoulli in 1727 [9] and demonstrated satisfactorily at least by Lagrange in 1763 [41].

The empirical validation of a principle in mechanics is a rather rare event. Galileo had done this by checking if the speed of a heavy body increases with the space traveled or the time passed [17].[56] If he had known Calculus, Galileo could have dissolved his doubt theoretically. Even if Smeaton had had a greater mathematical and/or physical skill he could have avoided experimentation. However at a time when Lagrange's demonstrations were not consolidated and could even be challenged in some way, showing that the principle of conservation of the living forces holds and in its most complete form, in which the static mechanical power is transformed into mechanical kinematic power, could have had a sense.

At the beginning of memoir Smeaton maintained it useful to clarify the correct use of the term *mechanical power*, or simply power [75].[57] Less emphasis had the meaning assigned to the effect, which is simply identified with the speed assumed by bodies under observation. The experimental set used by Smeaton is illustrated in Fig. 5.5. The weight s by its fall makes the two masses, K and L, of 3 pounds (1.35 kg) each, rotate around the axis BN. The rotation is due to the action of the rope connected to the plate of the scale that wounds on two barrels of diameter one the double of the other, large in M and small in N; clearly in the two cases there is a different relationship between the value of the descent of the weight and the number

---

[55] Vol. 2, preface, p, VII. A similar statement can be found in the *Discours préliminaire* to the *Traitéde dynamique*, written by d'Alembert one year before [21], p. XXIII.

[56] pp. 286–289.

[57] p. 458.

Table 5.1 Table of experiments on mechanical power

	1	2	3	4	5	6
1	8	M	W	5	14.25	29.00
2	8	N	W	10	28.25	29.25
3	8	N	W	2.5	14.25	58.50
4	32	M	W	5	7.00	14.00
5	32	N	W	10	14.00	14.75
6	32	N	W	2.5	7.00	28.75
7	8	M	H	5	7.00	14.75
8	8	N	H	10	14.00	15.00
9	8	N	H	2.5	7.00	30.25

of turns (one turn of the smaller barrel gives a descent of 2.525—~6,5 cm—inches while one turn of the larger barrel gives a double descent). The rope connected to the weight s has a fixed length; once it has reached the end the weight comes off and the two masses K and L continue to rotate by inertia, at a more or less constant speed for a certain number of revolutions, until the friction brings the whole to rest.

The results of the experiment are reported in Table 5.1. The first column shows the value of the weight s in ounces, in the second the barrel used, the large (M) and the small (N). The third column reports the distance of the rotating masses from the center of the axis, one equal to 8.25 in. (~21 cm) (W), the other to 3.92 in. (~10 cm) (H), more or less one half. The fourth column is relative to the turns the rope wound on the barrel, the fifth the time (t) of the descent of the weight s in second and the latter column the time (T) for the rotating masses make 20 turns in their free rotation, still in second. This value is inversely proportional to the angular velocity of the axis, and, for a given arrangement of the rotating masses, to their speed.

Table 5.1 allows to evaluate the (average) speed v of descent of the weight s and the speed V of the rotating masses in their free motion. Simple calculations shows that v is much less than V and because the two rotating masses also have a much greater weight than s, this means, using a modern language, that the potential energy lost by the weight s in its descent is entirely transformed into the kinetic energy of the rotating masses; in other words the speed of the weight s may be neglected and all occurs as if it would move 'statically': "It must, however, be always understood, that the descending body, when acting as a measure of power, is supposed to descend slowly, like the weight of a clock or a jack; for, if quickly descending, it is sensibly compounded with another law, viz. the law of acceleration by gravity" [75].[58]

The examination of the first two rows of the table shows how the same mechanical power (in both the two rows the weight of 8 ounces descends of 25.25 in.) is capable of producing the same final effect (that is speed) in a given body, whether it is applied in a greater (row 2) and lesser time (row 1). In the third row an half effect is obtained

[58]pp. 458–459.

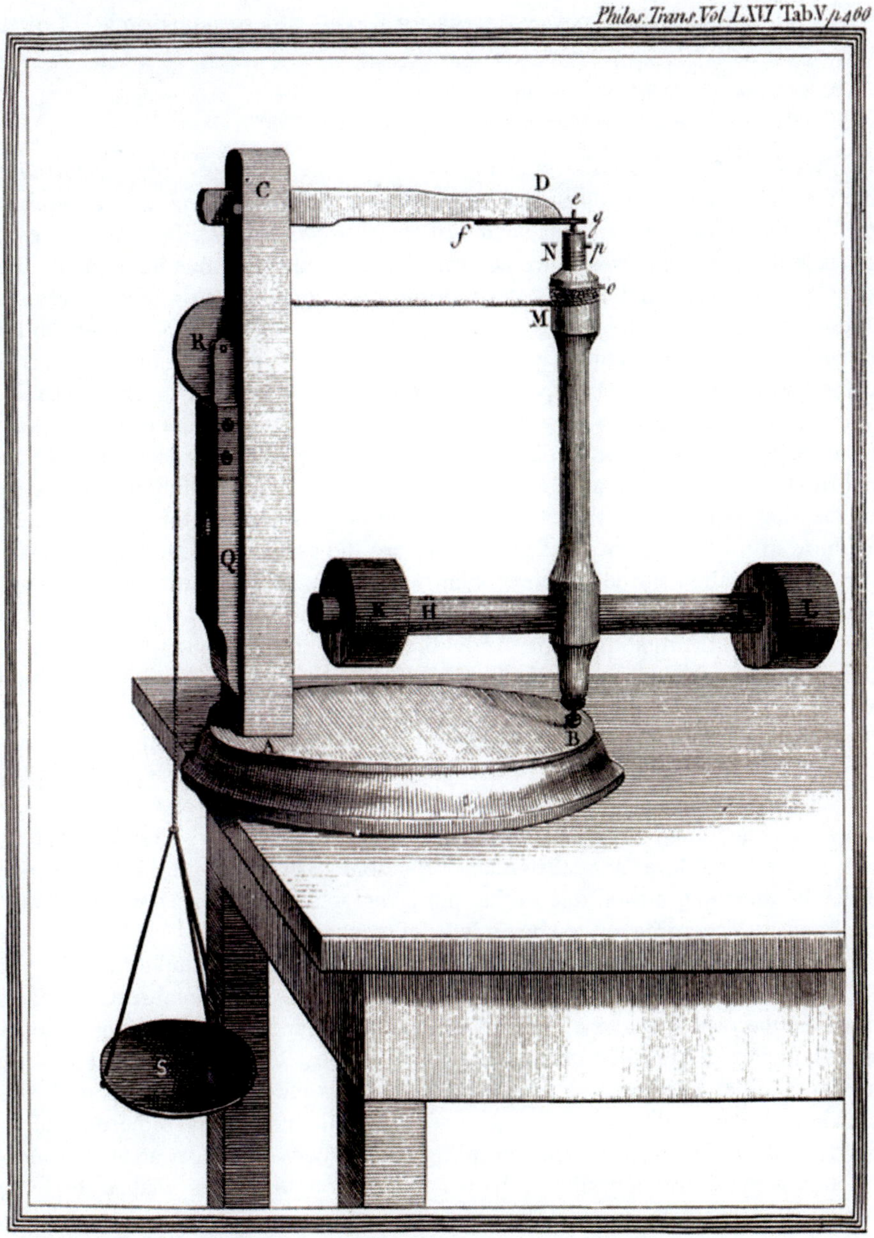

Fig. 5.5 Machine for experiments on mechanical power. From [75], after p. 460. Reproduced with the permission of Biblioteca Guido Castel Nuovo, Sapienza University of Rome

by 1/4 of the mechanical power (that is for 8 ounces descending only 1/4 of the former descent):

> We may conclude, in this instance that the mechanic power, employed in producing motion, is as the square of the velocity produced in the same body [75].[59]

In rows 4, 5 and 6 experiments as in row 1, 2, 3, are presented, only quadrupling the weight s, 32 ounces instead of 8. A double effect is obtaining, thus confirming the rule that power varies as the square of speed. Rows 7, 8, 9 differ from the other because the rotating masses have an arm of a reduced value (one half, about). The effect is apparently doubled with respect to rows 1, 2 and 3, as the angular velocity is double. It is quite interesting to see how Smeaton, who did not master the law of conservation of mechanical energy, justify the fact, in an interesting and may be consistent way (this is not however the point) using the law of Newtonian mechanics.

First, said Smeaton, actually the effect is the same for the cases of rows 1 and 7 for instance, because the effect is measured by the speed of the rotating masses and not by the angular speed around the axis BN and since the circumferences passed by the rotating masses in the case of row 7 are only one half of the circumferences as those of row 1, the speed of the masses are the same. This circumstance has a simple explanation inside the Newtonian mechanics; the time in which the power acts in the case of row 7 are nearly one half of those of the row 1, but the impulsive force acting on the rotating masses is double and "an impulsive power of double the intensity acting for half the time, produces the same effect in generating motion as an impulsive power of half the intensity acting for the whole time" [75].[60] That for the row 7 the force acting on the rotating masses is double than in the case of row 1 can be established with a statical reasoning. The barrel in row 7 was the same with the same number of turns as in row 1 and therefore the shorter arm of the lever by which the impelling power acts is the same, but the rotating masses, upon which the longer arm of the lever acts were placed upon only half the length from the center. Thus the impelling power, due to the first lever, acts upon the second with double the intensity, according to the known laws of mechanics.

A modern reader by browsing the table 5.1 is surprise for finding a quite perfect agreement of the theory he knows and Smeaton experimental results. Indeed a refined theoretical analysis, by using for instance the theorem of living forces, is quite complex. Besides the magnitudes considered by Smeaton one could take into account the weight of the scale, the inertia of the axis and the arm, the kinetic energy acquired by the weight s, friction. Smeaton was quite conscious of these problems and furnished some justifications for the agreement between theory and experiment. Friction is reduced to a minimum by his ability as an instrument maker, the living force of the weight s is very small. Even the inertia of the axis and the arms is taken in to account when deciding to locate the masses in their closest position to the axis at a distance of 3.92 in. instead of $8.25/2 = 4.125$ inches, with the following reasoning, more or less verbatim taken from Smeaton: When the bodies are at the

[59]p. 466.

[60]p. 468.

smallest distance from the axis of rotation, all goes as they were at half the greater distance from axis; for since the axis itself and the cylindric arms of wood, keep an unvaried distance from the center of rotation, the bodies themselves must be moved nearer than half their former distance, in order that, compounded with the invariable parts, they may be virtually at the half distance. To find the value of this distance, put in an arm of the same wood, that only went through the axis, without extending in the opposite direction; the two bodies, one at a time being at the distance of 8,25 in. and 3.92 in., the whole machine was inclined till the body and arm became a kind of pendulum. It is found that the pendulums vibrate one with the frequencies in the ration of $1 : \sqrt{2}$, as it should be with a pendulum of a half length [75].[61]

Smeaton ended his memoir with general considerations about the way of measuring force, or the quantity of motion produced, by stating that all depends if one takes in to account the time for which the force acts or the space passed by the force when acting.

> This then appears to be the foundation, not only of the disputes that have arisen, but of the mistakes that have been made, in the application of the different definitions of quantity of motion, that while those, that have adhered to the definition of Sir ISAAC NEWTON, have complained their adversaries in not considering the time in which effects are produced, they themselves have not always taken into the account the space that the impelling power is obliged to travel through in producing the different degrees of velocity. It seems, therefore, that, without taking in the collateral circumstances both of time and space the terms, quantity of motion, *momentum* and force of bodies in motion are absolutely indefinite; and that they cannot be so easily, distinctly and fundamentally compared as by having recourse to the common measure, *viz*, mechanical power [75].[62]

Impact of Soft, Elastic and Hard Bodies

The memoir *New fundamental experiments upon the collision of bodies* of 1782, concerns the study of the collision between bodies of different mechanical characteristics. The experimental apparatus consists of two wood pendulums which contain at their extremities bodies that are struck when the pendulums after being released meet. One of the pendulum is at rest in the vertical position, the other is raised and then left. The idea of using pendulums to study impact is not new, even Newton had done it and in the *Principia* he referred of similar experiences carried out by Edme Mariotte and Christopher Wren [56].[63] The system allows to measure the speed before and after the impact indirectly by measuring the angle of fall and raising. But Smeaton's ability as an instrument maker, makes the approach particularly interesting. The work is less rich than that of 1776 as regards the experimental aspects, because the use of experimental results is not essential. Unlike the work of 1776, Smeaton did not refer to any other author; nevertheless some considerations of an epistemological nature and of experimental philosophy reveal a good knowledge of scientific literature on the subject.

The paper starts with some considerations about the certitude of scientific knowledge: "It is universally acknowledged, that the first simple principles of science

[61] pp. 461–462.

[62] p. 473.

[63] p. 24.

cannot be too critically examined, in order to their being firmly established; more especially those which relate to the practical and operative parts of mechanics, upon which much of the active business of mankind depends" [76].[64] A similar sentiment was echoed by Lazare Carnot, more or less in the same period: "Sciences are as a beautiful river whose course is easy to follow, when it has acquired a certain regularity; but if one wants to sail to the source one cannot find it anywhere, because it is far and near; it is diffuse somehow in the whole earth surface. The same if one wants to sail to the origin of science, one finds nothing but darkness and vague ideas, vicious circles; and one loses himself in the primitive ideas" [19].[65]

Then Smeaton made a correlation between his work of 1776 with the present one, by asserting that the true doctrine of the collision of bodies (the object of the paper) hangs as it were upon the same hook as the doctrine of the gradual generation of motion, where the concept of mechanical power was fundamental. While there the mechanical power was defined statically as the product of height by weight, here it is introduced for body in motion and it is substantially the same as living force, which is assumed as the most meaningful magnitude to study the collision of bodies. This assumption was considered as arbitrary by the editor of the abridged Transactions of the Royal Society in a footnote to Smeaton's paper [68, 77].[66] He was not alone, in a period when the dispute on the way to measure the force was still heated.

Smeaton recalled that *mathematical philosophers* investigated principally three kinds of law of collision: that of bodies perfectly elastic, of bodies perfectly non-elastic and perfectly soft and of bodies perfectly non-elastic and perfectly hard. Notice the interesting, at least for the present book, the category of scholars Smeaton introduced, that of mathematical philosophers, to which most probably he thought to belong. It indicates mathematicians who have good training in natural and experimental philosophy.

For the sake of simplicity in the exposition of his both theoretical and experimental arguments, Smeaton assumed bodies which are equal in weight. For bodies perfectly elastic it is universally agreed, said Smeaton, because proved by experiments that when two of such bodies strike no motion is lost and hence if an elastic body in motion strikes another at rest, after the impact the former is reduced at rest and the latter assumes the speed of the former. Similarly, if two non-elastic perfectly soft bodies strike, one being at rest, they proceed together from the point of collision with exactly one half of the speed the moving body had before the impact.

With respect to the bodies non-elastic and perfectly hard there is no agreement among philosophers, continued Smeaton. Some maintain that there are no bodies in nature whereon to try experiments, some other do not agree and argue that if a non-elastic perfectly hard body strikes another of the same kind at rest, all happens as in the case of non-elastic soft bodies; that is the two bodies move together with one half of the speed of the striking body. However, noticed Smeaton, no experiment

[64]p. 338.

[65]p. 107.

[66]pp. 147–148, p. 299.

or any "fair deduction of reason" can prove the fact. So the question is whether it is true or not [76].[67]

According to Smeaton, solving this question is of relevance not only for scholars but also for practical men. Even though apparently the latter should not be worried if some argued about bodies they will never meet, in practice they may be mislead by reading considerations about the behavior of non-elastic perfectly hard bodies flanked to those of non-elastic perfectly soft bodies, which they have much to do with in their daily practice. Smeaton on this point said that for many years never had he doubted that non-elastic perfectly hard bodies and non-elastic perfectly soft bodies had the same behavior, but he changed his idea after his experiments on mills published in 1759.

Of course in practice there are only bodies of mixed nature. Some are mainly elastic, but also in part soft; other are mainly soft, but also in part elastic. So if there not exist perfectly hard non-elastic bodies, can imperfectly hard non-elastic bodies exist? They can, because as being imperfectly hard they are soft and being imperfectly elastic they are soft, so they are imperfectly elastic and imperfectly soft as in practice are all bodies; and as such are not a particularly interesting category.

For Smeaton this conclusion, however, was not entirely satisfactory to eliminate the difficulty of dealing with perfectly hard bodies. It must be shown that the idea itself is absurd. Smeaton did this by highlighting the differences between soft and hard bodies, using the laws of mechanics, even those not considered by the advocates of the concept of hard bodies. The difference existing between a soft and a hard material, is that the former is deformed during the impact the latter is not. And to deform a body, according to Smeaton, mechanical power must be spent. To be convinced of the fact Smeaton agued as follows. If the shape of a body can be changed, without a power, then one might be able to make a forge hammer work upon a mass of soft iron, without any other power than that necessary to overcome the resistance of the parts of the machine to be put in motion: for, as no progressive motion is given to the mass of iron by the hammer, no power can be expended that way. And if none is loft to the hammer from changing the shape of the iron, which is the only effect produced, then the whole power must reside in the hammer, and it would jump back. again to the place from which it fell, just in the same manner as if it fell upon a body perfectly elastic, upon which, if it did fall, the case would really happen. The power, therefore, to work the hammer would be the same, whether it fell upon an elastic or non-elastic body: "an idea so very contrary to all experience, and even apprehension of both. the philosopher and vulgar artist" [76].[68]

In the light of modern conceptions of thermodynamics Smeaton's suggestion about the loss of power appear as unsatisfactory, because what is now considered relevant is that during the deformation of a soft body there is a production of heat, and it is the conversion of living force into heat that explains the lost of mechanical power while the deformation is only an indirect index. But this possibly is a not historically correct criticism and Smeaton's intuition should be appreciated. Any way in most

---

[67] p. 340.

[68] p. 343.

cases the change of shape of a body may be an indicator of the lost of mechanical power during an impact, and modern design of objects able to resist the impact by absorbing energy is based in their capacity of deformation.

Smeaton assumed he could prove *bona fide* that one half of the original power is lost in the impact of soft bodies because of the change of shape ("as was suggested very strongly by the mills experiments") [76];[69] on the other hand no change in shape occurs for hard bodies and so no power is lost. For example let the speed of the impacting body be 20 and its mass 8 equal to that of the body at rest; then according the rule deduced from the experiment of 1766 for which the mechanical power can be measured with living force, the power of the impacting body is $20 \times 20 \times 8 = 3200$. If half of it is lost in the impact of soft bodies, it will reduce to 1660, to be shared between the impacted and impacting bodies, that is a mass of 16. This means that the two bodies will continue their motion with a speed of $\sqrt{1600/16} = 10$, that is just one half of the original speed. But for non-elastic perfectly hard bodies no power is lost in the impact, so the power remains 3200; if the two bodies move together as a body of mass 16, this power correspond to a speed $\sqrt{3200/16} = 10\sqrt{2} = 14.14$. But this contradicts what for Smeaton is a truth passible of strict demonstration, "that the velocity of the center of gravity of no system can be changed by any collision" [76].[70]

Indeed at the outset of the impact the center of gravity of the two bodies has a speed equal to one half of the speed of the impacting body—one body of mass m has speed zero, the other body still of mass m has ha speed v; the speed of the center of gravity is $(0 + mv)/(m + m) = 1/2v$; and thus the speed of the center of gravity is 10. This value remain the same immediately after the impact of two soft bodies, as both of them move with a speed 10. But the value increases in the impact between two perfectly hard bodies, as they move with a common speed of 14.14. Which is absurd.

Smeaton formulated the absurd in a formal way as follows. In the stroke of inelastic hard bodies they cannot possibly lose any mechanical power in the stroke; because no other impression is made than the communication of motion; and yet they must lost a quantity of mechanic power in the stroke; because, if they do not, their common center of gravity, as above shown, will acquire an *increase* of velocity-by their stroke, upon each other. "In a like manner the idea of a perpetual motion, perhaps, at first sight, may not appear to involve a contradiction in terms;" [76].[71] In conclusion: "an elastic hard body (perfectly so) is a repugnant idea and contains in itself a contradiction" [76].[72]

It remains to show that half of the mechanical power is lost in the collision between two soft bodies. This is done both experimentally and theoretically, with an analogical reasoning. As for the experimental demonstration, Smeaton could refer to the experiments available in the literature, but he preferred to rely on his own experiences. His

---

[69] p. 343.

[70] p. 352.

[71] p. 352.

[72] p. 352.

experimental results, which are not reported in detail in the article, confirmed those shared by all scholars of his time. In particular, for two equal soft bodies the speed after the impact of the assembly is one half the speed of the impacting and therefore one half of the power has been lost, as shown by simple calculations. As for analogical reasoning, imagine, said Smeaton, if one could construct a couple of bodies in such a way that they act as perfectly elastic bodies, but that their springs should at pleasure be hooked up, retained, or prevented from restoring themselves, when at their extreme degree of compression. Bodies under these circumstances observe the laws of collision of non-elastic soft bodies, then it is proved, that one half of the mechanical power, residing in the striking bodies, would be lost, because if the springs are released and the two bodies behave as elastic bodies they will recover a power equal to that stored [76].[73]

5.2.2 A Philosophical Engineer. James Watt

James Watt (1736–1819) was born at Greenock, the grandson of Thomas Watt, a teacher of mathematics, surveying, and navigation. He was always delicate, and suffered throughout his life from severe attacks of headache. When thirteen he began to study geometry and at once showed great interest in the subject. He then went to the Greenock Grammar school, where he acquired Latin and some Greek. During his boyhood he was a diligent worker in his father's shop, so far as regards the making of models, and gave early evidence of his great manual dexterity. At the age of seventeen to eighteen he was sent to Glasgow to live with his mother's relatives, then to London to improve himself as an instrument maker. But the atmosphere of London was unsuited for his delicate health and in less than a year he returned to Greenock to settle eventually in Glasgow, being then in his twenty-first year. He then endeavored to open a shop, as an instrument maker, but was prevented by the Corporation of Hammermen, on the ground that he had not served a proper apprenticeship.

Luckily Watt had for his most intimate schoolfellow Andrew Anderson, whose elder brother, John Anderson (1726–1796) was professor of natural philosophy at Glasgow University. The head of the university thus came to Watt's assistance by appointing him mathematical instrument maker to the university and by allowing him to establish a workshop within its precincts by 1757. Watt however besides preparing instruments for the university also continued his trade and also was busied in manufacturing musical instruments and fancy toys. At the university Watt continued to improve himself in various ways; here he made the acquaintance of many eminent men, such as Joseph Black (1728–1799), the discoverer of latent heat; Adam Smith (1723–1790) the famous economist and John Robinson (1739–1805), professor of natural philosophy.

In 1764, it occurred the well-known incident of the repair of the model of a Newcomen fire (steam) engine, belonging to the university, which had never acted

[73]pp. 345–348.

properly. It turned out that the model was not out of repair, in the ordinary sense of the word, for it had lately been put in order by a celebrated instrument maker in London, Jonathan Sisson [36].[74] It however exhibited a very poor performance; this fixed Watt's thoughts on the question of the economy of steam and laid the foundation of his first and greatest invention: the separate condenser. Watt prosecuted this invention so far as his limited means would admit, but nothing on a working scale seems to have been done, until he entered into an arrangement with John Roebuck, the founder of the Carron Works, to take a share in the invention, and an engine was made at Kinneil, near Linlithgow. But Roebuck fell into difficulties and this engine did not seem to have excited much attention; nor did the invention develop in the manner that might have been expected. Meantime, Watt became largely employed in making surveys and reports, in connection with canals, rivers, and harbors. Among the last of his engineering works of this character were an improvement of the harbor of his native place and a provision of water-works for that town.

In 1768 Watt knew Matthew Boulton, the founder of the Soho Works, near Birmingham. In 1769 Watt's invention was patented. In 1772 Roebuck failed, and Boulton offered to take a two-thirds share in Watt's engine patent, in lieu of a debt of 1 200 pounds. In 1775, nearly forty, Watt entered into partnership with Boulton at Soho. It was a happy partnership. The two men complemented one another ideally. Watt, the inventor and scientist, knew enough about organization of work; Boulton, the organizer and entrepreneur, was enough of an inventor to appreciate Watt's technical problems. His forthright and optimistic joviality contrasted with Watt's quick temper and morose outlook. The steam engine started to spread. Watt retired from the firm of Boulton & Watt in 1800, Boulton going out at the same time, leaving the business to their sons. After his retirement from Soho, Watt pursued at his residence, Heathfield Hall, near Birmingham, various inventions in the workshop which he had fitted up there. He died at Heathfield, in his eighty-fourth year, and was buried in St. Mary's Church at Handsworth [13].

Watt has been the subject of huge amount of papers and books; many biographies also have been written; of them it is still valid the *Life of James Watt*, by James Patrick Muirhead of 1858 [54]. A survey quite satisfactory of papers and book on Watt can be found in [25, 50]. Watt himself did not publish many writings. For instance he published only two papers (or better one divided into two parts) on a scientific journal, the Philosophical Transactions of the Royal Society of London, in 1784 [84, 85]. His ideas are otherwise spread in letters [66][75][65], in notebooks [65] and in his patents [66].[76] Many Watt's writings, now at the Birmingham central library, consist of Watt's personal papers, his extensive incoming correspondence and bound volumes of retained copies (made on Watt's copying-press) of his outgoing letters; notebooks, journals, personal and business accounts, surveying reports, memoranda, papers relating to the Act of Parliament of 1775 which extended his original patent, patent specifications and drawings, legal papers concerning court cases for

[74] p. 202.

[75] Vol. II.

[76] Vol. III.

infringement of his patents, and other miscellaneous papers. Watt's extensive correspondence is the rich core of the archive, documenting all aspects of his life and work and providing considerable information about his contemporaries. The overall quality and regularity of the correspondence with scientific and technological figures is exceptionally high. To give just two examples: Priestley's writings about phlogiston, inflammable air, the Lunar Society, and of his losses in the riots. Humphry Davy describes his galvanic experiments. The archive contains a considerable amount of juvenilia, apparently carefully preserved by his father, who was heartbroken by the early death of a favored son [86].

5.2.2.1 A Natural Philosopher and a Chemist

Today it seems strange to assimilate Watt to a chemist. Indeed a part from being an engineer, he is more easily seen as a physicist, because his works on heat and steam. Actually heat and steam in the 18th century were studied mostly by chemists. Heat sometimes referred to as *fire* and after Lavoisier *caloric*, was considered as a substance, or an element. And steam, water and ice, for instance, that today are considered as different states of a chemical substance, water (H_2O), were considered as different substances. Steam derived by water through a chemical combination with heat; ice from water by a separation of heat. In any case Watt was busied also with themes today classified as chemical. One of them was the nature and composition of water, object of many discussions until the end of the 18th century [88]. Watt was stimulated in this study by a letter he received from Priestley in 1782, who referred about the conversion of water into permanent air by first mixing it with quicklime and then exposing it to the red heat. In 1784 a mature Watt published two papers, one the sequel the other, in the Philosophical Transactions [84, 85], about his 'discovery' of the composition of water. In this same year and journal Cavendish published a his own paper on the subject and also Lavoisier shortly after announced a similar discovery. There was a controversy about priority, known a *water controversy*.

Here it is not a question of entering into the merits of the priority of discovery or even of analyzing in depth the theoretical value of Watt's writings. What matters is that Watt attached great importance to his discovery and demanded that his qualities as a chemist be appreciated. It should be remembered that we are in a period of study of gases in which not only the phenomena, but also the nomenclature was uncertain. The term *gas*, suggested by the Greek χάοζ, chaos, introduced by Van Helmont in 1652 [34],[77] was not yet widespread and it was generally spoken of air or airs. In particular for what concerns the composition of water, Watt used the terms *inflammable air* and *dephlogisticated air*. Scholars concerned with the chemistry of airs in the late 18th century were divided among *lumpers* and *splitters* [50].[78] Splitters inclined to see new discovered airs with distinct chemical identities. The lumpers, on

[77] "I have called this spirit gas, it being scarcely distinguishable from the Chaos of the ancients" p. 59.

[78] p. 86.

the other hand, whilst they recognized there were various kind of airs still believed in their fundamental unity. Lavoisier and new chemists were splitters, Watt and Priestley were lumpers. The common unifying agent of lumpers might be phlogiston or heat, the variability of airs being due to the amounts of heat or phlogiston combined with a 'base' of some sort. Another example of lumping devices was the *humor*, intended as the principle of humidity, introduced by Jean André de Luc (1727–1817), a friend and correspondent of Watt. In such a way the boundary between permanent air (gas) and vapor (for example steam) was recognized in the sense that the former would condense differently from the latter, but they were also lumped together by humor [50].[79]

In his paper of 1784 Watt began by saying that in 1783 he had written a letter to Priestley to have news about his experiment, "but before he had an opportunity of doing him that favor, he found, in the prosecution of his experiments, that the apparent conversion of water into air, by exposing it to heat in porous earthen vessels, was not a real transmutation, but an exchange of the elastic fluid for the liquid, in some manner not yet accounted for" [85].[80]

Watt then referred about other experiments by Priestley on calces and metals, and of an experiment mixing together certain proportions of pure dry dephlogisticated air and of pure dry inflammable air in a strong glass vessel, and then set them on fire by means of the electric spark,"in the fame manner as is done in the inflammable air pistol" [85].[81] As the glass grew cold a mist or visible vapor appeared in it, which was condensed on the glass in form of misture or dew [85].[82] Here in a footnote Watt recognized that Cavendish was the first who discovered that the combustion of dephlogisticated and inflammable air produced moistures. And also referred to experiments made in Paris on the subject, by which the essential point seems to be clearly proved, that the deflagration or union of dephlogisticated and inflammable air, by means of ignition, produces a quantity of water equal in weight to the airs; and that the water, thus produced, appeared, by every test, to be pure water [85].[83]

At this point Watt posed the question:

> Are we not then authorised to conclude, that water is composed of dephlogisticated air and phlogiston, deprived of part of their latent or elementary heat; that dephlogisticated or pure air is composed of water deprived of its phlogiston, and united elementary heat and light; and that the latter are contained in it in a latent state, so as not to be sensible to the thermometer or to the eye; and if light be only a modification of heat, or a circumstance attending it, or a component part of the inflammable air, then pure or dephlogisticated air is composed of water deprived of its phlogiston and united to elementary heat? [85].[84]

It is on the basis of this question raised in 1784 that Watt could claim to be the first to discover the composition of water.

---

[79] pp. 86–87.

[80] p. 330.

[81] p. 330.

[82] p. 332.

[83] p. 332.

[84] p. 332.

A modern is tempted to interpreter Watt's statement as an indication that inflammable air (that is hydrogen) reacted with dephlogisticated air (that is oxygen) to give water. Actually by reading the paper, it appears that Watt regarded the different airs and water as all being products of the union of dephlogisticated air and phlogiston in different degrees and by different means. This is the key insight that is missed when Watt's statement on the composition of water is extracted and taken in isolation [50].[85]

Watt declared that for many years he have entertained an opinion, that "air was a modification of water", originally founded on the facts that in most cases, wherein air was actually made, the substances were known to contain water as one of their constituent parts. This opinion, said Watt, arose from a discovery that the latent heat contained in steam diminished in proportion as the sensible heat (temperature) of the water, from which it was produced, increased (because of the higher pressure) or, in other words, "that the latent heat of steam was less when it was produced under a greater pressure, or in a more dense state, and greater when it was produced under a less pressure, or in a less dense state; which led me to conclude, that when a very great degree of heat was necessary for the production of the steam, the latent heat would be wholly changed into sensible heat; and that, in such cases, the steam itself might suffer some remarkable change" [85], [86] becoming air. This reasoning is at the basis of what since the 19th century will be known as *Watt's law*, referred to below.

Watt however abandoned this opinion in so far as relates to the change of water into air, as "I think that may be accounted for on better principles. In every case, wherein dephlogisticated air has been produced, substances have been employed, some of whose constituent parts have a strong attraction for phlogiston, and, as it would appear, a stronger attraction for that substance than humor has; they should, therefore, dephlogisticate the water or fixed air, and the humor thus set free should unite to the matter of fire and light and become pure air" [85].[87] Thus the conversion of water into air (steam) occurs not by the action of heat alone but by a process of dephlogistication of the water to produce the intermediate humor, which then becomes pure air by uniting with fire and light.

Watt returned on the subject with another paper. Here he advised the reader about precautions to be taken in repeating the experiments: "I think it necessary to resume the subject, in order to mention some necessary cautions to those who may chose to repeat the experiments mentioned there, and to point out some circumstances that may cause variations in the results" [84].[88]

As mentioned beforehand there was a controversy about the priority of the discovery of the composition of water, which saw Watt claiming his priority over Cavendish and Lavoisier, and not only for prestige, but also to accredit his image as a philosopher to spend in his trials of the 1790 s concerning the patent of the steam engine. The *water controversy* concerned one of the central discoveries of modern science,

---

[85] p. 104.

[86] pp. 33.

[87] pp. 335–336.

[88] p. 354.

that water is not an element but rather a compound. The allocation of priority in this discovery has occupied a number of 20th century historians. The matter is tied up with the larger issues of the so-called chemical revolution of the late eighteenth century. A case can be made for James Watt or Henry Cavendish or Antoine Lavoisier as having priority in the discovery depending upon precisely what the discovery is taken to consist of. However, neither the protagonists themselves in the 1780 s nor modern historians appear as strongly interested in the affair. In fact, a controversy, known as the *second water controversy*, attracted most attention in early Victorian Britain some fifty to seventy years after the actual work of Watt, Cavendish and Lavoisier.

This second water controversy arose long after Watt's death. It was prompted by the secretary of the Académie des sciences de Paris, François Jean Dominique Arago (1786–1853), who in his *Éloge* of James Watt asserted that Priestley was the first person to prove that air could be converted into water and that Watt was the first person to understand it [2].[89] The first to respond publicly to Arago and denying strongly his thesis was William Vernon Harcourt (1789–1871) in his presidential address at the British association for advancement of science meeting in 1839 [49].[90]

The water controversy was mainly driven by ideological struggles about the nature of science and its relation to technological invention and innovation in British society. More than credit for a particular discovery was at stake here. The controversy nourished large debates about the nature of science, about the relationship between science and technology and economic transformation, about the appropriate organization of scientific activity, and so on [49].[91] Watt and Cavendish were iconic characters in the Victorian era. Watt essentially represented a clear link between science and technology unmediated by an academic scientific elite. Men like Watt combined science and technology and were seen by many members of the industrial middle class, but not only, as the hero to be emulated. On the contrary, the aristocratic Cavendish was a more iconic character for large part of the scientific community. He represented a person who followed what they considered the right scientific approach, methodic cautious of highest quality and mainly riven by curiosity alone [49].[92]

Two groups of contenders are particularly worth considering. First the 'gentlemen of science'-who founded the British association for advancement of science and who denied Watt the status of a chemical discoverer-and second the 'Northern philosophers' including Rankine, Tait and William Thomson (Kelvin) who played such a signifiant role in the development of engineering science in the second half of the nineteenth century. The gentleman of science, such as Whewell, Harcourt, Herschel and Peacock, were mathematicians, or astronomers, or exponents of mathematical experimental philosophy, or even gentleman geologists. They bequeathed strong suspicions about mixing commercial and scientific objectives. The northern philosophers had no such reservations. William Thomson personified the amalga-

[89] pp. 80–87.

[90] p. 3.

[91] p. 4.

[92] pp. 4–5.

mation of roles as natural philosopher, engineer and businessman [50].[93] While the gentlemen of science tended to treat theory and practice as distinct though related realms, the Northern philosophers emphasized their unity. Consider, for example, Rankine's introductory lecture, published in 1856 on his accession to the Glasgow chair: "The subject of which, in this introductory, lecture, I propose to treat [...] is THE HARMONY OF THEORY AND PRACTICE IN MECHANICS", which was published in 1856 [61].[94]

5.2.2.2 The Steam Engine and the Separate Condenser

No doubt Watt cannot be considered as the inventor of the steam-engine. The grand principle of rendering the heat contained in steam available as a source of moving power may be traced so far back that we lose the clue altogether in the obscure, or impracticable, or simply puerile shapes in which the idea was contained. Papin, indeed, proposed a piston and cylinder in which the vacuum was produced by steam instead of by the air-pump; but Savery was the first who constructed a steam-engine, and applied it to the drainage of mines. His invention included the two capital properties of steam, its power of producing a vacuum by condensation and its elastic force at high temperatures. A few years later the piston-form was introduced or re-invented by Newcomen and Cawley, as well as the valuable expedient of producing condensation by a squirt of cold water injected into the cylinder. In this condition the atmospheric engine remained with slight improvement for above half a century, doing the work for which it was invented, the pumping of water out of shafts (the pump being moved by a chain attached to the end of a horizontal oscillating beam), wherever economy of fuel was unimportant. Such was the case at coal pits, but in other mines, usually situated remote from coal, it was of comparatively little use, on account of the enormous consumption of fuel [28].[95]

To reconstruct the history of Watt's contribution to the steam engine it is necessary to relay on Watt's account started not earlier than thirty years after his first investigation. There are two versions of it; one given in a footnote of Robinsons's dissertation, *A system of mechanical philosophy*, published in 1822 [66][96] but preceding 1814, and another referred to by Watt as *plain story*, in a document prepared in 1796 as a general answer to the objections that his opponents raised to the specification of his patents [54].[97] There are also accounts by Black and Robinson [54].[98]

According to his own narrative, Watt's attention was first directed in the year 1759 to the subject of the steam engines, by the "late Dr Robinson, then a student in the

[93] pp. 62–72.
[94] p. 4.
[95] p. 68.
[96] Vol. 2.
[97] pp. 83–91.
[98] Chap. 8.

university of Glasgow and nearly my own age" [66].[99] Watt and Robinson hoped to improve the operation of the Newcomen engine, then largely used to pump water from coal mines, whose basic elements are shown in Fig. 5.6; a stove fueled a boiler B that produced steam at atmospheric pressure. This steam was released from the bottom into the cylinder C and, aided by the rocker R that kept initially in equilibrium the pump P placed at the opposite end of the rocker arm, made the piston to be lifted. As soon as the steam had filled the cylinder, cold water was led into it through a valve originating condensation of the steam and a vacuum; soon the piston fell down because of atmospheric pressure. In this phase, the pump was operated for lifting water from the mine. At this point the cycle could start again. Systems of opening and closing of the valves for the entry and discharge of the steam (and water) were automated through the motion of the rod of the injection pump synchronized with the motion of the rocker.

Watt however was soon distracted in supervising his trade as an engineer. Any way in this period he frequently got in touch with Black. But in the winter of 1763–1764, being requested to repair the model of Newcomen's engine. Watt's mind was again directed to the steam engine. At that period, his knowledge on engines was derived principally from Desaguliers and partly from Belidor [66].[100]

Watt "set about repairing it [the Newcomen engine] as a mere mechanician" and when that was done and it was set to work, he was surprised to find that its boiler could not supply it with steam, though apparently large enough (the cylinder of the model being two inches in diameter, and six inches stroke, and the boiler about nine inches diameter). By blowing the fire it was made to take a few strokes, but required an enormous quantity of injection water [66].[101]

Since the beginning Watt was convinced that the way to improve the efficiency of the engine was to reduce the dissipation of heat. Watt acted here more as a chemist than as a physicist (a mechanician) by concentrating more on the consume and production of heat than in mechanical aspects [25].[102] A first idea shared by Robinson also was to replace the cylinder of brass with a less conducting material. Indeed the cylinder of the model being of brass, would conduct heat very well, and considerable advantage could be gained by making the cylinders of some substance that would receive and give out heat slowly. Of these, wood seemed to be the most likely, provided it should prove sufficiently durable. A small engine was therefore constructed with a cylinder six inches ($\sim$ 15 cm) diameter, and twelve inches stroke, made of wood, soaked in linseed oil, and baked to dryness. With this engine many experiments were made; but it was soon found that the wooden cylinder was not likely to prove durable and that the steam condensed in filling it still exceeded the proportion of that required [66].[103]

---

[99] Vol. 2, p. 113.

[100] Vol. 2, p. 113.

[101] Vol. 2, p. 113.

[102] p. 256.

[103] Vol. 2, p. 114.

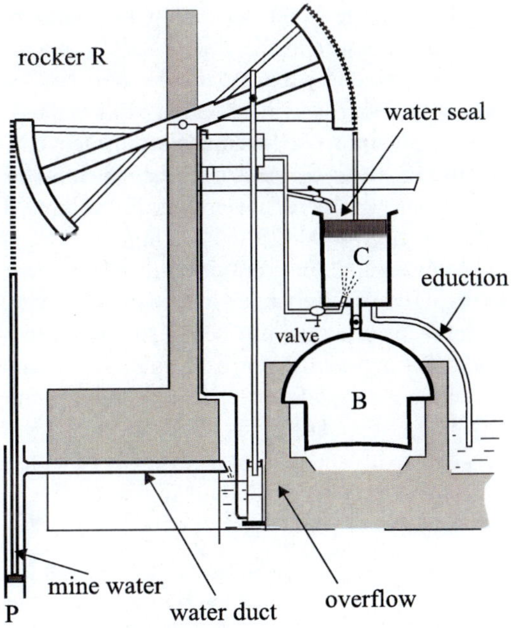

Fig. 5.6 Newcomen steam engine

Watt also concentrated by making the boiler performed with as little fuel as possible.

> I had often observed that the best way of heating bodies was to bring them in Contact with the burning fuel the great distance from the fire to the boyler in fire engines seemed in Consequence to be wrong after many fruitless thoughts on the subject *I saw no boyler so perfect in that Respect as the Common tea kitchen* [emphasis added] (an invention for which we are beholden to the Chinese) here the fuel is always in Contact with the sides of the boyler Containing the water the outside may be of wood with this advantage that very little heat will be able to penetrate it the Inside of very thin Iron which will Considerably diminish the Expence and being Constantly Covered with water it cannot burn [65].[104]

The reference to the tea kettle in this quotation and old biographies contributed to spread the myth of Watt and the kettle. According to this myth (not realistic indeed) Watt would have taken inspiration for his studies on steam engine by observing the steam coming from a tea kettle at his aunt Miss Muirhead, when a very young body. The spreading of this myth and what is at the basis of it are discussed in [47].

In studying the behavior of the steam in the cylinder of his Newcomen engine model Watt also found, that all attempts to produce a better exhaustion of the steam by throwing in more injection water, caused a disproportionate waste of steam. "On reflection, the cause of this seemed to be the boiling of water in vacuo at low heats, a discovery lately made by Dr Cullen, and some other philosophers, (below 100°F, as I was then informed)" [66].[105] This is the phenomenon of back pressure, which

[104] p. 435.

[105] Vol. 2, p. 114.

implies that, the water injected in the cylinder produces steam which would, in part, resist the pressure of the atmosphere with a reduction in efficiency.

Discouraged by the apparently inevitable heat loss, Watt gave temporarily up the Newcomen engine and passed to explore the Savery one; but he did not obtain better results. Meantime he had made experiments on which rate his boiler consumed water to produce steam by measuring also the ratio between the volumes of steam produced by a given volume of water and this volume. He found that the value furnished by Desaguliers, 14000 [23],[106] too high with respect the value 1800 he found [66].[107]

After his attempts in improving Newcomen and Savery engines, Watt turned back to his activity as a chemist, to study steam, with the hope of finding some suggestions to solve his practical problems. He knew from Black the phenomenon now known under the name of latent heat, that is that when water is at the boiling point it can receive a great addition of heat without increasing the temperature. But Watt was troubled, *astonished* in his words, by the converse, that is by the fact that the quantity of water to be injected in the cylinder of a Newcomen engine to cool the steam was much greater than one could think. To test this suspicion Watt, independently of Black, made his own experiments.

> A glass tube was bent at right angles, one end was inserted horizontally into the spout of a tea-kettle, and the other part was immersed perpendicularly in well-water contained in a cylindric glass vessel, and steam was made to pass through it until it ceased to be condensed, and the water in the glass vessel was become nearly boiling hot. The water in the glass vessel was then found to have gained an addition of about one-sixth part from the condensed steam. Consequently, water converted into steam can heat about six times its own weight of well-water to 212°F [the boiling point of water in Fahrenheit degrees], or till it can condense no more steam. Being struck with this remarkable fact, and not understanding the reason of it, I mentioned it to my friend Dr Black, who then explained to me his doctrine of latent heat, which he had taught for some time before this period, (summer 1764,) but having myself been occupied with the pursuits of business, if I had heard of it, I had not attended to it, when I thus stumbled upon one of the material facts by which that beautiful theory is supported [66].[108]

In Watt's account his first experiment dated 1764; he repeted the experiences some years later, but now on quantitative basis. In one of his notebook, under the date 23th February 1781, Watt reported experiments related to the measurement of latent heat. He measured the amount (m_0) of the water contained in the glass vessel, as referred to in the above quotation and measured its temperature (t_0) in Fahrenheit degrees. After some steam has passed from the tea kettle to the glass vessel he measured the amount (m_1) of the water, inclusive of the condensed steam, contained in the glass vessel and the relative temperature (t_1). The latent heat L was found by Watt with the relation:

$$L = \frac{m_0(t_1 - t_0)}{m_1 - m_0} + t_1 - 212$$

[106] p. 16.

[107] Vol. 2, p. 115.

[108] Vol. 2, p. 116.

Table 5.2 Watt's experiments on latent heat

N	m_0	t_0 °F	$m_1 - m_0$	t_1	$t_1 - t_0$	Total heat	Latent heat
1.	17500	43°.5	760	89°.5	46°.5	1159°.5	947°.5
2.	17500	44°.5	708.	86.5	42.5	1136.9	924.9
3°.	17500	44.5	899°.	98.	54.	1149.1	937.1
4°.	17500	44.5	467.5	73.5	29.5	1175.6	963.6
5°.	17500	44.5	369.	67.25	23.	1158.	946.
6.	17500	47.5	642.	87	40.	1177.3	965.3
7°.	17500	49.	588.5	84°.5	36.	1155.	943.
8.	17500	47.	675.	87.5	41.	1150.5	938.5
9°.	17500	45.	680.5	86.5	42.	1166.5	954.5
10.	17500	45.	664.25	8s.s	41.	1165.66	953.66
11	17500	45.	975.5	102	57°.5	1134.	922.

where $m_0(t_1 - t_0)$ is the heat that steam has passed to the water in the glass vessel; this is divided by the mass of steam $(m_1 - m_0)$. The term $(t_1 - 212)$ is the excess of temperature of the water in the vessel with respect to the boiling point of the water (that is the temperature lost by the steam in condensing), given by 212°F.[109]

Table 5.2 refers the experiments on the latent heat under atmospheric pressure in different experiments; masses are expressed in English grains (1 pound is 7000 grains) and temperature in Fahrenheit degrees. The mean value is 945.3, but Watt considered not reliable the values of row 1 and 11, which were very different from the others. By neglecting them, the mean value resulted to be 949.9, "which I believe is near the truth" [65].[110]

In the letter that preface *A system of mechanical philosophy* Watt resumed the results he had obtained in his preliminary studies:

> I had myself made experiments to determine the following facts:
>
> 1st The capacities for heat of iron, copper, and some sorts of wood, comparatively with water. Similar experiments had also subsequently been made by Dr Irvine, on these and other metals
>
> 2d The bulk of steam was compared with that of water.
>
> 3d The quantity of water which could be evaporated is a certain boiler by a pound of coals.
>
> 4th The elasticities of steam at various temperatures greater than that of boiling water, and an approximation to the law which it followed at other temperatures.
>
> 5th How much water, in the form of steam, was required every stroke by a small Newcomen's engine, with a wooden cylinder six inches diameter, and twelve inches long in the stroke.

[109] To justify previous relation see [64], pp. 480–481.

[110] p. 472.

6th I had measured the quantity of cold water required in every stroke to condense the steam in that cylinder, so as to give it a working power of about 7 lb. on the inch.

Here I was at a loss to understand how so much cold water could be heated so much by so small a quantity in the form of steam, and applied to Dr Black, and then first understood what was called Latent Heat [66].[111]

In 1765 Watt had a first inspiration on the way to improve Smeaton's engines. On his words "On reflecting further, I perceived, that in order to, make the best use of steam, it was necessary. First, that the cylinder should be maintained always as hot as the steam which entered it; and, Secondly, that when the steam was condensed, the water of which it was composed, and the injection itself, should be cooled down to 100°F, or lower, where that was possible" [66].[112] This is one of the first statements of Watt's definition of the *perfect engine*. This definition was made more explicit later, in the following quotations:

And laid it as axiom – *that to make a perfect steam-engine[emphasis added] it was necessary that the cylinder should be always as hot as the steam which entered it; and that the steam should be cooled down to* 100°F [54].[113]

I have some thoughts of writing a book, the 'Elements of the Theory of Steam-engines', in which, however, shall only give the enunciation of the *perfect engine* [emphasis added]. This book might do me and the scheme good. And would still leave the world in the dark as to the true construction of the engine. Something of this kind is necessary, as Smeaton is labouring hard at the subject; and if I can make no profit, I ought not to lose the honour of my experiments [53].[114]

A modern reader would be tempted to assimilate Watt's perfect engine with Sadi Carnot's ideal engine. But this is not a correct comparison. The perfection of Watt's engine is based on the avoidance of any loss of heat; the ideality of Carnot's engine is based in the optimization of the use of heat. Though both, Carnot and Watt, assumed heat as a substance, the former was a physicist and assumed some form of conservation of energy; the latter was a chemist and had no idea of thermal energy. In [50], much space has been spent to distance Watt from thermodynamics (too much may be) and to analyze the uncritical, Whig, statements of the scientists of the 19th century in their historical accounts of Watt's enterprise.

The means of accomplishing a perfect engine did not immediately present themselves; but early in 1765 it occurred to me, said Watt, that if a communication were opened between a cylinder containing steam, and another vessel which was exhausted of air and other fluids, the steam, as an elastic fluid, would immediately rush into the empty vessel, and continue so to do until it had established an equilibrium; and if that vessel were kept very cool by an injection, or otherwise, more steam would continue to enter until the whole was condensed. These is the basic idea of the separate condenser. Of course there were some practical difficulties in implementing the idea,

[111] Vol. 2, pp. VII–VIII.

[112] Vol. 2, p. 116.

[113] p. 87.

[114] Vol. 2, p. 59.

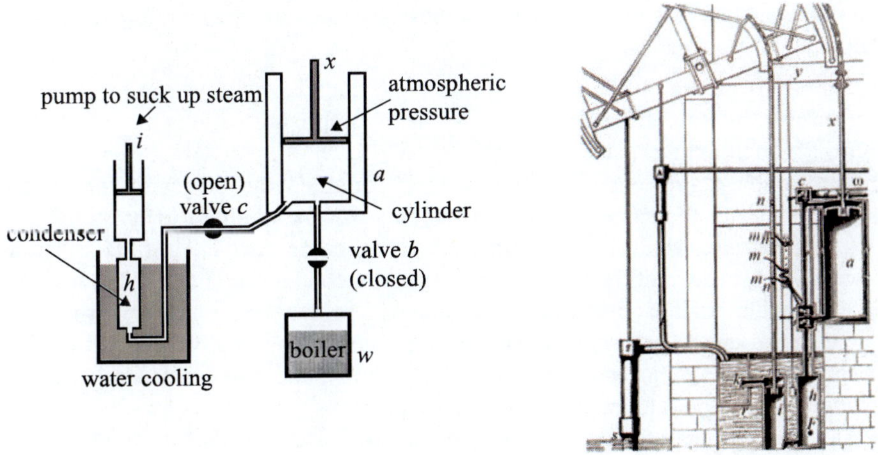

Fig. 5.7 **a** Basic version of Watt's steam engine. **b** A nearly complete scheme redrawn from [80], p. 114

but Watt soon solved them: "When once the idea of the separate condensation was started, all these improvements followed as corollaries in quick succession, so that in the course of one or two days, the invention was thus far complete in my mind, and I immediately set about an experiment to verify it practically" [66].[115]

Figure 5.7a shows the basic aspects of Watt's engine whose main elements were the great boiler w, the cylinder equipped with a steam jacket (a), the separate condenser h. The steam produced by the boiler enters the cylinder and lifts the piston (in this phase the valve b is open and c is closed). As soon as the piston has reached the top of the cylinder, c opens, b closes and a pump i sucks the steam from the cylinder. The cylinder goes down because of the atmospheric pressure. The sucked steam goes to the condenser to return to the liquid state. The valve b is reopened and the c closed, to begin a new cycle. Watt initially thought to directly connect the appliances to the piston x, but then he used the system of Newcomen's rocker. The rocker also drives the pump i that sucks the steam from the condenser. Figure 5.7b shows a more complete scheme.

Watt patented his ideas on separate condenser in 1769, under the title "A new method of lessening the consumption of steam and fuel in fire engines" [53].[116] In 1782 he presented the specifications of a new patent to introduce fundamental improvements to his engine. The title of the patent was "Specification of patent, March 12th, 1782, for certain new improvements upon steam or fire engines for raising water, and other mechanical purposes, and certain new pieces of mecha-

[115] Vol. 2, p. 117.

[116] Vol. 3, p. 10.

nism applicable to the same" [53].[117] In these specifications are comprehended the following improvements:

1. The use of steam on the expansive principle; together with various methods or contrivances, (six in number, some of them comprising various modifications), for equalizing the expansive power.
2. The double-acting engine which basically doubled the power of the simple machine with the same volume; in which steam is admitted to press the piston upwards as well as downwards; the piston being also aided in its ascent as well as in its descent by a vacuum produced by condensation on the other side.
3. The double-engine; consisting of two engines, primary and secondary, of which the steam-vessels and condensers communicate by pipes and valves, so that they can be worked either independently or in concert; and make their strokes either alternately or both together, as may be required.
4. The employment of a toothed rack and sector, instead of chains which was no longer usable, for guiding the piston-rod (Watt in 1788 realized a centrifugal regulator valve, the so-called Watt pendulum, a mechanism that regulated the supply of steam in order to keep the machine in motion with constant speed).
5. A rotative engine, or steam-wheel.

While from a technological point of view probably the most important innovation was the introduction of the double acting engine, from theoretical point of view the most interesting innovation is the introduction of the expansive principle; and may be Watt would have agree as he presented this innovation as first. Watt explained his idea with reference to Fig. 5.8. The steam entered thorough the opening J (top right), with a given pressure: "Then I say that the pressure or elastic power of the said steam on every square inch of the area or upper side of the piston, will be nearly fourteen pounds [...], and that if the said power were employed to act upon the piston through the whole length of its stroke, and to work a pump or pumps, either immediately by the piston rod beam or great lever, as is usual in steam engines, it would raise through the whole length of its stroke a column or columns of water, whose weight should be equal to ten pounds for each square inch of the area of the piston, besides overcoming all the frictions and vis inertiae of the water and the parts of the machine or engine" [53].[118]

But supposing that the steam from the boiler to be perfectly shut when the piston has descended to the point K two feet, or one-fourth of the length of the stroke or motion of the piston, then when the piston had descended four feet, or one-half of the length of the stroke, the elastic power of the steam would then be equal to seven pounds on each square inch of the area of the piston, or one-half of the original power; and that when the piston had arrived at the point P, the power of the steam would be one-third of the original power, or four pounds and two-thirds of a pound on each square inch of the piston's area [53].[119]

[117] Vol. 3, p. 55.

[118] Vol. 3. p. 62.

[119] Vol. 3. p. 62.

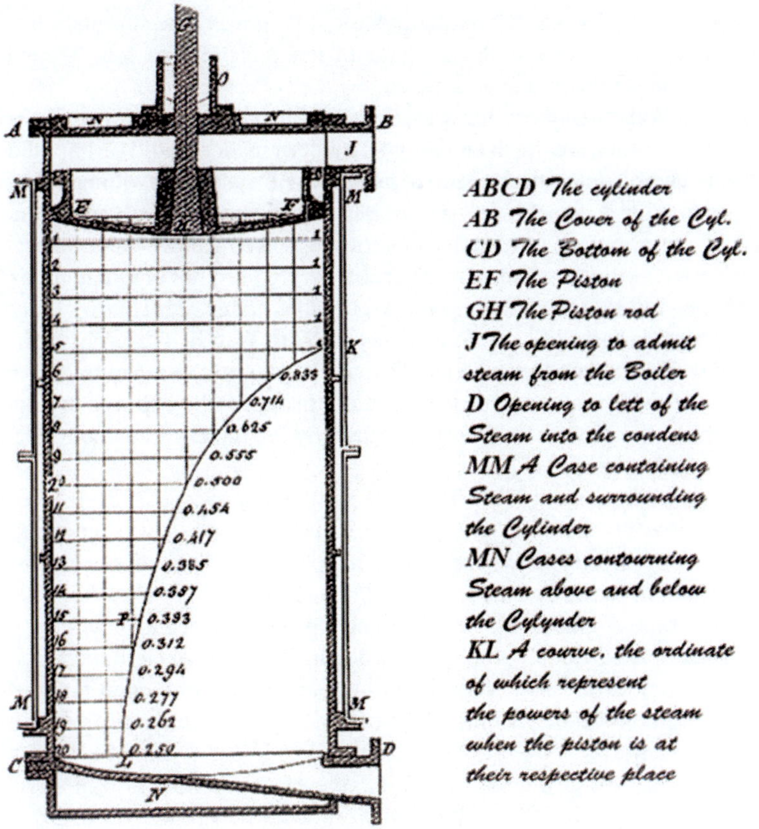

Fig. 5.8 Illustration of the expansive principle. Redrawn from [53], vol. 3, Plate VIII, Fig. 1

More generally the elastic powers, or pressure, of the steam are represented by the lengths of the horizontal lines or ordinates of the curve KL of Fig. 5.8, expressed in decimal fractions of the whole original power. opposite to the said ordinates or horizontal lines. Watt did not say how he obtained this curve; not experimentally however as a simple inspection shows that the values volume-pressure respect exactly the law of Boyle-Mariotte (volume multiplied by pressure is constant); therefore the curve is theoretical and according to Watt it has only the purpose to show at large the behavior of steam: "And I also say, that the sum of all these powers is greater than fifty-seven hundred parts of the original power multiplied by the length of the cylinder; whereby it appears that only one-fourth of the steam necessary to fill the whole cylinder is employed and that the effect produced is equal to more than one-half of the effect which would have been produced by one whole cylinder full of steam, if it had been admitted to enter freely above the piston during the whole length of its descent" [53].[120] In conclusion, though the total power (the summation of all the powers or pressure) is halved, the consume of steam is only one fourth, and thus the efficiency of the steam engine is doubled.

Notice that Watt did not explicitly attribute any physical meaning to the summation of the pressures of the steam, even though a modern reader would be tempted to read this value as a work, that is the integral of pressure with respect to volume. The concept of work as applied to machines was not very diffuse at the time, and even though the idea of the principle of virtual velocities was well known (and here concepts comparable with work are used), most probably Watt did not elaborate it, as Lazare Carnot (1753–1823), a contemporary of Watt, did with his *Essai sur les machines en général* published in 1783 [19]. In any way for sure Watt had an embryonic idea of power and work (modern meaning) and his use of pressure (and volume) to measure the efficacy of the steam machine was partially taken for the expression of power of water machines, were the effect was measured by the force impressed to the blames of wheel and their speed [15].

One more improvement Watt introduced in his engine, possibly in 1790 s, was the *steam indicator*; the precise origin of this invention are not clear and there are also questions of priority; these aspect discussed in [50][121] are not relevant here. The indicator of steam is a device made of a manometer which measures the pressure into the cylinder of the engine and drives a pen on a paper attached to a cylinder which is made to rotate proportionally the vertical displacement of the cylinder. In this way a plot is obtained of pressure versus volume; now an experimental one differently from that presented in the specifications of 1812 patent, which however was given for one stroke only.

Even this plot, when limited to a full cycle of the engine, to a modern reader, suggest that Watt is using the concept of work, as it is well known that the area of the surface delimited by the close curve produced by the indicator gives the value of the mechanical work produced by the steam machine and allows a comparison with the heat consumed to evaluate the efficiency of the engine. To Watt however the use of the indicator was mainly that of check the correct behavior of the engine; there is not indeed evidence of a thermodynamic interpretation in his writings. This notwithstanding the scientists of the 19th century interpreted the diagram of the indicator as the *diagram of energy* [50][122] and made Watt a proto thermodynamic scholar. A question still debated.

Watt thought about a further important idea to improve the efficacy of his steam engine, but that was not the object of a patent. From his studies on latent heat Watt knew, by his own experiments and by the scientific literature of the period, that the latent heat decreased with the pressure at which the water boils. This is in agreement with Watt's law; indeed the temperature for which water boils increases with the pressure, and thus the latent heat decreases.

[120] Vol.3, p. 63.

[121] Chapter 6.

[122] p. 153.

> That, in proportion as the sensible heat of steam increases, its latent diminishes, so, in the steam-engine, working with pressures above 15 lbs. must be more advantageous than below it; for not only the latent heat is diminished, but the steam is considerably expanded by the sensible heat, which is easily added [54].[123]

This means that a steam engine operating under high pressure is more efficient than an engine operating at the atmospheric pressure as a less amount of heat, and fuel, is necessary to produce the same amount of steam because of the lower latent heat. However the technology of time did not allowed the construction of high pressure engines.

Modern historian has long debated wether or not the scientific knowledge of physics and chemistry influenced the invention of the steam engine. This is part of a larger discussion about the influence of science over technology, especially before the 19th century. Some comment on this subject has been developed in previous sections; here I limit to what Watt himself though about this argument.

> But this theory [latent heat], though useful in determining the quantity of injection necessary where the quantity of water evaporated by the boiler, and used by the cylinder, was known, and in determining, by the quantity and heat of the hot water emitted by Newcomen's engines, the quantity of steam required to work them, did not lead to the improvements I afterwards made in the engine [66].[124]

These improvements proceeded upon the old-established fact, that steam was condensed by the contact of cold bodies and the later known one, that water boiled in vacuo at temperature below 100°F, and consequently that a vacuum could not be obtained unless the cylinder and its contents were cooled every stroke below that temperature.

> Although Dr Black's theory of latent heat did not suggest my improvements on the steam-engine, yet the knowledge upon Various subjects which he was pleased to communicate to me, and the correct modes of reasoning, and of making experiments of which he set me the example, certainly conduced very much to facilitate the progress of my inventions; and I still remember with respect and gratitude the notice he was pleased to take of me when I very little merited it, and which continued throughout his life [66].[125]

5.2.2.3 A Philosopher

Watt's image underwent a sensible change after his death. The scientists of the 19th century (the term *scientist* can be now used because introduced in this period), largely underestimated the scientific contribution by Watt as chemist. Moreover, chemists of the 19th century did not count Watt as one of them; they saw him as an engineer and the elitist scientists did the same.

---

[123] p. 166. Letter to John Roebuck (1718-1794), 23rd Aug. 1765.

[124] Vol. 2, p. VIII.

[125] Vol. 2, p. VIII.

However he was considered a very special kind of engineer; he was characterized as a *philosophical engineer*—The term was not Watt's but used in [48]—because of his deep understanding of natural philosophy. This is an estimation of Watt by James David Forbes (1809–1868) one of the founder of the British association for advancement of science

> James Watt may be considered as the most distinguished practical man of science of the last century, or even for a much longer period. But this is not all. Few men achieve such a reputation as his without having done more than originate a great invention for the use and benefit of mankind in all ages : He also taught men to raise the useful arts to a new dignity, to marry them to genuine, unpretending, and inductive science, to disparage ignorance and empiricism, and to render the labours of the workshop subservient to intellectual progress [28].[126]

Forbes, in the first chapter of his *A review of the progress of mathematical and physical science in more recent times* of 1858 [28], placed the scientific part of engineering in its due relation to pure physics and compared the relation between physics and engineering to that subsisting between mathematics and physics; the one as an instrument, the other as an end. According to Forbes the masterpieces of civil engineering: the steam engine, the locomotive engine and the tubular bridge were only experiments of the powers of nature on a gigantic scale, and were not to be compassed without inductive skill as remarkable and as truly philosophic as any effort which the man of science exerts, save only the origination of great theories, of which one or two in a hundred years may be considered as a liberal allowance.

That is the activity of engineers consist in the application of science. The engineer is inferior to pure scientist because the former cannot create great theories. Forbes continued by asserting that, whilst then one claims for Watt a place amongst the eminent contributors to the progress of science in the eighteenth century:

> We must reserve a similar one for the Stephensons and the Brunels of the present: and whilst we are proud of the changes wrought by the increase of knowledge during the last twenty-five years on the face of society, we must recollect that these very changes, and the inventions which have occasioned them, have stamped perhaps the most characteristic feature its intense practicalness on the science itself of the same period [28].[127]

Rankine, like Forbes, presented Watt's experiments with the model steam engine as a model of the relations of theory and practice. For Rankine, who however saw engineering and science more close to each other than Forbes, Watt became the model of the engineering scientist:

> Watt set to work scientifically from the first. He studied the laws of the pressure of elastic fluids, and of the evaporating action of heat, so far as they were known in his time; he ascertained as accurately as he could, with the means of experimenting at his disposal, the expenditure of fuel in evaporating a given quantity of water, and the relations between the temperature, pressure and volume of the steam. Then reasoning from the data which he had thus obtained, he framed a body of principles expressing the conditions of the efficient and

[126] p. 67.

[127] p. 3.

> economic working of the steam engine, which are embodied in an invention described by himself [...], in the specification of his patent of 1769 Miller p. 795 [62].[128]

So Rankine had Watt moving smoothly from the puzzle of the Newcomen engine model to experiments, to data, to principles and thence to specifications of the improved engine.

To note that at Watt's time the role of philosophical engineering was object of discussion. From this point of view it is interesting to quote a letter by Robinson, where the latter discussed about the opportunity to create a doctorate in engineering (sciences).

> And this brings another thing in to my head Would not a description and even a good Model of your Engine be a becoming present from You to the Museum of the University of Glasgow. I know that it would be received with great Affection and Respect. Think of this at your Leisure, as also of a new doctorate that I am scheming with some hopes of Success, if we can find a proper Name for it—Doctor of Arts—a Collegium or Corporation of Scientific Engineers, with three degrees of Bachelor Master and Doctor—not merely academical honours, of no more value than the offices of a Mason Lodge, but to have Civil Consequences. As a Man must have a diploma to entitle him to a consulting fee, so should an Engineer etc. etc.-I had more to say-but the Wright has come to close the Box-so farewell [65].[129]

James Watt engineer was the term Watt used as his identification in the two scientific papers published in the Philosophical Transactions. The term was however not fully adequate for the times in England, when *engineer* mainly indicated practical men, also involved in trade activity. Watt was for sure also an engineer in this sense, but not only this.

David Miller, who devoted in-depth researches in the role of Watt as engineer [48] [50],[130] suggests that when Watt, unusually and deliberately, designated himself *engineer* in his publications in the Philosophical Transactions, or elsewhere, did it with a mixture of deference and defiance. He was saying, yes I am an engineer but not a mere engineer. I am superior to the 'run of the mill' engineer because I am an experimentalist and a philosopher. "This was not an entirely personal act in the sense that Watt was creating space for the idea of the philosophical engineer. Yet it was, deeply personal too, going to the heart of Watt's sense of identity" [50].[131]

After the patent of 1769 was extended by Act of parliament in 1775, and after he in the 1780 s took out a number of patents on the improvement of the steam engine, Watt was involved with Boulton in a series of patent trials [48]. Eventually January 1799 the trials ended with a victory of our 'heroes', however only one year before the expiry of the 1769/1775 patent.

The various patent trials of the 1790 s offer a discussion on the role of the engineer and of Watt himself. The question was raised: Was the engineer characterized by extensive hands-on experience, or was it a combination of such experience with a philosophical understanding? In presenting the specifications for his patent of

---

[128] p. XXII.

[129] p. 273, letter of Robinson to Watt of M ay 3rd 1797.

[130] Chapter 3.

[131] p. 61.

1769 on the separate condenser, Watt decided to relate about ‘principles’ only. By principles he indented of course not principles of natural philosophy, but simply the most general idea about his invention. For instance he neither intended to patent the discovery of the latent heat nor that the fact that steam condenses when cooled, but only the idea that steam should be condensed in a separate vessel and not in the cylinder of the engine.

In 1769 William Small, (1734–1775), a Scottish professor of natural philosophy at the College of William and Mary in Virginia, and Matthew Boulton wrote to Watt on the specifications for the patent of the separate condenser.

> Mr. Boulton and I have considered your paper, and think you should neither give drawings nor descriptions of any particular machinery, (if such omissions would be allowed at the office) but specify in the clearest manner that you can, that you have discovered some principles, and thought of new applications of others, by means of both which joined together, you intend construct steam engines of much greater powers [...].
>
> As to your principles, we think they should be enunciated (to use an hard word) as generally as possible, to secure you as effectually.[132]

The main idea for presenting only general principles is stressed in the final part of the quotation. The specifications are in general terms; it avoids any particular description; it includes no drawings; the purpose is to avoid piracy being perpetrated by those who would make minor modifications.

This however was not only an expedient to avoid piracy. Watt was also convinced that the main role of a philosophical engineer is to produce general ideas which play a similar role of theories in natural philosophy. He, Small and Boulton, were convinced that any skillful mechanic may construct the engine with separate condenser, and there is any oblige to teach any blockhead in the nation to construct masterly engine [64].[133] The philosophical engineer so is justified to patent his ideas.

5.2.2.4 An Inventor

Watt always gave much importance to his manual skills which he always exercised giving an imprint of his acumen to the objects produced. His friend Robinson reported in a his memoir of the way in which Watt experimented with the construction of an organ in the first 1760, although he could not distinguish one note from another.

> He began by building a very small one for his intimate friend Dr. Black, which is now in my possession. In doing this, a thousand things occurred to him which no organ—builder ever dreamed of—nice indicators of the strength of the blast, regulators of it, &c. &c. He began to the great one. He then began to study the philosophical theory of music. Fortunately for me, no book was at hand but the most refined of all, and the only one that can be said to contain, any theory at all, ‘Smith’s Harmonics. Before Mr. Watt had half finished this organ, he and I were completely masters of that most refined and beautiful theory of the beats of imperfect consonances [53].[134]

[132] From Watt’s Birmingham central library archive as reported in [64] pp. 119–120.

[133] p. 126.

[134] Vol. 1, pp. CVIII–CIX.

Another example of Watt's skill and curiosity is provided by the construction of a device to facilitate perspective drawings. Still another example is the construction of a micrometer, or in modern term something like a tachometer, useful for topographic surveys; they are described both in [53][135] and [54].

The practical spirit of Watt made possible a civil engineering career that could guarantee him to keep the family since he had married his cousin Miss Miller, occurred in 1764. In the last year of 1760 and the beginning of 1770 he carried out as an engineer the survey of some canals. But the last and most remarkable of his civil engineering works was probably that for which he was called on, also in 1773, to perform for his employers the Commissioners of Police; viz. a survey and estimate for a navigable canal, to pass through the chain of rivers and lakes in the wild and remote tract of country between Fort-William and Inverness [53].[136]

Watt patented many inventions. Below a list of the specifications he presented for a patent:

1. January 5th, 1769, for a new method of lessening the consumption of steam and fuel in fire engines.
2. February 14th, 1780, for a new method of copying letters and other writings expeditiously.
3. October 25th, 1781, for certain new methods of applying the vibrating or reciprocating motion of steam or fire engines, to produce a continued rotative or circular motion round an axis or centre, and thereby to give motion to the wheels of mills or other machines
4. March 12th, 1782, for certain new improvements upon steam or fire engines for raising water, and other mechanical purposes, and certain new pieces of mechanism applicable to the same.
5. April 28th, 1784, for certain new improvements upon fire and steam engines, and upon machines worked or moved by the same.
6. June 14th, 1785, for certain newly improved methods of constructing furnaces or fire-places for heating, boiling, or evaporating of water and other liquids which are applicable to steam engines and other purposes; and also for heating, melting, and smelting of metals and their ores, whereby greater effects are produced from the fuel, and the smoke is in a great measure prevented or consumed.

Of these patent I will comment only briefly some of them, while for its relevance and the relative shortness I will report in full the first, that of 1769, related to the invention of the separate condenser:

> SPECIFICATION OF PATENT, JANUARY 5th, 1769, FOR A NEW METHOD OF LESSENING THE CONSUMPTION OF STEAM AND FUEL IN FIRE ENGINES.
>
> To all to whom these presents shall come, I, James Watt, of Glasgow, in Scotland, Merchant, send greeting.
>
> Whereas His Most Excellent Majesty King George the Third, by his Letters Patent, under the Great Seal of Great Britain, bearing date the fifth day of January, in the ninth year

[135] Vol. 1.

[136] Vol. 1, p. CXXIV.

of his said Majesty's reign, did give and grant unto me, the said James Watt, his special licence, full power, sole privilege and authority, that I, the said James Watt, my executors, administrators, and assigns, should, and lawfully might, during the term of years therein expressed, use, exercise, and vend throughout that part of his Majesty's Kingdom of Great Britain called England, the Dominion of Wales, and Town of Berwick upon Tweed, and also in his Majesty's Colonies and Plantations abroad, my new invented "Method of Lessening" the Consumption of Steam and Fuel in "Fire Engines", in which said recited Letters Patent is contained a Proviso, obliging me, the said James Watt, by writing under my hand and seal, to cause a particular description of the nature of the said invention to be inrolled in his Majesty's High Court of Chancery, within four calendar months after the date of the said recited Letters Patent, as in and by the said Letters Patent and the Statute in that behalf made, relation being thereunto respectively had, may more at large appear.

Now know ye, that in compliance with the said Proviso, and in pursuance of the said Statute, I, the said James Watt, do hereby declare that the following is a particular description of the nature of my said invention and the manner in which the same is to be performed (that is to say): my Method of lessening the consumption of steam, and consequently fuel, in fire engines, consists of the following principles:

- First, that vessel in which the powers of steam are to be employed to work the engine, which is called the Cylinder in common fire engines, and which I call the Steam Vessel, must, during the whole time the engine is at work, be kept as hot as the steam that enters it; first, by inclosing it in a case of wood, or any other materials that transmit heat slowly; secondly, by surrounding it with steam or other heated bodies ; and, thirdly, by suffering neither water nor any other substance colder than the steam to enter or touch it during that time.
- Secondly, in Engines that are to be worked wholly or partially by condensation of steam, the steam is to be condensed in vessels distinct from the steam vessels or cylinders, although occasionally communicating with them: these vessels I call Condensers; and, whilst the engines are working, these condensers ought at least to be kept as cold as the air in the neighbourhod of the engines, by application of water, or other cold bodies.
- Thirdly, whatever air, or other elastic vapour, is not condensed by the cold of the condenser, and may impede the working of the engine, is to be drawn out of the steam vessels or condensers by means of pumps, wrought by the engines them selves, or otherwise.
- Fourthly, I intend in many cases to employ the expansive force of steam to press on the pistons, or whatever may be used instead of them, in the same manner as the pressure of the atmosphere is now employed in common fire engines : in cases where cold water cannot be had in plenty, the engines may be wrought by this force of steam only, by discharging the steam into the open air after it has done its office.
- Fifthly, where motions round an axis are required, I make the steam vessels in form of hollow rings, or circular channels, with proper inlets and outlets for the steam, mounted on horizontal axles, like the wheels of a water-mill; within them are placed a number of valves, that suffer any body to go round the channel in one direction only: in these steam vessels are placed weights, so fitted to them as entirely to fill up a part or portion of their channels, yet rendered capable of moving freely in them by the means hereinafter mentioned or specified. When the steam is admitted in these engines, between these weights and the valves, it acts equally on both, so as to raise the weight to one side of the wheel, and by the re-action on the valves successively, to give a circular motion to the wheel, the valves opening in the direction in which the weights are pressed, but not in the contrary: as the steam vessel moves round, it is supplied with steam from the boiler, and

that which has performed its office may either be discharged by means of condensers, or into the open air.[137]

- Sixthly, I intend, in some cases, to apply a degree of cold not capable of reducing the steam.
- Lastly, instead of using water to render the piston, or other parts of the engines, air and steam-tight, I employ oils, wax, resinous bodies, fat of animals, quicksilver, and other metals, in their fluid state.

In Witness whereof I have hereunto set my hand and seal this twenty-fifth day of April, in the year of our Lord one thousand seven hundred and sixty-nine.

James Watt.

Sealed and delivered in the presence of

Coll. Wilkie.
Geo. Jardine.
John Roebuck.

Be it remembered, that the said James Watt doth not intend that anything in the Fourth Article shall be understood to extend to any engine where the water to be raised enters the steam vessel itself, or any vessel having an open communication with it.[138]

James Watt.

Witnesses. Coll. Wilkie.
Geo. Jardine [53].[139]

The patent of 1780, for a new method of copying letters and other writings expeditiously, is particularly interesting to document Watt's versatility. The method anticipated the modern concept of photocopier by making practicable an approach already known. In substance it consists of superposing to the sheet to be copied a pieces of thin paper of the same dimension "which contain no size, or glue, or gummy or mucilaginous matter, or which at least does not contain so much size or other matter as would make it fit for being written upon" [53].[140] This thin paper should be moisten or wet with water, or other liquid, by means of a sponge or brush, or otherwise. Having moistened or wet the thin paper, it should be located between two thick unsized spongy papers, or between two cloths, or other substances capable of absorbing the superfluous moisture from the thin paper. Then the original sheet and the thin paper are pressed together, for example by means of a rolling press. Some cautions should be taken; first the original text should be written with inks that allows to be wet whiteout danger; secondly the liquor with which to wet the thins paper should have been prepared with appropriate recipe. In any case the copying machine worked very

[137]This device, a rotatory engine, is known as *steam-wheels*; for its description and comments see [30], pp. 76–78.

[138]This comment was added by Watt, to prevent it being supposed that he had any intention of claiming the principle of Savery's engine, in which the water was raised by the elastic force of steam, but without the intervention of a piston.

[139]Vol. 3, pp. 10–15.

[140]Vol. 3, p. 30.

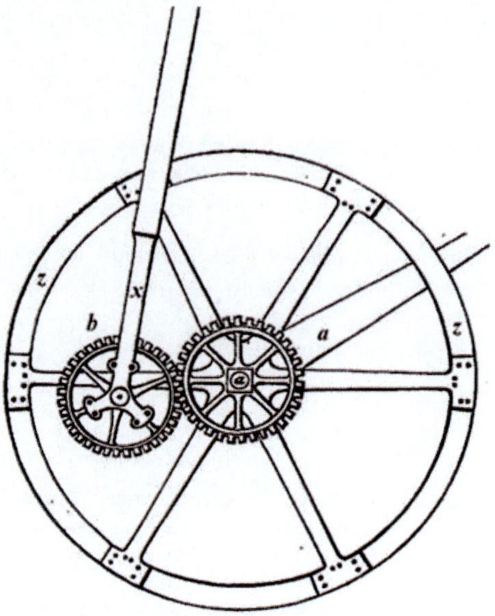

Fig. 5.9 Sun and planet gear system. Redrawn from [80], Fig. 29, p. 136

well and in the great establishment at Soho, there was the practice to retain copies of the drawings of all the engines sent out from that manufactory.

In the patent of 1781, Watt presented five methods to transform the alternating motion of the piston in to a circular one. The application for this patent, for new methods of producing a rotative motion from a reciprocating one in the steam-engine, and thereby giving motion to mill-work, was rendered necessary on the one hand by the difficulties experienced in working the steam-wheels such as described in the patent specification of 1769; on the other hand by Mr. Watt having been unfairly anticipated by Matthew Wasbrough (1753–1781), in patenting the crank to produce rotatory motion. Five different methods are enumerated, by any one of which the proposed end might be attained without the intervention of a crank; all of them admitting, as mentioned in the specification, of many varieties. The last one, known as sun and planet gear wheels, was the one most used by Watt, and it is described below. Notice that though the invention apparently did not required a deep knowledge of mathematics and kinematics, it required some; or at least the knowledge on the literature about gears. The way the system works is made clear from Fig. 5.9. The alternating motion transmitted by the thread x, is transformed into a rotatory motion through the gear b (the planet) and a (the sun).

The content of the patent of 1782, related to the double action engine, has already be discussed. The last Watt's patent dates 1784, and contains many improvements. According to his biographer Muirhead "this specification may probably be viewed as second in importance to none of those prepared by Mr. Watt subsequent to that of the Separate Condenser in 1769" [53].[141] Besides many improvements now of minor

[141] Vol. 3, p. 88.

Fig. 5.10 Watt's drawing of parallel motion, redrawn from [54], vol. 3, p. 89

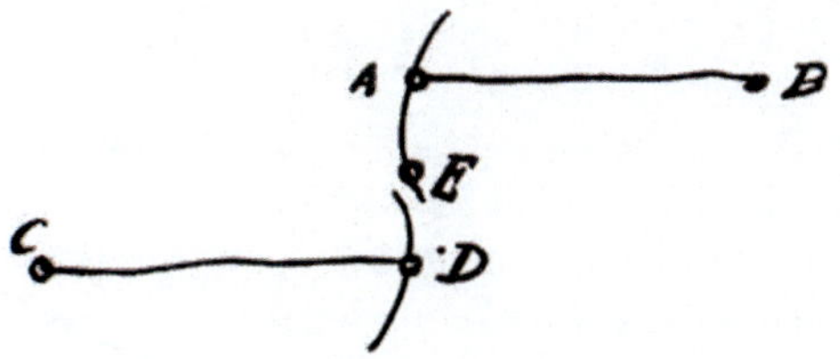

consequence, such balancing of pump-rods, new steam-wheels, communication of motion from the same engine to two separate primary axes, and apparatus for opening the regulating valves with rapidity, it contains various methods of converting a circular or angular motion into a perpendicular or rectilineal motion—one of those methods being the well-known and much-admired *parallel motion*; a method of working a tilt-hammer for forging iron, making steel, &c, by steam—and the application of the steam-engine to give motion to wheel carriages for carrying persons or goods [53].[142]

Watt was particularly proud of the invention of the parallel motion that served to rectify the motion along an arc of circle of the extremity of his rockets beam; the solution with the chain was no longer practicable for his double action engines the chain; the solution with the chain was no longer practicable for his double action engines. Here is what Watt wrote to his son James in 10th November 1808:

> The idea originated in this manner. On finding double chains, or racks and sectors, very inconvenient for communicating the motion of the piston-rod to the angular motion of the working-beam, I set to work to try if I could not contrive some means of performing the same from motions turning upon centres, and after some time it occurred to me that AB, CD, being two equal radii revolving on the centres B and C, and connected together by a rod AD, in moving through arches of certain lengths, the variations from the straight line would be nearly equal and opposite, and that the point E would describe a line nearly straight, and that if for convenience the radius CD was only half of AB, by moving the point E nearer to D, the same would take place, and from this the construction, afterwards called the parallel motion, was derived. *Though I am not over anxious after fame, yet I am more proud of the parallel motion than of any other" mechanical invention I have ever made* [emphasis added] [54].[143]

In Fig. 5.10, the line CD represents the rocket beam, which rotates around C in its oscillating motion B is a fixed point and AB a thread. Point E it is know today to describe a lemniscate, but because the small angular excursion of CD it approximates a straight line.

Actually what is described in Fig. 5.10 is more exactly what is known as *Watt linkage*. Indeed the use of the scheme of Fig. 5.10 would have made the engine machine an awkward shape, with B a long way from the end of the rocket beam. To avoid this, Watt added the parallelogram linkage BADGH to form a pantograph as shown in Fig. 5.11, and this is properly what is known as *Watt parallel motion.* This system guarantees that E always lies on a straight line and that the motion of

[142] Vol. 3, p. 88.

[143] Vol. 3, p. 89.

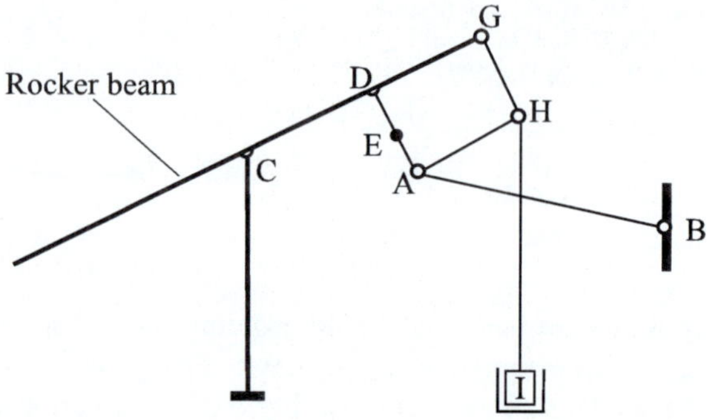

Fig. 5.11 Watt's improved version of parallel motion

H, the point to which the piston rod HI is attached, is a magnified version of the motion of E. The addition of the pantograph made the mechanism shorter and so the building containing the engine could be smaller. As already noted, the path of E is not a perfect straight line, but merely an approximation. Watt's design produced a deviation of about one part in 4000 from a straight line. Later, in the 19th century, perfect straight-line linkages were invented, beginning with the Peaucellier-Lipkin linkage of 1864 [81]; however Watt's mechanism of parallel motion in many cases should be preferred because of its greater simplicity.

5.3 Quotations

E.1 L'emploi d'ingénieur exige beaucoup d'étude, de talens, de capacite & de génie. Les sciences fondamentales de cet état sont l'Arithmétique, la Géométrie, la Méchanique & l'Hydraulique. Un ingénieur doit avoir quelqu'usage du dessein. La physique lui est nécessaire pour juger de la nature des matériaux qu'on emploie dans les bâtimens, de celle des eaux, & des différentes qualités de l'air des lieux qu'on veut fortifier. Il est très-utile qu'il ait des connoissances générales & particulieres de l'Architecture civile, pour la construction des bâtimens militaires, comme casernes, magazins, arsenaux, h'pitaux, logemens de l'état-major, &c. dont les ingénieurs sont ordinairement chargés. M. Frézin recommande aux ingénieurs de s'appliquer à la coupe des pierres.

E.2 Mais nous reprenons insensiblement le dessus, & l'on peut dire qu'aux yeux mêmes de la multitude, les bornes de cette prétendue magie naturelle se rétrécissent tous les jours; parce qu'éclairés du flambeau de la Philosophie, nous faisons tous les jours d'heureuses découvertes dans les secrets de la nature, & que de bons systèmes soutenus par une multitude de belles expériences annoncent àl'humanité dequoi elle peutêtre capable par elle-même & sans magie. Ainsi la boussole, les thélescopes, les microscopes, &c. & de nos jours, les polypes, l'électricité; dans la Chimie, dans la Méchanique & la Statique, les découvertes les plus belles & les plus utiles, vont immortaliser notre siecle; & si l'Europe retomboit jamais dans la barbarie dont elle est enfin sortie, nous passerons chez de barbares successeurs pour autant de magiciens.

References

1. Agrippa HC (1676) The vanity of arts and sciences. Speed, London
2. Arago FJD (1839) Éloge historique de James Watt. Firmin, Paris
3. Aristotle (2018) Analytica posteriora. The internet Classical Archive, Translated into English by Mure GRG
4. Aristotle (2018) Ethica nichomachea. The internet Classical Archive, Translated into English by Ross WD
5. Aristotle (2018) Metaphysica. The internet Classical Archive, Translated into English by Ross WD
6. Auyang SY (2004) Engineering-An endless frontier. Harvard University Press, Cambridge
7. Belhoste B (1989) Les origines de l'École polytechnique. Des anciennes écoles d'ingénieurs àl'École centrale des travaux publics. Histoire de l'Éducation 42:13–53
8. Belhoste B (2003) La formation d'une technocratie. L'Ècole polytechnique et ses élèves de la révolution au second empire, Belin, Paris
9. Bernoulli J (1727) Theoremata selecta pro conservatione virium vivarum demonstranda et esperimenta confirmanda. In: Bernoulli J (ed) Opera omnia (1742, 4 vol), vol. 3, Bousquet, Lausanne and Geneva, pp. 124–130
10. Bo-cong L (2010) The rise of philosophy of engineering in the East and the West. In: Van de Poel I, Goldberg DE (eds) Philosophy of engineering. An emerging agenda. Springer, Dordrecth, pp 31–40
11. Böhme G, van de Daele W, Krohn W (1976) Finalization in science. Soc Sci Inf 15:307–330
12. Boon M (2011) In defense of engineering sciences: On the epistemological relations between science and technology. Thechné: Research in Philosophy and Technology 15(1):49–71
13. Bramwell FJ (1885–1900) James W (1736–1819). Dictionary of National Biography 60:51–62
14. Bunge M (1966) Technology as applied science. Technol Cult 7(3):329–347
15. Capecchi D (2013) Over and undershot waterwheels in the 18th century. Science-technology controversy. Adv Hist Stud 2(3):131–139
16. Capecchi D (2014) The problem of motion of bodies. Springer, Cham
17. Capecchi D (2018) The path to post-Galilean epistemology. Springer, Cham
18. Capecchi D, Ruta G (2015) Strength of materials and theory of elasticity in 19th century Italy. Springer, Dordrecht
19. Carnot L (1786) Essai sur les machines en général. de Defay, Dijon
20. Channell DF (2009) The emergence of the engineering sciences: An historical analysis. In: Meijers A (ed) Philosophy of technology and engineering sciences, vol 9. Elsevier, Amsterdam, pp 117–154
21. d'Alembert J (1743) Traité de dynamique. David, Paris
22. Della Porta GB (1677) Della magia naturale del sig. Gio. Battista Della Porta linceo napolitano. Libri 20 (2nd edn). Bulifon, Naples
23. Desaguliers JT (1730) An attempt to solve the phaenomenon of the rise of vapours, formation of clouds and descent of rain. Philos Trans R Soc Lond 36(407):6–22
24. Desaguliers JT (1734–1744) A course of experimental philosophy (2 vol). Longman et als, London
25. Donovan A (1975) Philosophical chemistry in the Scottish Enlightement. Edinburgh University, Edinburgh

26. Encyclopédie (1751–1772) Encyclopédie ou dictionnaire raisonné des sciences, des arts et des métiers, par une societè de gens de lettres (17 vols). Briasson-David-Le Breton-Durand, Paris
27. Fioravanti L (1571) Il reggimento della peste. Sessa, Venice
28. Forbes JD (1858) A review of the progress of mathematical and physical science in more recent times, and particularly between the years 1775 and 1850. Black, Edinburgh
29. Galen (1819–1833) Claudii Galeni opera omnia (20 vol). In: Kühn CG (ed) Officina Libraria Cnoclochii, Lipsia
30. Galloway E (1831) History and progress of the steam engine. Kelly, London
31. Gille B (1964) Les ingénieur de la Renaissance. Hermann, Paris
32. Grafton A (2002) Magic and technology in early modern Europe. Smithsonian Institution Libraries, Washington
33. Hall AR (1961) Engineering and the scientific revolution. Technol Cult 2(4):333–341
34. van Helmont JB (1652) Ortus medicinae id est initia physicae inaudita. Elsevier, Amsterdam
35. Houkes W (2009) The nature of technological knowledge. In: Meijers A (ed) Philosophy of technology and engineering sciences, vol 9. Elsevier, Amsterdam, pp 309–350
36. Jones R (1970) The 'plain story' of James Watt: the Wilkins lecture 1969. Notes Records R Soc Lond 2(2):194–220
37. Khattab A (2109) Institution of civil engineers ICE, history of the institution of civil engineers. https://www.designingbuildings.co.uk/wiki/Institution_of_Civil_Engineers_ICE
38. Knowles E (2019) Engineering timelines. John Smeaton edited by Jane Joyce. http://www.engineering-timelines.com/who/Smeaton_J/smeatonJohn.asp
39. Krohn W, Schäfer W (1983) Agricultural chemistry. The origin and structure of a finalized science. In: Schäfer W (ed) Finalization in science. The social orientation of scientific progress. Springer, Dordrecht, pp 17–52
40. Kuhn T (1962) The structure of scientific revolution. The University of Chicago Press, Chicago
41. Lagrange JL (1763) Recherches sur la libration de la lune. In: Serret JA, [Darboux G] (1867–1892) (ed) Oeuvres de Lagrange (14 vol), vol 6, Gauthier-Villars, Paris, pp 5–61
42. Latour B (1987) Science in action. How to follow scientists and engineers through society. Open University Press, Cambridge
43. Lelas S (1993) Science as technology. Br J Philos Sci 44(3):423–442
44. Maclaurin C (1748) An account of Sir Isaac Newton's philosophical discoveries: In four books. Murdoch, London
45. Mammola S (2012) La ragione e l'incertezza. Franco Angeli, Milan
46. Meijers A (2009) General introduction. In: Meijers A (ed) Philosophy of technology and engineering sciences, vol 9. Elsevier, Amsterdam, pp 1–22
47. Miller DP (2004) True myths: James Watt's kettle, his condenser, and his chemistry. Hist Sci 42:333–360
48. Miller DP (2006) Watt in court: specifying steam engines and classifying engineers in the patent trials of the 1790s. Hist Technol 27:43–76
49. Miller DP (2016a) Discovering water: James Watt, Henry Cavendish and the nineteenth-century 'water controversy'. Routhledge, London
50. Miller DP (2016b) James Watt, chemist: Understanding the origins of the steam age. University of Pittsburgh, London
51. Mitcham C (1994) Thinking through technology. The path between engineering and philosophy. The University of Chicago, Chicago
52. Mitcham C, Schatzberg E (2009) Defining technology and the engineering sciences. In: Meijers A (ed) Philosophy of technology and engineering sciences, vol 9. Elsevier, Amsterdam, pp 27–64
53. Muirhead JP (1854) The origin and progress of the mechanical invention of James Watt (3 vols). Murray, London
54. Muirhead JP (1858) The life of James Watt. Murray, London
55. Musson A, Robinson E (1969) Science and technology in the industrial revolution. University Press of Manchester, Manchester
56. Newton I (1726) Philosophia naturalis principia mathematica, 3rd edn. Innys, London

57. Parent A (1700) Elemens de mechanique et de physique. Ou l'on donne geometriquement les principes du choc & des equilibres entre toutes sortes de corps. Delaulne, Paris
58. Parent A (1704) Sur la plus grande perfection possible des machines. Mémoires de l'Académie Royale des Sciences de Paris, pp 323–338
59. Parent A (1713) Analyse de la plus avantageuse disposition de l'axe des moulins à vent verticaux à l'égard du cours du vent. In: Parent A (ed) Essais et recherches de mathématique et de physique (2 vol), de Nully, Paris, pp 530–536
60. Radder H (2009) Science, technology and the science-technology relationship. In: Meijers A (ed) Philosophy of technology and engineering sciences, vol 9. Elsevier, Amsterdam, pp 65–92
61. Rankine WM (1856) Introductory lecture on the harmony of theory and practice in mechanics. Griffin and company, London
62. Rankine WM (1859) A manual of the steam engine and other prime movers. Griffin and company, London
63. Reynolds TS (1979) Scientific influences on technology: the case of the overshot waterwheel, 1752–1754. Technol Cult 20(2):270–295
64. Robinson E (1972) James Watt and the law of patents. Technol Cult 13(2):115–139
65. Robinson E, McKie D (1970) Partners in science. Letters of James Watt and Joseph Black. Harvard University, Cambridge
66. Robinson J (1822) A system of mechanical philosophy (4 vols). Murray, Edinburgh
67. Schäfer W (ed) (1983) Finalization in science. The social orientation of scientific progress. Springer, Dordrecht
68. Scott WL (1970) The conflict between atomism and conservation theory, 1644–1860. Elsevier, New York
69. Shepherd DG (1990) Historical development of the windmill. Technical Report 4337, NASA
70. Singer C, Holmyard E, Hall A, Williams TI (1954–1978) A history of technology (5 vol). Clarendon Press, Oxford
71. Smeaton J (1753) Description of a new pyrometer, with a table of experiments made therewith. Philos Trans R Soc Lond 48:598–613
72. Smeaton J (1759) An experimental enquiry concerning the natural powers of water and wind to turn mills, and other machines, depending on a circular motion. Philos Trans R Soc Lond 51:100–174
73. Smeaton J (1768) Description of a new method of observing the heavenly bodies out of the meridian. Philos Trans R Soc Lond 58:170–173
74. Smeaton J (1768b) A discourse concerning the menstrual parallax, arising from the mutual gravitation of the earth and moon; it's influence on the observations of the sun and planets; with a method of observing it. Philos Trans R Soc Lond 58:156–169
75. Smeaton J (1776) An experimental examination of the quantity and proportion of mechanic power necessary to be employed in giving different degrees of velocity to heavy bodies from a state of rest. Philos Trans R Soc Lond 66:450–475
76. Smeaton J (1782) New fundamental experiments upon the collision of bodies. Philos Trans R Soc Lond 72:337–354
77. Smeaton J (1809) New fundamental experiments on the collision of bodies. In: Hutton C, Shaw G, Pearson R (eds) The Philosophical Transactions of the Royal Society of London, from their commencement, in 1665, to the year 1800; abridged, with notes and biographic illustrations (18vols), vol 15. Baldwin, London, pp 295–305
78. Smeaton J (1812) Reports of the late John Smeaton: made on various occasions, in the course of his employment as a civil engineer (3 vols). Longmans, London
79. Stewart R (2017) John Smeaton and the fire engine: 1765–1785. Int J Hist Eng Technol 87(2):190–225
80. Stuart R (1824) A descriptive history of the steam engine. Knight, John and Lacey, Henry, London
81. Taimina D (2008) Geometry and motion links mathematics and engineering in collections of 19th century kinematic models and their digital representation. In: Symposium on the occasion of the 100th anniversary of ICMI, Rome

82. Truesdell CA (1976) History of classical mechanics. Die Naturwissenschaften 63: Part I 53–62, Part II 119–130
83. VanDyck M, Vermeir K (2014) Varieties of wonder: John Wilkins' mathematical magic and the perpetuity of invention. Hist Math 41(4):463–489
84. Watt J (1784) Sequel to the thoughts on the constituent parts of water and dephlogisticated air. Philos Trans R Soc Lond 74:354–357
85. Watt J (1784) Thoughts on the constituent parts of water and of dephlogisticated air; with an account of some experiments on that subject. Philos Trans R Soc Lond 74:329–353
86. Watt J (2109) Industrial revolution: a documentary history. Series three. The papers of James Watt and his family, Birmingham Central Library. http://www.ampltd.co.uk/collections_az/IndRev-3-2/description.aspx
87. Wilkins J (1680) Mathematicall magick, or, the wonders that may be performed by mechanical geometry in two books (2nd edn). Gellibrand, London
88. Wisniak J (2004) The nature and composition of water. J Chem Techn 11(May):434–444
89. Wood DH (1999) Smeaton laws? Wind Eng 23(6):373–375

Index

D. Capecchi, *Epistemology and Natural Philosophy in the 18th Century*, History of Mechanism and Machine Science 39,
https://doi.org/10.1007/978-3-030-52852-2

Y